LEWIS-CLARK STATE COLLEGE
D0531918

Series Editor: Rhodes W. Fairbridge
Columbia University

Published Volumes and Other Volumes in Preparation

RIVER MORPHOLOGY, Stanley A. Schumm
ENVIRONMENTAL GEOMORPHOLOGY AND LANDSCAPE CONSERVATION, VOLUME I: Prior to 1900, Donald R. Coates
SPITS AND BARS, Maurice L. Schwartz
ENVIRONMENTAL GEOMORPHOLOGY AND LANDSCAPE CONSERVATION, VOLUME II: Urban, Donald R. Coates
GEOCHRONOLOGY, C. T. Harper
BARRIER ISLANDS, Maurice L. Schwartz
VERTEBRATE PALEONTOLOGY, J. T. Gregory
PALEOMAGNETISM, Kenneth Creer
GLACIAL ISOSTASY, John T. Andrews
SLOPES, Stanley A. Schumm and M. Paul Mosley
EVAPORITES, Douglas Kirkland and Robert Evans
PALEOECOLOGY, John Chronic, Jr.
TEKTITES, Virgil and Mildred Barnes
ENVIRONMENTAL GEOMORPHOLOGY AND LANDSCAPE CONSERVATION, VOLUME III: Non-urban, Donald R. Coates
ROCK CLEAVAGE, Dennis S. Wood
SOLID STATE GEOPHYSICS, Thomas J. Ahrens
EARTHQUAKES: Geological Effects, Clarence R. Allen
OCEANOGRAPHY, Takashi Ichiye

SPITS AND BARS

Edited by
MAURICE L. SCHWARTZ
Western Washington State College

To my loving wife, Norma, for her many years
of confidence and devotion.

Library of Congress Catalog Card Number: 72–88983
ISBN: 0–87933–012–0

Manufactured in the United States of America.

Exclusive distributor outside the United States and Canada:
John Wiley & Sons, Inc.

Acknowledgments and Permissions

ACKNOWLEDGMENTS

Government Printing Office—*U.S. Geological Survey Monograph I*
"Lake Bonneville"

Government Printing Office—*U.S. Army Corps of Engineers, Seattle District, Public Brochure, Alternatives and Their Pros and Cons, 4th Ed.*
"Erosion Control, Ediz Hook, Port Angeles, Washington"

PERMISSIONS

The following papers have been reprinted with the permission of the authors and the copyright owners.

The University of Chicago Press—*Journal of Geology*
"The Origins of Spits, Bars, and Related Structures"
"A Simulation Model of a Complex Recurved Spit"

Coastal Engineering Research Center—*BEB Technical Report*
"An Experimental Study of Submarine Sand Bars"
"Longshore Bars and Longshore-Troughs"

American Association of Petroleum Geologists—*Geometry of Sandstone Bodies*
"Laboratory Experiments in Form and Structure of Longshore Bars and Beaches"

Gebrüder Borntraeger—*Zeitschrift fur Geomorphologie,* N.F.B.d.7
"The Growth of Sand and Shingle Spits Across Estuaries"

Institute of British Geographers—*Institute of British Geographers Transactions*
"Shingle Spit and River Mouth: Short Term Dynamics"

UNESCO—*Humid Tropic Research, Scientific Problems of the Humid Tropical Zone Deltas and Their Implications*
"Hydrology and Formation of Rivermouth Bars"

Mr. F. J. Meistrell—*Master's Thesis*
"The Spit-Platform Concept: Laboratory Observation of Spit Development"

Editions Technip—*Photo Interpretation*
"Tropical Littoral Accumulation Shapes"

American Geophysical Union—*Journal of Geophysical Research*
"Shifting Offshore Bars and Harbor Shoaling"

George Allen & Unwin Ltd.—*Oceanography and Marine Biology: Volume 5*
"Stages in Island Linking"

Reinhold Publishing Corporation—*Encyclopedia of Geomorphology*
"Bars"

Society of Economic Paleontologists and Mineralogists—*Journal of Sedimentary Petrology*
"Depositional and Dispersal Dynamics of Intertidal Sand Bars"

The University of Nottingham—*East Midland Geographer*
"Changes in the Spit at Gibraltar Point, Lincolnshire, 1951 to 1969"

Series Editor's Preface

The philosophy behind the "Benchmark Papers in Geology" series is one of collection, sifting and rediffusion. Scientific literature today is so vast, so dispersed and, in the case of old papers, so inaccessible for readers not in the immediate neighborhood of major libraries, that much valuable information has become ignored, by default. It has become just so difficult, or time consuming, to search out the key papers in any basic area of research that one can hardly blame a busy man for skimping on some of his "homework."

This series of volumes has been devised, therefore, to make a practical contribution to this critical problem. The geologist, perhaps even more than any other type of scientist, often suffers from twin difficulties—isolation from central library resources and an immensely diffused source of material. New colleges and industrial libraries simply cannot afford to purchase complete runs of all the world's earth science literature. Specialists simply cannot locate reprints or copies of all their principal reference materials. So it is that we are now making a concentrated effort to gather into single volumes the critical material needed to reconstruct the background to any and every major topic of our discipline.

We are interpreting "Geology" in its broadest sense: the fundamental science of the Planet Earth, its materials, its history and its dynamics. Because of training and experience in "earthy" materials, we also take in astrogeology, the corresponding aspect of the planetary sciences. Besides the classical core disciplines such as mineralogy, petrology, structure, geomorphology, paleontology, or stratigraphy, we embrace the newer fields of geophysics and geochemistry, applied also to oceanography, geochronology, and paleoecology. We recognize the work of the mining geologists, the petroleum geologists, the hydrologists, the engineering and environmental geologists. Each specialist needs his working library. We are endeavoring to make his task a little easier.

Each volume in the series contains an Introduction prepared by a specialist, the volume editor—and a "state-of-the-art" opening or a summary of the objects and content of the volume. The articles selected, usually some 30–50 reproduced either in their entirety or in significant extracts, attempt to scan the field from the key papers of the last century until fairly recent years. Where the original references may be in foreign languages, we have endeavored to locate or commission translations. Geologists, because of their global subject, are often acutely aware of the oneness of our world. Its literature, therefore, cannot be restricted to any one country and, whenever possible, an attempt has been made to scan the world literature.

To each article, or group of kindred items, some sort of "Highlight Commentary" is usually supplied by the volume editor. This should serve to bring that article into historical perspective and to emphasize its particular role in the growth of the field. References or citations, wherever possible, will be reproduced in their entirety; for by this means the observant reader can assess the background material available to that particular author, or, if he wishes, he too can double check the earlier sources.

A "benchmark," in surveyor's terminology, is an established point on the ground, recorded on our maps. It is usually anything that is a vantage point, from a modest hill to a mountain peak. From the historical viewpoint, these benchmarks are the bricks of our scientific edifice.

Rhodes W. Fairbridge

Contents

Contents by Author

Introduction

As industrial, residential, and recreational pressures are intensified in the coastal zone our knowledge of the processes operating there must keep ahead, or at least abreast, of man's intrusion. Spits and bars in populated areas throughout the world are being encroached upon by some form of human exploitation. Too often after the fact, when the morphology of the structure continues its relentless change, does the populace become aware of what is then called an emergency condition. Suddenly, countermeasures are taken to remedy what was inevitable to begin with.

If this opening statement seems overly dramatic, take a look at the coastal zone nearest you. Consider all the spits and bars that you can identify in this area, then recall what you have read in your newspaper about them and man's eternal battle with the sea. A classic case in point exists only a few dozen miles from where this piece is being written. Ediz Hook, Washington, is a narrow spit extending out into the Strait of Juan de Fuca on the northeasterly coast of the Olympic Peninsula. The distal third of the spit is the site of a U.S. Coast Guard Station, while all but the outer foreshore of the landward two-thirds is covered by industrial construction. Migration of the former and erosion of the latter is presently wreaking havoc upon these man-made facilities. The history of Ediz Hook's geomorphic and economic development, and alternatives considered to remedy the present situation, are outlined in the last paper of this volume. That study in applied coastal geomorphology concludes this revue of almost a hundred years of spit and bar literature.

The purpose of this volume in the Benchmark Papers in Geology series is to present the editor's considered choice of spit and bar literature that covers the historical growth of studies on this topic and highlights recent reports. You, the reader, may not find a particular paper that you seek here; but then, it is doubtful that any two bibliographic listings on the subject would be the same. Space limitations necessitate that the volume be a selection of papers.

A word concerning the introduction to each paper is appropriate here. In these introductions there is no attempt to repeat the abstract or summary of the paper, but rather to briefly focus upon the significance of the work and place it in proper perspective with relation to the included papers in this and other volumes of the series. Short biographical sketches, furnished for the most part by the authors, are offered to give the reader some idea of the background of the people whose works he is perusing.

It is hoped that no single conclusion will be drawn from a reading of this volume. On the contrary, a concerted effort has been made to include papers representing the divergent viewpoints inherent in prolonged studies of such complex structures as spits and bars. For example, there seems to be little agreement on what constitutes a bar, and whether are not submarine bars can build up through sea level.

Evans, in his classic paper, includes as bars structures that rise above water level; bay bars and looped bars, among others. On the other hand Price considers only subaqueous forms to be true bars. Price's classification includes longshore, transverse, and river-mouth bars, which are treated in this collection by Shepard, Neidoroda and Tanner, and Mikhailov, respectively.

On the matter of bar emergence, McKee and Sterrett model and describe processes for building through sea level. Shepard and Price maintain though, that barrier bars are not developed by bar emergence.

Such is the dichotomy that prevails in coastal studies today. There are many unknowns; and for each of these, various schools of thought to explain them. I do not presume to tell you what the *right* or *correct* answer is. Rather, we present here the background and important present works on spits and bars and suggest that you, the reader, pursue your own intelligent conclusions.

In order to delay no longer the reading of these works, a brief acknowledgment of thanks is offered to the kind and helpful people whose cooperation made this volume possible: my good friend Rhodes W. Fairbridge, editor of the series; my most cooperative publisher; and all of the publishers and authors who provided the articles included here.

MAURICE L. SCHWARTZ
November, 1972

Editor's Comments on Paper 1

During the span of his lifetime, from May 6, 1843 to May 1, 1918, Grove Karl Gilbert devoted half a century to geology and geological organizations. He started his unique career as a museum worker under Professor Henry A. Ward, founder of the famous natural history supply house. Gilbert then became an assistant on the Ohio Geological Survey where, among other duties, he reported on elevated beach terraces bordering Lake Erie. Following this he moved west with the Wheeling Survey and visited the Lake Bonneville region; thus began a major study which led to the pioneer monograph reproduced here in part. Gilbert later joined the Powell Survey and with the establishment of the U.S. Geological Survey in 1879 became a Senior Geologist on its staff: an active and prolific association that continued until his death.

Though G. K. Gilbert was interested in a great many aspects of geologic science, his Lake Bonneville paper is a particularly fitting point of departure for this volume of selected papers. Named by Gilbert to honor a military explorer of the early West, Lake Bonneville had once covered twenty thousand square miles in northwestern Utah and attained a depth of nine hundred feet. As the surface level fluctuated with Pleistocene glacial cycles (though not necessarily in phase), wave action along its shores developed the stepped terraces seen today on the mountain slopes bordering the present Great Salt Lake. It was on these terraces that Gilbert found a subaerial inventory of coastal forms ideally preserved, and where he was among the first to report on the development of shoreline features.

For the sake of brevity only a portion of chapter II is reproduced here. In this chapter Gilbert deals with the topographic features of lake shores. The originality of his work is attested to by the modest statement on page 25 of the monograph. The following treatment of shore erosion, transportation, and deposition, stands as a primer to the serious student of coastal processes. Gilbert's description of spit and bar development provides an interesting model for comparison with the subsequent papers in this collection.

UNITED STATES GEOLOGICAL SURVEY

J. W. POWELL, DIRECTOR

1

LAKE BONNEVILLE

BY

GROVE KARL GILBERT

WASHINGTON
GOVERNMENT PRINTING OFFICE
1890

CHAPTER II.

THE TOPOGRAPHIC FEATURES OF LAKE SHORES.

It has been assumed in the preceding pages that valleys from which lakes have recently disappeared are characterized by certain features whereby that fact can be recognized. Perhaps no one observant of natural phenomena will dispute this. But there is, nevertheless, some diversity of opinion as to what are the peculiar characters to which lakes give rise; and especially has the true interpretation of certain local topographic features been mooted, some geologists ascribing them to waves, and others to different agencies.

In the investigation of our ancient lake, it has been found necessary not only to discriminate from all other topographic elements the features created by its waves, but also to ascertain the manner in which each was produced, so as to be able to give it the proper interpretation in the reconstruction of the history of the lake. It is proposed in this chapter to present the more general results of this study, describing in detail the various elements which constitute shore topography, explaining their origin, so far as possible, and finally contrasting them with topographic features of other origin which so far simulate them as to occasion confusion.

The play of meteoric agents on the surface of the land is unremitting, so that there is a constant tendency to the production of the forms characteristic of their action. All other forms are of the nature of exceptions, and attract the attention of the observer as requiring explanation. The shapes wrought by atmospheric erosion are simple and symmetric and need but to be enumerated to be recognized as normal elements of the sculpture of the land. Along each drainage line there is a gradual and gradually increasing ascent from mouth to source; and this law of increasing acclivity applies to all branches as well as to the main stem. Between each pair of adjacent drainage lines is a ridge or hill, standing midway and rounded at the top. Wherever two ridges join there is a summit higher than the adjacent portion of either ridge; and the highest summits of all are those which, measuring along lines of drainage, are most remote from the ocean. The crests of the ridges are not horizontal but undulate from summit to summit. There are no sharp contrasts of slope; the concave profiles of the drainage lines change their inclination little by little and merge by a gradual transition in the convex profiles of the crests and summits.

The factor which most frequently, and in fact almost universally, interrupts these simple curves is heterogeneity of terrane. Under the influence of this factor, just as in the case of a homogeneous terrane, the declivities adjust themselves in such way as to oppose a maximum resistance to erosion; and with diversity of rock texture this adjustment involves diversity of form. Hard rocks survive, while the soft are eaten away. Peaks and cliffs are produced. The apices are often angular instead of rounded. Profiles exhibit abrupt changes of slope. Flat-topped ridges appear, and the distribution of maximum summits becomes in a measure independent of the length of drainage lines.

A second factor interrupting the continuity of erosion profiles is upheaval; and this produces its effects in two distinct ways. First, the general uprising of a broad tract of land affects the relation of the drainage to its point of discharge or to its base level, causing corrasion by streams to be more rapid than the general waste of the surface and producing canyons and terraces. Second, a local uprising by means of a fault produces a cliff at the margin of the uplifted tract; and above this cliff there is sometimes a terrace.

A third disturbing factor is glaciation, the cirques and moraines of which are distinct from anything wrought by pluvial erosion; and a fourth is found in eruption.

The products of all these agencies except the last have been occasionally confused with the phenomena of shores. The beach-lines of Glen Roy have been called river terraces and moraine terraces. The cliffs of the Downs of England have been ascribed to shore waves. Glacial moraines in New Zealand have been interpreted as shore terraces. Beach ridges in our own country have been described as glacial moraines, and fault terraces as well as river terraces have been mistaken for shore-marks.

In the planning of engineering works for the improvement and protection of harbors, it is of prime importance to understand the natural processes by which coast features are produced and modified, and this necessity has led to the production by engineers of a large though widely scattered literature on coast-forming agencies. Geologists also require for the interpretation of strata originating as coast deposits an understanding of the methods of coastal degradation and coastal deposition, and from their point of view there has arisen an independent literature on the subject. The physical theory of water waves required alike by engineers and geologists has been developed by physicists, and has its own literature. The three groups of writers have so thoroughly traversed the subject of shore processes that the present chapter would have need to demonstrate its raison d'être were it not that the general subject has as yet received no compendious and systematic treatment in the English language.

It happens, moreover, that the present treatment of the subject has its own peculiar point of view, and is in large part independent. During the progress of the field investigation I was unaware of the greater part of the literature mentioned above, having indeed met with but one important paper, that in which Andrews describes the formation of beaches at the head of Lake Michigan, and I was induced by the requirements of my work to develop the philosophy of the subject ab initio. The theories here presented had therefore received approximately their present form and arrangement before they were compared with those of earlier writers. They are thus original without being novel, and their independence gives them confirmatory value so far as they agree with the conclusions of others.

The peculiarity of the point of view lies in the fact that the phenomena chiefly studied are fossil shore-lines instead of modern. The bodies of water to which they pertain having disappeared, the configuration of the submerged portion is directly seen instead of being interpreted from laborious soundings. There are, moreover, natural sections of the deposits, exposed by subsequent erosion, and these reveal features of internal structure or anatomy quite as important to the geologist as the features of morphology.

The literature of shore-lines is so feebly connected by cross reference, and portions of it have been discovered in places so unexpected, that the writer fears many important contributions have escaped his attention. Within the range of his reading, the earliest discussions of value are by Beaumont[1] and De la Beche,[2] and it must be admitted that the writers of geologic manuals now in use have improved very little upon their presentation. Fleming, in an essay on the origin and preservation of the harbor of Toronto,[3] set forth the process of littoral transportation with admirable clearness; and Andrews, who appears to have reached his conclusions by independent observation, added to the theory of littoral transportation an important factor in the theory of littoral deposition.[4] Mitchell, in an essay on tidal marshes,[5] incidentally describes the growth of the protecting barrier. A general treatise by Cialdi[6] gives a systematic discussion of coast processes from the engineer's point of view, and reviews the Italian literature of the subject; and a shorter paper by Keller[7] has a similar scope. Richthofen, in his manual of instruction to scientific travelers, treats analytically and at length of the work of waves in conjunction with tides, and discusses a subsiding continent.[8] The theory of waves has been developed experimentally by a committee of the British Association, with J. Scott Russell as reporter;[9] and it is analytically treated by Airy[10] and Rankine.[11]

---

[1] Leçons de geologie pratique. Par Elie de Beaumont. Vol. 1, pp. 221–253, Paris, 1845.

[2] A Geological Manual. By Henry T. De la Beche. 3d edition, enlarged, London, 1833, pp. 67–91. The Geological Observer. By the same. London, 1851, pp. 49–117.

[3] Toronto Harbor—its formation and preservation. By Sandford Fleming, C. E.: Canadian Journal, vol. 2, 1854, pp. 103–107, 223–230. Reprinted with additions as Report on Preservation and Improvement of Toronto Harbor. In Supplement to Canadian Journal, 1854, pp. 15–29.

[4] The North American Lakes considered as chronometers of post-Glacial time. By Dr. Edmund Andrews. Trans Chicago Acad. Sci., Vol. 2, pp. 1–23.

[5] On the recamation of tide-lands and its relation to navigation. By Henry Mitchell. Appendix No. 5, to Rept. U. S. Coast Survey for 1869. Washington, 1872, pp. 75–104.

[6] Sul moto ondoso del mare e su le correnti di esso specialmente su quelle littorali. Alessandro Cialdi, Roma, 1866.

[7] Studien uber die Gestaltung der Sandkusten, etc., H. Keller, Berlin, 1881.

[8] Führer für Forschungsreisende, von Ferdinand Freiherr von Richthofen. Berlin, 1886, pp. 336–365.

[9] Report of the Committee on Waves, by Sir John Robinson, and John Scott Russell, Reporter: Rept. British Ass. Adv. Sci., 7th meeting, 1837, pp. 417–496.

[10] G. B. Airy, Vol. V, Ency. Metrop.

[11] W. J. McQ. Rankine, Philos. Trans. Royal Soc. London, vol. 153, 1863, pp. 127–138.

In the following treatment of the subject the description and analysis of the elements of shore topography will be followed by a comparison of certain of these elements with simulating features of different origin. First, however, a few words will be devoted to the consideration of shore shaping as a division of the more general process of earth shaping.

The earth owes its spheroidal form to gravity and rotation. It owes its great features of continent and ocean bed to the unequal distribution of the heterogeneous material of which it is composed. Many of its minor inequalities can be referred to the same cause, but its details of surface are chiefly molded by the circulation of the fluids which envelope it. This shaping or molding of the surface may be divided into three parts—subaerial shaping (land sculpture), subaqueous shaping, and littoral shaping. In each case the process is threefold, comprising erosion, transportation, and deposition.

In subaerial or land shaping the agents of erosion are meteoric—rain, acting both mechanically and chemically, streams, and frost. The agent of transportation is running water. The condition of deposition is diminishing velocity.

In subaqueous shaping, or the molding of surface which takes place beneath lakes and oceans, currents constitute the agent of erosion. They constitute also the agent of transportation; and the condition of deposition is, as before, diminishing velocity.

In littoral shaping, or the modeling of shore features, waves constitute the agent of erosion. Transportation is performed by waves and currents acting conjointly, and the condition of deposition is increasing depth.

On the land the amount of erosion vastly exceeds the amount of deposition. Under standing water erosion is either nil or incomparably inferior in amount to deposition. And these two facts are correlatives, since the product of land erosion is chiefly deposited in lakes and oceans, and the sediments of lakes and oceans are derived chiefly from land erosion. The products of littoral erosion undergo division, going partly to littoral deposition and partly to subaqueous deposition. The material for littoral deposition is derived partly from littoral erosion and partly from land erosion.

That is to say, the detritus worn from the land by meteoric agents is transported outward by streams. Normally it is all carried to the coast, but owing to the almost universal complication of erosion with local uplift, there is a certain share of detritus deposited upon the basins and lower slopes of the land. At the shore a second division takes place, the smaller portion being arrested and built into various shore structures, while the larger portion continues outward and is deposited in the sea or lake. The product of shore erosion is similarly divided. A part remains upon the shore, where it is combined with material derived from the land, and the remainder goes to swell the volume of subaqueous deposition.

The forms of the land are given chiefly by erosion. Since the wear by streams keeps necessarily in advance of the waste of the intervening surfaces, and since, also, there is inequality of erosion dependent on diversity of texture, land forms are characterized by their variety.

The forms of sea beds and lake beds are given by deposition. The great currents by which subaqueous sediments are distributed sweep over the ridges and other prominences of the surface and leave the intervening depressions comparatively currentless. Deposition, depending on retardation of current, takes place chiefly in the depressions, so that they are eventually filled and a monotonous uniformity is the result.

The forms of the shore are intermediate in point of variety between those of the land and those of the sea bed; and since they alone claim parentage in waves, they are sui generis.

Ocean shores are genetically distinguished from lake shores by the cooperation of tides, which modify the work accomplished by waves and wind currents.

The phenomena of ocean shores are therefore more complicated than those of lake shores, and an exhaustive treatment of the subject would include the discussion of their distinguishing characteristics. They fall, however, without the limits of the present investigation, and in the analysis which follows, the influence of tides is not considered. It is perhaps to be regretted that the systematic treatment here proposed could not be so extended as to include all shores, but there is a certain compensation in the fact that the results reached in reference to lake shores have an important negative bearing on tidal discussions. It was long ago pointed out by Beaumont[1] and Desor[2] that many of the more important features ascribed by hydraulic engineers to tidal action, are produced on the shores of inland seas by waves alone; and the demonstration of wave work pure and simple should be serviceable to the maritime engineer by pointing out those results in explanation of which it is unnecessary to appeal to the agency of tides.

The order of treatment is based on the three-fold division of the process of shore shaping. Littoral erosion and the origin of the sea-cliff and wave-cut terrace will be first explained, then the process of littoral transportation with its dependent features, the beach and the barrier, and finally the process of littoral deposition, resulting in the embankment, with all its varied phases, and the delta.

WAVE WORK.

LITTORAL EROSION.

In shore sculpture the agent of erosion is the wave. All varieties of wave motion which affect standing water are susceptible of producing erosive effect on the shore, but only those set in motion by wind need be considered here. They are of two kinds: the wind wave proper, which exists only during the continuance of the wind; and the swell, which continues after the wind has ceased. It is unnecessary to discriminate the effects of these upon the shore further than to say that the wind wave is the more efficient and therefore the better deserving of special consideration. In the wind wave two things move forward, the undulation and the water. The velocity of the undulation is relatively rapid; that of the water is slow and rhythmic. A particle of water at or near the surface, as each undulation passes, describes an orbit in a vertical plane, but does not return to the starting point. While on the crest of the wave it moves forward, and while in the trough it moves less rapidly backward, so that there is a residual advance.[3]

---

[1] Leçons de géologie pratique, vol. 1, p. 232.

[2] E. Desor, Geology of Lake Superior Land District by Foster & Whitney, Washington, 1851, vol. 2, pp. 262, 266.

[3] The theory of wave motion involved in this and the following paragraphs is based partly on observation but chiefly on the discussions of J. S. Russell, Airy, Cialdi, and Rankine.

This residual advance is the initiatory element of current. By virtue of it the upper layer of water is carried forward with reference to the layer below, being given a differential movement in the direction towards which the wind blows. This movement is gradually propagated to lower aqueous strata, and ultimately produces movement of the whole body, or a wind-wrought current. So long as the velocity of the wind remains constant, the velocity of the current is less than that of the wind; and there is always a differential movement of the water, each layer moving faster than the one beneath. The friction is thus distributed through the whole vertical column, and is even borne in part by the lake bottom. The greater the depth the smaller the share of friction apportioned to each layer of water and the greater the velocity of current which can be communicated by a given wind.[1] The height of waves is likewise conditioned by depth of water, deep water permitting the formation of those that are relatively large.

When the wave approaches a shelving shore its habit is changed. The velocity of the undulation is diminished, while the velocity of the advancing particles of water in the crest is increased; the wave length, measured from trough to trough, is diminished, and the wave height is increased; the crest becomes acute, with the front steeper than the back; and these changes culminate in the breaking of the crest, when the undulation proper ceases. The return of the water thrown forward in the crest is accomplished by a current along the bottom called the *undertow*. The momentum of the advancing water contained in the wave crest gives to it its power of erosion. The undertow is efficient in removing the products of erosion.

The retardation of the undulation by diminishing depth of water changes the direction of its axis or crest line—excepting when the axis is parallel to the contours of the shoaling bottom—and the phenomena are analogous to those of the refraction of light and sound. As a wave passes obliquely from deep water to a broad shoal of uniform depth, the end first entering shoal water is first retarded and the crest line is for the moment bent. When the entire crest has reached shoal water it is once more straight, but with a new trend, a trend making a narrower angle with the line of separation

[1] This is a matter of observation rather than theory. It implies that the friction between contiguous films of water increases in more than simple ratio with the differential velocity of the films.

between deep and shallow water. The wave has been refracted. When a wave passes obliquely from deep water to shoal water whose bottom gradually rises to a shore, the end nearer the shore is the more retarded at all stages of progress and the crest line is continuously curved. When the wave breaks and the undulation ceases, the crest line is nearly parallel to the shore. It results that for a wide range of wind direction there is but small range in the direction of wave trend at the shore. It results also, as has been often noted, that when the wind blows normally into a circling bay, the waves it brings are diversely turned, so as to beat against both sides as well as the head of the bay.

When the land at the margin of the water consists of unconsolidated material or of fragmental matter lightly cemented, the simple impact of the water is sufficient to displace or erode it. The same force is competent also to disintegrate and remove firmer rock that has been superficially weakened by frost or is partially divided by cracks, but it may be doubted whether it has any power to wear rock that is thoroughly coherent. The impact of large waves has great force, and its statement in tons to the square foot is most impressive; but, so far as our observation has extended, the erosive action of waves of clear water beating upon firm rock without seams is practically nil. On the shores of Lake Bonneville, not only was there no erosion on the faces of cliffs at points where the waves carried no detrital fragments, but there was actually deposition of calcareous tufa; and this deposition was most rapid at points specially exposed to the violence of the waves.

The case is very different when the rock is divided by seams, for then the principle of the hydrostatic press finds application. Through the water forced into the seams, and sometimes through air imprisoned and compressed by the water, the blow struck by the wave is applied not merely to large surfaces but in directions favorable to the rending and dislocation of rock masses.

It rarely happens, however, that the impact of waves is not reinforced by the impact of mineral matter borne by them. The detritus worn from the shore is always at hand to be used by the waves in continuance of the attack; and to this is added other detritus carried along the shore by a process presently to be described.

The rock fragments which constitute the tool of erosion are themselves worn and comminuted by use until they become so fine that they no longer lie in the zone of breakers but are carried away by the undertow.

The direct work of wave erosion is restricted to a horizontal zone dependent on the height of the waves. There is no impact of breakers at levels lower than the troughs of the waves; and the most efficient impact is limited upward by the level of the wave crests, although the dashing of the water produces feebler blows at higher levels. The indirect work has no superior limit, for as the excavation of the zone is carried landward, masses higher up on the slope are sapped so as to break away and fall by mere gravity. Being thus brought within reach of the waves, they are then broken up by them, retarding the zonal excavation for a time but eventually adding to the tool of erosion in a way that partially compensates.

Let us now consider what goes on beneath the surface of the water. The agitation of which waves are the superficial manifestation is not restricted to their horizon, but is propagated indefinitely downward. Near the surface the amount of motion diminishes rapidly downward, but the rate of diminution itself diminishes, and there seems no theoretic reason for assigning any limit to the propagation of the oscillation. Indeed, the agitation must be carried to the bottom in all cases where the depth operates as a condition in determining the magnitude of waves, for that determination can be assigned only to a resistance opposed by the bottom to the undulation of the water.

During the passage of a wave each particle of water affected by it rises and falls, and moves forward and backward, describing an orbit. If the passing wave is a swell, the orbit of the particle is closed,[1] and is either a circle or an ellipse; but in the case of a wind wave the orbit is not closed. The relative amounts of horizontal and vertical motion depend on the depth of the particle beneath the surface, and on the relation of the total depth of the water to the size of the wave. If the water is deep as compared to the wave-length, the horizontal and vertical movements are sensibly equal, and their amount diminishes rapidly from the surface downward. If the depth

[1] This is strictly true only while the swell traverses deep water. It is pointed out by Cialdi that in passing to shoal water the swell is converted into a wave of translation, and the particles no longer return to their points of starting.

is small, the horizontal motion is greater than the vertical, but diminishes less rapidly with depth. Near the line of breakers, the vertical motion close to the bottom becomes inappreciable, while the horizontal oscillation is nearly as great as at the surface. This horizontal motion, affecting water which is at the same time under the influence of the undertow, gives to that current a pulsating character, and thus endows it with a higher transporting power than would pertain to its mean velocity. Near the breaker line, the oscillation communicated by the wave may even overcome and momentarily reverse the movement of the undertow. Inside the breaker line no oscillation proper is communicated. The broken wave crest, dashing forward, overcomes the undertow and throws it back; but the water returns without acceleration as a simple current descending a slope.

It should be explained that the increment given by pulsation to the transporting power of the undertow depends upon the general law that the transporting power of a current is an increasing geometric function of its velocity. Doubling the velocity of a current more than doubles the amount it can carry, and more than doubles the size of the particles it is able to move.

The transporting power of the undertow diminishes rapidly from the breaker line outward. That part of its power which depends on its mean velocity diminishes as the prism of the undertow increases; that part which depends on the rhythmic accelerations of velocity diminishes as the depth of water increases.

The pulsating current of the undertow has an erosive as well as a transporting function. It carries to and fro the detritus of the shore, and, dragging it over the bottom, continues downward the erosion initiated by the breakers. This downward erosion is the necessary concomitant of the shoreward progress of wave erosion; for if the land were merely planed away to the level of the wave troughs, the incoming waves would break where shoal water was first reached and become ineffective at the water margin. In fact, this spending of the force of the waves where the water is so shallow as to induce them to break, increases at that point the erosive power by pulsation, and thus brings about an interdependence of parts. What may be called a normal profile of the submerged terrace is produced, the parts of which are adjusted to a harmonious interrelation. If some exceptional temporary condition produces abnormal wearing of the outer margin of the terrace, the greater depth of water at that point permits the incoming waves to pass with little impediment and perform their work of erosion upon portions nearer the shore, thus restoring the equilibrium. If exceptional resistance is opposed by the material at the water margin, erosion is there retarded until the submerged terrace has been so reduced as to permit the incoming waves to attack the land with a greater share of unexpended energy. Conversely, if there is a diminution of resistance at the water margin, so as to permit a rapid erosion, the landward recession of that margin causes it to be the less exposed to wave action. Thus the landward wear at the water margin and the downward wear in the several parts of the submerged plateau are adjusted to an interdependent relation.

The Sea-Cliff.—Wave erosion, acting along a definite zone, may be rudely compared to the operation of a horizontal saw; but the upper wall of the saw cut, being without support, is broken away by its own weight and falls in fragments, leaving a cliff at the shoreward margin of the cut. This wave-wrought cliff requires a distinctive name to avoid confusion with cliffs of other origin, and might with propriety in this discussion be called a lake-cliff; but the term *sea-cliff* is so well established that it appears best to retain it.

One of the most noteworthy and constant characters of the sea-cliff is the horizontality of its base. Being determined by wave erosion the base must always stand at about the level of the lake on which the waves are formed. The material of the cliff is the material of the land from which it is carved. Its declivity depends partly on the nature of that material and partly on the rate of erosion. If the material is unconsolidated, the inclination cannot exceed the normal earth slope; if it is thoroughly indurated, the cliff may be vertical or may even overhang. If the rate of wave erosion is exceedingly rapid, the cliff is as steep as the material will permit; if the rate is slow, the inclination is diminished by the atmospheric waste of the cliff face.

Figure 1 represents a cliff on the shore of Great Salt Lake. The material in this case is arenaceous limestone. At the base of the cliff may be seen a portion of the accompanying wave-cut terrace, and the foreground exhibits a portion of the associated beach. The large bowlders of the foreground have an independent origin, but the shingle and other material of the beach were derived from the erosion of the cliff and transported to their present position by the waves. Sheep Rock is overlooked by the northern face of the Oquirrh mountain range, on which the Bonneville shores are traced, and the partial view of the mountain face given in the frontispiece shows a line of ancient sea-cliffs, originally as precipitous as Sheep Rock but now shattered by frost and partially draped by talus.

Fig. 1.—Sheep Rock, a Sea-Cliff on the shore of Great Salt Lake. From a photograph by C. R. Savage.

It will appear in the sequel that the distribution of sea-cliffs is somewhat peculiar, but this cannot be described until the process of littoral transportation has been explained.

The Wave-Cut Terrace.—The submerged plateau whose area records the landward progress of littoral erosion, becomes a terrace after the formative lake has disappeared, and, as such, requires a distinctive name. It will be called the *wave-cut terrace.*

Its prime characteristics are, first, that it is associated with a cliff; second, that its upper margin, where it joins the cliff, is horizontal; and, third, that its surface has a gentle inclination away from the cliff. There is an exceptional case in which an island or a hill of the mainland has been completely pared away by wave action, so that no cliff remains as a companion for the wave-cut terrace; but this exception does not invalidate the rule. The lakeward inclination is somewhat variable, depending on the nature of the material and on the pristine acclivity of the land. It is greater where the material is loose than where it is coherent; and greater where the ratio of terrace width to cliff height is small. It is probably conditioned also by the direction of the current associated with the wind efficient in its production; but this has not been definitely ascertained.

The width of the terrace depends on the extent of the littoral erosion, and is not assignable. Its relative width in different parts of a given continuous coast depends entirely on the conditions determining the rapidity of erosion, and the discussion of these at this point would be premature.

Sometimes a portion of the eroded material gathers at the outer edge of the terrace, extending its profile as indicated in Figure 4.[1]

Figures 2 and 3 show ideal sections of cliffs and terraces, carved in one case from soft material, in the other from hard. The station of the artist

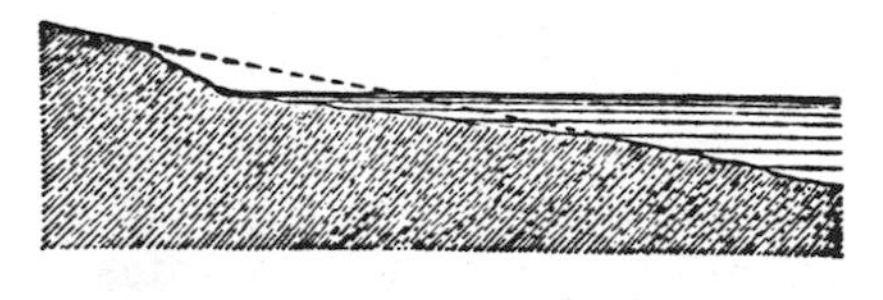

FIG. 2.—Section of a Sea Cliff and Cut-Terrace in Incoherent Material.

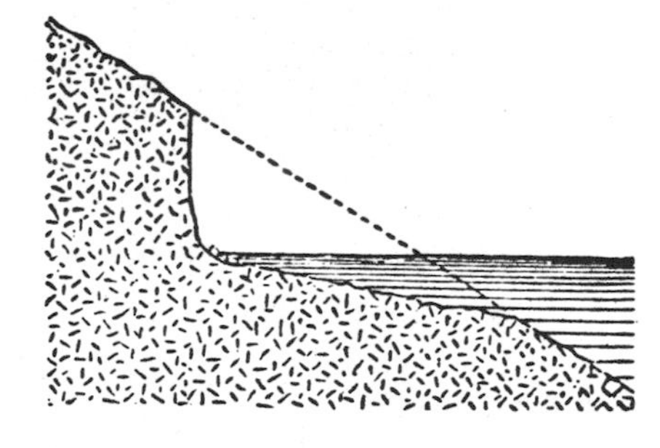

FIG. 3.—Section of a Sea Cliff and Cut-Terrace in Hard Material.

in sketching the view represented in the frontispiece was on a cut-terrace, and a portion of it appears in the foreground.

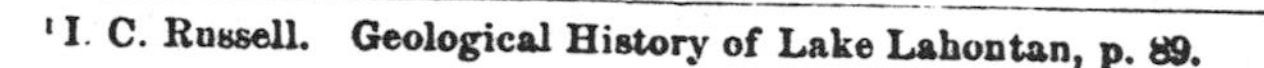

[1] I. C. Russell. Geological History of Lake Lahontan, p. 89.

LITTORAL TRANSPORTATION.

Littoral transportation is performed by the joint action of waves and currents. Usually, and especially when the wind blows, the water adjacent to the shore is stirred by a gentle current flowing parallel to the water margin. This carries along the particles of detritus agitated by the waves. The waves and undertow move the shallow water near the shore rapidly to and fro, and in so doing momentarily lift some particles, and roll others forward and back. The particles thus wholly or partially sustained by the water are at the same moment carried in a direction parallel to the shore by the shore current. The shore current is nearly always gentle and has of itself no power to move detritus.

When the play of the waves ceases, all shore action is arrested. When the play of the waves is unaccompanied by a current, shore action is nearly arrested, but not absolutely. If the incoming waves move in a direction normal to the shore, the advance and recoil of the water move particles toward and from the shore, and effect no transfer in the direction of the shore; but if the incoming waves move in an oblique direction the forward transfer of particles is in the direction of the waves, while the backward transfer, by means of the undertow, is sensibly normal to the shore, and there is thus a slow transportation along the shore. If there were no currents a great amount of transportation would undoubtedly be performed in this way, but it would be carried on at a slow rate. The transporting effect of waves alone is so slight that only a gentle current in the opposite direction is necessary to counteract it. The concurrence of waves and currents is so general a phenomenon, and the ability of waves alone is so small, that the latter may be disregarded. The practical work of transportation is performed by the conjoint action of waves and shore currents.

In the ocean the causes of currents are various. Besides wind currents there are daily currents caused by tides upon all coasts, and it is maintained by some physicists that the great currents are wholly or partly due to the unequal heating of the water in different regions. But in lakes there are no appreciable tides, and currents due to unequal heating have never been discriminated. The motions of the water are controlled by the wind.

A long-continued wind in one direction produces a set of currents harmoniously adjusted to it. A change in the wind produces a change in the currents, but this adjustment is not instantaneous, and for a time there is lack of harmony. The strong winds, however, bring about an adjustment more rapidly than the gentle, and since it is to these that all important littoral work is ascribed, the waves and currents concerned in littoral transportation may be here regarded as depending on one and the same wind.

A wind blowing directly toward a shore may be conceived of as piling the superficial water against the shore, to be returned only by the undertow, but, in fact, so simple a result is rarely observed. Usually there is some obliquity of direction, in virtue of which the shoreward current is partially deflected, so as to produce as one of its effects a flow parallel to the shore, or a littoral current. The littoral current thus tends in a direction harmonious with the movement of the waves, passing to the right if the waves tend in that direction, to the left if the waves tend thither.

To this rule there is a noteworthy exception. The undertow is not the only return current. It frequently occurs that part of the water driven forward by the wind returns as a superficial current somewhat opposed in direction to the wind. If this current follows a shore it constitutes a littoral current whose tendency is opposed to that of the waves. Thus the littoral current may move to the right while the waves tend to the left, and vice versa. In every such case the direction of transportation is the direction of the littoral current.

The waves and undertow accomplish a sorting of the detritus. The finer portion, being lifted up by the agitation of the waves, is held in suspension until carried outward to deep water by the undertow. The coarser portion, sinking to the bottom more rapidly, can not be carried beyond the zone of agitation, and remains as a part of the shore. Only the latter is the subject of littoral transportation. It is called *shore drift.*

With the shifting of the wind the direction of the littoral current on any lake shore is occasionally, or it may be frequently, reversed, and the shore drift under its influence travels sometimes in one direction and sometimes in the other. In most localities it has a prevailing direction, not necessarily determined by the prevailing direction of the shore current, but rather by the direction of that shore current which accompanies the greatest waves. This is frequently but not always the direction also of the shore current accompaning the most violent storms.

The source of shore drift is two-fold. A large part is derived from the excavation of sea-cliffs, and is thus the product of littoral erosion. From every sea-cliff a stream of shore drift may be seen to follow the coast in one direction or the other.

Another part is contributed by streams depositing at their mouths the heavy part of their detritus, and is more remotely derived from the erosion of the land. The smallest streams merely reinforce the trains of shore drift flowing from sea-cliffs, and their tribute usually cannot be discriminated. Larger streams furnish bodies of shore drift easily referred to their sources. Streams of the first magnitude, as will be explained farther on, overwhelm the shore drift and produce structures of an entirely different nature, known as deltas.

The Beach.—The zone occupied by the shore drift in transit is called the *beach.* Its lower margin is beneath the water, a little beyond the line where the great storm waves break. Its upper margin is usually a few feet above the level of still water. Its profile is steeper upon some shores than others, but has a general facies consonant with its wave-wrought origin. At each point in the profile the slope represents an equilibrium in transporting power between the inrushing breaker and the outflowing undertow. Where the undertow is relatively potent its efficiency is diminished by a low declivity. Where the inward dash is relatively potent the undertow is favored by a high declivity. The result is a sigmoid profile of gentle flexure, upwardly convex for a short space near its landward end, and concave beyond.

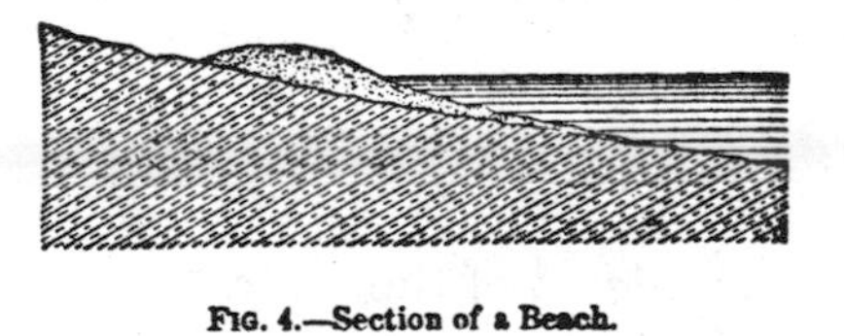

FIG. 4.—Section of a Beach.

In horizontal contour the beach follows the original boundary between land and lake, but does not conform to its irregularities. Small indentations are filled with shore drift, small projections are cut away, and smooth, sweeping curves are given to the water margin and to the submerged contours within reach of the breakers.

The beach graduates insensibly into the wave-cut terrace. A cut-terrace lying in the route of shore drift is alternately buried by drift and swept bare, as the conditions of wind and breaker vary. The cut-and-built terrace (Figure 5), which owes its detrital extension to the agencies determining the beach profile, may be regarded as a form intermediate between the beach and the cut terrace.

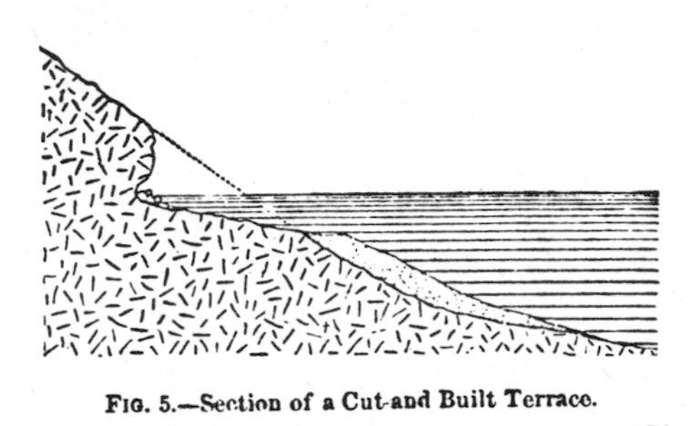

FIG. 5.—Section of a Cut-and Built Terrace.

The Barrier.—Where the sublittoral bottom of the lake has an exceedingly gentle inclination the waves break at a considerable distance from the water margin. The most violent agitation of the water is along the line of breakers; and the shore drift, depending upon agitation for its transportation, follows the line of the breakers instead of the water margin. It is thus built into a continuous outlying ridge at some distance from the water's edge. It will be convenient to speak of this ridge as a *barrier*.

FIG. 6.—Section of a Barrier.

The barrier is the functional equivalent of the beach. It is the road along which shore drift travels, and it is itself composed of shore drift. Its lakeward face has the typical beach profile, and its crest lies a few feet above the normal level of the water.

Between the barrier and the land a strip of water is inclosed, constituting a lagoon. This is frequently converted into a marsh by the accumulation of silt and vegetable matter, and eventually becomes completely filled, so as to bridge over the interval between land and barrier and convert the latter into a normal beach.

The beach and the barrier are absolutely dependent on shore drift for their existence. If the essential continuous supply of moving detritus is cut off, not only is the structure demolished by the waves which formed it, but the work of excavation is carried landward, creating a wave-cut terrace and a cliff.

The principal elements of the theory of shore-drift deposits here set forth are tacitly postulated by many writers on the construction of harbor and coast defenses. According to Cialdi[1] the potency of currents in connection with waves was first announced by Montanari; it has been concisely and, so far as appears, independently elucidated by Andrews.[2]

Still water level is the datum with which all vertical elements of the profile of the beach and barrier are necessarily compared; and, referred to this standard, not only does the maximum height of the beach or barrier vary in different parts of the same shore, but the profile as a whole stands at different heights.

The explanation of these inequalities depends in part on a principle of wide application, which is on the one hand so important and on the other so frequently ignored that a paragraph may properly be devoted to it, by way of digression. There are numerous geologic processes in which quantitative variations of a causative factor work immensely greater quantitative variations of the effect. It is somewhat as though the effect was proportioned to an algebraic power of the cause, but the relation is never so simple. Take, for example, the transportation of detritus by a stream. The variable cause is the volume of water; the variable effect is the amount of geologic work done—the quantity of detritus transported. The effect is related to the cause in three different ways: First, increase of water volume augments the velocity of flow, and with increase of velocity the size of the maximum particle which can be moved increases rapidly. According to Hopkins, the size of the maximum fragment which can be moved varies as the sixth power of the velocity, or (roughly) as the $\frac{3}{2}$ power of the volume of water. Second, the increase of velocity enlarges the capacity of the water to transport detritus of a given character; that is, the per cent of load to the unit of water is increased. Third, increase in the number of unit volumes of water increases the load pro rata. The summation of these three tendencies gives to the flooded stream a transporting power scarcely to be compared with that of the same stream at its low stage, and it gives to the exceptional flood a

[1] Loc. cit., p. 394, et seq. Cialdi himself maintains at great length that the work is performed by waves, and that the so-called shore current, a feeble peripheral circulation observed in the Mediterranean, is qualitatively and quantitatively incompetent to produce the observed results. Whether he would deny the efficiency of currents excited by the same winds which produce the waves is not clearly apparent.

[2] Trans. Chicago. Acad. Sci., vol. 2, p. 9.

power greatly in excess of the normal or annual flood. Not only is it true that the work accomplished in a few days during the height of the chief flood of the year is greater than all that is accomplished during the remainder of the year, but it may even be true that the effect of the maximum flood of the decade or generation or century surpasses the combined effects of all minor floods. It follows that the dimensions of the channel are established by the great flood and adjusted to its needs.

In littoral transportation the great storm bears the same relation to the minor storm and to the fair-weather breeze. The waves created by the great storm not only lift more detritus from each unit of the littoral zone, but they act upon a broader zone, and they are competent to move larger masses. The currents which accompany them are correspondingly rapid, and carry forward the augmented shore drift at an accelerated rate. It follows that the habit of the shore, including not only the maximum height of the beach line and the height of its profile, but the dimensions of the wave-cut terrace and of various other wave products presently to be described, is determined by and adjusted to the great storm.

It should be said by way of qualification that the low-tide stream and the breeze-lifted wave have a definite though subordinate influence on the topographic configuration. After the great flood has passed by, the shrunken stream works over the finer debris in the bed of the great channel, and by removing at one place and adding at another shapes a small channel adjusted to its volume. After the great storm has passed from the lake and the storm swell has subsided, the smaller waves of fair weather construct a miniature beach profile adapted to their size, superposing it on the greater profile. This is done by excavating shore drift along a narrow zone under water and throwing it up in a narrow ridge above the still water level. Thus, as early perceived by De la Beche[1] and Beaumont,[2] it is only for a short time immediately after the passage of the great storm that the beach profile is a simple curve; it comes afterward to be interrupted by a series of superposed ridges produced by storms of different magnitude.

Reverting now to the special conditions controlling the profiles of beach or barrier at an individual locality, it is evident that the chief of these is the

[1] Manual of Geology, Philadelphia, 1832, p. 72.

[2] Leçons, p. 226 and plate IV.

magnitude of the largest waves breaking there. The size of the waves at each locality depends on the force of the wind and on its direction. A wind blowing from the shore lakeward produces no waves on that shore. One from the opposite shore produces waves whose height is approximately proportional to the square root of the distance through which they are propagated, provided there are no shoals to check their augmentation. For a given force of wind, the greatest waves are produced when the direction is such as to command the broadest sweep of water before their incidence at the particular spot, or in the technical phrase, when the *fetch* is greatest.

A second factor is found in the configuration of the bottom. Where the off-shore depth is great the undertow rapidly returns the water driven forward by the wind, and there is little accumulation against the shore; but where the off-shore depth is small the wind piles the water against the shore, and produces all shore features at a relatively high level.

The Subaqueous Ridge.—Various writers have mentioned low ridges of sand or gravel running parallel to the shore and entirely submerged. As the origin of such ridges is not understood, they have no fixed position in the present classification, and they are placed next to the barrier only because of similarity of form. The following description was published by Desor in 1851:

An example of this character occurs on the northern shore of Lake Michigan, not far from the fish station of Bark Point (Pointe aux Écorces), under the lee of a promontory, designated on the map as Point Patterson. Here, the shore, after running due east and west for some distance, bends abruptly to the northeast. The voyageur coming from the west, after having passed Point Patterson, is struck by the appearance of several bands of shallow water, indicated by a yellowish tint. These bands, which appear to start from the extremity of the point, are caused by subaqueous ridges, which spread, fan-like, to the distance of nearly half a mile to the east, being from three to ten yards wide, and from five to ten feet above the general bed of the lake, at this point. They are not composed, like the flats, of fine sand, but of white limestone pebbles, derived from the adjacent ledges, with an admixture of granitic pebbles, some of which are a foot in diameter. It is difficult to conceive of currents sufficiently powerful to transport and arrange such heavy materials, and yet we know of no other means by which this aggregation could have been accomplished.

These subaqueous ridges afford, on a small scale, an interesting illustration of the formation of similar ridges now above water. If the north coast of Lake Michigan were to be raised only twenty feet, such a rise would lay dry a wide belt of almost level ground, on which these ridges would appear conspicuously, not unlike those which occur on the south shores of lakes Erie and Ontario, and thus confirm the views of Mr. Whittlesey, that most of these ridges are not ancient beaches, but have been formed under water, by the action of currents.[1]

[1] Foster and Whitney's "Geology of the Lake Superior Land District." Part 2, p. 258.

Whittlesey describes no examples on existing coasts, but refers to them as familiar features and relegates to their category numerous inland ridges associated with earlier water surfaces in the basins of Lakes Erie, Ontario, and Michigan. He says that "their composition is universally coarse water-washed sand and fine gravel", while beaches consist of "clean beach sand and shingle"; and also that beaches are distinguished from subaqueous ridges by the fact "that the former are narrow and are steepest on the lake side, resembling miniature terraces."[1]

Having personally observed many of the inland ridges described by Whittlesey and recognized them as barriers, having failed or neglected to observe ridges of this subaqueous type in the Bonneville Basin, and having independent reason to believe that the waters of Lakes Michigan, Erie, and Ontario have recently advanced on their coasts, I leaped to the conclusion that the ridges seen by Desor beneath the water of Lake Michigan, as well as the subaqueous ridges mentioned without enumeration by Whittlesey, were formed as barriers or spits at the water surface and were subsequently submerged by a rise of the water.[2] In so doing I ignored an important observation by Andrews, who, writing of the beach at the head of Lake Michigan, describes "a peculiarity in the contour of the deposit, which is uniform in all the sand shores of this part of the coast. As you go out into the lake, the bottom gradually descends from the water line to the depth of about five feet, when it rises again as you recede from the shore, and then descends toward deep water, forming a subaqueous ridge or 'bar' parallel to the beach and some ten or twenty rods from the shore."[3] It is impossible to regard this sand ridge as a beach or barrier submerged by the rise of the lake, for it stands within the zone of action of storm waves, and no mole of loose debris can be assumed to successfully oppose their attack. It is to be viewed rather as a product of wave action, or of wave and current action, under existing relations of land and lake.

The subject is advanced by Russell, who visited the eastern shore of Lake Michigan in 1884. He says:

> Bars of another character are also formed along lake margins, at some distance from the land, which agree in many ways with true barrier bars, but differ in being

[1] Fresh-water Glacial Drift of the Northwestern States. By Charles Whittlesey. Smithsonian Contribution No. 197. Washington, 1866, pp. 17, 19.

[2] Fifth Ann. Rept. U. S. Geol. Survey, p. 111.

[3] Trans. Chicago Acad. Sci., vol. 2, p. 14.

composed of homogeneous, fine material, usually sand, and in not reaching the lake surface.

The character of structures of this nature may be studied about the shores of Lake Michigan, where they can be traced continuously for hundreds of miles. There are usually two, but occasionally three, distinct sand ridges; the first being about 200 feet from the land, the second 75 or 100 feet beyond the first, and the third, when present, about as far from the second as the second is from the first. Soundings on these ridges show that the first has about 8 feet of water over it, and the second usually about 12; between, the depth is from 10 to 14 feet. From many commanding points, as the summit of Sleeping Bear Bluff, for example, these submerged ridges may be traced distinctly for many miles. They follow all the main curves of the shore, without changing their character or having their continuity broken. They occur in bays as well as about the bases of promontories, and are always composed of clean, homogeneous sand, although the adjacent beach may be composed of gravel and boulders. They are not shore ridges submerged by a rise of the lake, for the reason that they are in harmony with existing conditions, and are not being eroded or becoming covered with lacustral sediments.

In bars of this character the fine debris arising from the comminution of shore drift appears to be accumulated in ridges along the line where the undertow loses its force; the distance of these lines from the land being determined by the force of the storms that carried the waters shoreward. This is only a suggested explanation, however, as the complete history of these structures has not been determined.[1]

In the survey of these lakes by the U. S. Engineers, numerous inshore soundings were made, and while these do not fall near enough together to determine the configuration of subaqueous ridges, they serve to show whether the profile of the bottom descends continuously from the beach lakeward. A study of the original manuscript sheets, which give fuller data than the published charts, discovers that bars similar to those described by Russell occur along the eastern coast of Lake Michigan wherever the bottom is sandy, being most frequently detectible at a depth of 13 feet, but ranging upward to 3 feet and downward to 18 feet. At the south end of the lake they are not restricted to the 5-foot zone indicated by Andrews, but range to 13 feet. A single locality of occurrence was found on the shore of Lake Erie, but none on Lake Ontario.

These ridges constitute an exception to the beach profile, and show that the theory of that profile given above is incomplete. Under conditions not yet apparent, and in a manner equally obscure, there is a rhythmic action along a certain zone of the bottom. That zone lies lower than the trough between the greatest storm waves, but the water upon it is violently oscil-

[1] Geol. Hist. of Lake Lahontan. pp. 92–93.

lated by the passing waves. The same water is translated lakeward by the undertow, and the surface water above it is translated landward by the wind, while both move with the shore current parallel to the beach. The rhythm may be assumed to arise from the interaction of the oscillation, the landward current, and the undertow.

LITTORAL DEPOSITION.

The material deposited by shore processes is, first, shore drift; second, stream drift, or the detritus delivered at the shore by tributary streams Increasing depth of water is in each case the condition of littoral deposition. The structures produced by the deposit of shore drift, although somewhat varied, have certain common features. They will be treated under the generic title of *embankments*. The structures produced by the deposit of stream drift are *deltas*.

EMBANKMENTS.

The current occupying the zone of the shore drift and acting as the coagent of littoral transportation has been described as slow, but it is inseparably connected with a movement that is relatively rapid. This latter, which may be called the off-shore current, occupies deeper water and is less impeded by friction. It may in some sense be said to drag the littoral current along with it. The momentum of the off-shore current does not permit it to follow the sinuosities of the water margin, and it sweeps from point to point, carrying the littoral current with it. There is even a tendency to generate eddies or return currents in embayments of the coast. The off-shore current is moreover controlled in part by the configuration of the bottom and by the necessity of a return current. The littoral current, being controlled in large part by the movements of the off-shore current, separates from the water margin in three ways: first, it continues its direction unchanged at points where the shore-line turns landward, as at the entrances of bays; second, it sometimes turns from the land as a surface current; third, it sometimes descends and leaves the water margin as a bottom current.

In each of these three cases deposition of shore drift takes place by reason of the divorce of shore currents and wave action. The depth to which wave agitation sufficient for the transportation of shore drift extends is small, and when the littoral current by leaving the shore passes into deeper waters the shore drift, unable to follow, is thrown down.

When the current holds its direction and the shore-line diverges, the embankment takes the form of a *spit*, a *hook*, a *bar*, or a *loop*. When the shore-line holds its course and the current diverges, whether superficially or by descent, the embankment usually takes the form of a *terrace*.

The Spit.—When a coast line followed by a littoral current turns abruptly landward, as at the entrance of a bay, the current does not turn with it, but holds its course and passes from shallow to deeper water. The water between the diverging current and coast is relatively still, although there is communicated to the portion adjacent to the current a slow motion in the same direction. The waves are propagated indifferently through the flowing and the standing water, and reach the coast at all points. The shore drift can not follow the deflected coast line, because the waves that beat against it are unaccompanied by a littoral current. It can not follow the littoral current into deep water, because at the bottom of the deep water there is not sufficient agitation to move it. It therefore stops. But the supply of shore drift brought to this point by the littoral current does not cease, and the necessary result is accumulation. The particles are carried forward to the edge of the deep water and there let fall.

In this way an embankment is constructed, and so far as it is built it serves as a road for the transportation of more shore drift. The direction in which it is built is that of the littoral current. It takes the form of a ridge following the boundary between the current and the still water. Its initial height brings it just near enough to the surface of the water to enable the wave agitation to move the particles of which it is constructed; and it is narrow. But these characters are not long maintained. The causes which lead to the construction of the beach and the barrier are here equally efficient, and cause the embankment to grow in breadth and in height until the cross-profile of its upper surface is identical with that of the beach.

The history of its growth is readily deduced from the configuration of its terminus, for the process of growth is there in progress. If the material is coarse the distal portion is very slightly submerged, and is terminated in the direction of growth by a steep slope, the subaqueous "earth-slope" of the particular material. If the material is fine the distal portion is more deeply submerged, and is not so abruptly terminated. The portion above water is usually narrow throughout, and terminates without reaching the extremity of the embankment. It is flanked on the lakeward side by a submerged plateau, at the outer edge of which the descent is somewhat steep. The profile of the plateau is that normal to the beach, and its contours are confluent with those of the beach or barrier on the main shore. Toward the end of the embankment its width diminishes, its outer and limiting contour turning toward the crest line of the spit and finally joining it at the submerged extremity.

The process of construction is similar to that of a railroad embankment the material for which is derived from an adjacent cutting, carted forward along the crest of the embankment and dumped off at the end; and the symmetry of form is often more perfect than the railway engineer ever accomplishes. The resemblance to railway structures is very striking in the case of the shores of extinct lakes.

As the embankment is carried forward and completed, contact between the current and the inshore water is at first obstructed and finally cut off, so that there is practically no communication of movement from one to the other at the extremity of the spit. At the point of construction the moving and the standing water are sharply differentiated, and there is hence no uncertainty as to the direction of construction. The spit not only follows the line between the current and still water, but aids in giving definition to that line, and eventually walls in the current by contours adjusted to its natural flow.

The Bar.—If the current determining the formation of a spit again touches the shore, the construction of the embankment is continued until it spans the entire interval. So long as one end remains free the vernacular of the coast calls it a *spit;* but when it is completed it becomes a *bar.* Figure 7 gives an ideal cross-section of a completed embankment.

The bar has all the characters of the spit except those of the terminal end. Its cross-profile shows a plateau bounded on either hand by a steep slope. The surface of the plateau is not level, but has the beach profile, is

BAR ON THE SHORE OF LAKE MICHIGAN.

From a photograph by I. C. Russell.

slightly submerged on the windward side and rises somewhat above the ordinary water level at the leeward margin. At each end it is continuous with a beach or barrier. It receives shore drift at one end and delivers it at the other.

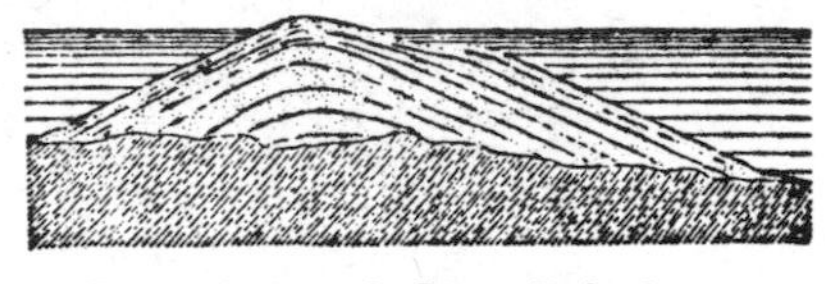

Fig. 7.—Section of a Linear Embankment.

The bar may connect an island with the shore or with another island, or it may connect two portions of the same shore. In the last case it crosses the mouth either of a bay or of a river. If maintained entire across the entrance to a bay it converts the water between it and the shore into a lagoon. At the mouth of a river its maintenance is antagonized by the outflowing current, and if its integrity is established at all it is only on rare occasions and for a short time. That is to say, its full height is not maintained; there is no continuous exposed ridge. The shore drift is, however, thrown into the river current, and unless that current is sufficient to sweep it into deep water a submerged bar is thrown across it, and maintains itself as a partial obstruction to the flow. The site of this submerged bar is usually also the point at which the current of the stream, meeting the standing water of the lake, loses its velocity and deposits the coarser part of its load of detritus. If the contribution of river drift greatly exceeds that of shore drift, a delta is formed at the river mouth, and this, by changing the configuration of the coast, modifies the littoral current and usually determines the shore drift to some other course. If the contribution of river drift is comparatively small it becomes a simple addition to the shore drift, and does not interrupt the continuity of its transportation. The bars at the mouths of small streams are constituted chiefly of shore drift, and all their characters are determined by their origin. The bars at the mouths of large streams are constituted chiefly of stream drift, and belong to the phenomena of deltas.

On a preceding page the fact was noted that the horizontal contours of a beach are more regular than those of the original surface against which it rests, small depressions being filled. It is now evident that the process of filling these is identical with that of bar construction. There is no trenchant line of demarkation between the beach and the bar. Each is a carrier of shore drift, and each employs its first load in the construction of a suitable road.

Plate IV represents a part of the east shore of Lake Michigan seen from the hill back of Empire Bluffs. In the extreme distance at the left stand the Sleeping Bear Bluffs, and somewhat nearer on the shore is a timbered hill, the lakeward face of which is likewise a sea-cliff. A bar connects the latter with the land in the foreground and divides the lagoon at the right from the lake at the left. The symmetry of the bar is marred by the formation of dunes, the lighter portion of the shore-drift being taken up by the wind and carried toward the right so as to initiate the filling of the lagoon.

Figure 8 is copied from the U. S. Engineer map of a portion of the south shore of Lake Ontario west of the mouth of the Genesee River. The orig-

FIG. 8.—Map of Braddock's Bay and vicinity, N. Y., showing headlands connected by Bars.

inal contour of the shore was there irregular, consisting of a series of salient and reentrant angles. The waves have truncated some of the salients and have united them all by a continuous bar, behind which several bays or ponds are inclosed. The movement of the shore drift is in this case from northwest to southeast, and the principal source of the material is a point of land at the extreme west, where a low cliff shows that the land is being eaten by the waves.

The map in Figure 9 is also copied from one of the sheets published by the U. S. Engineers, and represents the bars at the head of Lake Superior. These illustrate several elements of the preceding discussion. In the first place they are not formed by the predominant winds, but by those which bring the greatest waves. The predominant winds are westerly, and produce no waves on thi scoast. The shore drift is derived from the south coast, and its motion is first westerly and then northerly. Two bars are exhibited, the western of which is now protected from the lake waves, and must have been completed before the eastern was begun. The place of deposition of shore drift was probably shifted from the western to the eastern by reason of the shoaling of the head of the lake. The converging shores should theoretically produce during easterly storms a powerful undertow, by which a large share of the shore drift would be carried lakeward and distributed over the bottom. The manner in which the bars terminate against the northern shore without inflection is explicable likewise by the theory of a strong undertow. If the return current were superficial the bars would be curved at their junctions with both shores.

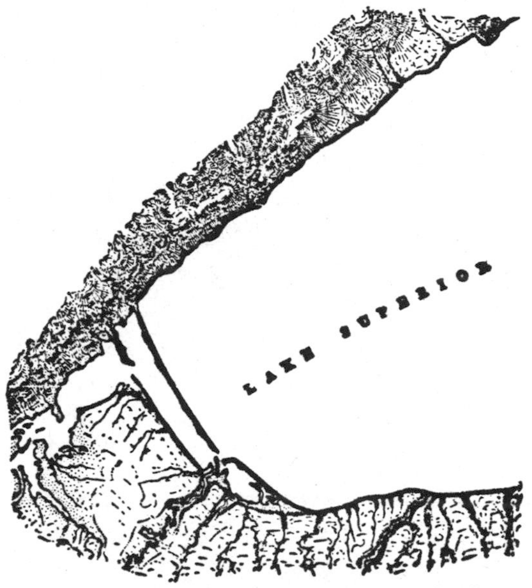

FIG. 9.—Map of the head of Lake Superior, showing Bay Bars.

An instructive view of an ancient bar will be found in Pl. IX, representing a portion of the Bonneville shore line. The town of Stockton, Utah, appears at the right. The plain at the left was the bed of the lake. The storm waves, moving from left to right, carved the sea-cliff which appears at the base of the mountain at the left, and drifting the material toward the right built it into a great spit and a greater bar. The end of the spit is close to the town. The bar, which lies slightly lower, having been formed by the lake at a lower stage of its water, sweeps in a broad curve across the valley to the rocky hill on the opposite side, where the artist stood in making the sketch.

The Hook.—The line of direction followed by the spit is usually straight, or has a slight concavity toward the lake. This form is a function of the littoral current, to which it owes origin. But that current is not perpetual; it exists only during the continuance of certain determining winds. Other winds, though feebler or accompanied by smaller waves, nevertheless have systems of currents, and these latter currents sometimes modify the form of the spit. Winds which simply reverse the direction of the littoral current retard the construction of the embankment without otherwise affecting it; but a current is sometimes made to flow past the end of the spit in a direction making a high angle with its axis, and such a current modifies its form. It cuts away a portion of the extremity and rebuilds the material in a smaller spit joining the main one at an angle. If this smaller spit extends lakeward it is demolished by the next storm; but if it extends landward its position is sheltered, and it remains a permanent feature. It not infrequently happens that such accessory spits are formed at intervals during the construction of a long embankment, and are preserved as a series of short branches on the lee side.

It may occur also that a spit at a certain stage of its growth becomes especially subject to some conflicting current, so that its normal growth ceases, and all the shore drift transported along it goes to the construction of the branch. The bent embankment thus produced is called a *hook*.

The currents efficient in the formation of a hook do not cooperate simultaneously, but exercise their functions in alternation. The one, during the prevalence of certain winds, brings the shore drift to the angle and accumulates it there; the other, during the prevalence of other winds, demolishes the new structure and redeposits the material upon the other limb of the hook.

In case the land on which it is based is a slender peninsula or a small island, past which the currents incited by various winds sweep with little modification of direction by the local configuration, the hook no longer has the sharp angle due to the action of two currents only, but receives a curved form.

Hooks are of comparatively rare occurrence on lake shores, but abound at the mouths of marine estuaries, where littoral and tidal currents conflict.

Plate V represents a recurved spit on the shore of Lake Michigan, seen from a neighboring bluff. The general direction of its construction is from left to right, but storms from the right have from time to time turned its end toward the land and the successive recurvements are clearly discernible near the apex.

The mole enclosing Toronto harbor on the shore of Lake Ontario is a hook of unusual complexity, and the fact that its growth threatens to close the entrance to the harbor has led to its thorough study by engineers. Especially has its history been developed by Fleming in a classic essay to which reference has already been made. A hill of drift projects as a cape from the north shore of the lake. The greatest waves reaching it, those having the greatest fetch, are from the east (see Fig. 10), and the cooperating current flows from east to west. As the hill gradually yields to the waves, its coarser material trails westward, building a spit. The waves and currents set in motion by southwesterly winds carry the spit end northward, producing a hook. In the past the westward movement has been the more powerful and the spit has continued to grow in that direction, its northern edge being fringed with the sand ridges due to successive recurvements, but the shape of the bottom has introduced a change of conditions. The water at the west end of the spit is now deep, and the extension of the embankment is correspondingly slow. The northward drift, being no longer subject to frequent shifting of position, has cumulative effect on the terminal hook and gives it a greater length than the others. In the chart of the harbor (Fig. 11) the composite character of the mole is readily traced. It may

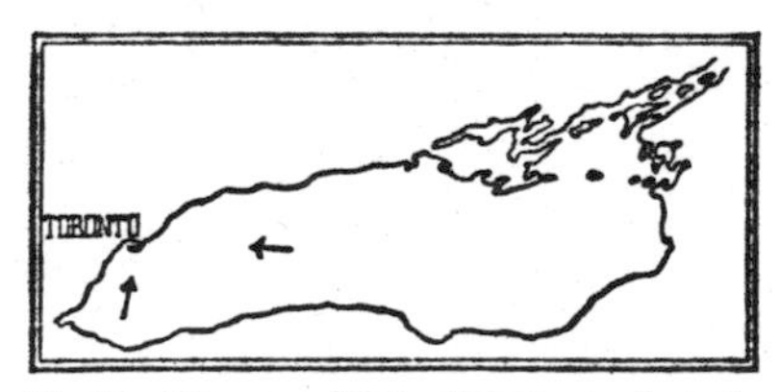

Fig. 10.—Diagram of Lake Ontario, to show the Fetch of Waves reaching Toronto from different directions.

A HOOK. DUTCH POINT, GRAND TRAVERSE BAY, LAKE MICHIGAN.

From a photograph by I. C. Russell.

also be seen that the ends of the successive hooks are connected by a beach, the work of waves generated within the harbor by northerly winds.[1] It will be observed furthermore that while the west end of the spit is continuously fringed by recurved ridges its eastern part is quite free from them. This does not indicate that the spit was simple and unhooked in the early stages of growth, but that its initial ridge has disappeared. As the cliff is eroded,

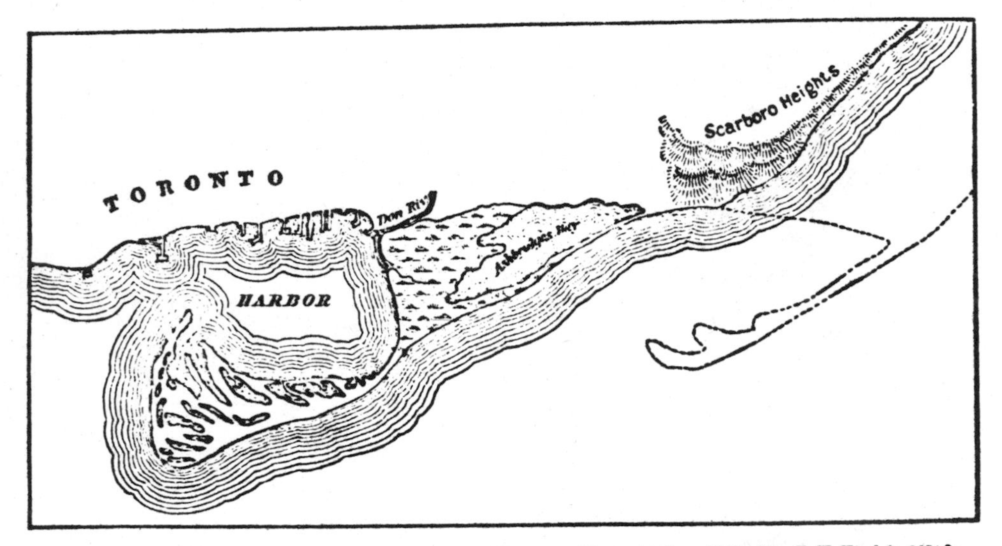

FIG. 11.—Map of the harbor and peninsula (Hook) at Toronto. From charts published by H. Y. Hind, in 1854.[2]

its position constantly shifts landward, the shore current follows, and the lakeward face of the spit is carried away so that the waves break over it, and then a new crest is built by the waves just back of the line of the old one.[3] By this process of partial destruction and renewal the spit retreats, keeping pace with the retreating cliff. At an earlier stage of the process the spit may have had the position and form indicated by the dotted outline, but whatever hooks fringed its inner margin have disappeared in the process of retreat.

[1] The marsh occupying part of the space between the spit and the mainland (Fig. 11) is only incidentally connected with the feature under discussion. A small stream, the Don, reaches the shore of the lake within the tract protected from waves by the hook and is thus enabled to construct a delta with its sediment.

[2] Report on the preservation and improvement of Toronto Harbor. In Supplement to Canadian Journal, 1854.

[3] At the present time the spit is divided near the middle, a natural breach having been artificially prevented from healing. The portion of the peninsula fringed by successive hooks stands as an island.

CUP BUTTE, A FEATURE OF THE BONNEVILLE SHORE-LINE.

The landward shifting illustrated by the Toronto hook affects many embankments, but not all. It ordinarily occurs when the embankment is built in deep water and the source of its material is close at hand. Wherever it is known that an embankment has at some time been breached by the waves, it may be assumed with confidence that retreat is in progress.

As retreat progresses the layers constituting the embankment are truncated at top, and new layers are added on the landward side. In the resulting structure the prevailing dip is landward (Fig. 12), and it is thereby distinguished from all other forms of lacustrine deposition. This structure was first described and explained by Fleming, who observed it in a railway cutting through an ancient spit.[1]

The Loop.—Just as the spit, by advancing until it rejoins the shore, becomes a bar, so the completed hook may with propriety be called a *loop* or a *looped bar*. There is, however, a somewhat different feature to which the name is more strikingly applicable. A small island standing near the main-land is usually furnished on each side with a spit streaming toward the land. These spits are composed of detritus eroded from the lakeward face of the island, against which beat the waves generated through the broad expanse. The currents accompanying the waves are not uniform in direction, but vary with the wind through a wide angle; and the spits, in sympathy with the varying direction of currents, are curved inward toward the island. If their extremities coalesce, they constitute together a perfect loop, resembling, when mapped, a festoon pendent from the sides of the island.

Such a loop in the fossil condition, that is, when preserved as a vestige of the shore of an extinct lake, has the form of a crater rim, the basin of the original lagoon remaining as an undrained hollow. The accompanying illustration (Pl. VI) represents an island of Lake Bonneville standing on the desert near what is known as the "Old River Bed." The nucleus of solid rock was in this instance nearly demolished before the work of the waves was arrested by the lowering of the water.

The Wave-built Terrace.—It has already been pointed out that when a separation of the littoral current from the coast line is brought about by a divergence of the current rather than of the coast line, there are two cases, in the

[1] Notes on the Davenport gravel drift. Canadian Journal, New Series, vol. 6, 1861, pp. 247–253.

first of which the current continues at the surface, while in the second it dives beneath the surface. It is now necessary to make a further distinction. The current departing from the shore, but remaining at the surface, may continue with its original velocity or it may assume a greater cross-section and a diminished velocity. In the first case the shore drift is built into a spit or other linear embankment. In the second case it is built into a terrace. The quantity of shore drift moved depends on the magnitude of the waves; but the speed of transit depends on the velocity of the current, and wherever that velocity diminishes, the accession of shore drift must exceed the transmission, causing accumulation to take place. This accumulation occurs, not at the end of the beach, but on its face, carrying its entire profile lakeward and producing by the expansion of its crest a tract of new-made land. If afterward the water disappears, as in the case of an extinct lake, the new-made land has the character of a terrace. A current which leaves the shore by descending, practically produces at the shore a diminution of flow, and the resulting embankment is nearly identical with that of a slackening superficial current.

The wave-built terrace is distinct from the wave-cut terrace in that it is a work of construction, being composed entirely of shore drift, while the wave-cut terrace is the result of excavation, and consists of the pre-existent terrane of the locality. The wave-built terrace is an advancing embankment, and its internal structure is characterized by a lakeward dip (Fig. 13). It is thus contrasted with the retreating embankment (Fig. 12).

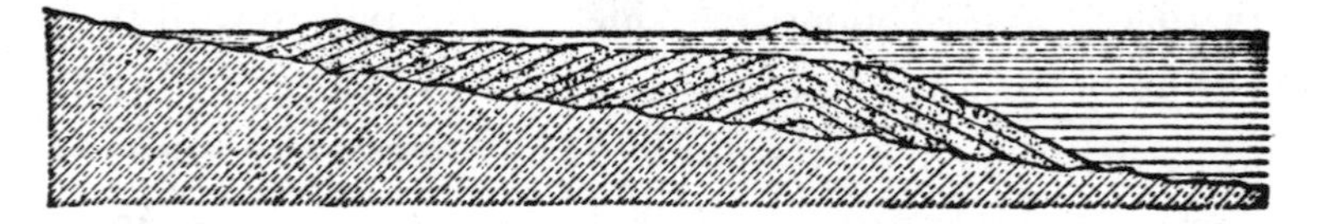

FIG. 12.—Section of a Linear Embankment retreating landward. The dotted line shows the original position of the crest.

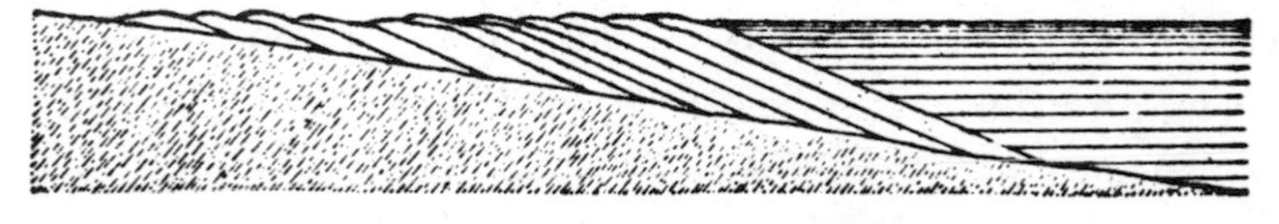

FIG. 13.—Section of a Wave-built Terrace.

The surface of the wave-built terrace, considered as a whole, is level, but in detail it is uneven, consisting of parallel ridges, usually curved. Each of these is referable to some exceptional storm, the waves of which threw the shore drift to an unusual height.

Where the shore drift consists wholly or in large part of sand, and the prevailing winds are toward the shore, the wave-built terrace gives origin to dunes, which are apt to mask its normal ribbed structure.

The locality most favorable for the formation of a wave-built terrace is the head of a triangular bay, up which the waves from a large body of water are rolled without obstruction. The wind sweeping up such a bay carries the surface of the water before it, and the only return current is an undertow originating near the head of the bay. The superficial advance of the water constitutes on each shore a littoral current conveying shore drift toward the head of the bay, and as these littoral currents are diminished and finally entirely dissipated by absorption in the undertow, the shore drift taken up along the sides of the bay is deposited. If the head of the bay is acute, the first embankment built is a curved bar tangent to the sides and concave toward the open water. To the face of this successive additions are made, and a terrace is gradually produced, the component ridges of which are approximately parallel. The sharpest curvature is usually at the extreme head of the bay.

The converging currents of such a bay give rise to an undertow which is of exceptional velocity, so that it transports with it not only the finest detritus but also coarser matter, such as elsewhere is usually retained in the zone of wave action. In effect there is a resorting of the material. The shore drift that has traveled along the sides of the bay toward its head, is divided into two portions, the finer of which passes out with the reinforced undertow, while the coarser only is built into the terrace.

The V-Terrace and V-Bar.—It remains to describe a type of terrace for which no satisfactory explanation has been reached. The shores of the ancient Pleistocene lakes afford numerous examples, but those of recent lakes are nearly devoid of them, and the writer has never had opportunity to examine one in process of formation. They are triangular in ground plan, and would claim the title of delta were it not appropriated, for they simulate the Greek letter more strikingly than do the river-mouth structures. They are built against coasts of even outline, and usually, but not always, upon slight salients, and they occur most frequently in the long, narrow arms of old lakes.

One side of the triangle rests against the land and the opposite angle points toward the open water. The free sides meet the land with short curves of adjustment, and appear otherwise to be normally straight, although they exhibit convex, concave, and sigmoid flexures. The growth is by additions to one or both of the free sides; and the nucleus appears always to have been a miniature triangular terrace, closely resembling the final structure in shape. In the Bonneville examples the lakeward slope of the terrace is usually very steep down to the line where it joins the preexistent slope of the bottom.

There seems no reason to doubt that these embankments, like the others, were built by currents and waves, and such being the case the formative currents must have diverged from the shore at one or both the landward angles of the terrace, but the condition determining this divergence does not appear.

In some cases the two margins appear to have been determined by currents approaching the terrace (doubtless at different times) from opposite directions; and then the terrace margins are concave outward, and their confluence is prolonged in a more or less irregular point. In most cases, however, the shore drift appears to have been carried by one current from the mainland along one margin of the terrace to the apex, and by another current along the remaining side of the terrace back to the mainland. The contours are then either straight or convex.

In Lake Bonneville it happened that after the best defined of these terraces had attained nearly their final width the lake increased in size, so that they were immersed beneath a few feet of water. While the lake stood at the higher level, additions were made to the terraces by the building of linear embankments at their outer margins. These were carried to the water surface, and a triangular lagoon was imprisoned at each locality. The sites of these lagoons are now represented by flat triangular basins, each walled in by a bar bent in the form of a V. These bars were at first observed without a clear conception of the terrace on which they were founded, and the name V-*bar* was applied. The V-bar, while a conspicuous feature of

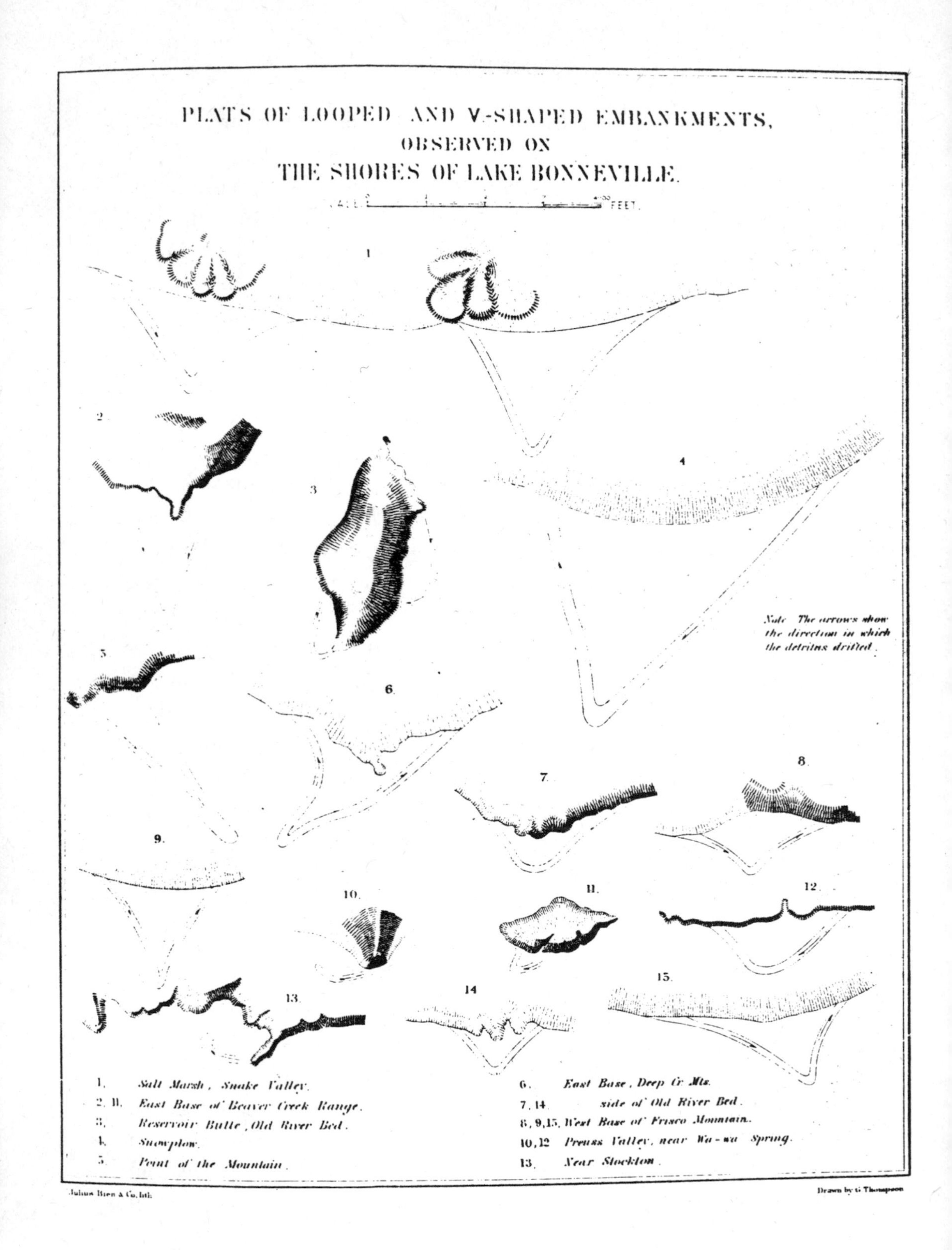
PLATS OF LOOPED AND V-SHAPED EMBANKMENTS,
OBSERVED ON
THE SHORES OF LAKE BONNEVILLE.
SCALE
FEET.
1
2
3
4
5
6.
7.
8.
9.
10.
11.
12.
13.
14
15.
Note. The arrows show the direction in which the detritus drifted.
1. Salt Marsh, Snake Valley.
2, 11. East Base of Beaver Creek Range.
3, Reservoir Butte, Old River Bed.
4. Snowplow.
5. Point of the Mountain.
6. East Base, Deep Cr. Mts.
7, 14. side of Old River Bed.
8, 9, 15. West Base of Frisco Mountain.
10, 12 Preuss Valley, near Wa-wa Spring.
13. Near Stockton.
Julius Bien & Co. lith.
Drawn by G. Thompson

the Bonneville shores, is not believed to be a normal feature of lakes maintaining a constant level.

DRIFTING SAND; DUNES.

The dune is not an essential shore feature, but is an accessory of frequent occurrence.

Dunes are formed wherever the wind drifts sand across the land. The conditions essential to their production are wind, a supply of sand, and sterility or the absence of a protective vegetal growth. In arid regions sterility is afforded by the climatic conditions, and the sand furnished by river bars laid bare at low water, and by the disintegration of sand rocks, is taken up by the wind and built into dunes; but where rain is abundant, accumulations of such sort are protected by vegetation, and the only sources of supply are shores, either modern or ancient.

Shore drift nearly always contains some sand, and is frequently composed exclusively thereof. The undertow carries off the clay, which might otherwise hold the sand particles together and prevent their removal by the wind; and pebbles and bowlders, which, by their superior weight oppose wind action, are less able to withstand the attrition of littoral transportation, and disappear by disintegration from any train of shore drift which travels a considerable distance. Embankments are therefore apt to be composed largely of sand; and the crests of embankments, being exposed to the air during the intervals between great storms, yield dry sand to the gentler winds.

The sand drifted from the crests of free embankments, such as barriers, spits, and bars, quickly reaches the water on one side or the other. What is blown to the lakeward side falls within the zone of wave action, and is again worked over as shore drift. What is blown to the landward side extends the area of the embankment, correspondingly encroaching on the lagoon or bay.

Sand blown from the crests of embankments resting against the land, such as beaches and terraces, will spread over the land if the prevailing wind is favorable. In cases where the prevailing wind is toward the lake the general movement of sand is, of course, in that direction, and it is merely returned to the zone of the waves and readded to the shore drift; but where the prevailing winds are toward the land, dunes are formed and slowly rolled forward by the wind. The supply of dry sand afforded by beaches is comparatively small, and dunes of magnitude are not often formed from it. The great sand magazines are wave-built terraces, and it is from these that the trains of sand so formidable to agriculture have originated.

The sands accumulated on the shores of lakes and oceans now extinct are sometimes so clean that vegetation acquires no foothold, and the wind still holds dominion. The "oak openings" of Western States are usually of this nature; and in the Great Basin there are numerous trains of dunes conveying merely the sand accumulated on the shores of the Pleistocene lakes.

One product of littoral deposition—the delta—remains undescribed; but this is so distinct from the embankment, not only in form but in process of construction, that its consideration will be deferred until the interrelations of the three processes already described have been discussed.

THE DISTRIBUTION OF WAVE-WROUGHT SHORE FEATURES.

Upon every coast there are certain tracts undergoing erosion; certain others receive the products of erosion, and the intervals are occupied by the structures peculiar to transportation. Let us now inquire what are the conditions determining these three phases of shore shaping.

It will be convenient to consider first the conditions of transportation. In order that a particular portion of shore shall be the scene of littoral transportation, it is essential, first, that there be a supply of shore drift; second, that there be shore action by waves and currents; and in order that the local process be transportation simply, and involve neither erosion nor deposition, a certain equilibrium must exist between the quantity of the shore drift on the one hand and the power of the waves and currents on the other. On the whole this equilibrium is a delicate one, but within certain narrow limits it is stable. That is to say, there are certain slight variations of the individual conditions of equilibrium, which disturb the equilibrium only in a manner tending to its immediate readjustment. For example, if the shore drift receives locally a small increment from stream drift, this increment, by adding to the shore contour, encroaches on the margin of the littoral current and produces a local acceleration, which acceleration leads to the removal of the obstruction. Similarly, if from some temporary cause there is a local defect of shore drift, the resulting indentation of the shore contour slackens the littoral current and causes deposition, whereby the equilibrium is restored. Or if the force of the waves is broken at some point by a temporary obstruction outside the line of breakers, as for example by a wreck, the local diminution of wave agitation produces an accumulation of shore drift whereby the littoral current is narrowed and thus accelerated until an adjustment is reached.

Outside the limits thus indicated everything which disturbs the adjustment between quantity of shore drift and capacity of shore agents leads either to progressive local erosion or else to progressive local deposition. The stretches of coast which either lose or gain ground are decidedly in excess of those which merely hold their own.

An excessive supply of shore drift over and above what the associated current and waves are competent to transport leads to deposition. This occurs where a stream of some magnitude adds its quota of debris. A moderate excess of this nature is disposed of by the formation of a wave-built terrace on the lee side of the mouth of the stream, that is, on the side toward which flows the littoral current accompanying the greatest waves. A great excess leads to the formation of a delta, in which the stream itself is the constructing agent and the influence of waves is subordinate.

On the other hand, there is a constant loss of shore drift by attrition, the particles in transit being gradually reduced in size until they are removed from the littoral zone by the undertow. As a result of the defect thus occasioned, a part of the energy of the waves is expended on the subjacent terrane, and the work of transportation is locally accompanied by a sufficient amount of erosion to replenish the wasting shore drift. For the maintenance of a continuous beach in a permanent position, it appears to be necessary that small streams shall contribute enough debris to compensate for the waste by attrition.

Theoretically, transportation must be exchanged for erosion wherever there is a local increase in the magnitude of waves, and for deposition where there is a local decrease of waves; but practically the proportions of waves are so closely associated with the velocities of the accompanying currents that their effects have not been distinguished.

The factor which most frequently, by its variation, disturbs the equilibrium of shore action is the littoral current. It has already been pointed out that wherever it leaves the shore, shore drift is deposited; and it is equally true that wherever it comes into existence by the impinging of an open-water current on the shore, shore drift is taken up and the terrane is eroded. It has been shown also that the retardation of the littoral current produces deposition, and it is equally true that its acceleration causes erosion. Every variation, therefore, in the direction or velocity of the current at the shore has a definite effect in the determination of the local shore process.

Reentrant angles of the coast are always, and reentrant curves are usually, places of deposition. The reason for this is twofold: first, currents which follow the shore move with diminished velocity in passing reentrants; second, currents directed toward the shore escape from reentrants only by undertow, and, as heretofore explained, build terraces at the heads of the embayments.

Salient angles are usually eroded, and salient curves nearly always, the reasons being, first, that a current following the shore is relatively swift opposite a salient, and, second, that a current directed toward the shore is apt to be divided by a salient, its halves being converted into littoral currents transporting shore drift in opposite directions *away* from the salient.

Some salient angles, on the contrary, grow by deposition. This occurs where the most important current approaches by following the shore and is thrown off to deep water by a salient. The most notable instances are found on the sides of narrow lakes or arms of lakes, in which case currents approaching from the direction of the length are accompanied by greater waves than those blown from the direction of the opposite shore, and therefore dominate in the determination of the local action.

It thus appears that there is a general tendency to the erosion of salients and the filling of embayments, or to the simplification of coast outlines. This tendency is illustrated not only by the shores of all lakes, but by the coasts of all oceans. In the latter case it is slightly diminished by the action of tides, which occasion currents tending to keep open the mouths of estuaries, but it is nevertheless the prevailing tendency. The idea which sometimes appears in popular writings that embayments of the coast are eaten out by the ocean is a survival of the antiquated theory that the sculpture of the land is a result of "marine denudation." It is now understood that the diversities of land topography are wrought by stream erosion.

Figure 8, representing about seven miles of the shore of Lake Ontario, illustrates the tendency toward simplification. Each bluff of the shore marks the truncation by the waves of a cape that was originally more salient. Each beach records the partial filling of an original bay. Each bar is a wave-built structure partitioning a deep reentrant from the open lake. The lagoons receive the detritus from the streams of the land and are filling; partly for this reason there is a local defect of shore drift, and the coast is receding by erosion; and by this double process the original reentrants are suffering complete effacement. For the original coast line—a sinuous contour on a surface modeled by glacial and fluvial agencies—will be substituted a relatively short line of simple curvature.

The simplification of a coast line is a work involving time, and the amount of work accomplished on a particular coast affords a relative measure of the time consumed. There are many modifying conditions—the fetch of waves, the off-shore depth, the material of the land, the original configuration, etc.—and these leave no hope of an absolute measure; but it is possible to distinguish the young coast from the mature. When a water level is newly established against land with sinuous contour, the first work of the waves is the production of the beach profile. On the gentlest slopes they do this by excavating the terrane at the point where they first break and throwing the material shoreward so as to build a barrier. On all other slopes they establish the profile by carving a terrace with its correlative cliff. The coarser products of terrace-cutting gather at the outer edge of the terrace, helping to increase its breadth; the finer fall in deeper water and help to equalize the off-shore depth. The terrace gradually increases by the double process of cutting and filling until it has attained a certain minimum width essential to the transportation of shore drift. This width is for each locality a function of the size of the greatest waves. Before it is reached, the fragments detached from the cliff linger but a short time on the face of the terrace; after a few excursions up and down the slope they come to rest at the edge of the deeper water. When it is reached—when the beach profile is complete—the excavated fragments torn from the cliff no longer escape from the zone of wave action, but are rolled to and fro by the waves of every storm, lose their angles by attrition, and are drifted along by the shore current. It may happen that the material of the cliff is a gravel, already rounded by some earlier and independent process, but when this is not the case, the cut-terraces of adolescent and mature coasts are distinguished by the angular forms on the one hand and the rounded forms on the other of the associated detritus. When the formation of shore drift has once been begun, its further development and the development of efficient shore currents are gradual and by reciprocation. The spanning of minor recesses of the coast-line by its beach helps to smooth the way for the shore current, and the current promotes the beach. Embankments come later, when ways have been straightened for the current and shore drift, and those first constructed usually attempt the partition of only small embayments. The more extended and powerful shore currents, competent to span the bays between the greater headlands, become possible only after minor rugosities of coast and bottom have disappeared.

Low but nearly continuous sea-cliffs mark the adolescent coast; simple contours and a cordon of sand, interspersed with high cliffs, mark the mature coast. As a result of the inconstancy of the relations of land and water, it is probable that all coasts fall under these heads, but Richthofen has sketched the features of the theoretic senile coast.[1] As sea-cliffs retreat and terraces grow broader the energy of the waves is distributed over a wider zone and its erosive work is diminished. The resulting defect of shore drift permits

[1] Führer für Forschungsreisende, p. 338.

the erosion of embankments, and the withdrawal of their protection extends the line of cliff; but eventually the whole line is driven back to its limit and erosion ceases. The cliffs, no longer sapped by the waves, yield to atmospheric agencies and blend with the general topography of the land. Shore drift is still supplied by the streams and is spread over the broad littoral shoal, where it lies until so comminuted by the waves that it can float away.

The length of the period of adolescence varies with local conditions. Where the waves are powerful, maturity comes sooner than where they are weak. It comes sooner, too, where the material to be moved by the waves is soft or incoherent than where it is hard and firm; and it comes early where the submerged contours and the contour at the water's edge have few irregularities. Different parts of the same coast accordingly illustrate different stages of development. The shores of Lake Bonneville are in general mature, but in small sheltered bays they are adolescent. The shore of Lake Ontario is in general mature, being traced on a surface of glacial drift, but near the outlet is a region of bare, hard rock disposed in promontories and islands, and there much of the coast is adolescent.

The classic "parallel roads" of Glen Roy in Scotland illustrate the adolescent type, and this although the local conditions favor rapid development. The smooth contours of the valley gave no obstruction to shore currents, depth and length of lake permitted the raising of large waves, and a mantle of glacial drift afforded material for shore drift; but the beach profile was not completed, the bowlders of the narrow terraces are still subangular, and there are no embankments. It is fairly inferred that the time represented by each shore-line was short.

STREAM WORK; THE DELTA.

The detritus brought to lakes by small streams is overwhelmed by shore drift and merges with it. The tribute of large streams, on the contrary, overwhelms the shore drift and accumulates in deltas. In the formation of a normal delta the stream is the active agent, the lake is the passive recipient, and waves play no essential part.

Editor's Comments on Paper 2

This is the classic paper on spits and bars. While O. F. Evans recognized the earlier work of G. K. Gilbert (Lake Bonneville: *U.S. Geol. Survey Mono.*, vol. 1, 1890), F. P. Gulliver (Cuspate Forelands: *Bull. Geol. Soc. Amer.*, vol. 7, pp. 399–422, 1896), and D. W. Johnson, "Shore Processes and Shoreline Development" (New York, Wiley & Sons, 1919), he undertook a systematic investigation of spits and bars in the process of development. Much of what has followed in this field stems from his studies.

Foremost in the paper reproduced here are Evans' conclusions on spit and bar development. Significant too are his opinions on the building of these structures through, and above, sea level; a topic of debate that is still going on to this day. While admitting that bar emergence is possible, he states that evidence is lacking. In addition the reader may be interested to note what Evans had to say about looped bars, those features which puzzled G. K. Gilbert in the preceding paper.

Oren F. Evans was born near Shelly, Michigan, in 1878. He received his bachelor's degree from Albion College, Michigan, in 1910, and his master's and Ph.D. degrees from the University of Michigan in 1920 and 1939, respectively. Evans joined the University of Oklahoma as an Assistant Professor in 1920 and remained there as an Associate Professor, and later Professor, until his retirement in 1951. During this time he initiated a course in petroleum geology at the university, wrote a textbook on meteorology, published a great many papers, and counted as one of the leading specialists in the study of shore processes in the United States. Dr. Evans, then University of Oklahoma Professor Emeritus of Geology, died at his home in Norman, Oklahoma, in August 1965.

The following article is reprinted with the permission of the *Journal of Geology*, as edited by R. T. Chamberlin and published by the University of Chicago Press.

THE ORIGIN OF SPITS, BARS, AND RELATED STRUCTURES[1]

O. F. EVANS
University of Oklahoma

2

ABSTRACT

Building and lengthening of the superaqueous portion of a spit or bar is caused by the transportation of material from the shore toward the distal end of the structure by the swash and backwash when the waves break obliquely on the shore. If the water surface has a constant elevation, the submerged part of a spit or bar is not brought above the surface by waves moving in a direction perpendicular to its axis. Such wave action results only in widening the submerged ridge.

Hooks are caused by the work of refracted waves which cause transportation of material across the end of the spit. A change of wind direction or the presence of tidal or hydraulic currents is not necessary to their building, although if such currents are present they may aid the process. Looped bars are built by refracted waves in the same way as are hooks. However, in the looped bar the process operates more uniformly and continues until the bar rejoins the shore. The windward end of a looped bar originates in shallow water on a gently sloping bottom. The location of the point at which the return to shore begins depends on the depth of water, the amount of sediment being carried by beach drifting, and the energy of the waves.

Extensive observations indicate that the lengthening of spits and bars takes place only when the waves move in a direction such that they cause the swash and backwash to move material along the shore toward the end of the spit. Waves from any other direction cause it to be widened and shortened.

INTRODUCTION

A "spit" is a ridge or embankment of sediment attached to the land at one end and terminating in open water at the other. It is younger than the land mass to which it is attached. The crest of the spit from the land outward for some distance rises above the water.

A "bar" is a completed or extended spit which encloses, or nearly encloses, a portion of the water body into which it extends. It may be attached only at one end, or it may be the result of two spits building from opposite directions. If such a bar extends across a bay it is called a "bay bar." If it departs from a relatively straight shoreline and then swings back it is a "looped bar."

An "offshore bar" is a ridge or embankment of sediment lying approximately parallel to a shoreline and reaching above the water. It is not attached to the shore at either end. Like the spit and the bar, it is younger than the land body along which it lies.

[1] With the support of a grant from the American Philosophical Society.

Any of these structures may vary greatly in size and age. Some spits are only a few feet long and have a life of only a few hours. Examples of these can be seen where small streams empty along sandy shores which are acted upon by waves produced under the influence of variable winds. Others may be large and have a long life, such as Sandy Hook or Cape Cod. Likewise, bars may shut off lagoons a few feet across or may close the entrance to great bays large enough to form important harbors. Unlike spits and bars, offshore bars of small size are not very common. The offshore bars usually described in print are such as those along the coast of New Jersey or the great offshore bars of the Gulf Coast or the Carolinas.

In this article the processes by which spits and bars are formed will first be considered. Offshore bars will be discussed only with regard to their possible relation to spits and bars.

Our geological textbooks give a somewhat indefinite and hazy explanation of the methods of formation of spits and bars. The most common statement is that they are the result of transportation by the longshore currents, but usually no attempt is made to give a complete picture of the process. Of sixteen modern geological textbooks of college grade which were examined, five stated that spits and bars are built by shore currents, ten that they are the result of shore currents combined with the work of the waves as they encounter the submerged part of the ridge, and one that they are built by the shore current which is sometimes helped by "wave wash from opposite directions."

D. W. Johnson[2] gives an excellent summation of the literature of spits and bars which is still fairly complete. The earlier studies of spits, bars, and related structures were confined chiefly to descriptions of their physical forms. Attempts at explanation of origin were largely deductive. A good example of this is F. P. Gulliver's work.[3] Little was known at that time of the processes by which sediments are transported in water bodies, and very few systematic subaqueous observations had been attempted. Gulliver recognizes this weakness in his own work, for he states that "its confirmation, extension, or re-

[2] *Shore Processes and Shoreline Development* (New York: John Wiley & Sons, Inc., 1919), pp. 333–39.

[3] "Cuspate Forelands," *Bull. Geol. Soc. Amer.*, Vol. VII (1896), pp. 399–422.

jection awaits the local observer in the field." In field work emphasis was at first placed largely on the study of fossil structures. An outstanding example of this is G. K. Gilbert's classical work.[4]

Later work by various investigators made it clear that the energy of longshore currents alone is hardly sufficient in itself to explain the formation of all spits and bars, especially those containing coarse material. Also, it was soon realized that spits and bars are numerous along the shores of small bodies of water where eddy and tidal currents are either absent or negligible. A few years after the publication of Gilbert's studies, Tarr, Woodman, and Wilson, each as the result of independent work, suggested wave action as a possible cause of the formation of spits and related structures. However, they failed to make any adequate investigations of processes of sediment transportation. Consequently, their explanations do not give a clear picture of the exact processes by which the structures are built.

Johnson, in his excellent book,[5] appears clearly to recognize the incompleteness of the explanations that have been offered and emphasizes the importance of beach drifting and wave-formed currents in the formation of shore structures. He gives an analysis of the process of beach drifting[6] and seems to be the first to have seen the great importance of the swash and backwash in the transportation of sediment. However, as will appear later, even he probably failed to appreciate fully the extremely important role played by the beach-drifting process.

Since spits and bars extend above the water, any genetic account of their origin must explain the building of the exposed portion as well as that part below the surface. This is the weakest point of the explanation usually given in the textbooks. It is easy to understand how currents, if strong enough to pick up sediment, may again deposit it under water in the form of ridges; but it is evidently impossible for such currents operating entirely within the water to deposit above its surface.

The necessity for explaining this point seems to be thoroughly understood by Johnson, who gives an extended discussion of the subject in describing the building of offshore bars.[7] However, he follows

[4] "Lake Bonneville," *U.S. Geol. Surv. Mono. I* (1890).

[5] *Op. cit.*, pp. 335–39. [6] *Ibid.*, p. 94. [7] *Ibid.*, pp. 356–70.

De Beaumont, Shaler, and Davis in assuming that the action of the waves is capable of building a submerged offshore bar above the surface of a body of water having a constant elevation. The writer has not had access to De Beaumont's original article, but Shaler, Davis, and Johnson[8] give rather complete and essentially identical descriptions of the building of offshore bars and of the process by which they are supposed to be brought above the water, although none of them except Johnson appears to have seen the process in operation. He says:

> The writer has seen a very perfect miniature off-shore bar formed in a few hours by waves raised on the surface of a small lake at Lakehurst, New Jersey, during a fresh breeze. The bar was a few inches in width, and separated a shallow lagoon one or two feet broad from the gently sloping shore which it paralleled for some yards.[9]

However, there is a possibility that even as experienced an observer as Johnson may have been deceived in this instance. The level of the water on the lee side of a small lake is often raised several inches during strong onshore winds. At such times it is not unusual for small subaqueous ridges of sediment to be built up by longshore currents nearly to the surface of the water and then to be driven shoreward by the waves of translation across them. Such a ridge may appear above the water surface as the wind goes down, as a result of the fall in the water-level. Such a process would be entirely different from that of building the ridge above the water surface by the work of the waves.

But extensive inquiry among commercial fishermen, members of the coast guard, and others who have lived along the shore has failed to discover anyone who has ever seen an offshore deposit of sand or mud built above the surface except when the water surface has been lowered or the growth of weeds has checked the water movements sufficiently to cause deposition. It seems probable that, if offshore bars can be formed by the work of the waves, they should be numerous along the shores of some parts of the Great Lakes because of the many subaqueous sand ridges or "balls" found in many places in the Great Lakes and especially along the east side of Lake Michigan.

[8] *Ibid.*, pp. 365–66.

[9] *Ibid.*, p. 388.

According to my studies,[10] it is impossible for a subaqueous ridge which is not connected with the shore to be built above a water surface of constant level by wave action. As soon as such a ridge is brought near the surface the waves of oscillation break against it. This causes waves of translation which carry the sediment across its top, and its further upward growth is prevented. Thus the material on the top of the ridge is planed off and deposited on the lee side, and the ridge migrates as a subaqueous dune.[11]

Johnson appears to recognize clearly the necessity of explaining how spits and bars are brought above the water surface, but he seems to accept completely the hypothesis of wave-built ridges as it has been applied to offshore bars and attempts to apply it also to spits and bars. He says: "The seaward side of the narrow embankment is acted upon by the ocean waves, which build its crest above normal sea-level and establish a profile of equilibrium, similar to that of an ordinary beach."[12] He also says:

> The super-aqueous portion [of the spit] owes its height primarily to the waves, but in the case of sand spits wind action may locally raise the level a number of feet by forming dunes. Disregarding the disturbing effect of the wind, the height of a spit will depend upon the exposure to wave action; big waves will cast the debris many feet above mean water level, while small waves will raise the surface but slightly above the lake or sea.[13]

Again, in discussing compound spits, he says:

> Large quantities of debris, borne by a current which departs from the former shoreline and advances into open water, must be built into an embankment which elongates rapidly in the direction of current advance. Waves raise the surface of the new embankment into a beach ridge; and by repetitions of this process there are formed successive beach ridges separated by lagoons of considerable breadth.[14]

In describing the building of bars Johnson says:

> There is, however, an entirely different process by which bars, indistinguishable in surface form from those developed from growing spits, may be pro-

[10] Evans, "The Low and Ball of the Eastern Shore of Lake Michigan," *Jour. Geol.*, Vol. XLVIII (1940), p. 497.

[11] *Ibid.*, pp. 498–99. The ridges of the low and ball formed by the plunging breakers are never brought into this upper zone of migration except by accident, but there are numerous ridges nearer shore, which are formed in other ways, that do commonly migrate shoreward in this way.

[12] *Op. cit.*, p. 287. [13] *Ibid.*, pp. 295–97. [14] *Ibid.*, p. 298.

duced. Waves entering shallow water may break before reaching the coast, and cast up the bottom debris into a narrow ridge, in the manner discussed more fully in connection with "Offshore Bars." The irregular bottom of a typical young shoreline of submergence is usually highly unfavorable to this process; but whenever the initial form or later deposition does give a fairly uniform slope to the bottom near the shore, wave action may produce a bar independently of longshore transportation. Such a bar may form a short distance offshore and be driven in until the portion opposite a headland becomes a headland beach, and the portion opposite the bay remains a typical bay bar extending from headland to headland and nearly or quite closing the bay mouth; or the waves may construct the bar just at the mouth of the bay in the first place; or they may break on the gently sloping bottom well within the bay and produce a bar near the middle or even near the head of the bay. It is possible that some supposed sandspits are really the beginnings of, or last remnants of, bars formed in this manner.[15]

He frequently refers, however, to the importance of "beach drifting" and "wave currents." For example, he says:

There can be no doubt that wave currents and the associated longshore beach drifting play a very important role in the formation of various types of beaches, spits, bars, tombolos, and forelands.[16]

But nowhere does he speak of beach drifting as a direct factor in building the structures above the water surface, while he does, as indicated above, repeatedly refer to the efficiency of breaking waves in bringing subaqueous embankments above the water surface. Therefore, it seems evident that Johnson's hypothesis of spit- and bar-building includes two processes: first, the building of subaqueous embankments by longshore currents and, second, the elevation of these embankments above the water by direct wave action in a way similar to that in which beach ridges are thrown up.

Since direct wave work is incapable of bringing offshore subaqueous embankments above a water surface of constant elevation, it is not probable that spits and bars are built above the water by the processes heretofore supposed. This study was undertaken, therefore, for the purpose of determining the exact processes involved in the building of such structures.

THE REGIONS STUDIED

The first observations were directed toward discovering, if possible, any instance where breaking waves were effective in building the

[15] *Ibid.*, p. 301. [16] *Ibid.*, p. 334.

submerged end of the spit above water. It is evident that if subaqueous embankments can be brought above the water by the direct action of the waves, the most effective direction of wave travel is perpendicular to the axis of the submerged end of the spit, as at *A* in Figure 1. With the waves as indicated at *B* the conditions would be less favorable than at *A*, but there should still be some building, since one component of the wave direction at *B* is perpendicular to the axis of the spit. With the wind as shown at *C* we would not expect an extension of the spit, but there should be no decrease in its length. Should the wind be from the direction shown at *D*, the observations would be indeterminate, since in this case it would be impossible to distinguish between the effects of wave-building and beach drifting. If wave-building is the effective agent by which embankments are brought above the water, there might well be built occasionally, with the wind as shown at *A* or *B*, small islands of sediment along the crest of the ridge because of inequalities in wave energy and in the amount of sediment available. These islands, if later joined, would become a part of the spit built above the water.

For this study three spits were selected. Two of these are permanent, but one is temporary, in that it alternately grows and is destroyed, depending on the wind direction. One of the two permanent spits is located on the south side of Blue Lake a few miles northeast of Whitehall, Michigan. It points east and is at the end of a small cape which partly encloses a bay about 200 feet across. The spit was first measured on June 12, 1941, following several days of fairly strong northeast and east winds. At that time it was 27½ feet long and 18 feet wide. Its outer end was almost semicircular in shape and had the appearance of having been blunted and cliffed by the recent waves. On June 17, following three days of strong northwest winds, the spit had been extended east by south 4 feet and was 31½ feet long (see Fig. 2). A strong beach ridge had developed along the north side of the structure, evidencing that the material added to its outer end had come by the process of beach drifting from the sandy shore to the west. Although the wind had dropped somewhat at the time of the visit, transportation and building were still going on, the waves on reaching the end of the spit being refracted in the shallow water and moving around its end toward the bay. As a consequence, beach

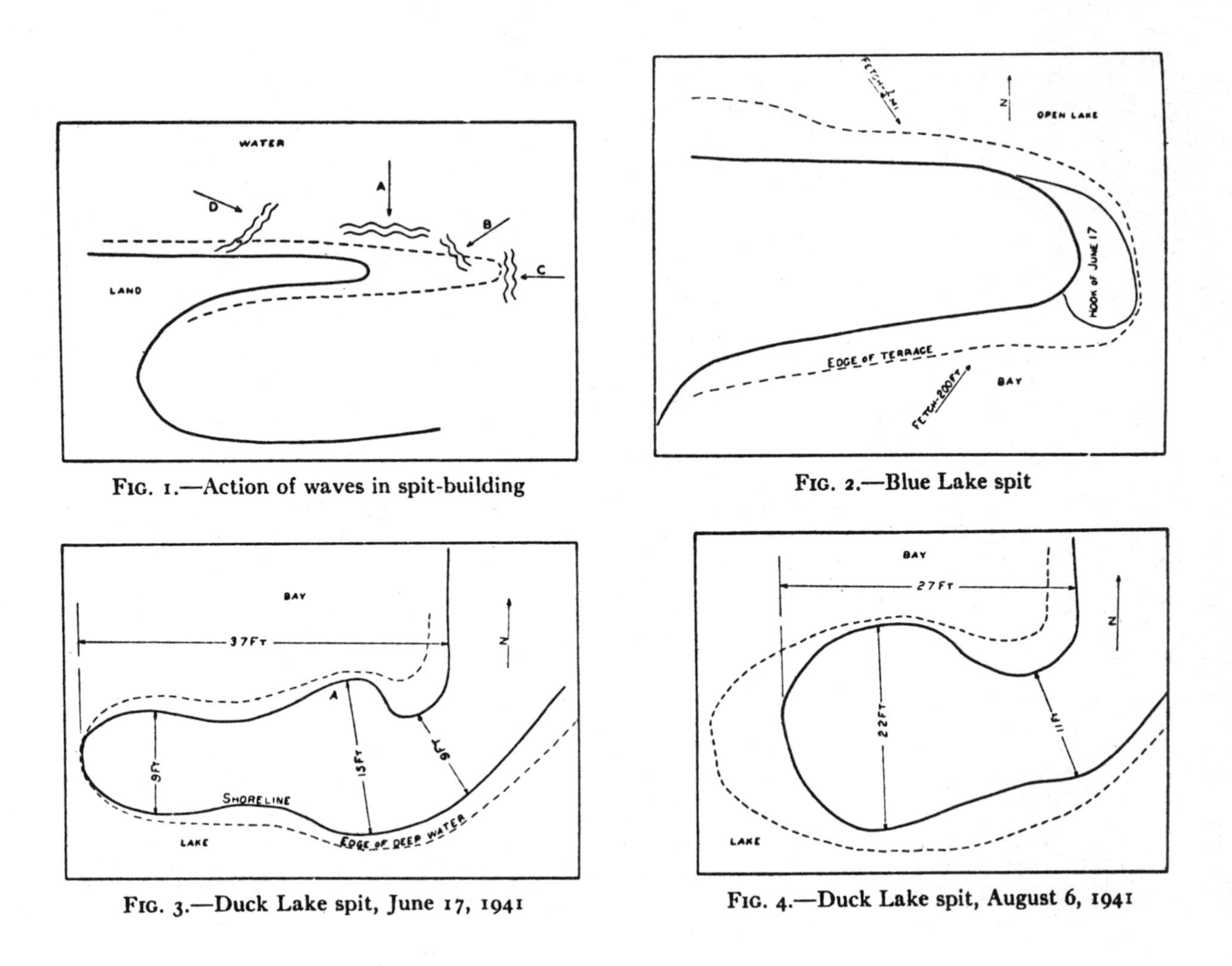

FIG. 1.—Action of waves in spit-building

FIG. 2.—Blue Lake spit

FIG. 3.—Duck Lake spit, June 17, 1941

FIG. 4.—Duck Lake spit, August 6, 1941

drifting across the end of the spit was forming a small hook pointing south. For some days following, the waves were gentle, and little change took place. During the night of July 1 the winds were fresh west-northwest and by mid-forenoon of July 2 a hook at the end of the spit was forming. Some cliffing had occurred on the north side and across the end of the point, and the spit was now shortened to 29 feet. On July 19 strong northwest winds were again blowing, and the spit was again lengthened to nearly 31 feet. The wind was a little more northerly than on June 17, and the waves were being refracted more strongly around the end of the spit and were tending to broaden it. Thus this spit was lengthened by material brought along the beach ridge during northwest winds and was shortened and rounded with a strong tendency toward hook formation when the wind was north or northeast.

The second permanent spit is a deposit of sand on the north side of Duck Lake about 2 miles south of White Lake Harbor entrance. It is an extension of a geologically recent sand point which partly separates the west arm of Duck Lake from the main body of water. The first visit to this structure on June 17 followed a considerable period of strong northeast and east winds, and its general appearance and relation to the shore was as shown in Figure 3. There was a strong beach ridge along its south side, and it had every appearance of having made a recent rapid growth to the west. The end terminated abruptly in relatively deep water. The material of the spit clearly had come from the abundant sand to the northeast. Several ensuing days of gentle to moderate southwest winds with a fetch of about one-fourth of a mile shortened the spit from 37 to $33\frac{1}{2}$ feet, increased its width at the end from 9 to 15 feet, and turned it more toward the north shore. There was now a subaqueous shelf 3 feet wide at the end and along the south side of the spit where none was present on June 17. At *A*, erosion was taking place and material was moving eastward toward shore. By July 10, after a further period of fairly strong westerly winds, the shelf at the end of the spit had widened to 5 feet, and the length of the spit above water measured only 31 feet. The shelf on the south side had widened to 8 or 9 feet. On the north side at *A* the shoreline was now straight. Here, again, is evidence that winds which cause beach drifting along the axis of the spit from the land to which it is attached cause it to grow, while winds from the other directions destroy it. By August 6, after a long period of southwest, west, and northwest winds, it had the form and dimensions shown in Figure 4. At this time the winds were moderate from the west, and material was moving eastward by beach drifting on both sides of the spit.

The third study was made on the west side of Silver Lake. Here the Little Point Sable dunes are moving eastward and slowly filling the lake. This sand is shaped by the waves into a shallow, gently

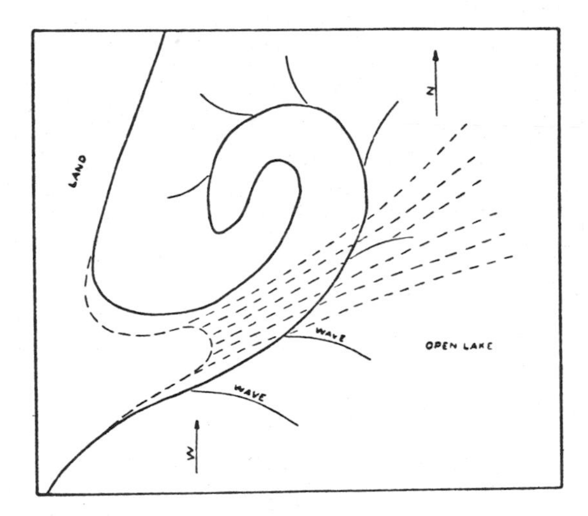

FIG. 5.—Hooked spit, Silver Lake

sloping underwater terrace, and the shoreline is being re-worked with every change of wind. The terrace is so shallow and has such a gentle slope that numerous cusps form, some of which extend almost straight out from shore and terminate in long subaqueous ridges that often become subaqueous migrating dunes.[17] The shoreline here lies north by east. On June 12, with the wind east-southeast, one of these low ridges attached to the shore at its south end was being driven toward shore. The waves were 3–4 inches high, and the water was very shallow on the ridge. By the next day the wind had shifted to the south, the subaqueous ridge had become more nearly parallel with the shore, and a spit had formed at the shoreward end of the

[17] Evans, "Mass Transportation of Sediments on Subaqueous Terraces," *Jour. Geol.*, Vol. XLVII (1939), pp. 327–28.

ridge. The part projecting above the water was 8 feet long and 28 inches wide (see Fig. 5). Beyond where it became submerged, it gradually diverged from shore as the water became deeper, and at about 60 feet it could no longer be distinguished as a ridge. By June 21, with continued south and southeast winds, the spit had become 25 feet long and 4 feet wide, and a hook had formed that turned directly back into the little bay on the landward side of the spit (Fig. 5). This hook was not the result of currents caused by change in wind direction but was the result of beach drifting caused by refraction and diffraction of waves coming from the south and southeast. This process of hook formation will be discussed in greater detail under the subject of "looped bars."

Streams which enter Lake Michigan across the broad sandy beaches of the east side of the lake have their outlets shifted to the south or the north as the direction of the wind changes. When such shifting occurs, it is always accompanied by the building of spits which have their shore attachment on the side from which the waves come. The growth is always in the direction of beach drifting, and the more material carried forward by the process the more rapidly the spit grows. When the winds happen to be almost directly onshore, the process is very slow, and spits already formed are sometimes shortened. However, this depends on the relation between the sand-transporting energy of the waves and that of the stream. In no case is the mouth of the stream ever closed simultaneously throughout its width by a ridge thrown up by the waves. The outlets at Duck Lake and at Flower Creek were especially kept under observation during the summer of 1941; and these, with others, have been under occasional observation for a much longer time. At no time was there observed a tendency toward spit-building by any process except that of shore drifting, although the rapidity of the process was greatly increased by the amount of sand brought down by the stream.

Thus it seems clearly indicated that spits and bars increase in length above the water only when the waves move in such a direction that material is moved in the swash and backwash outward from the attached end. Waves coming from the other direction or perpendicular to the axis of the structure tend to shorten it. However, beach drifting along the side of a spit does not necessarily lengthen the spit. Because of wave refraction in the shallow water the sand may be carried around the end of the spit and thus tend to form a hook (Fig. 5). The spit is increased in length only when the amount of sediment carried along its windward side by beach drifting is more than can be forced around its end by the work of the refracted waves. The processes of transportation involved are largely those which have been described by Johnson[18] under the head of "wave currents." Of course, wind-caused longshore currents are also usually present. These aid in the process by building up the bottom and thus shallowing it over the submerged end of the spit.

If Johnson did not accept so completely the theories of De Beaumont in explaining the building of subaqueous embankments above the water, he would seem to be in complete accord with the above statements. He says:

> I can say that I have found many forelands and embankments which seemed to me demonstrably due, principally if not wholly, to wave currents; but none which seemed undoubtedly the product of other types of currents. I am therefore inclined to believe that wave currents have played the most important part in the construction of all the sandy forelands and embankments of our coasts.[19]

Elsewhere he accepts beach drifting as an important process in the formation of beaches, spits, and bars.

An interesting corollary of the above-described process of spit formation is that the building of hooks is a normal accompaniment of spit-building, and in accounting for them it is not at all necessary to postulate tidal or hydraulic currents across the end of the spit. The shore drifting which accompanies wave refraction is entirely competent to form hooks, and it is not necessary to postulate any other change of wave or current direction to explain the process. The factors involved in hook formation under these conditions are wind direction and velocity, slope of bottom, and amount, size, and texture of material available.

The observations described above were made on relatively small structures for convenience and economy of time in doing the work. With large structures it requires a much longer period of time for the waves and currents to produce easily measured results. However,

[18] *Op. cit.*, pp. 90–106.

[19] *Ibid.*, p. 337.

effects in the two cases are the same. It is well known that, as waves approach a shore at an acute angle, refraction may cause them to change direction by several degrees, and the more gradual the slope of the bottom the greater is the change. Also, waves coming into a bay between headlands have their courses changed by refraction, so that as they approach the beach they are nearly parallel with the shoreline. This is discussed by Johnson[20] in explaining the concentration of waves in the erosion of headlands. The same effect can be observed where the entrance to a harbor is so situated that waves drive directly into it during a storm. In discussing this action Jeans says: "In the same way sea-waves which fall on the entrance to a harbour do not travel in a straight line across the harbour but bend round the edges of the breakwater, and make the whole surface of the water in the harbour rough."[21]

In studying the building of hooks or recurved spits along ocean shores it is difficult to distinguish between the work of the refracted waves and that of tidal currents; but in the Great Lakes tides are not a factor. Presque Isle is a compound recurved spit which forms the harbor at Erie, Pennsylvania. At its eastern end the process of hook-building is still active. The records of the United States Weather Bureau show that the prevailing winds at Erie are from the southwest, west, and south; and the winds of maximum velocity are also from those directions. There are occasional winds from the north or east, but their effect is not great. The growth of Presque Isle has been from southwest to northeast, and its eastward progression has been marked by the building of a series of hooks across its eastern end. The last of these extensions had its beginning about 1866 and is still building. It is 3,000 feet long and about 700 feet wide. The last 1,000 feet of the south end of the hook is itself a series of hooks whose axes lie approximately northeast-southwest. The testimony of several employees of the park indicates that both the principal hook and the smaller ones at its south end have been built by waves that refracted around the end of the structure in the manner shown in Figure 5.

[20] *Ibid.*, pp. 74–76.

[21] Sir James Jeans, *The Mysterious Universe* (London: Macmillan & Co., 1934), pp. 42–43.

Obviously, the same principles apply to the formation of cuspate bars and looped bars, since a spit which has its distal end turned back until it again joins the shore constitutes a looped bar. Thus it is not at all necessary to postulate two currents circling in opposite directions in the formation of cuspate bars or cuspate forelands. Johnson[22] agrees with this and also says[23] that cuspate bars sometimes form as the result of a spit turning back toward the shore. He does not discuss the process by which it is brought about, and in his discussion of the formation of hooks[24] he seems to place chief emphasis on currents running athwart the end of the spit. Of course, it is entirely possible that cuspate forelands may sometimes form, as has been stated frequently, by two converging spits growing out from the shore; but those which I have observed in the process of building have been recurved spits caused entirely by wave refraction. They are most numerous in lakes having broad, shallow sandy terraces. Along the shore of Silver Lake near Little Point Sable, Michigan, several of them can usually be seen either completed or in the process of formation. They seem to develop best with a wind blowing at an angle of 30°–45° with the shoreline. Studies of looped bars, both completed and in the process of formation on Silver Lake, as well as a consideration of certain fossil looped bars on the shores of Crystal Lake near Frankfort, Michigan, indicate that this is the usual method of formation.

For several days during the latter part of July, 1941, winds from the south over Silver Lake were fairly strong. On the west side of the lake, where the beach is of pure dune sand, a series of five looped bars formed. The shoreline here lies about north-northeast. The bars started as spits building toward the northeast and were later swung back to the shore by the refracted waves. Three of them were complete by July 29, but two had not completely joined the shore at the north end.

Figure 6 is a diagram of one of these looped bars and is typical of them all. A spit began at *A*, and the distal end of it passed beneath the water as a subaqueous ridge as shown at *B*. It continued to build out above the water until it reached a depth such that material brought forward by beach drifting was carried around the end by the

[22] *Op. cit.*, p. 337. [23] *Ibid.*, p. 318. [24] *Ibid.*, p. 289.

refracted waves and a hook began to form at *C* and to grow toward the shore. The refraction of the waves was increased by the continued broadening of the subaqueous dune at *B*. At *D*, to the lee of this subaqueous ridge, the water was deeper. Here the waves, which were partly translation waves across *B*, reformed into oscillation waves and continued the beach drifting along *C* until it joined the shore. In all five bars the ridges became somewhat narrower and lower from south to north.

A good example of combined spit and hooked-bar formation was seen at Old Mission Point on the east arm of Grand Traverse Bay. On July 12, 1941, its extreme tip to the south-southeast reached out about 200 feet. Below the 40-foot bluff here the gravel and sand beach extends around the point (see Fig. 7). The bay lies to the south and southwest of the point. The longest fetch and probably the strongest winds to reach the point are from the north and northeast. Sand yielded by undercutting of the bluff is moved by the shore current and the beach-drifting process toward the point where it is constantly being built and rebuilt into beach ridges, spits, hooks, and looped bars by the waves and currents. Each change of wind having considerable duration probably results in a change of form of the structures. The hook at *B* is clearly older than the ridge at *C*. It was probably built by waves from the northeast or north-northeast which were somewhat refracted as they moved in along the shore. *C* was then built by larger waves which were perhaps from a more northerly direction, and such a large quantity of material was brought by the shore drift that it could no longer be carried westward to follow the outline of *B* but was deposited in the spit *C* to the south. As the water deepened and the waves were more refracted, so as to travel in a more westerly direction, the hook *A* was built. This was practically complete at the time of my visit, but there was a small outlet at *b* from the enclosed water to the lake. More recent southwest winds then caused the formation of the spit at *D*. All this was clearly shown in the relations of the various beach ridges.

Crystal Lake near Frankfort, Michigan, was lowered about 7 feet in the early seventies, exposing many shore structures and also starting new adjustments along the newly formed emergent shoreline. The lake is now surrounded by a broad sandy terrace which was for-

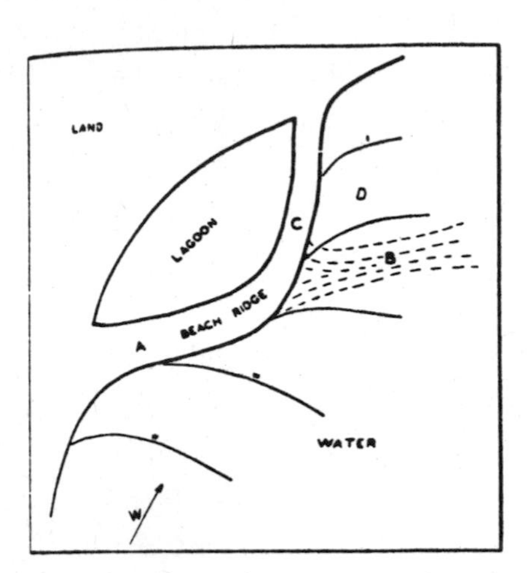

Fig. 6.—Looped bar, Silver Lake

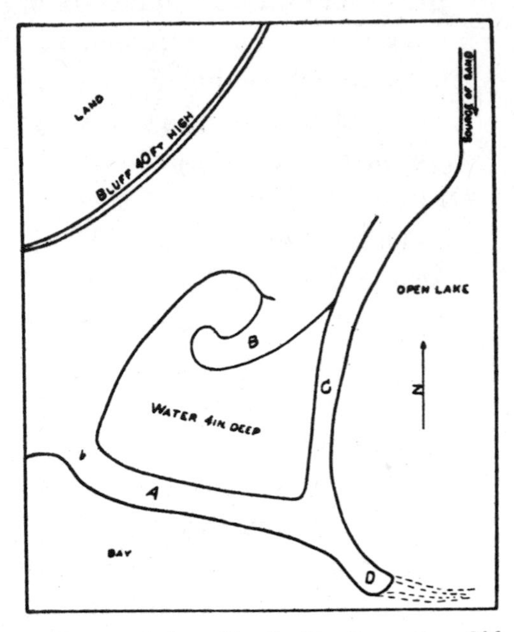

Fig. 7.—Spit and hooked bar formation, Old Mission Point.

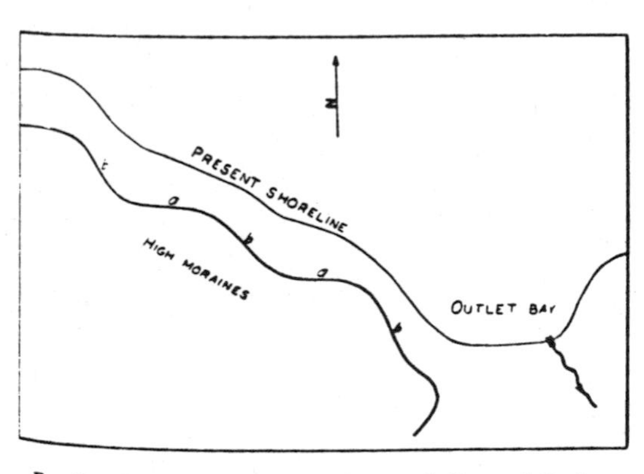

Fig. 8.—Portion of south shore of Crystal Lake

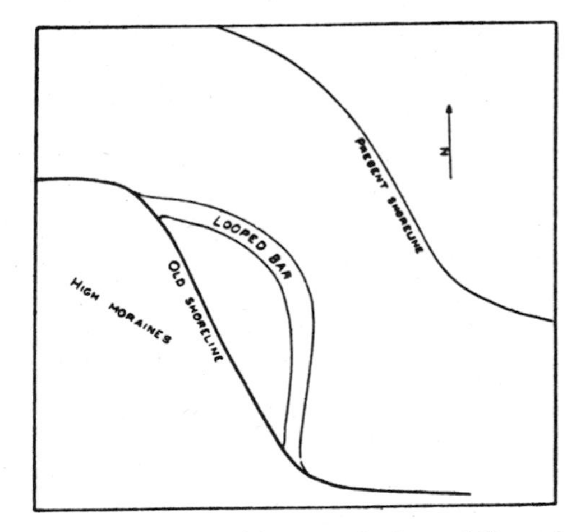

Fig. 9.—Looped bar, south shore of Crystal Lake.

merly lake bottom. On the south shore, west of Outlet Bay, the former bluffs of morainic material form a sawtooth arrangement of three projecting points (see Fig. 8). According to the United States Weather Bureau records, the strongest winds on this shore are from the west and northwest. Before the lowering of the lake the waves had cut deeply into the moraines, thus forming a terrace of glacial material extending out beneath the water 200 or 300 feet with a depth of 9 or 10 feet at the "dropoff." The material of this old terrace, the upper part of which is now above the water, consists of sand containing a considerable amount of gravel and large boulders. The sand and fine gravel were worked out to form part of the terrace, and some was carried southeast by the currents and shore drift. Some of this was laid down as looped bars in front of the old shore at *b*. These ridges, from 4 to 7 feet high and 20–40 feet wide, leave the old shores near the end of the promontories on their east side and gradually diverge from the cliffs until they are as much as 100 feet from them in some places. Then they swing back and again join the old shore. It is certain that one of them, which is shown roughly in Figure 9, formed a complete looped bar, but the other two may have been partly submerged at their southeast ends. These bars are, in general, steeper on the land side, showing that, after being deposited by the shore currents and before being brought above the water surface, they were worked in toward the shore by waves of translation, as so commonly happens with structures of this kind. No ridges are present on the old terrace in front of the cliffs at *a*, where the waves drove directly toward shore. The old terraces contain considerable sand which would be good material for the building of shore structures, but these features are found only along the cliffs at *b*, which lie nearly parallel to the direction of the prevailing winds. This is a strong indication that spits, bars, and related structures result from the combined work of the shore current and beach drifting rather than being thrown up by waves driving directly on the beach.

Since the lowering of the lake, considerable shifting of sediment has occurred in adjusting the shoreline to its new environment. This has resulted along this south shore in the building of large subaqueous dunes wherever the shoreline turns sharply toward the land. The largest, Figure 10, is about 900 feet long and nearly half as wide. Along their lee sides to the southeast the depths increase abruptly from about 4 to 12 feet, with the sand generally lying at the angle of repose. According to De Beaumont's theory, these should be built up above the water by the waves, and since they are connected to the shore at one end they should become spits, but instead they are continuing to extend in length and to migrate shoreward beneath the water.

An examination of the shoreline shows the reason for the absence of spits. There is so much sand on the submerged part of the terrace that considerable quantities are brought into suspension by wave agi-

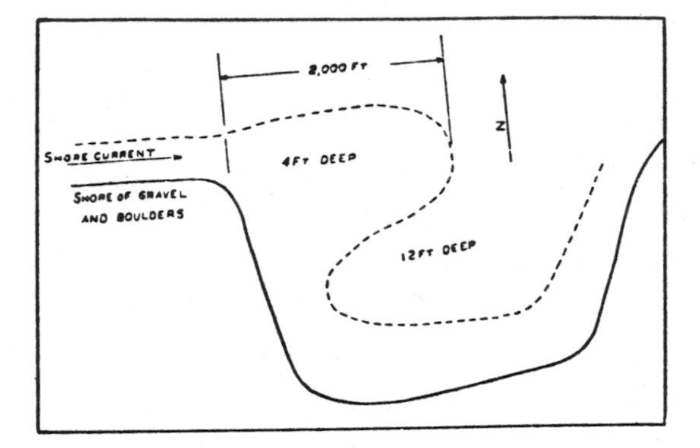

FIG. 10.—Sand deposit in Outlet Bay, Crystal Lake

tation and are transported readily by the shore currents. But at the new shoreline on the old terrace so much coarse debris occurs that very little beach drifting is possible. Because of this lack of beach material it is not possible for the submerged ridges to be built up above the water where they leave the shore, and as a result no spits form.

CONCLUSIONS

The above studies and observations indicate:

1. The subaqueous ridges which constitute the submerged portion of spits and bars are not built above a water surface of constant elevation by the work of the waves bringing up material from the bottom.

2. Spits increase in length above the water only when the waves and the shore drift move from the direction of their land connection. Waves from any other direction tend to destroy them.

3. Hooks are a normal accompaniment of spit-building, usually as a result of wave refraction. Neither change of wind direction nor the presence of tidal or hydraulic currents is necessary to their building.

4. The part of a spit above the water receives its material as the result of beach drifting. The direction the spit takes is determined by relations among the variables of current direction, wave direction, wave energy, amount of sediment available, and depth of water.

5. A looped bar is built by the same processes that produce spits and hooks. In fact, a looped bar is only a hooked spit that has continued to build until it has rejoined the shore.

Because the above summary disagrees somewhat radically with the conclusions of some previous investigators, it seems advisable to further simplify and explain some of the points. As far as my work shows, no ridges of fine sediment such as sand are brought by the work of the waves alone above a water surface having a constant elevation. However, it is possible, where rapid variations in water-level occur, as with tides, that a ridge formed at high tide by the plunging breakers might be exposed at low tide. A beach ridge might then be formed by the waves and be further elevated by dune-building; and so, under a fortuitous combination of conditions, an offshore bar might be built above the water. However, as far as I know, this process has never been observed.

Since offshore bars are not brought above the surface by wave work, it seems most logical to consider them as long spits that have been broken through by the waves. However, as stated by Johnson,[25] it is difficult to see how such great quantities of material could be transported along the shore the necessary distances for the building of such structures. It is possible that some of the material in offshore bars may be brought from the bottom and added to the spit by waves in the same way they sometimes add to a beach ridge from a shallow terrace. This sand would then be subject to beach drifting and thus be added to the spit along with that brought directly from the land. However, the whole subject of offshore bars needs further study.

The statement that hooks and looped bars are usually built by

[25] *Ibid.*, p. 355.

waves coming from the same direction as those that built the spit does not mean that tidal or hydraulic currents across the end of the spit might not result in hook formation. Also, an onshore wind blowing perpendicular to the axis of a spit causes beach drifting across the end of it and so aids in the formation of the hook. However, refracted waves are usually the chief agent in the process.

Johnson[26] states that bay bars are sometimes formed and brought above the water by the breakers. While, according to my observations, these ridges are never brought above the water surface by direct wave work, the foundation of the bar in the form of a subaqueous ridge might be formed by the plunging breakers as described in the building of the low and ball, and this might be followed by the actual building of the ridge above water by the processes of shore drifting from one or both sides.

[26] *Ibid.*, p. 301.

Editor's Comments on Paper 3

While previous workers had investigated subaerial fossil shore features or modern subaqueous structures in the field, G. H. Keulegan turned to model simulation for his study of submarine sand bars. Thus quantitative treatment of the problem was added to the qualitative approach by comparing these early experimental studies with field observations.

In this model experiment Keulegan was mainly concerned with the mechanism of sand transport in bar movement, and bar stability as related to beach slope, water depth, wave height, and wave steepness. Only under such controlled conditions could all but one factor be held constant while the effect of a single variable was tested. Discrepancies in form between his model bars and those observed on the Pomeranian Coast and in Lake Michigan may possibly be attributed to difficulties inherent in scale-model procedures in use at the time. For a more recent treatment of this topic a later paper on laboratory experiments in bar formation by McKee and Sterrett is included after the following paper by Francis Shepard.

Garbis H. Keulegan was born in Sebastia, Asia Minor, in 1890 and came to the United States in 1912. He attended the Ohio State University from 1912 to 1917, then served in the U.S. Army in World War I. Keulegan was with the National Bureau of Standards from 1921 to 1960, mostly as a staff member of the National Hydraulics Laboratory, and received his doctorate from Johns Hopkins University in 1928. Dr. Keulegan is presently a consultant to the Hydraulics Division of the U.S. Army Corps of Engineers Waterways Experiment Station in Vicksburg, Mississippi.

DEPARTMENT OF THE ARMY CORPS OF ENGINEERS, U. S. ARMY

AN EXPERIMENTAL STUDY OF SUBMARINE SAND BARS

TECHNICAL REPORT NO. 3

BEACH EROSION BOARD
OFFICE OF THE CHIEF OF ENGINEERS

DEPARTMENT OF THE ARMY

OFFICE OF THE CHIEF OF ENGINEERS

BEACH EROSION BOARD

The Beach Erosion Board of the Department of the Army was established by section 2 of the River and Harbor Act approved 3 July 1930 (Public, 520, 71st Cong.), to cause investigations and studies to be made in cooperation with the appropriate agencies of various States on the Atlantic, Pacific, and Gulf coasts, and on the Great Lakes, and the Territories, with a view to devising effective means of preventing erosion of the shores of the coastal and lake waters by waves and currents. The duties of the Board were modified by an act approved 31 July 1945 (Public, 166, 79th Cong.), "to make investigations with a view to preventing erosion of the shores of the United States by waves and currents and determining the most suitable methods for the protection, restoration, and development of beaches; and to publish from time to time such useful data and information concerning the protection of the beaches as the Board may deem to be of value to the people of the United States * * *."

(III)

TECHNICAL REPORTS

AUGUST 1948.

This paper was prepared for the Beach Erosion Board during World War II by Dr. Garbis H. Keulegan of the Hydraulics Laboratory of the National Bureau of Standards. Dr. Keulegan completed work for this paper in 1944 while detailed to the Beach Erosion Board as a consultant in hydrodynamics. While this paper was prepared primarily for departmental use and the opinions and conclusions expressed are those of the author, it is believed that it will be of material value to engineering and research agencies engaged in work involving submarine sand bars.

(IV)

CONTENTS

LIST OF ILLUSTRATIONS

LIST OF SYMBOLS

A = Cross-sectional area.
a = Measured wave height in front of wave generator.
a_o = Corresponding deep water wave height.
C = Crest of the submarine sand bar.
γ_s = Specific weight of sand.
D = Distance traveled by a sand ripple.
d_{GM} = Median diamteer of sand.
Δf_1 = Elevation of the wave crest above the undisturbed water surface.
Δf_2 = Depression of the wave trough below the undisturbed water surface.
ΔH_1 = Maximum elevation of the surface above the undisturbed water surface during the passage of waves.
ΔH_2 = Depression of the trough at the start of wave break.
H_1 = Undisturbed water depth.
H_2 = Depth of water at the point where wave reformation begins.
H_B = Depth from the undisturbed water surface to the bar base when the bar becomes relatively stable.
H'_B = Depth from the undisturbed water surface to the bar base when the bar is in the process of formation.
H_c = Depth of the bar crest below the undisturbed water surface.
H_t = Depth of the bar trough below the undisturbed water surface.
i = Slope of the beach.
λ = Measured wave length in front of the wave generator.
λ_o = Corresponding deep water wave length.
ν = Kinematic viscosity.
0 = Reference point for measuring and identifying stations.
Q = Rate of sand transportation.
Q_r = Weight of sand transported per hour per foot of width.
ρ_s = Density of sand.
ρ_w = Density of water.
s = Distance between the point where the wave begins to break and the breaker is completed.
S_1 = Point of impending wave break.
σ_φ = Logarithmic standard deviation (Krumbein's notation).
T = Period of the wave.
t_1 = Time required for the bar to become relatively stable.
y = Depth of any point with respect to the undisturbed water surface.
z = Horizontal distance from a vertical line passing through the center of the crest of the bar to any other point at depth y.

AN EXPERIMENTAL STUDY OF SUBMARINE SAND BARS

Section I. INTRODUCTION

Submarine sand bars are frequently found as characteristic features of ocean and lake beaches. The bars may occur singly or in series and are usually associated with sand beaches and offshore areas. An individual bar formation consists of a crest or ridge to seaward and a trough or depression, to shoreward. A series of bars is a number of related crests and troughs. These bar formations have been studied by geographers and geologists, and the literature on the subject has a history of almost three quarters of a century. The earliest descriptions of these underwater formations were those of the German investigator Hagen, whose work appeared in 1863 (reference 1) and was followed by that of Otto, Lehmann, Hartnack, and Evans (references 2, 3, 4, and 5) among others. Although considerable data on the material aspects (form, dimensions, and number) of bars is thus available, almost no authoritative information on the mechanism of bar formation and movement has been obtained.

Admittedly the formation and migration of offshore sand bars are hydrodynamical phenomena of a complex nature difficult to study in a natural environment; under these circumstances laboratory experimentation can prove of considerable value. Economy of time, the limitations of available apparatus, and the nature of the problem require that the general case be resolved into components. Each component can then be considered individually as involving only a limited number of variables. The present investigation concerns, as a first step, the metrical aspects of bars; i. e., the shape and disposition of bars, as they are influenced by the size characteristics of waves.

This paper reports the results of experiments made to determine the existence of basic relationships governing bar phenomena. Observations were made of the form, dimensions, and number of bars; wave characteristics; ripple formation; and nature and volume of sand movement involved in bar phenomena. During the investigation certain qualitative observations on some of the factors affecting the mechanism of bar formation and movement were made. Although these qualitative observations are limited, they will be discussed briefly because of their implications for further study.

806229—48——2

2

Section II. PRELIMINARY CONSIDERATIONS

A. *Bar formation.*—If an experimental beach be assumed having initially a smooth sand surface of constant slope and subject to the action of waves, the part of the beach which lies between the point of impending wave break and that of reformation following the breaker is the area of most active change. This region may be called the bar environment since it is here that the bar is ultimately formed. As will be discussed more fully later, the breaker is the most important element of the region and is the genetic cause of the bar itself.

The deformation of the beach surface is in the form of a ridge, which is relatively flat in the initial stages and moves toward the shore in an observable manner. In the course of time the ridge is enlarged, its form is established or stabilized and its motion decreases to an imperceptible value. When the changes in the position and shape of the bar become minute, the bar can be considered relatively stable. Thus, beginning with an initial smooth beach of slope i at time $t=0$, the bar will reach a stable position and assume a stable shape at time t_1, such that after that time the changes in the bar position and the bar shape take place at an imperceptible rate. Hence, t_1 may be said to be the time required for the bar to become relatively stable. The first problem in studying bar formation therefore is to determine the factors which affect the position of the bar at the time that it becomes relatively stable.

The dimension defining the position of bars in a hydrodynamical sense must necessarily be selected. Inasmuch as the types of bars considered here are associated with breaking waves, the relevant dimension is the depth of the bar with respect to the undisturbed water surface rather than the distance between the bar and the shore. The depth of the bar can be represented by the bar base, defined as the straight line joining the seaward and shoreward toes of the bar (fig. 1.) The depth of the bar base is measured from the undisturbed water surface to the point directly below the crest, C. This depth will be denoted by H'_B when the bar is in the process of formation and by H_B after the bar becomes relatively stable.

Suppose that the variables affecting H'_B are the deep water wave length λ_0, wave height a_o, time t, the sand size d_{GM}, the sand size distribution coefficient σ_φ (Krumbein notation), the kinematic viscosity ν, the sand and water densities P_s and P_w, respectively, and the slope of beach i. The usual considerations of dimensional analysis then lead to the general relation:

$$f\left(\frac{a_o}{H'_B}, \frac{a_o}{\lambda_0}, \frac{d_{GM}\sqrt{ga_o}}{\nu}, \frac{\lambda_0}{d_{GM}}, \sigma_\varphi, i, \frac{P_s}{P_w}, \frac{T}{\sqrt{g\lambda_0}}\right)=0 \qquad (1)$$

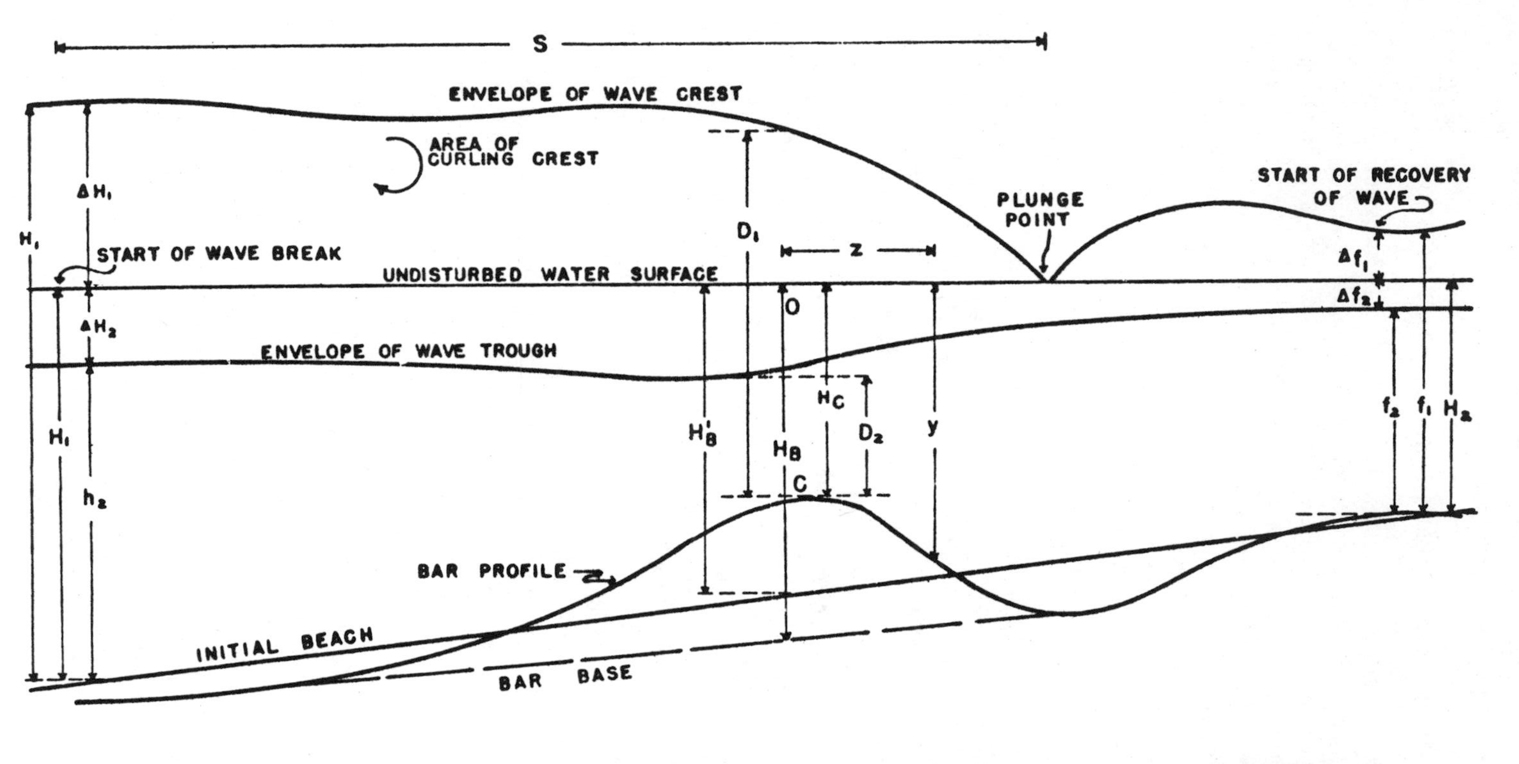

DEFINITIONS OF TERMS DESCRIBING THE BAR ENVIRONMENT

Figure 1.

The dimensionless terms are the transfer parameters relating phenomena of different scale. The multiplicity of these parameters makes the possibility of a true model experiment of bar processes questionable, if not negative. It is possible, however, that approximate models which ignore the least important terms are sufficient. The term, $\frac{\rho_s}{\rho_w}$, may be omitted because the densities are constant. If it is assumed that hydrodynamic effects on the motion of sand are independent of the Reynolds number, the parameter containing the kinematic viscosity also may be eliminated. With these simplifications,

$$f_1\left(\frac{a_o}{H'_B}, \frac{a_o}{\lambda_0}, \frac{\lambda_0}{d_{GM}}, \sigma_\varphi, i, \frac{T}{\sqrt{g\lambda_0}}\right)=0 \tag{2}$$

To obtain the equation for the bar after it is relatively stable, H_B may replace H'_B and the term involving time be omitted. Accordingly,

$$f_1\left(\frac{a_o}{H_B}, \frac{a_o}{\lambda_0}, \frac{\lambda_0}{d_{GM}}, \sigma_\varphi, i,\right)=0 \tag{3}$$

or omitting σ_φ, which may be a constant for a series of tests.

$$\frac{a_O}{H_B}=f_2\left(\frac{a_o}{\lambda_0}, \frac{\lambda_0}{d_{GM}}, i\right) \tag{4}$$

The last expression suggests that the effect of wave height, a_o, and the bar base depth H_B can be conveniently related by plotting a_o/H_B against the wave steepness ratio a_o/λ_0.

B. *Depth and form of bars.*—The depth of the bar crest below the water surface (H_c in fig. 1) is the highest point of the bar and therefore is probably a critical feature. It may then be assumed that a relationship exists between the bar depth, the wave characteristics and the nature of the sand. The general expression

$$f_2\left(\frac{H_c}{a_o}, \frac{a_o}{\lambda_0}, \frac{\lambda_0}{d_{GM}}, \sigma_\varphi, i\right)=0 \tag{5}$$

is obtained from dimensional analysis by assuming that the bar is relatively stable and the effect of kinematic viscosity is negligible. In view of equation 3, H_c/a_o may be replaced by H_c/H_B, giving

$$\frac{H_c}{H_B}=f_3\left(\frac{a_o}{\lambda_0}, \frac{\lambda_0}{d_{GM}}, \sigma_\varphi, i\right) \tag{6}$$

If $\sigma\varphi$ is the same in all tests it suffices to write

$$\frac{H_c}{H_B}=f_3\left(\frac{a_o}{\lambda_0}, \frac{\lambda_0}{d_{GM}}, i\right) \tag{7}$$

Accordingly the depth and form of the bar can be related to wave characteristics by considering the dependence of H_c/H_B upon a_o/λ_0.

A more complete description of a bar is obtained from an examination of variations in the form, considering the bar profile as the form. The general relation defining the bar profile is obtained as follows. Any point on the profile may be determined by the parameters y and z (see fig. 1) in which y is the depth of the point with respect to the undisturbed water surface and z is the horizontal distance from a vertical line passing through the crest, C, of the bar. This distance will be regarded as positive if the point is shoreward of C. By dimensional reasoning as in Equation 6, we obtain the relationship:

$$\frac{\gamma}{H_B}=f\left(\frac{z}{H_B}, \frac{a_o}{\lambda_0}, \frac{\lambda_0}{d_{GM}}, \sigma_\varphi, i\right) \tag{8}$$

and again neglecting σ_φ,

$$\frac{\gamma}{H_B}=f\left(\frac{z}{H_B}, \frac{a_o}{\lambda_0}, \frac{\lambda_0}{d_{GM}}, i\right) \tag{9}$$

which is the desired general relation for the form or configuration of bars.

Section III. EQUIPMENT AND PROCEDURE

Experiments were made in two wave tanks, one 1.5 feet wide, 2 feet deep and 30 feet long; the other 14 feet wide, 4 feet deep and 85 feet long. Schematic views of the small and large tanks are shown in figure 2A and 1B, respectively. The conditions illustrated are those for a beach having a slope of 1 on 70. The experimental beach for each sketch is to the left of the mark *0;* the section to the right of the mark is a transition. In a more suitable tank this transition section would not have been necessary; instead the beach slope would have extended to the bottom. The distorted proportions of the approach, or transition segment, restrict the normal motion of the sand toward the experimental beach and thus might alter the dimensions of the bar eventually formed. It is considered that the test results for the 1/70 slope may be affected by these conditions but those for the steeper slopes are believed to be reliable. The mark *0* of the figure is solely a reference point for measuring distances and identifying stations. The depth of water in front of the wave generator is variable, the depth being adjusted, depending on the height of the waves, to control the position of the breaking waves on the experimental beach.

Tests in the small tank were made with beach slopes of 1 on 70, 1 on 30, and 1 on 15. Those slopes were established always to the left of *0,* with the transition segment to the right of *0* the same for all

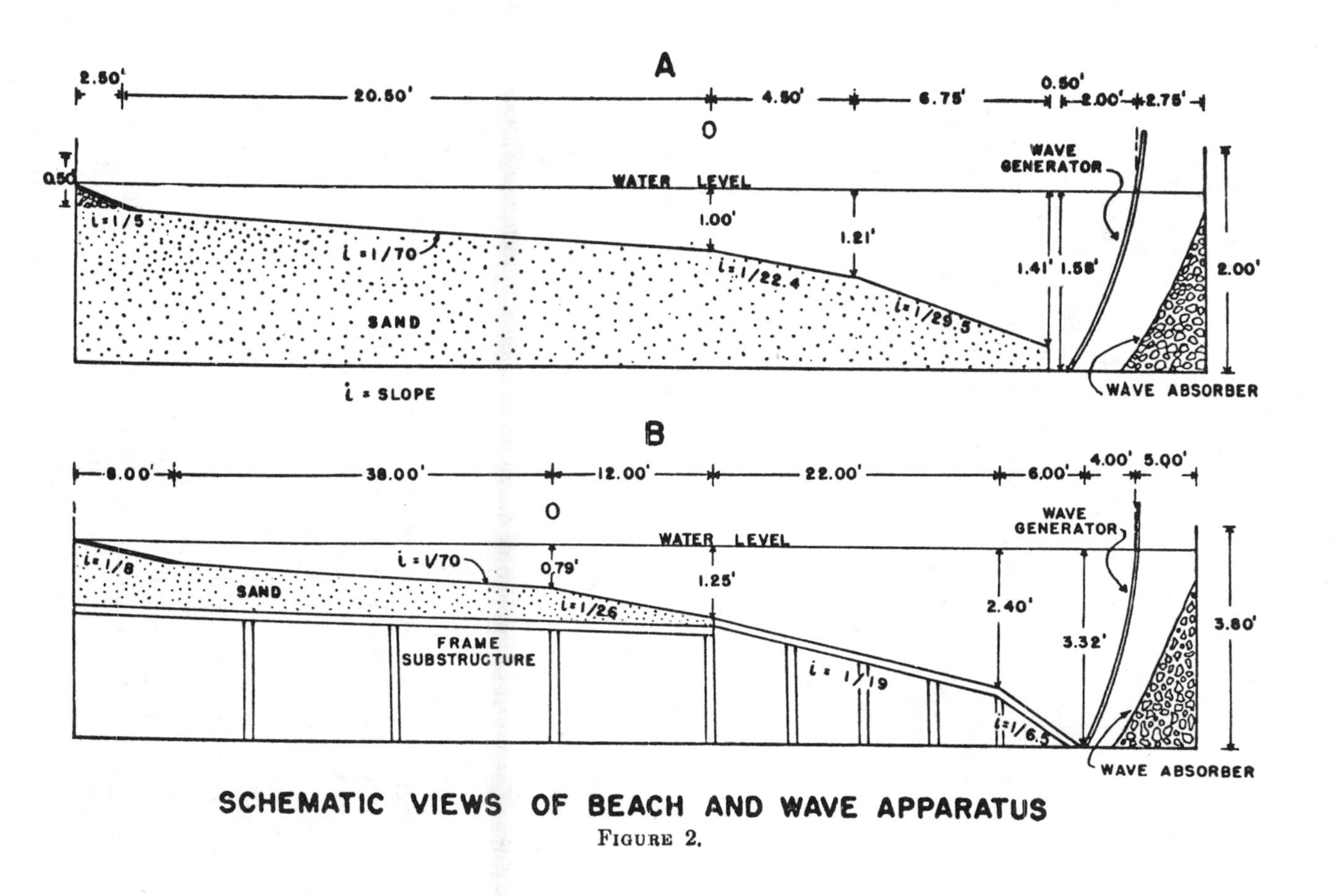

SCHEMATIC VIEWS OF BEACH AND WAVE APPARATUS

FIGURE 2.

three slopes. The small tank is provided with a glass wall allowing observation of the experimental beach.

The large tank was subdivided into three parallel compartments with beach slopes of 1 on 30, 1 on 50, and 1 on 70, respectively. The experimental beach of slope 1 on 70 is shown in figure 2B. The sand of the experimental beach and of a part of the approach segment is placed on a platform. The remaining major part of the approach segment, consisting of a wooden platform, is firmly attached to the concrete bottom of the tank. In the sketch the experimental beach is to the left of mark *0*, and the approach segment of the right of *0*. In the three compartments the approach conditions were alike.

The sand used for the experiments was commercial Potomac River sand. Sieve analysis for the distribution of grain size gave the following percentage values:

Grain size (millimeter)	*Percent larger than*
2.38	1.01
1.168	6.68
0.589	27.08
0.297	72.93
0.149	95.66
0.074	99.30

The defining parameters of the experimental sand in Krumbein's notation are, d_{GM}=0.42 millimeter and σ_{φ}=0.96. (See fig. 3.)

The quantities observed during a test are shown in figure 4, which is representative of the basic data. The envelopes of the wave crest and trough, the undisturbed water surface, and the initial beach profile were marked on the glass tank wall, then transferred to a graph. The point of impending wave break, point S_1 in figure 4, was believed to be significant and was noted. At the point of impending wave break the wave front in the immediate vicinity of the crest is almost vertical and the water particles on the crest are moving with a velocity slightly below the velocity of wave propagation. As motion continues the crests deform rapidly, then break. It will be shown later that the transport of sand at the point of impending wave break is a maximum.

In the tests made with the small tank, the gradual formation and stabilizing of the bar was observed through the glass wall. Roughly speaking, stability was reached in one hour with a slope of 1 on 15, in two hours with a slope of 1 on 30, and in 4 hours with a slope 1 on 70, the time intervals being the duration of the respective tests made in the small tank. It was not possible to observe the formation of the bars during the tests made in the larger tank; therefore, the time intervals for those tests were determined from the values obtained for tests in the smaller tank by comparing wave lengths and using a transference equation based on the Froude number. The minimum

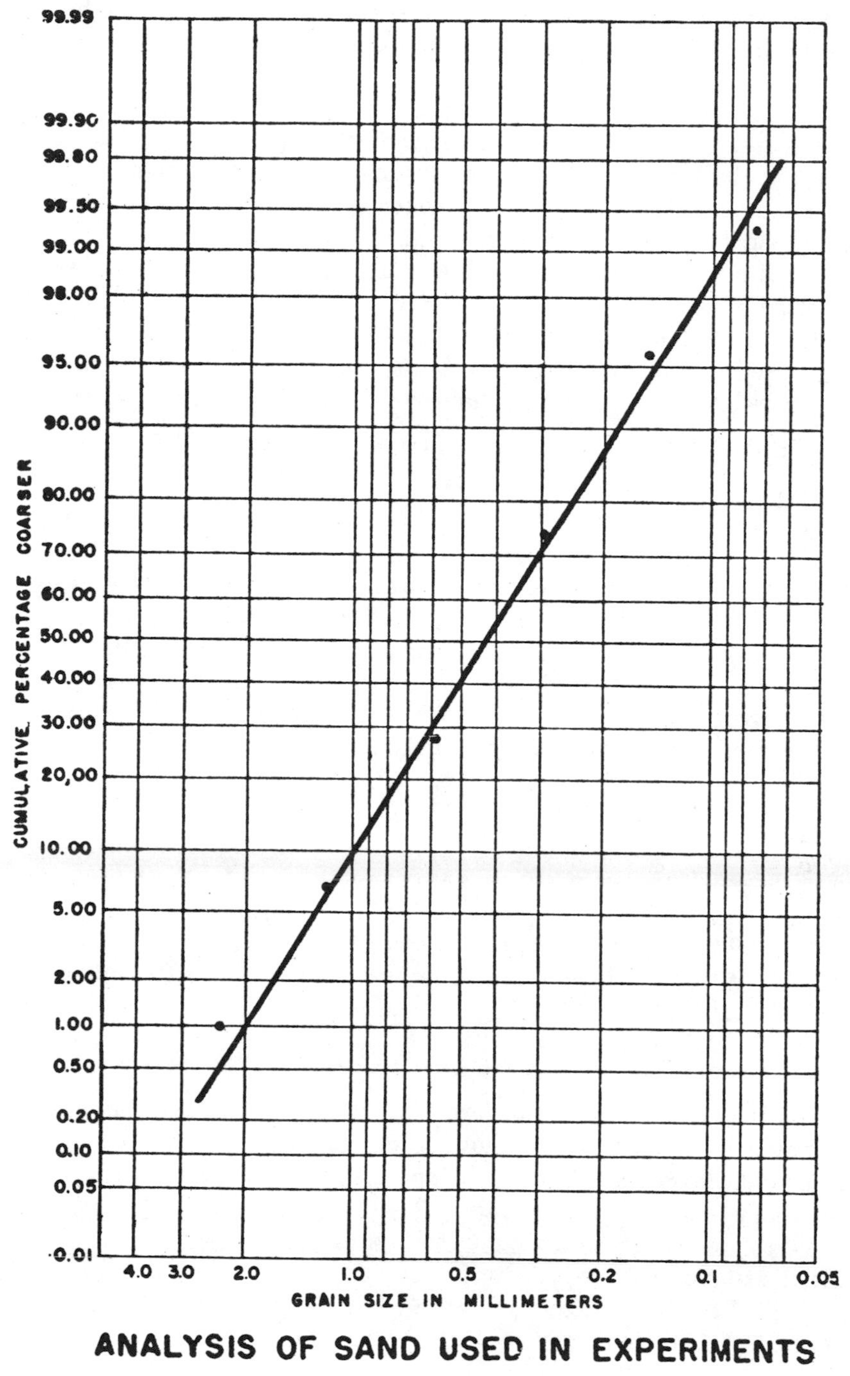

ANALYSIS OF SAND USED IN EXPERIMENTS

FIGURE 3.

test durations adopted for the larger tank on this basis were 4 hours for the slope 1 on 30, 8 hours for the slope 1 on 50, and 12 hours for the slope 1 on 70.

The configuration of the bar shown in figure 4 is indicated by a smooth curve. This is an idealization in that the bar surface and the adjacent bottom actually were covered with ripple-marks. The curve shown was derived from master traces of the actual surface; it and similar data are the average of two or more runs.

It is desirable to define the waves reaching the experimental beach in terms of their deep-water characteristics but the waves generated in both the small and large tanks were shallow water waves. Since the period is constant, the wave length, λ, and the wave height, a, may be related to the corresponding deep water characteristics λ_o and a_o by the following approximate expressions: $\frac{\lambda}{\lambda_0} = tanh\,\frac{2\pi H}{\lambda}$

$$\frac{a_o}{a} = \left(\frac{2\pi H}{\lambda} + sinh\,\frac{2\pi H}{\lambda}\,cosh\,\frac{2\pi H}{\lambda}\right) \Big/ cosh\,\frac{2\pi H}{\lambda} \qquad (10)$$

where λ and a are measured at a point having the undisturbed water depth H (reference 7). The elimination of H between equations 1 and 2 gives an expression for the dependence of a_o/a upon λ/λ_o. This expression is shown graphically in figure 5 and is the basis of the desired reductions. Taking the observed value of the wave period, T, the deep water wave length λ_o may be computed from

$$\lambda_0 = gT^2/2\pi \qquad (11)$$

The value a_o/a corresponding to the ratio λ/λ_o where λ is the measured wave length in front of the wave generator, may be read from the curve in figure 5. Knowing the value of the ratio a_o/a, the deep water wave height a_o, is readily obtained since the wave height a measured in front of the wave generator is known. These computations were made for all the tests.

Section IV. RESULTS OF LABORATORY EXPERIMENTS

The initial changes in the form and movement of the bar are illustrated in figure 6 which shows a typical bar development. The form and movement changes do not progress uniformly with time, but oscillate about mean values. The inset on figure 6 is a smooth curve representing the average position of the bar with time. The initial assumption as set forth in section II*a*, to the effect that the motion of the bar is decreased to an imperceptible value is thereby shown to be substantiated by the laboratory observations.

806229—48——3

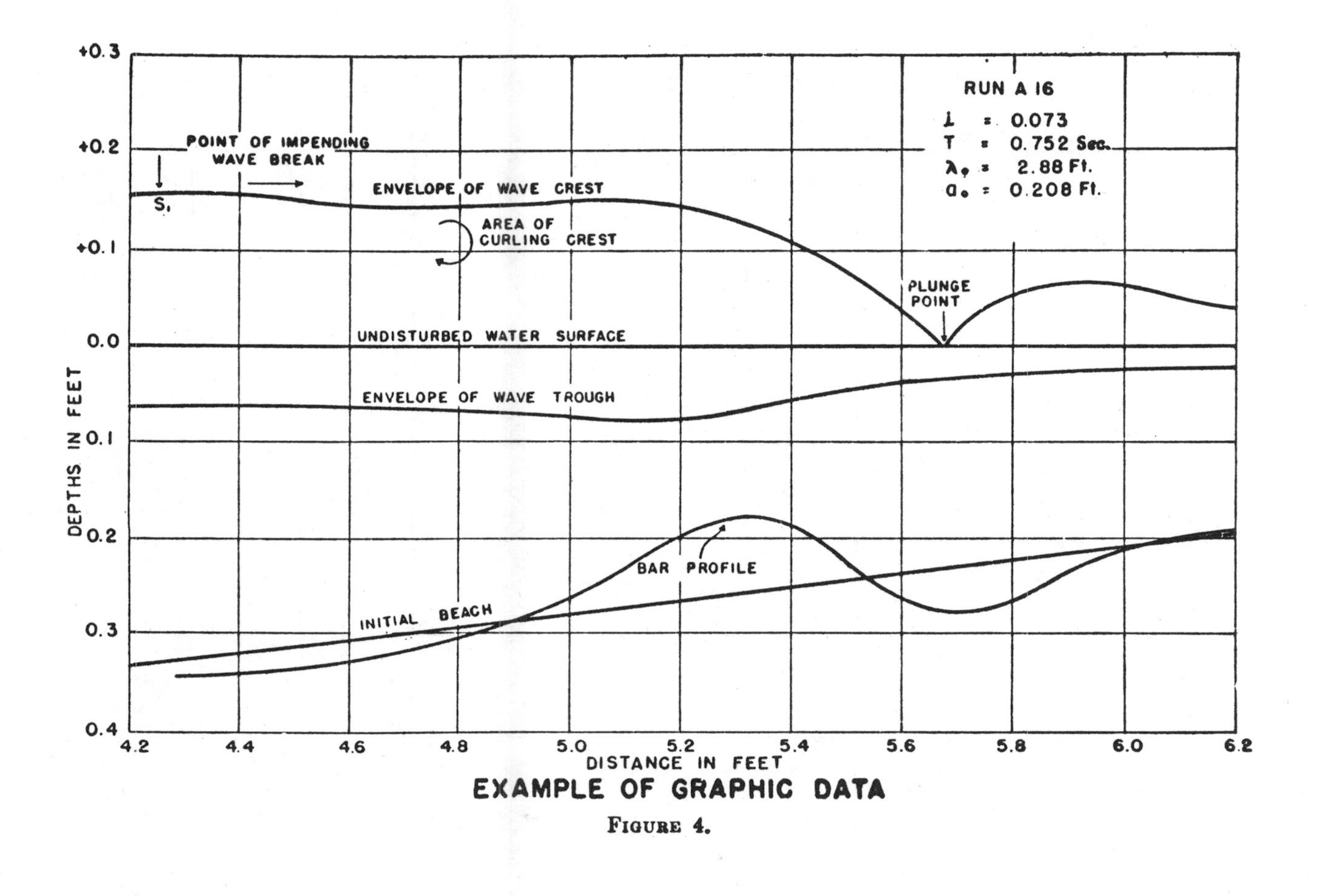

EXAMPLE OF GRAPHIC DATA

FIGURE 4.

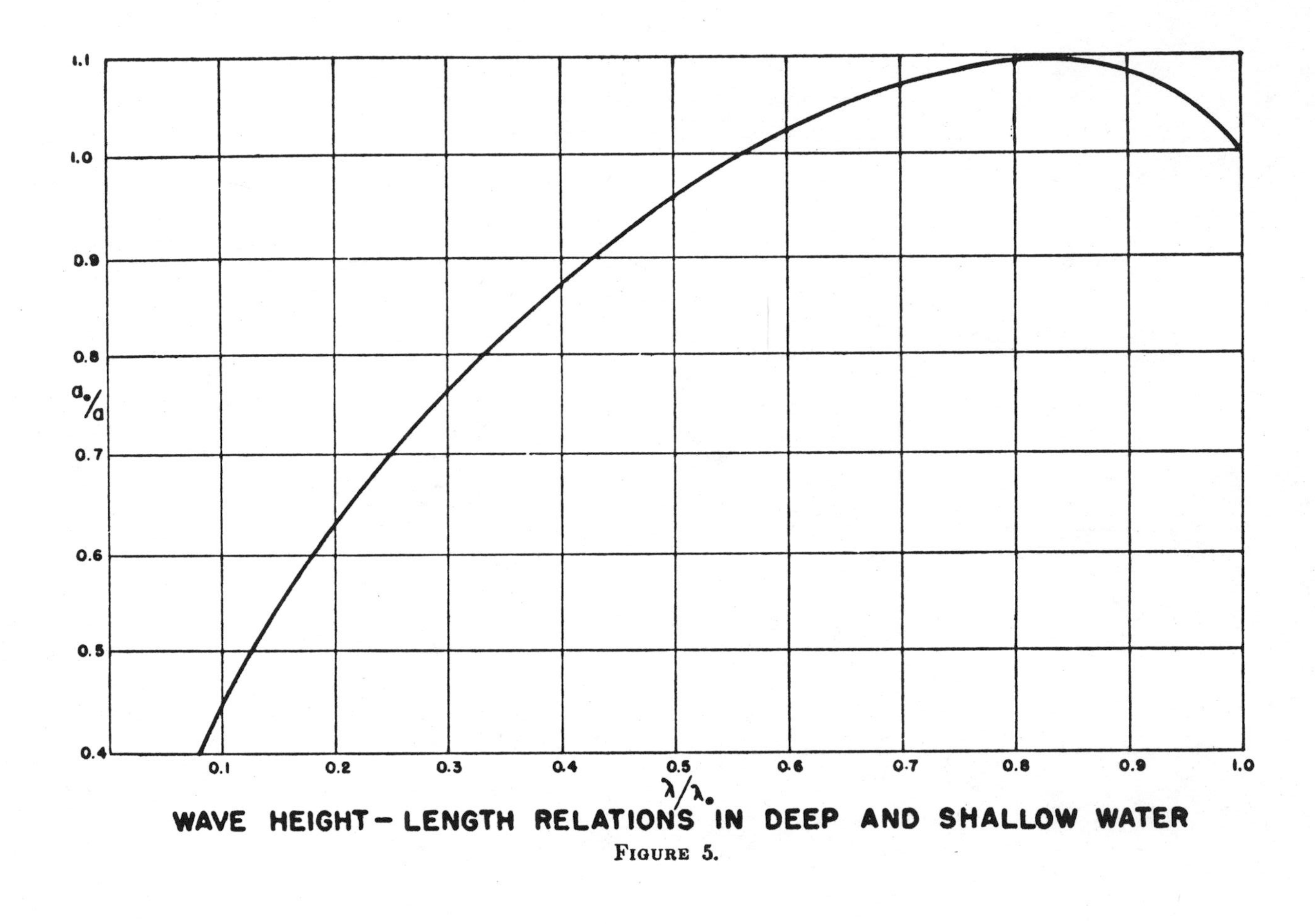

WAVE HEIGHT – LENGTH RELATIONS IN DEEP AND SHALLOW WATER

FIGURE 5.

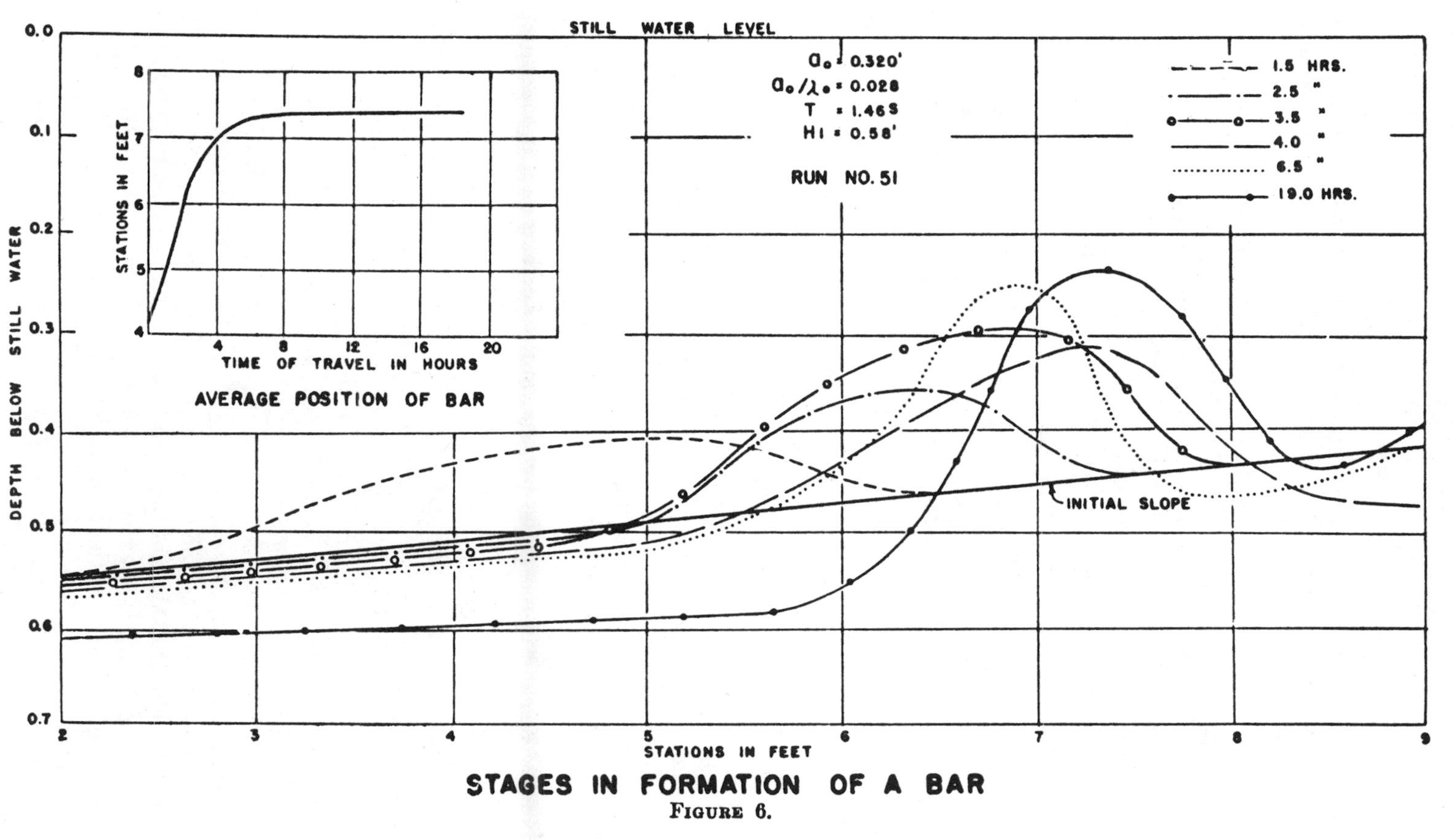

STAGES IN FORMATION OF A BAR

FIGURE 6.

The factors controlling the position of the bar at the time it becomes relatively stable are stated in equation 4. The results of the tests to evaluate the terms in equation 4 are plotted on figure 7. Figure 7A shows the relation between a_o/H_B i. e., the ratio of wave height to the depth of bar base; and the wave steepness, $a_o\lambda_0$.

It appears that for the same value of wave steepness, a_o/λ_0, the initial slope i, of the beaches did not affect the ratio a_o/H_B. If an effect is present, it is of the order of the error of observation.

Since the points are plotted without differentiating between the values of the ratio $\frac{\lambda_0}{d_{GM}}$ and do not scatter to any great extent, it may be concluded that the effect of this ratio, if present, also cannot be larger than the observation error. For example, considering the following pair of data taken from the records:

$$\lambda_0=48 \text{ ft.}, a_o/\lambda_0=0.06, a_o/H_B=0.50$$

$$\lambda_0=18 \text{ ft.}, a_o/\lambda_0=0.06, a_o/H_B=0.45$$

a 300 percent variation is indicated in the $\frac{\lambda_0}{d_{GM}}$ values, yet the variation in the values of the ratio a_o/H_B is only 10 percent.

Interpretation of the curve in figure 7A leads to the conclusion that if the water depth remains constant, the depth of bar base and consequently, the position of the bar, formed by a single system of waves is a function of the wave height, a_o, and wave steepness a_o/λ_o. If the water depth and wave steepness are held constant, an increase in the wave height will move the bar seaward. Furthermore, if the water depth and wave height are held constant, an increase in the wave steepness will move the bar shoreward. Further, if the wave height and wave length are held constant, H_B, will likewise remain constant, and any increase in the depth of water moves the bar shoreward.

In figure 7B the data from the beaches of different slopes are plotted individually. No distinction is made, however, between the tests made in the large and small tank. Similarly, no differentiation is made between points associated with the variable sand parameter, λ_0/d_{GM}. It should be noted that the ratio of the depth of bar crest to the depth of bar base H_c/H_B, is practically independent of the wave steepness ratio a_o/λ_0, and of the slope i. Since the data covers a wide variation in the values of the ratio λ_0/d_{GM}, it may also be inferred that the ratio H_c/H_B is likewise independent of the ratio λ_0/d_{GM} within the order of the error of the observations. The data as plotted in figure 7B give a constant value

$$H_c/H_B=0.58 \tag{12}$$

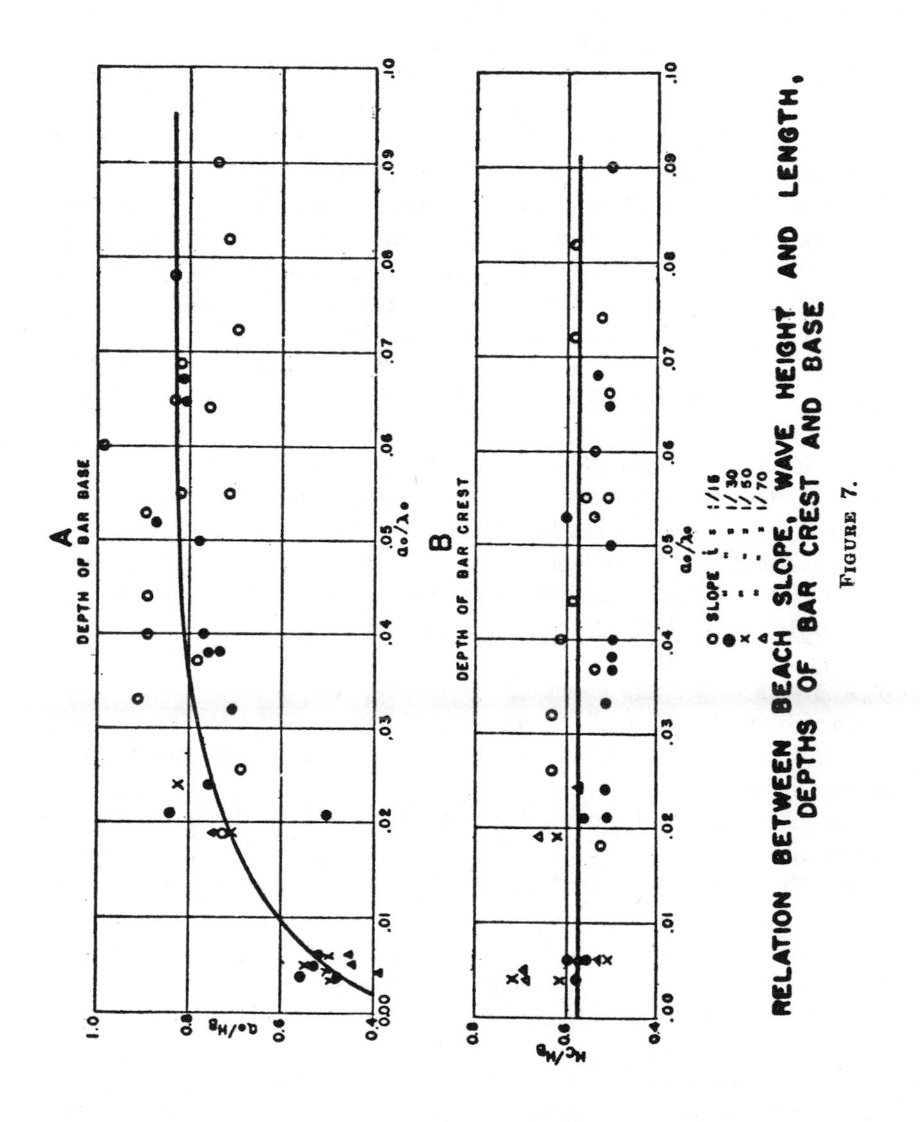

RELATION BETWEEN BEACH SLOPE, WAVE HEIGHT AND LENGTH, DEPTHS OF BAR CREST AND BASE

FIGURE 7.

In examining the forms of the bars obtained in the laboratory on the basis of equation 9 to determine the importance of a_o/λ_0, it is desirable to replace the bar base depth H_B with the depth of the initial beach surface H'_B measured at the point just below the crest. See figure 1. This substitution is useful because H'_B is more accurately determined than H_B and is permissible because the ratio H_B/H'_B approximates a constant quantity, which is a function of the slope i.

Figure 8 gives the results of data obtained in the small tank for a slope of 1/15. The values of λ/H'_B as plotted in figure 8 were grouped and averaged for certain intervals of the wave steepness ratio. No systematic variation in the configuration of the bars with wave steepness is discernible. Therefore a single curve passing through the points may be considered as defining the form of the bar for this slope ($i=1/15$). The results indicate that the form of the bar is apparently independent of the size of the generating waves. The bar base and the initial beach surface also are shown in the same figure. It is seen that here $H_B/H'_B=1.2$. This ratio differs from unity due to the fact that the slope of the initial beach surface is steep and the bar troughs have descended relatively far below the initial surface. A similar analysis of the bar form developed in tests made on beaches with a 1/30 slope leads to the same conclusion; i. e., the form of the bar is independent of the wave steepness ratio. The ratio H_B/H'_B in this case has the value of 1.09, indicating that although the troughs of the bar descend below the initial surface, the effect is not so pronounced as that obtained with the steeper slope, $i=1/15$.

The results of the test with 1/70 beach slopes are shown in figure 10. In this case the effect of wave steepness is noticeable. Bars formed by waves of small steepness ratio have wider crests and longer bar bases. With an increasing steepness ratio the bars become slender and pointed.

An examination of figures 8 and 9 shows that in the bar environment of beaches having steep slopes the troughs descend markedly below the level of the initial undisturbed beach. In the bar environment of beaches having flat slopes the troughs hardly descend below the level of the undisturbed beaches. As a result of this behavior, the trough depth below the undisturbed water surface manifests a remarkably uniform relationship with the depth of water over the bar crest.

Let H_t denote the depth of the trough below still water level. The dependence of H_t/H_c on a_o/λ_o is indicated in figure 10. To facilitate the comparison the data for the different slopes are shown separately. It is seen, making due allowance for errors of observation, that H_t/H_c is practically independent of wave steepness and of beach slope. The ratio is found to have an average value of 1.69.

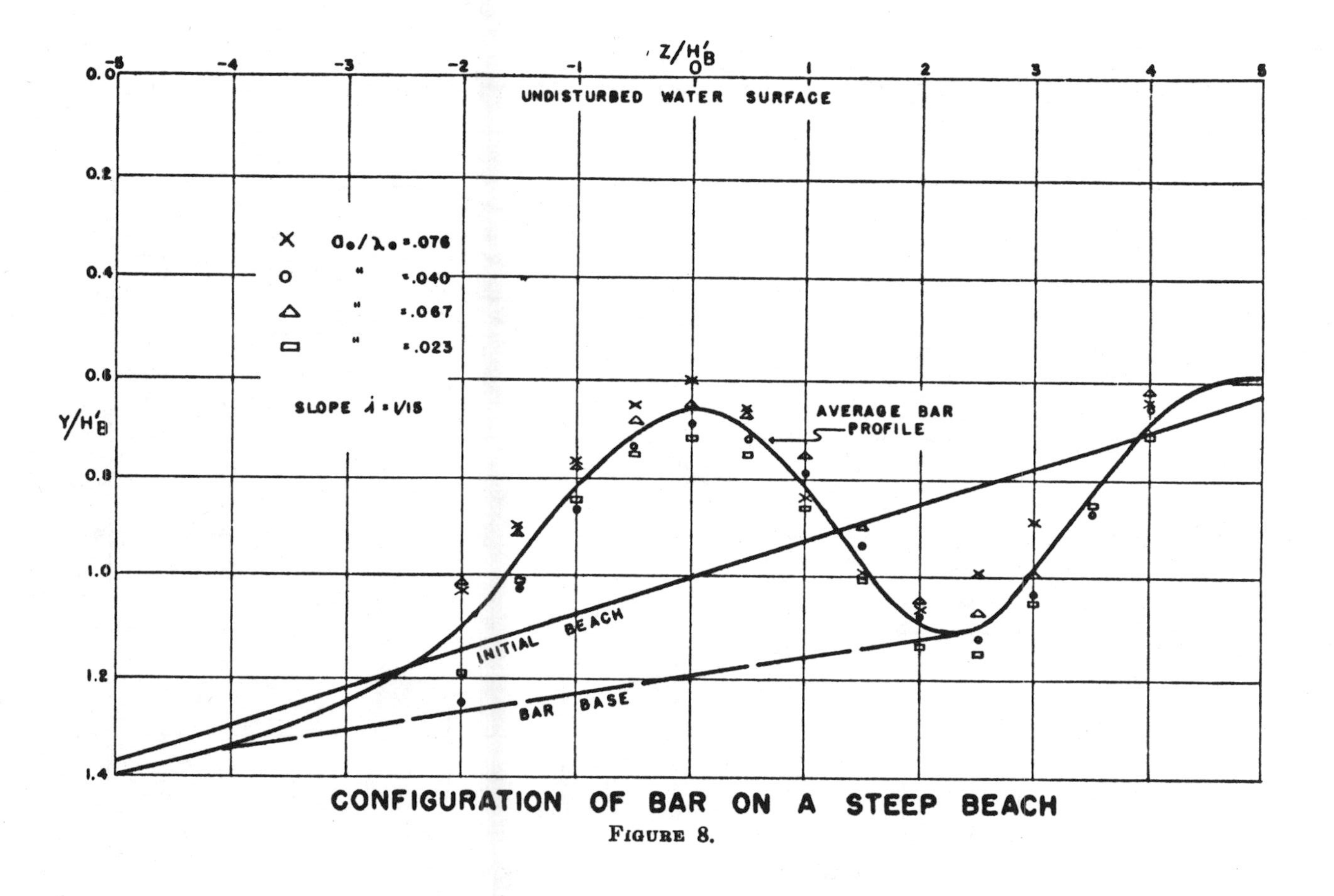

CONFIGURATION OF BAR ON A STEEP BEACH

Figure 8.

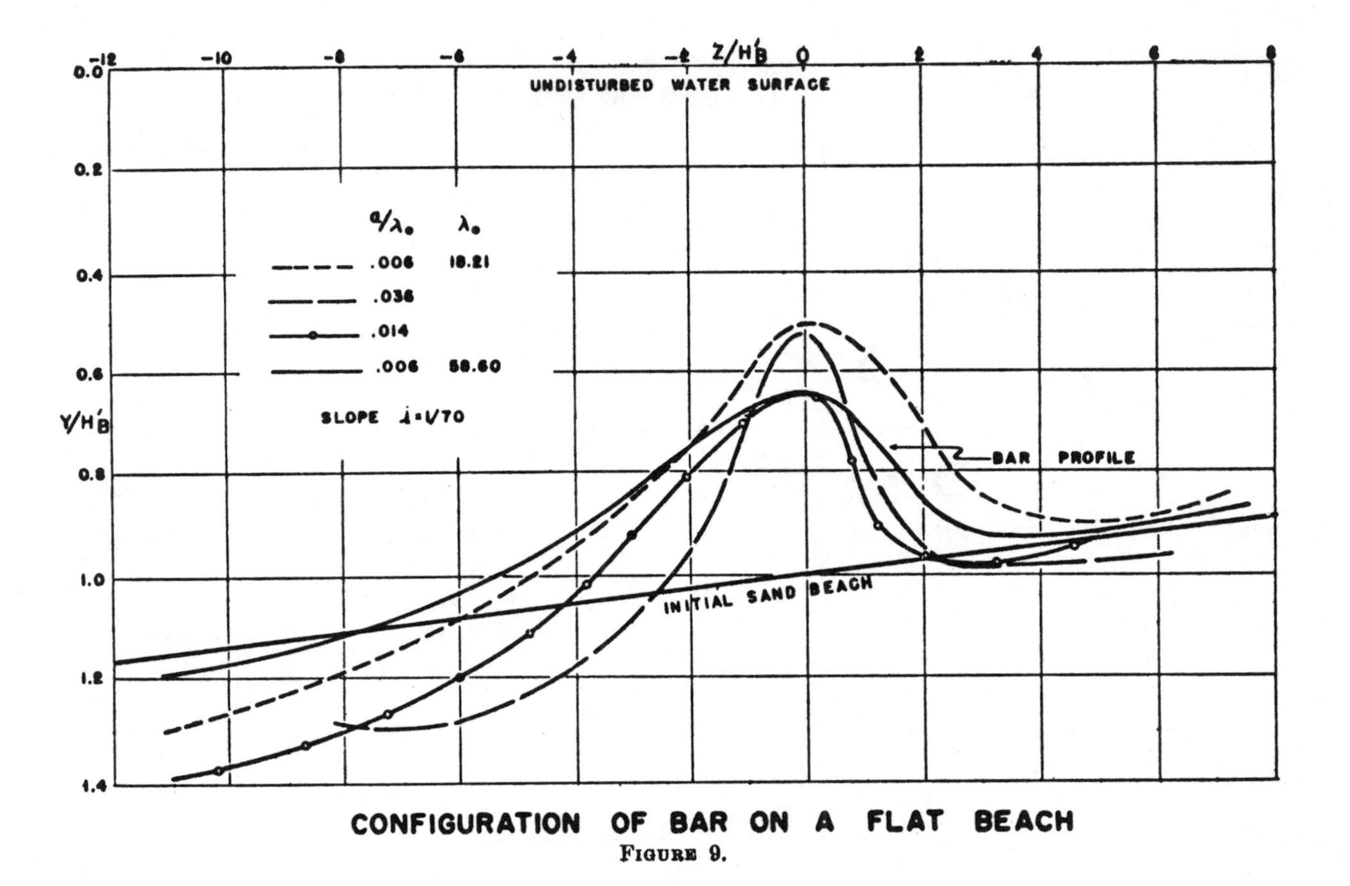

CONFIGURATION OF BAR ON A FLAT BEACH

FIGURE 9.

806229—48——4

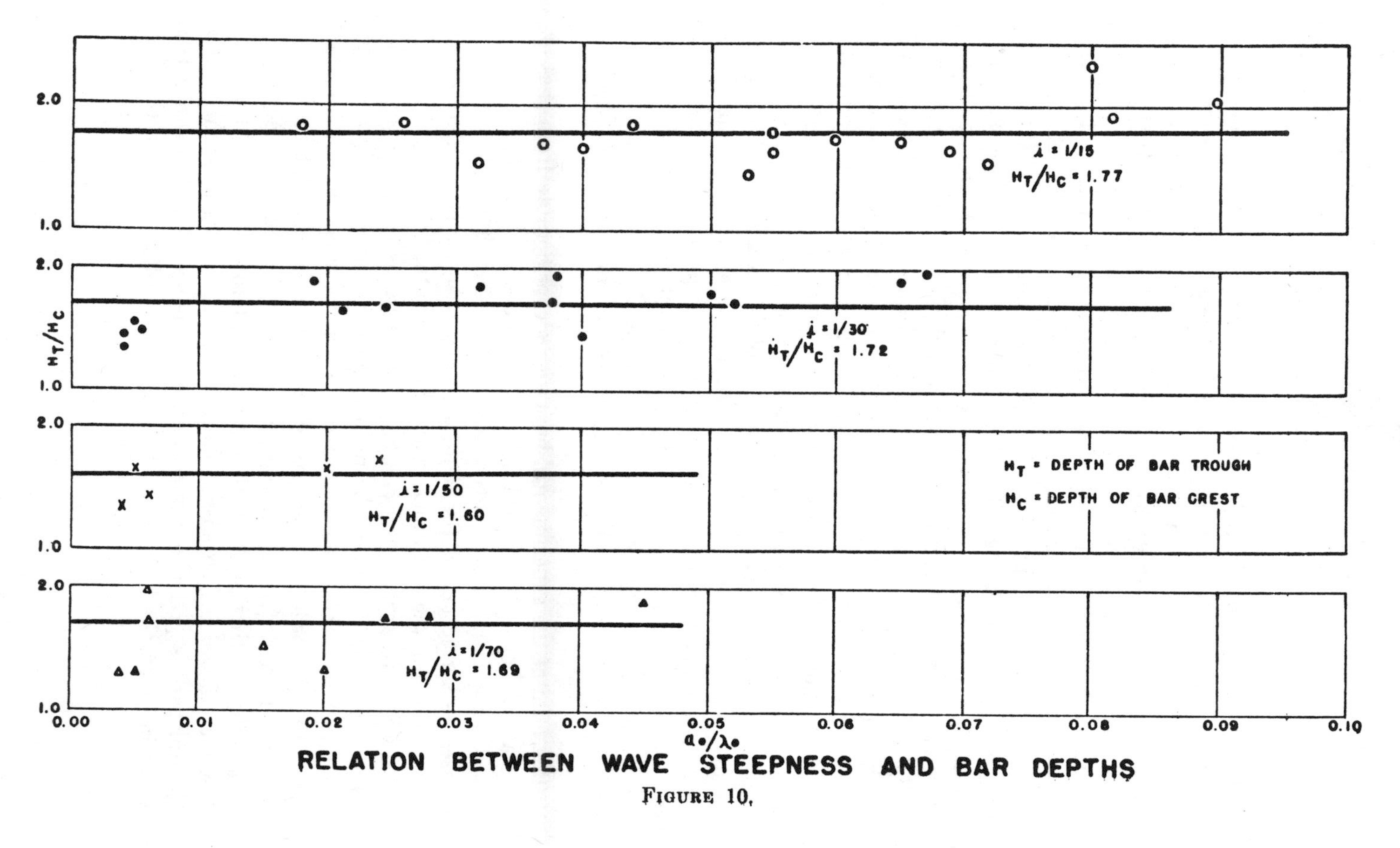

RELATION BETWEEN WAVE STEEPNESS AND BAR DEPTHS

FIGURE 10.

Section V. FIELD OBSERVATIONS

A. *Form and dimensions of natural bars.*—The bars of the Pomeranian coast have been sounded by Otto and Hartnack (references 2 and 4), and the latter's paper includes records of soundings by Lehmann and others. Evans (reference 5) measured the bars of Lake Michigan. The data from these investigations will be examined here to ascertain whether any similarity exists between the form of natural bars and those developed in laboratory tests.

Consider the profiles of Lake Michigan bars shown in figure 11, from which two groups of bars are selected; one consists of the bars in a first zone of the four profiles, i. e., the bars closest to the shore and the other, those in a second zone the next seaward. The bar base depth H_B of each bar was determined, and the distance y and z of profile points of a given bar obtained, from which the space parameters y/H_B and z/H_B are formed. The averaged and smoothed values of these parameters for the bars of the two groups are shown in figure 12. The values are plotted as rectangles for the first zone and circles for the second zone. Since the distribution of points for the two zones is approximately the same, a mean curve can be drawn as shown in the figure. The curve may be representative of the bars of Lake Michigan. In preparing the Pomeranian coast data, the average for the given locality was taken irrespective of the zones of bars. These data are represented by dots, triangles and crosses and the mean curve through the points is representative of the bars of the Pomeranian coast. The lowest curve represents the average value for the bar determined by the experiments based on the data shown in figure 9.

The form of the experimental bars varies considerably from that of the natural bars. The natural bars are flatter and longer than the experimental bars; significantly, however, the depth of the crests expressed as a fraction of the depths of bar base have practically the same value for the natural and the experimental bars.

No adequate explanation can be offered now as to the cause of the difference in the forms of the natural and experimental bars, except to suggest that the waves producing the natural bars are of a mixed type, and the order and extent of turbulence is of a different scale resulting in a difference in the suspension and motion of particles. It is believed that such factors would tend to produce flatter bars.

A more detailed comparison of the crest and trough depths of the natural bars is given in table 1. The sequence of bars defines the zones in which the bars occur, the bar nearest shore lying in the first zone.

TABLE I.—*Crest and trough depths of natural bars*

Locality and investigator	Average slope	Factor	Zone			
			I	II	III	IV
Pomeranian Coast (Otto)	0.013	H_B(ft.)	....	3.8	6.5	....
	.013	H_C/H_B	....	.46	.57	....
	.013	H_t/H_C	....	1.79	1.56	....
Pomeranian Coast (Lehmann)	.017	H_B(ft.)	3.12	5.74	10.40	14.8
	.017	H_C/H_B	.54	.55	.57	.58
	.017	H_t/H_C	1.71	1.50	1.66	1.56
Pomeranian Coast (Otto)	.014	H_B (ft.)	....	7.02	12.20	20.8
	.014	H_C/H_B	....	.46	.52	.52
	.014	H_t/H_C	....	1.87	1.77	1.81
Pomeranian Coast (Hartnack)	.017	H_B (ft.)	2.22	6.56	9.88	13.1
	.017	H_C/H_B	.42	.49	.55	.62
	.017	H_t/H_C	1.86	1.77	....	....
Lake Michigan (Evans)	.017	H_B (ft.)	....	7.12	13.37	20.6
	.017	H_C/H_B	....	.55	.62	.62
	.017	H_t/H_C	....	1.45	1.42	1.55

The dimensions shown are average values derived from different profiles for natural bars found in similar zones. The bar nearest the shore is designated as being in the first zone, but at times it was difficult to recognize that zone. Therefore, in some cases the zone designation may not be correct. The data in table I, as would be expected, indicate that the depth of the bar base increases with the order of zones. The ratio of the crest depth to the bar base depth tends to increase as the distance the bars lie from the shore increases. This may indicate that the bars that are farther away from the shore have lost material from their surface. It is possible, also, that the larger waves do not operate long enough in this area for the complete formation of bars. The mean value of H_C/H_B when computed from the table is 0.54 which compares favorably with the corresponding ratio, H_C/H_B of 0.58, obtained from the laboratory tests. For the different determinations the ratio of trough depth to the bar crest depth varies between 1.42 and 1.86. The average value H_tH_C of 1.66 compares well with the average of the laboratory determination; H_t/H_C of 1.69.

B. *Lake Michigan bars and observed waves.*—The profiles of the bars of Sylvan Beach, near the White Lake piers on the eastern shore of Lake Michigan, were measured by Evans in 1939 (reference 5). Some of these measurements are reproduced in figure 11. The essential bars in a given profile are three in number, ignoring the indefinite forms near the shore. An application of the results shown in figure 7A might be to determine the magnitude of the waves necessary to produce those bars. The underlying supposition is, of course, that each bar in a given profile is created under the action of a singular system

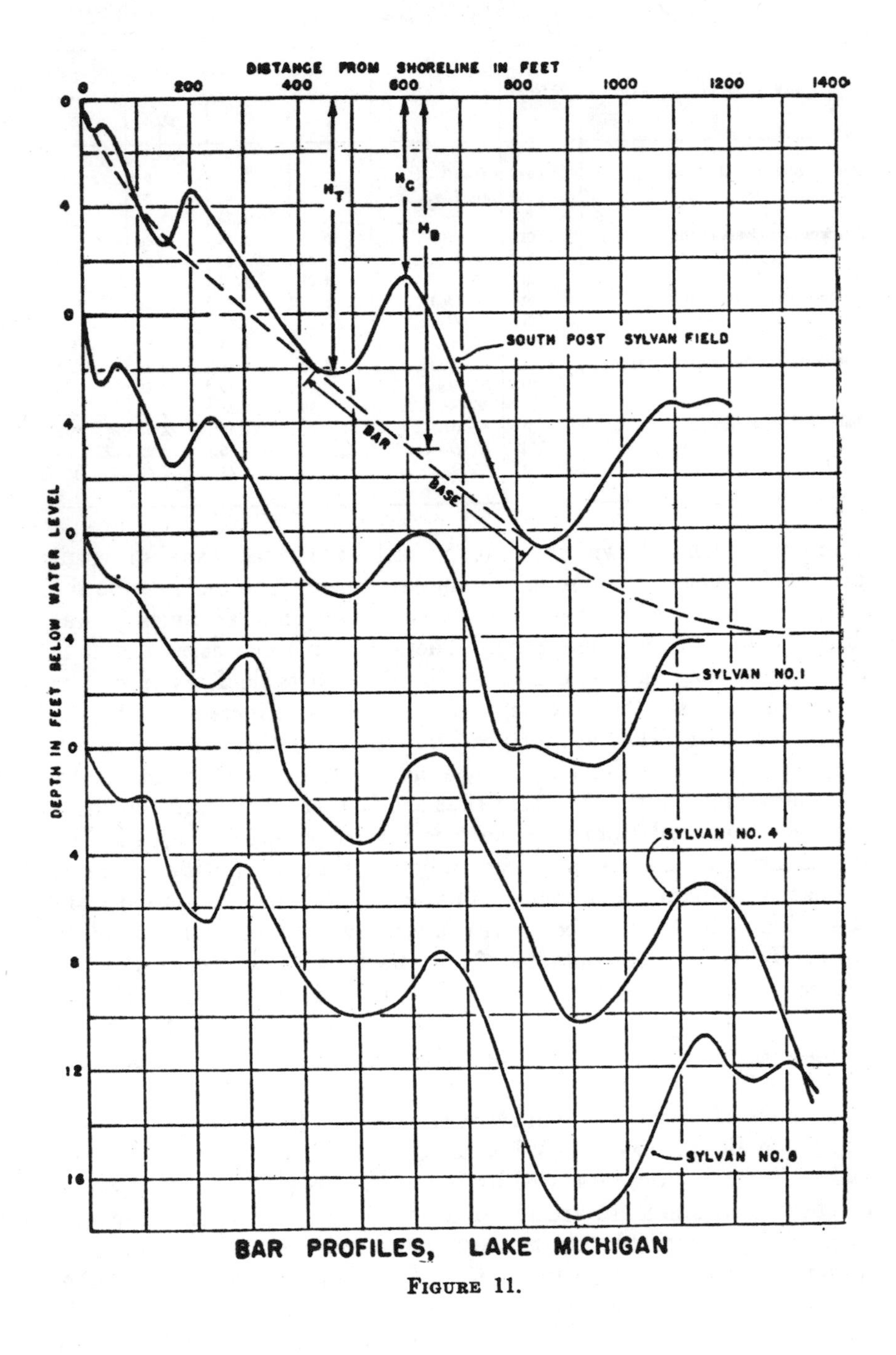

BAR PROFILES, LAKE MICHIGAN

FIGURE 11.

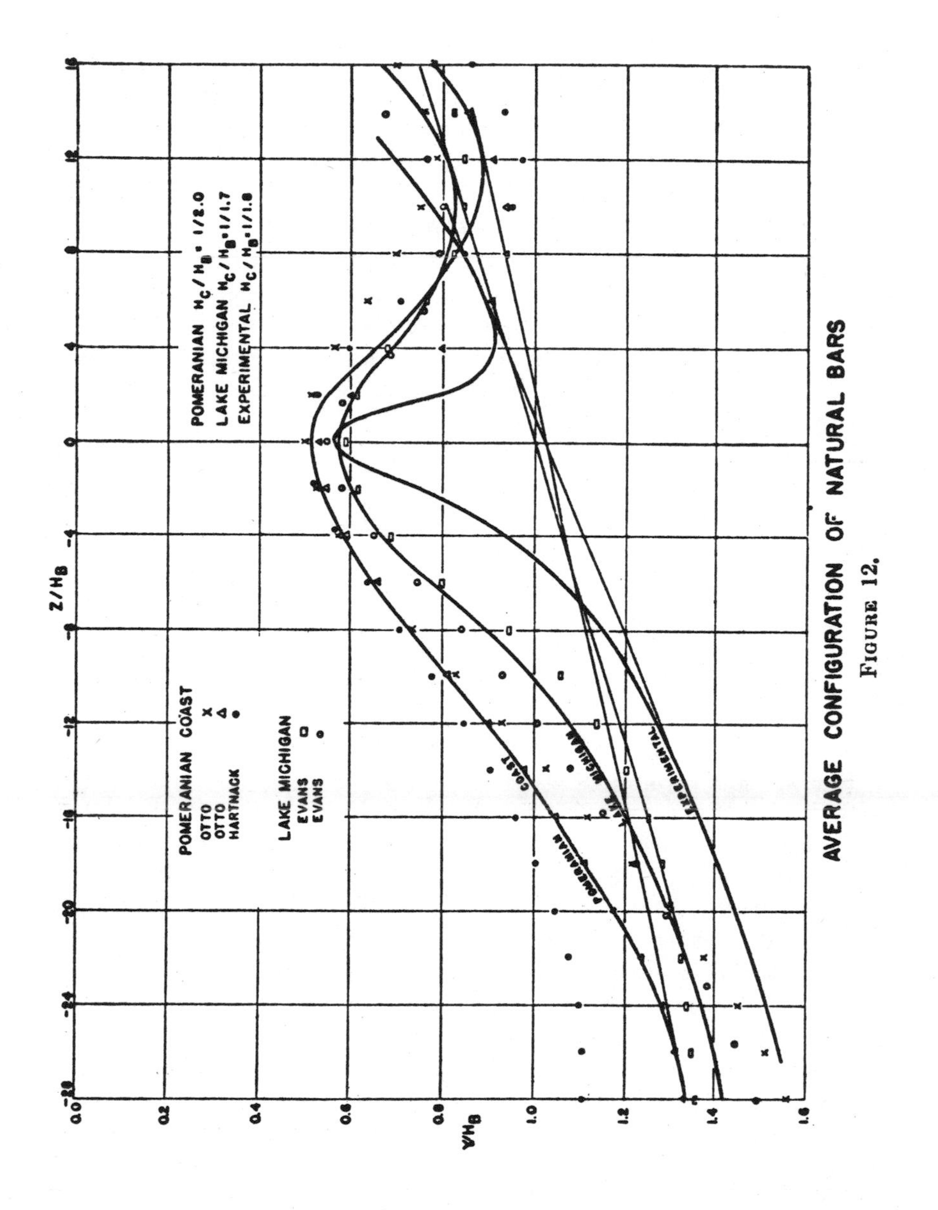

AVERAGE CONFIGURATION OF NATURAL BARS

FIGURE 12.

of waves, and the respective wave systems are different. It is recognized that natural conditions may not approximate this supposition. The bar base depths of the bars in the different zones are defined by lines joining the troughs of the bars. According to the definition, the segment of the line under the body of a given bar becomes the bar base. From figure 11 the average value of the bar base depth, H_B for the bars in the first zone of the profiles is 7.2 feet; in the second zone, 13.2 feet; and in the third zone, 19.5 feet. If it be assumed, for example, that the bars are formed under waves of steepness ratio $a_o/\lambda_0=0.03$, the required value a_o/H_B from figure 7A is 0.75. Accordingly, the bar of the first zone required 5.4-foot waves for its formation, that of the second zone required 10.5-foot waves, and the third zone 14.5-foot waves. H. A. Montgomery observed wave heights in Lake Michigan, near Milwaukee, from 10 April 1931 to 28 September 1932 (reference 7). The maximum wave heights recorded over the first 12 months of the investigation were:

Maximum wave height (in feet)	*Number of days on which recorded*
14	1
13–14	0
12–13	0
11–12	1
10–11	2
9–10	7
8–9	6
7–8	5
6–7	22
5–6	32
4–5	48
3–4	71

The average period of the waves of lowest height was 4.3 seconds, with a minimum period of 3.5 and a maximum of 5.0 seconds. Since the water at the location of the measurements was quite deep, the inference is that the low waves were about 95 feet long. The wave steepness ratio was about 0.03. The above measurements were made on the western shore of the Lake. If they apply to the eastern shore near White Lake, the waves are of the order of magnitude expected to create bars of the dimensions observed in the first and second zones.

The above comparison of laboratory experiments and natural processes would be of greater value were it possible to discuss simultaneously the effect of a spectrum of waves operating over a beach. Actually it may be surmised that waves in nature are far from a singular system.

Section VI. FACTORS AFFECTING MECHANISM OF BAR FORMATION AND MOVEMENT

As stated in the introduction, certain observations of the mechanism of bar formation and movement were made during the course of the experiments. While these observations are somewhat limited it is believed that they are of sufficient value to justify some discussion.

A. *Sand transportation on smooth beaches.*—A qualitative measure of the bed motion of sand in the initial and the final stages of bar formation was obtained from an experiment in the small tank with a 1 on 70 slope, wave height of 0.32 foot, and wave period of 1.4 seconds. The undisturbed water depth H_1, at the point where waves began to deform was 0.58 feet. In figure 13 are shown the initial condition of the beach, and the envelopes of the wave crests and troughs. The breaker was of the spilling type and was completed at station 4. In figure 14 are shown the last stages of the bar formation and the corresponding envelopes of crests and troughs. Very marked changes are noticeable both in the beach surface and the water surface. Ripple marks in the area in front of the bar are in evidence. The general slope of this area is almost horizontal. The breaker, now of the plunging types, has moved toward the shore. It is important to note that the bar was formed, not where the breaker was initially present, but appreciably nearer the shore.

Determinations of the rate of transportation of sand were made for the initial and the final conditions of the beach. Initially measurements were made with galvanized rectangular traps, one-half inch wide, one inch deep, and extending across the tank. Four of these traps were set level with the sand surface at one-foot intervals to collect the sand in motion along the beach. Samples were collected for selected intervals and the locations of the traps were changed periodically so as to obtain the rate of sand transportation over a stretch of 13 feet between the stations *0* and 13. The data thus obtained are shown graphically in figure 15 on which the rate of sand transportation, Q, is plotted against the station distances. The rate is expressed as pounds per hour per foot of width. On the graph are shown the location of the breaker as it initially occurred and the final position of the bar.

The maximum rate of transportation of sand on the initial smooth beach in the area investigated occurred at the point of impending wave break. Considerable movement of sand was observed along the beach surface where the waves were breaking. The breaker was of the spilling type, and the distinctive discontinuity of the plunging type of breaker was lacking. The bar is eventually formed at a point where the rate of sand transportation is nearly a minimum. Movement of sand occurs on the beach up to the limit of wave action.

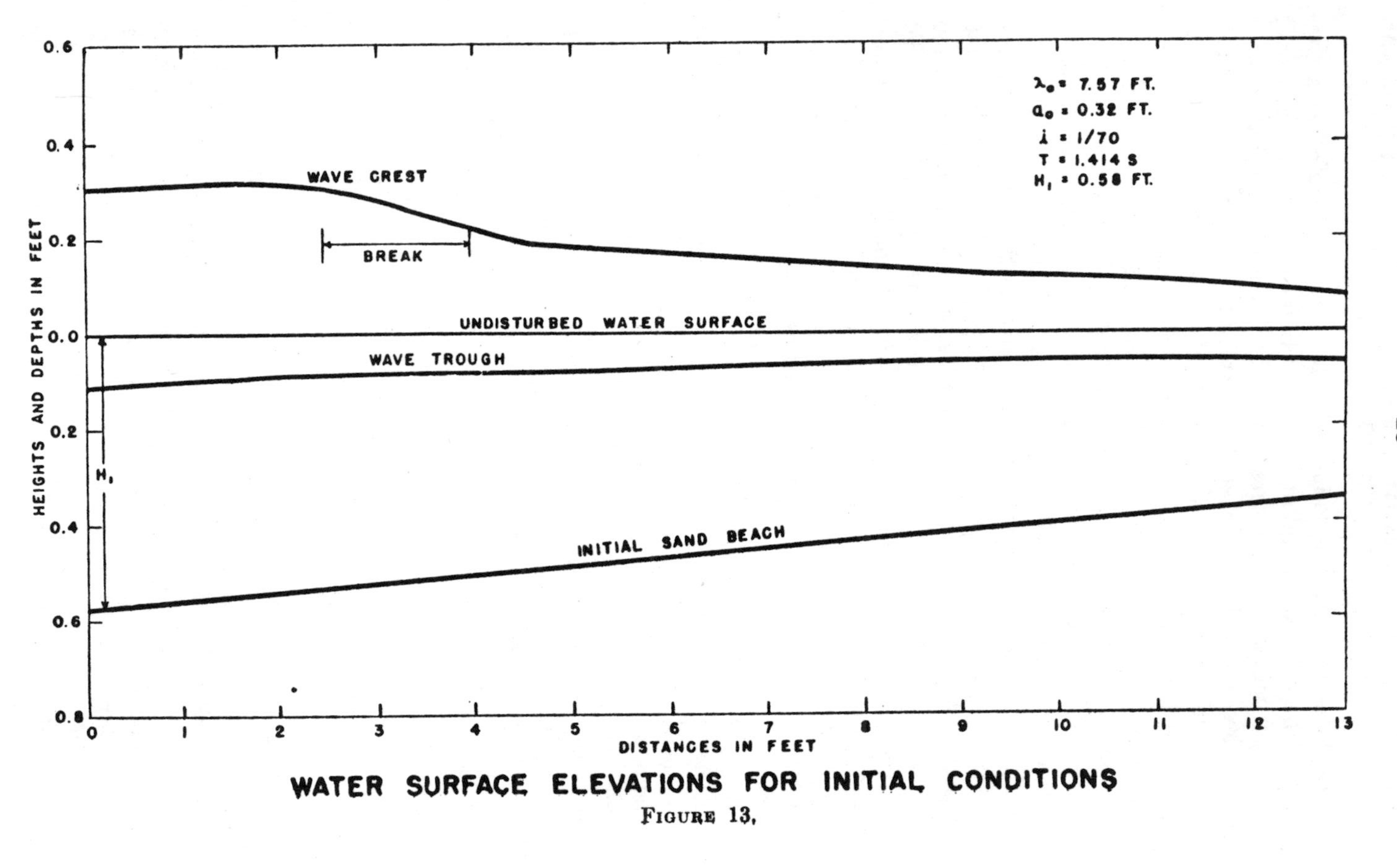

WATER SURFACE ELEVATIONS FOR INITIAL CONDITIONS

FIGURE 13,

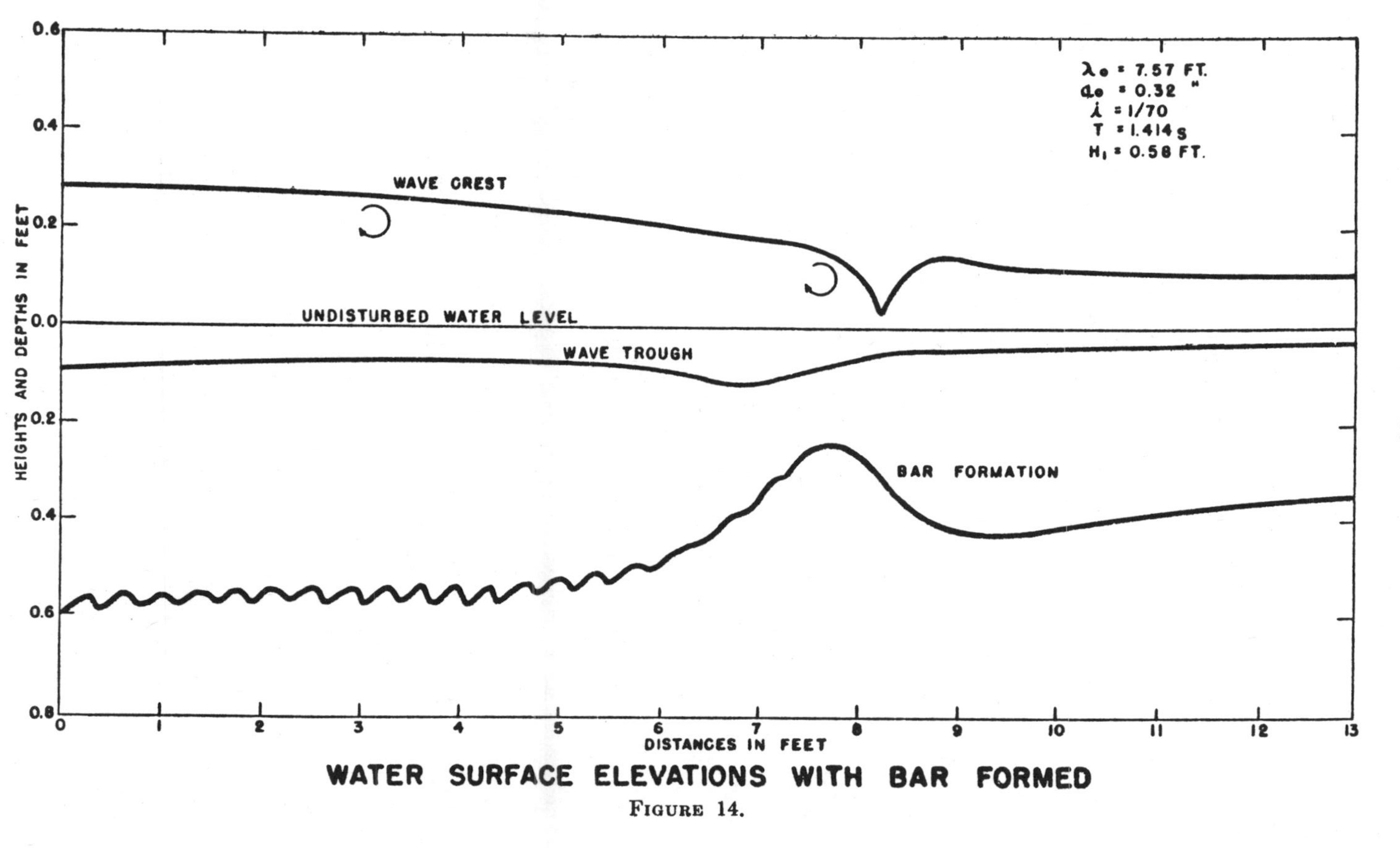
λo = 7.57 FT.
do = 0.32 "
i = 1/70
T = 1.414 s
H1 = 0.58 FT.
WAVE CREST
UNDISTURBED WATER LEVEL
WAVE TROUGH
BAR FORMATION
HEIGHTS AND DEPTHS IN FEET
0.6
0.4
0.2
0.0
0.2
0.4
0.6
0.8
0
1
2
3
4
5
6
7
8
9
10
11
12
13
DISTANCES IN FEET
WATER SURFACE ELEVATIONS WITH BAR FORMED

FIGURE 14.

Obviously, the non-uniform transportation of sand is effected by the total displacement of the water surface at a given point during the passage of waves. If ΔH_1 is the maximum elevation of the surface above the undisturbed water level during the passage of the waves and ΔH_2 the maximum depression, the total displacement is given by $\Delta H = \Delta H_1 + \Delta H_2$. The relation between the rate of initial sand transportation and the fall of the water surface is shown in figure 15. It is apparent that correlation exists between the rate of sand transportation and the total displacement of the water surface during the wave passage. A clear picture of the correlation of surface displacement with the rate of sand transport is obtained by considering the dependence of the rate of sand transportation Q, upon ΔH expressed in terms of dimensionless quantities. Three distinct regions must be kept in mind in dealing with the problem of sand transportation under wave action. The most important region has been referred to previously as the bar environment, which is bounded seaward by the point of impending wave break and shoreward by the point of reformation of waves beyond the breaker; the limits of this region are readily obtained in laboratory experiments. The remaining two regions are: the one extending indefinitely seaward from the point of impending wave break; the other extending to the shore line from the point of reformation of waves. The laws of transportation of sand in these regions are expected to be different.

In this investigation only the movement of material in the region of bar environment under initial conditions will be considered. The assumption is made that the rate of sand transportation Q depends on the total surface displacement ΔH, the depth of water H_1 at the point of impending wave break; the period of the wave T; the characteristic sand grain size d_{GM}, the kinematic viscosity ν; the densities of water and sand ρ_w and ρ_s, respectively; the gravity constant g; the slope i; and the sand dispersion coefficient σ_φ. The ordinary arguments of dimensional analysis result in the general relationship

$$\frac{QT}{\rho_s g H_1^2} = f\left(\Delta H/H_1,\ \rho_w/\rho_s,\ \frac{\sqrt{gH_1}\,d_{GM}}{\nu},\ H_1/gT^2,\ d_{GM}/H_1,\ i,\ \sigma_\varphi\right) \quad (13)$$

defining the law of sand transportation. In the equation the dimensions of the initial deep water waves are missing, since the wave characteristics are determined by T and ΔH. The last six quantities may be considered constants for a given test hence it suffices to write

$$\frac{QT}{\rho_s g H_1^2} = f(\Delta H/H_1) \quad (14)$$

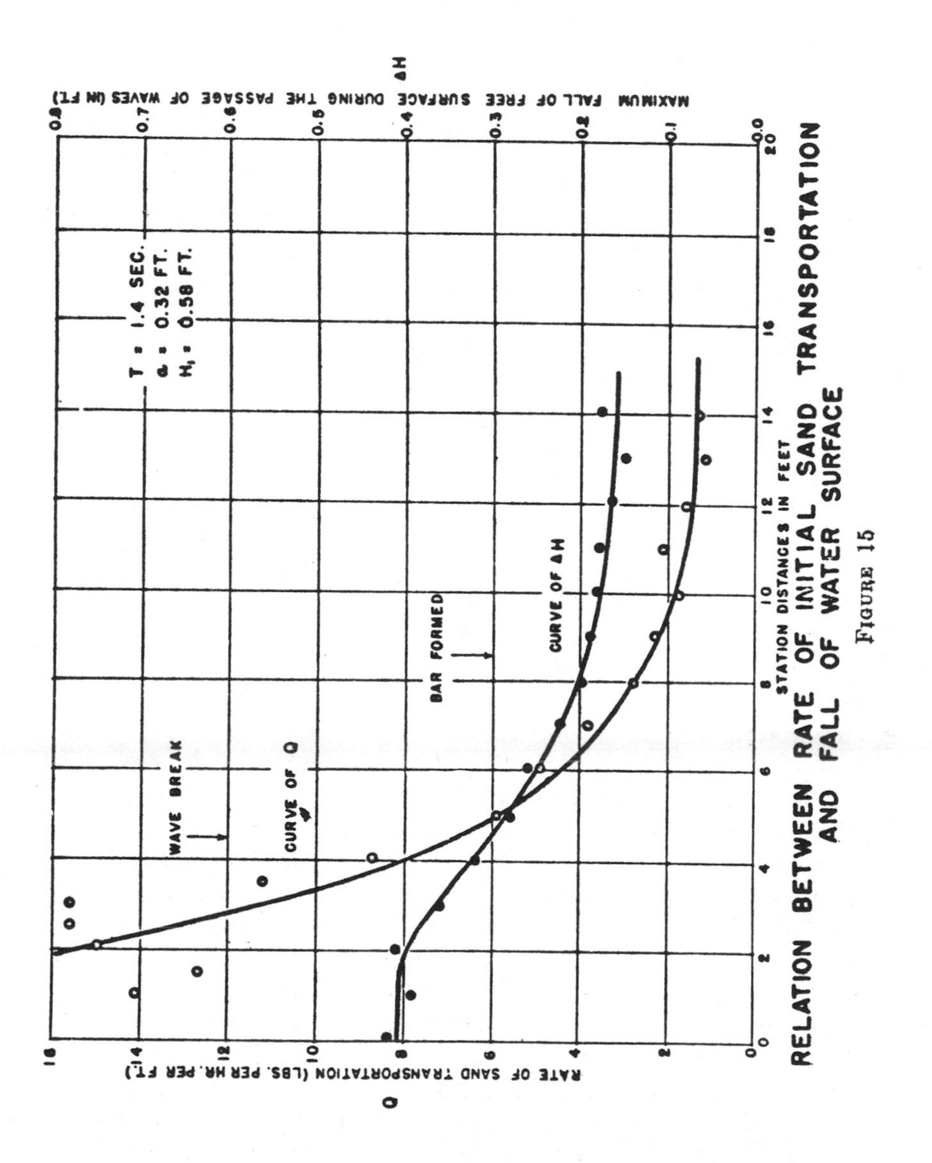

RELATION BETWEEN RATE OF INITIAL SAND TRANSPORTATION AND FALL OF WATER SURFACE

Figure 15

Figure 16 is plotted in accordance with the latter equation from figure 15. The product $\rho_s g$, being the specific weight of sand, is taken to equal 137 pounds (weight) per cubic foot. The quantities Q and ΔH are read from the curves in figure 15. Since Q was expressed as pounds (weight) per foot width per hour and T is given in seconds, the numerical value of QT involves the ratios of two different units of time. Interpretation of the curve reveals that the rate of sand transportation in the bar environment is controlled, among other things, by the total displacement of the water surface, ΔH and that there is a critical displacement value below which no appreciable transportation of sand is possible.

Further conclusions cannot be drawn on the basis of the meager data available, but the problem is important enough to warrant a separate and a more complete study. In an extended study the effects of all the parameters entering into the right hand side of equation 14, should be examined.

B. *Sand transportation and sand ripples.*—The discussion of sand transportation given above refers to the initial conditions when the beach is smooth and the slope of the surface is constant. Certain changes in the beach surface occur as transportation continues. See figure 14. Sand accumulates in the process of bar development and sand ripples appear seaward of the bar. The effects of these two changes are quite significant. With the formation of the bar the breaker characteristics are changed, the total displacement of water surface in the area between the breaker and the point of impending break of the waves becoming nearly constant everywhere in this area. The sand ripples signify a new mode of sand movement and a very considerable reduction in the rate of sand transport, as will be shown.

Sand ripples begin to form in the area just under the breaking wave, apparently from two causes. One is the secondary undulations, which manifest themselves at the surface of the breaker and are transmitted to the bottom to form corrugations. The other is an accidental unevenness of the bottom surface due to nonuniformity of resistance of the sand or some other discontinuity of the bottom geometry. Once the sand ripples have been started, in one way or another, a continuous series of them results under favorable conditions. Their final dimensions are controlled by the depth of water, the period of the waves, the total displacement of the water surface during the passage of waves, and the size and dispersion coefficient of the sand.

Movement of sand in the bar environment after sand ripples have covered the entire beach surface appears to take place in the following manner. At the instant that a crest is moving over a sand ripple, the motion of water particles just above the ripple is toward the shore. At this moment the sand of the ripple surface is moved toward the

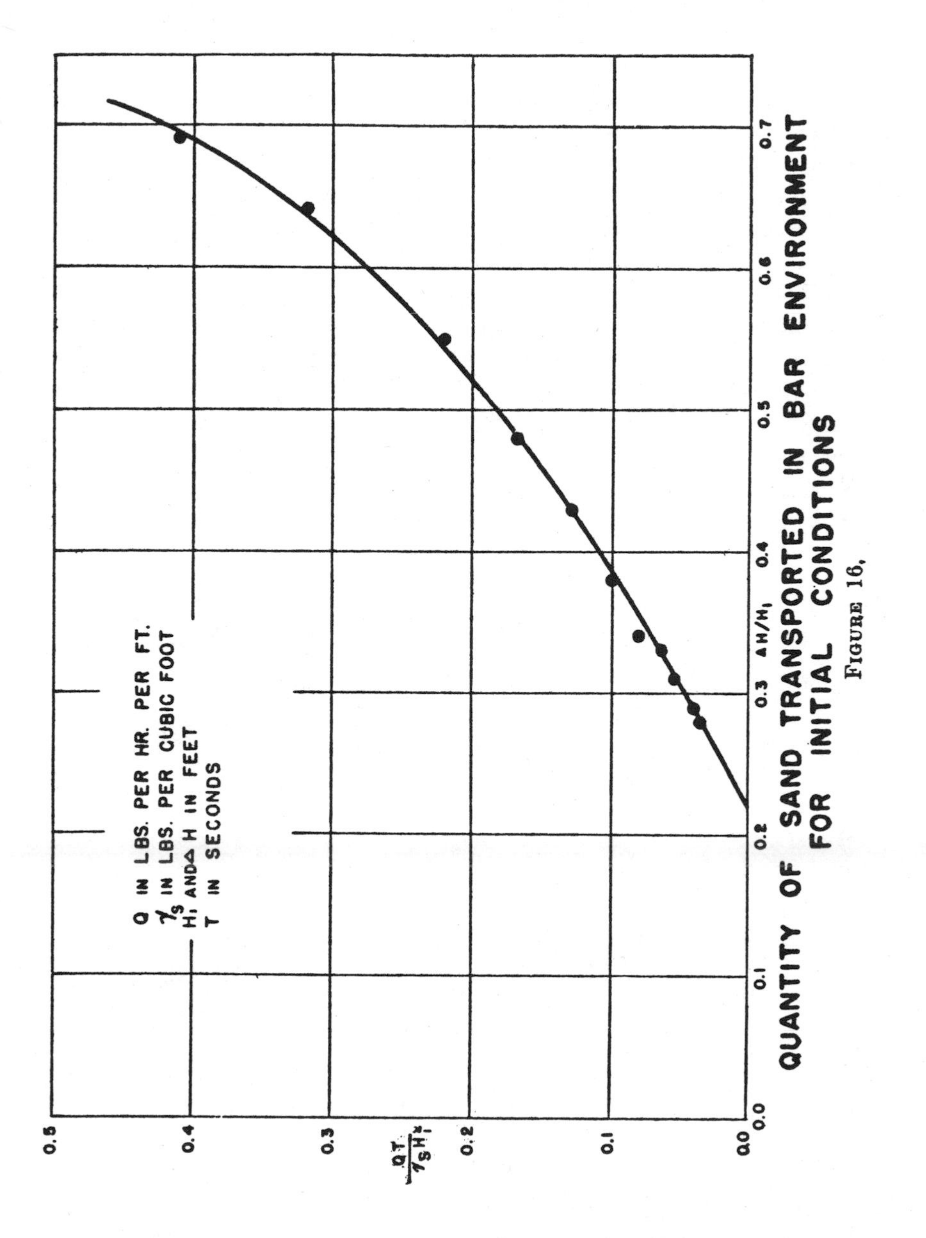

QUANTITY OF SAND TRANSPORTED IN BAR ENVIRONMENT FOR INITIAL CONDITIONS

FIGURE 16,

crest of the ripple and then downward into the depression of the next ripple. See figure 17. Simultaneously an eddy is formed in the depression and is loaded with small sand particles picked up from the depression. The hydrodynamic reaction between the rotating eddy and the sand surface ejects the eddy together with its sand particles upward some distance. With the trough of the wave following, the movement of water particles near the sand ripple surface is reversed. This moving mass of water carries with it toward the sea the suspended sand particles of the previous eddy. In this reversed motion the major portion of the suspended particles is deposited on the surface of the sand ripple from which it initially derived and a minor amount is taken over the next ripple on the seaward side. The cumulative effect of this cyclic process is that a ripple moves gradually by small steps toward the shore. In the process the larger particles of sand of the environment are taken to the shore and the lighter particles, through suspension, are taken to sea. Thus the sand ripple, rather than its motion, becomes the sorting agent.

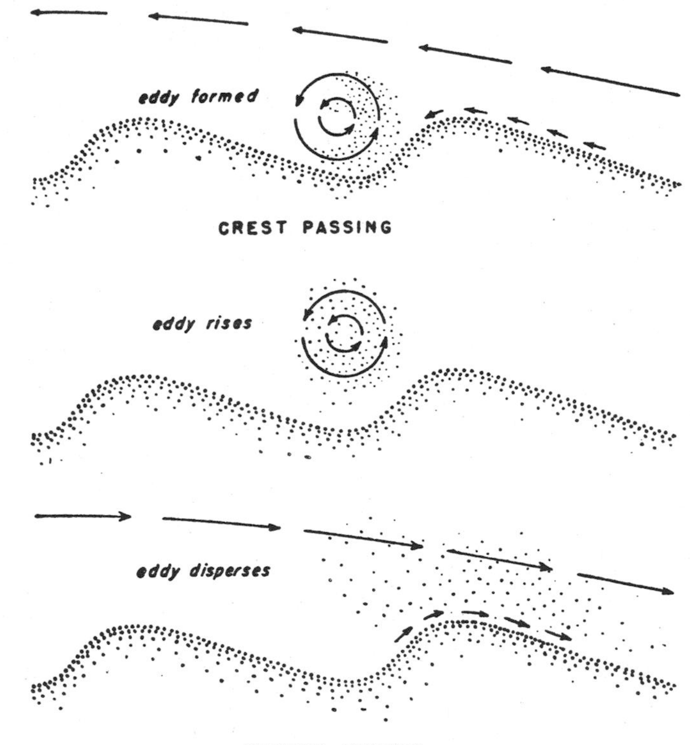

MECHANISM OF MOVEMENT OF SAND OVER SAND RIPPLES

FIGURE 17.

The net rate of sand transportation in the form of ripples can be ascertained once the distance traveled by the ripple and the area of the vertical section of the ripple is known. Let Q_r denote the weight of sand transported per hour per foot of width. Denoting the distance traveled by a sand ripple per hour by D and the sectional area by A,

$$Q_r = \gamma_s DA \tag{15}$$

where γ_s is the specific weight of sand. For the sand ripples in test 51

$$A = 0.54hl \tag{16}$$

where h is the height of the ripple mark and l the length of the base, both quantities being expressed in feet. Accordingly,

$$Q_r = 0.54hlD\gamma_s \tag{17}$$

Table 2 lists the dimensions of some thirteen sand ripples of test 51 together with their rate of travel. The first ripple mark was located at station 2, and the remainder covered the distance between the bar and that station. See figure 14. It can be seen from the table that with a few exceptions the ripples have nearly equal sizes and travel with nearly equal velocities. This can be explained on the basis that the depth of water over the ripples is constant and the displacement of the surface uniform during the passage of waves. On the average the rate of transportation of each is $Q_r = 1.21$ pounds per hour per foot of width. Comparing this with the data of figure 15 it is obvious that sand transportation in the form of ripples is much smaller than the sand movement under initial conditions.

TABLE II.—*Transportation of Sand by Ripples*

No.	l	h	D	Q
	Ft.	*Ft.*	*Ft./hr.*	*Lbs./hr./ft.*
1	0.31	0.06	0.75	0.97
2	.37	.05	1.50	1.93
3	.34	.06	.90	1.25
4	.35	.06	.60	.87
5	.34	.07	.45	.74
6	.34	.06	.60	.83
7	.33	.06	.90	1.23
8	.36	.06	.90	1.34
9	.32	.07	.90	1.41
10	.37	.05	.90	1.10
11	.36	.05	1.35	1.68
12	.40	.04	1.50	1.62
13	.28	.02	2.40	.75
Average				1.21

The general expression for the law of sand transportation in the form of sand ripples remains to be obtained. As it is assumed on the basis of table 2 that in the bar environment sand transportation is independent of the position of sand ripples, the desired law can be written immediately from equation 15 with the term $\Delta H/H_1$ omitted. Accordingly,

$$\frac{Q_r T}{P_s g H_1^2} = f_2\left(P_w/P_s, \frac{\sqrt{gH_1}d_{GM}}{\nu}, \frac{H}{gT^2}, \frac{d_{GM}}{H_1}, i, \sigma_\phi\right) \tag{18}$$

In the test under study the parameters on the right hand side are constants and from the data for the test we obtain

$$\frac{Q_r T}{P_s g H_1^2} = 0.039 \tag{19}$$

The data used to establish the relation are: $Q_r = 1.21$ pounds (weight) per foot per hour, $T = 1.4$ seconds, $P_s g = 137$ pounds (weight) per cubic foot and $H_1 = 0.58$ foot. As an application to a natural condition, let us suppose that $H_1 = 6$ feet, $T = 3$ seconds and other conditions are similar to those of the test. The rate of transportation of sand would be: $Q_r = 0.039 \times 137 \times 6 \times 6/3$, or 64 pounds per foot width per hour.

The problem of sand transportation in this form of ripples on sea beds is an important one and can be treated with sufficient detail only by an independent investigation. The investigation should be made for the bar environment, and for the seaward region adjacent to the bar environment. The law probably assumes different forms in the two regions.

C. *Extreme water surface variations in the bar environment.*—The bar environment, as was mentioned previously, is limited seaward by waves which are beginning to break and shoreward by waves which reform following the breaker. Variations in the water surface elevations at these points determine the flow of energy into and out of the bar environment. In studying breaking of waves, the first question to consider is the determination of the locality where breaking is impending. The upper curve in figure 18 shows the dependence of the ratio $\Delta H_1/H_1$, upon the wave steepness, a_o/λ_0. Here H_1 is the depth of water at the locality where the waves are beginning to break, and ΔH_1 is the elevation of the crests at the point of impending wave break, measured from the level of the undisturbed water. Thus, if ΔH_1 is known the curve just given will enable us to evaluate H_1. Obviously, ΔH_1 is a function of a_o and $a_o\lambda_0$ in the form

$$\Delta H_1/a_o = f(a_o/\lambda_0) \tag{20}$$

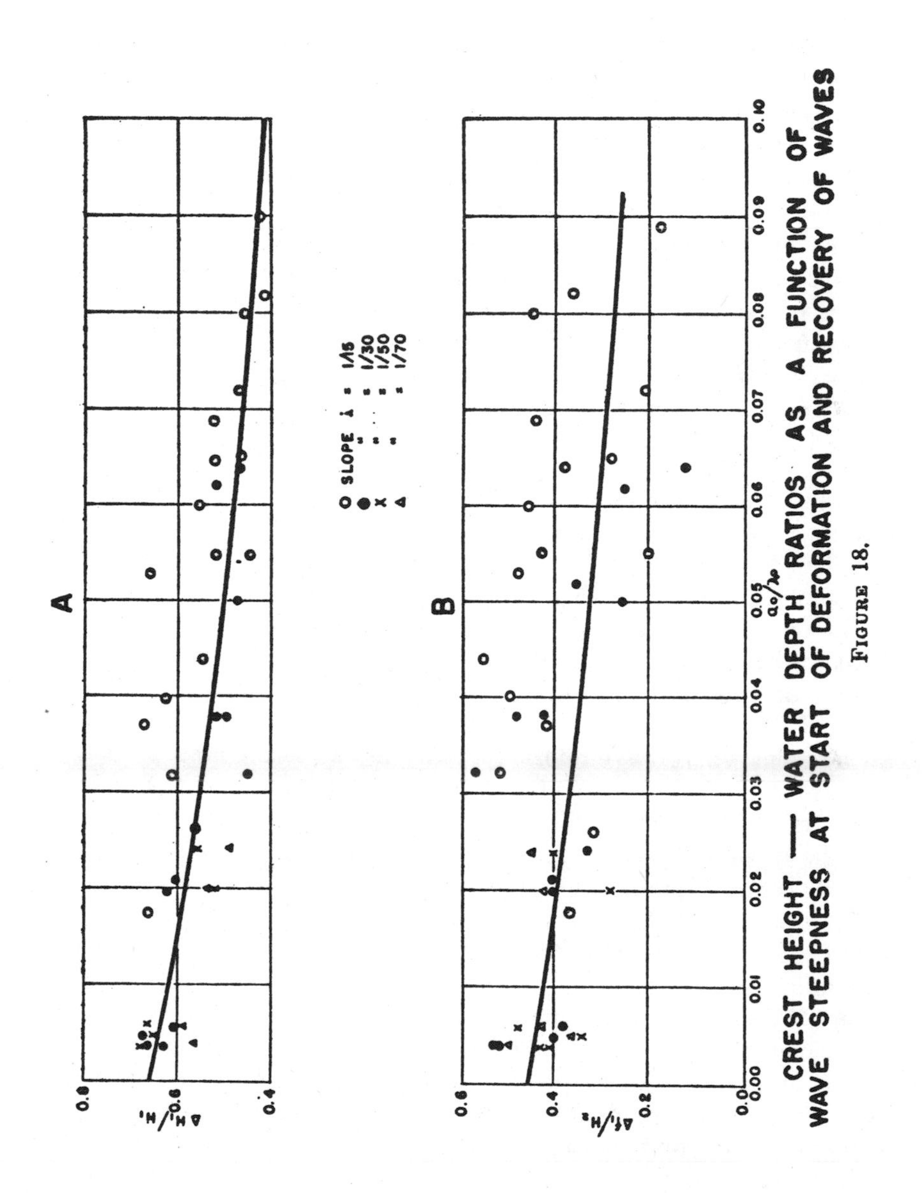

CREST HEIGHT — WATER DEPTH RATIOS AS A FUNCTION OF WAVE STEEPNESS AT START OF DEFORMATION AND RECOVERY OF WAVES

FIGURE 18.

if the effect of the slope, i, can be neglected. This relationship is shown in the upper curve of figure 19, and it is noted that the effect of slope can be neglected within the order of experimental error.

The second question to consider is the depression of the trough ΔH_2 at the start of wave break. The desired information can be obtained conveniently if we consider the ratio $\Delta H_2/\Delta H_1$ as a function of $a_o\lambda_0$. The upper curve in figure 20 approximates the relationship.

Evaluation of the energy flowing into the bar environment can be made on the basis of the quantities H_1, ΔH_1, and $\Delta H_2/\Delta H_1$. Unfortunately, the theory of shallow water waves is not complete enough for an appropriate analysis, there being available no general energy formula covering the case of breaking waves.

It is necessary to justify the absence of a_o and a_o/λ_0 in the general and transportation formulas in equations 13 and 18. It was established that $a_o/H_1=f(a_o/\lambda_0)$. By multiplying both sides of the equation by λ_0/a_o, it can be shown that there is a one to one correspondence between a_o/λ_0 and λ_0/H_1. Inasmuch as $\lambda_0=\frac{gT^2}{2\pi}$, it may be stated that a_o/λ_0 and H_1/gT^2 are uniquely related and the latter may replace the former. Because of these relations the quantities H_1 and H_1/gT^2 are sufficient to characterize, in a functional manner, the magnitude and the shape of waves entering the bar environment.

The movement of sand from the shoreward limit of the bar environment to the shore is controlled by the energy content of waves at the instant of reformation. For the evaluation of the energy the quantities H_2, Δf_1, and $\Delta f_2/\Delta f_1$ are fundamental. H_2 is the depth of water at the point where wave reformation begins, Δf_1 is the crest elevation above the undisturbed level and Δf_2 the depression of the trough below that same level. The relationship between these quantities and the wave steepness ratio a_o/λ_0 is given in figures 18, 19, and 20. These relationships should be studied further.

D. *Energy distribution in the bar environment.*—The ease with which the mathematical expressions for the form and position of bars is established results from the simplicity of the experimental procedure. The tests represent conditions under a singular system of waves and in the absence of tidal flow. Each test began with a smooth sloping surface and a bar was allowed to form and attain a relatively fixed position in association with the breaker; then, the quantities having a bearing on the form of the bar were measured. The resulting data is of the most elementary nature and really contains little information on the beach processes involved in the production of the bar. During the tests, however, attempts were made to observe these processes qualitatively.

Sand movement along the bar appears to take place in the following

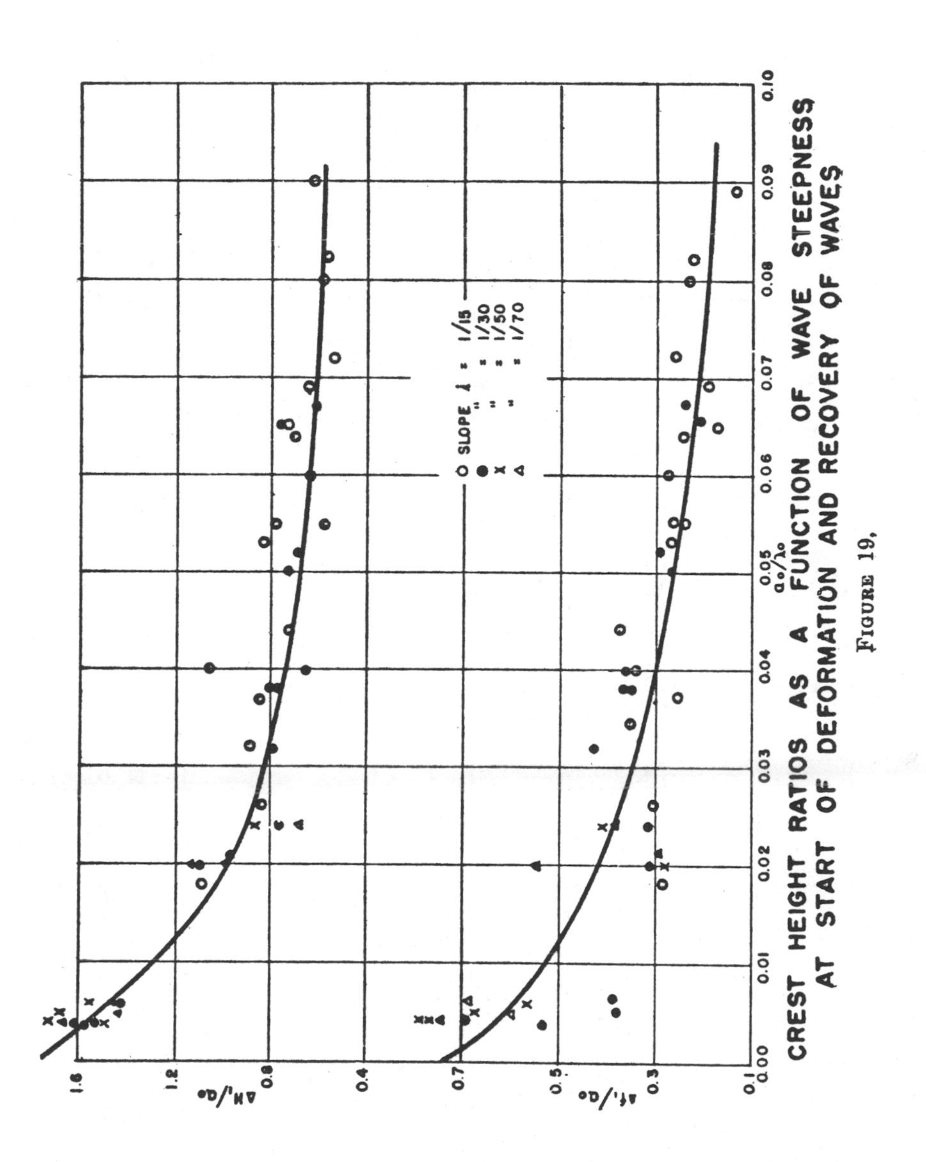

CREST HEIGHT RATIOS AS A FUNCTION OF WAVE STEEPNESS AT START OF DEFORMATION AND RECOVERY OF WAVES

FIGURE 19,

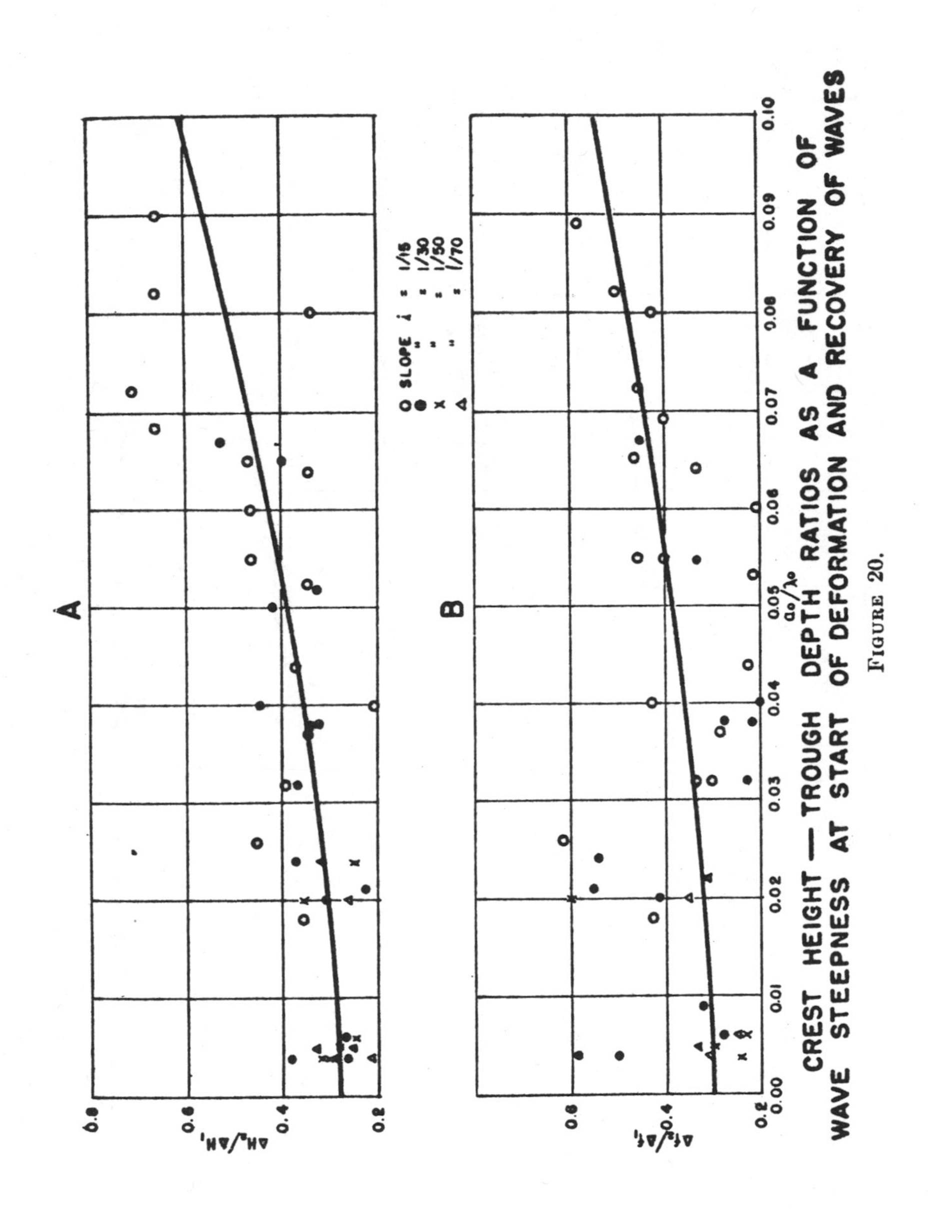

A
B
0.8
0.6
0.4
0.2
0.00
0.01
0.02
0.03
0.04
0.05
0.06
0.07
0.08
0.09
0.10
SLOPE i = 1/15
" = 1/30
" = 1/50
" = 1/70
CREST HEIGHT — TROUGH DEPTH RATIOS AS A FUNCTION OF
WAVE STEEPNESS AT START OF DEFORMATION AND RECOVERY OF WAVES

FIGURE 20.

manner. Initially, in the absence of the bar, a certain distance exists between the points where the wave begins to break and the breaker is completed; this distance may be referred to as the breaker distance and is denoted by s in figure 1. On flat beaches the breaker initially is of the spilling type. As the bar is formed it changes into the plunging type and the distance, s, increases. The magnitude of s is a function of the slope of the beach and of the depth of water at the point of impending wave break. The functional dependence may be expressed as

$$s/H_1 = f(a_s/\lambda_0, i) \tag{21}$$

or

$$s/H_1 = f\left(\frac{H_1}{gT^2}, i\right)$$

which probably assumes a different form when no bar is present. When bars are present, it is found that, to a rough approximation $s/H_1 = 5$, 8, 9, and 11 when $i = 1/15$, 1/30, 1/50, and 1/70, respectively.

It is the interpretation of the writer that the two changes mentioned above have a bearing on the energy distribution in the bar environment. This may be explained by reference to the two curves in figure 4 which represent the maximum and minimum elevation of the water surface with reference to the still water level during the passage of waves. In a sense these curves represent the paths of travel of the crests and troughs. In tracing the travel path of the crests, it is seen that the crest deforms rapidly and develops a central curl somewhere between the point where the wave starts to break and the plunge point. The position of the fully developed curl is shown in figure 14. The curl, or rotating body of water, causes a greater internal dissipation of energy than otherwise would be possible. After the appearance of the central curl when the water elevation is maximum at the bar crest, a strong current moves parallel to the seaward slope of the bar over the crest and is directed shoreward. The current in passing over the crest develops a curl and falls into the trough of the bar. At the instant of the fall of the curl its linear momentum is imparted to the waters ahead and the rotary momentum is consumed locally. The impact of the curl reforms the wave. In following the path of the wave troughs seaward it is seen that the current attains a maximum at the crest of the bar. The trough path at this point is a concave surface just above the seaside face of the bar, as shown in figure 14; here the hydraulic gradient is quite steep. This together with the fact that depths are small indicate the existence of a strong seaward current. The net result is that in all probability the main role of the bar is to reduce the energy of the incoming waves causing them to impart a lesser amount of energy to the reforming waves than would be the case if the bar were absent. It is believed that the longer breaker distance signifies a greater dissipation of energy over the sand ripples as a result of increased turbulence. In such cases the rotary motions of the central curl and plunging curl are locally damped. Maximum currents over the bar are associated with loss of energy in the boundary layers over the sand.

The currents naturally affect the movement of sand in the bar environment. When the water depth over the crest is greatest the strong currents directed toward the shore move the layers of sand from that part of the seaside slope of the bar surface and the crest area into the trough of the bar. Subsequently, when the plunging curl is falling in the bar trough, the rotating water picks up sand from the trough and suspends it in the main body of water. At the instant that the depth of water over the crest is minimum, the seaward current transports the suspended sand particles and at the same time induces a movement of sand over the bar surface toward the seaside foot of the bar. In fact, the cyclic movement of sand over the bar is quite similar to that over a sand ripple, except that the advance of the main body of the bar is imperceptible. In the case of bars, the sand transported to the seaside foot of the bar by sand ripples is collected at the shore side of the bar by the recovering and reforming waves and is transported to the shore.

The formation of the main body of the bar is the final step in the sorting of sand. The coarser sand reaches the shore through the cyclic creeping movement on the surface of the sand ripples and the bar; the finer particles return to the sea through the turbulent action of the eddies of the sand ripples and the bar. This interpretation is based on incomplete data and a detailed investigation on this particular subject as an independent problem is necessary.

Section VII. ACKNOWLEDGMENT

The author acknowledges gratefully the numerous suggestions made by Dr. M. A. Mason, Chief, Engineering and Research Branch of the Beach Erosion Board, during the conduct of the tests. The importance and significance of bar structures were brought to the author's attention by Prof. W. W. Williams of Cambridge University, and his general discussions of different problems relating to the subject have been very valuable. The work of Messrs. W. H. Vesper and D. G. Dumm, who operated the models and assisted in the computations, was invaluable.

REFERENCES

1. Hagen, Gotthilf: Handbuch der Wasserbaukunst, 3 Bde. 1863.
2. Otto, Theodore: Der Darss und Zingst. Jahresber. Geogr. Ges. Greisswald, XIII, 1911–12.
3. Lehmann, F. W. Paul: Das Kustengebiet Hinterpommers. Zschr. d. Ges. f. Erdk. Berlin 19, p. 391, 1884.
4. Hartnack, Wilhelm: Uber Sandriffe, Jahresber. Geogr. Ges. Greisswald, XL–XLII, 1924.
5. Evans, O. F.: The Low and Ball of the Eastern Shore of Lake Michigan, The Journal of Geology, Vol. XLVIII, p. 476, 1940.
6. Burnside, W.: On the Modification of a Train of Waves as it advances into Shallow Water, Proc. London Math. Soc., Vol. 14, p. 131, 1915.
7. Montgomery, Lt. H. A.: Lake Michigan Wave Measurements at Milwaukee, U. S. Engineers, Milwaukee Office, Nov. 4, 1933.

U. S. GOVERNMENT PRINTING OFFICE: 1946

Editor's Comments on Paper 4

In this paper, Shepard deals with the submerged sand bars and troughs found parallel to the shoreline along most long sandy beaches. He is concerned with the origin and characteristics of the bar and trough, and their relationship to waves, currents, and tides. In this he continues Keulegan's quantification of their parameters.

An important part of the paper is given over, under a discussion of bar and trough origins, to the aforementioned problem of bar growth through and above sea level. In this argument, Shepard sides with an earlier study by Evans (The low and ball of the eastern shore of Lake Michigan: *J. Geol.*, vol. 48, pp. 476–511, 1940); holding that barrier islands are not built from submerged bars. In a hopefully forthcoming volume on barrier islands in this selected reprint series we will take up this matter again and examine the relevant literature in more complete detail. The interested reader is advised to look for that volume to further his reading on this topic.

Francis P. Shepard was born in Brookline, Massachusetts, in 1897. He received the A.B. degree from Harvard in 1919 and the Ph.D. from Chicago University in 1922. He then taught geology at the University of Illinois for a number of years. First Research Associate, then Professor of Submarine Geology at the Scripps Institution of Oceanography, he became Professor Emeritus there in 1967. Dr. Shepard has specialized in submarine geology, continental shelf and bay sedimentation, submarine canyons, and coastal changes. He is a fellow or honorary member of many professional organizations, and was president from 1958 to 1963 of the International Association of Sedimentology. His textbook entitled "Submarine Geology" (New York, Harper and Row, 1963) is now in its second edition.

4

LONGSHORE-BARS AND LONGSHORE-TROUGHS

TECHNICAL MEMORANDUM NO. 15
BEACH EROSION BOARD
CORPS OF ENGINEERS

JANUARY, 1950

Contribution from the Scripps Institution of Oceanography, New Series, No. 454

FOREWORD

This paper was prepared by Dr. Francis P. Shepard of the Scripps Institution of Oceanography and represents in part the results of research carried out for the U. S. Navy and the Beach Erosion Board. The report first appeared in limited issue as Submarine Geology Report No. 6 of the Scripps Institution of Oceanography, University of California. It is believed that the findings of the investigations are of sufficient value to merit publication at this time.

The opinions and conclusions expressed by the author are not necessarily those of the Board.

The authority for publication of this report was granted by an Act for the improvement and protection of beaches along the shores of the United States, Public Law No. 166, Seventy-ninth Congress, approved July 31, 1945.

TABLE OF CONTENTS

LONGSHORE-BARS AND LONGSHORE TROUGHS

Abstract

The submerged longshore-bars and longshore-troughs which skirt the shores off most sandy beaches are described. The troughs which lie landward of the bars are explained as the result of plunging breakers and the longshore currents which are feeders to rip currents. The bars are thought to be partly the result of the excavation of the troughs and partly due to landward migration of sand outside the breakers and seaward migration from the troughs. The depths of the bars and troughs are shown to be related to wave and breaker heights. The elimination of some bars is seen to be the effect of a long continued period of small waves during which the bar moves landward filling the trough. In many areas the deeper bars persist undisturbed through long periods of quiet seas. The analysis of thousands of profiles, mostly taken along California open ocean piers, is the chief basis for the preceding conclusions.

Introduction

Because of their instability, submerged sand bars are a source of navigational difficulty to every type of craft from row boats to ocean liners. There appear to be three principal types of these shifting sand bars. One is crescentic in shape and is found off many river mouths and narrow entrances to bays (figure 1A). A second type is oval in shape and occurs in groups of somewhat parallel dune-like masses. Many of these are elongate parallel to the sides of straits and estuaries where there are strong currents and where abundant sediment is available (figure 1B). Others are more irregularly distributed. The third type of bar is long and narrow, extending essentially parallel to the shore line along most long sand beaches. The present discussion will deal only with this third type. This type presumably includes the cusp-like bars which were reported by King and Williams (9, p.75) from the shallow water along the margin of bays in the Mediterranean. King and Williams refer to the latter as crescent bars but they are decidedly different from the large crescents off bay entrances and are apparently more akin to beach cusps.

The term <u>ball</u> has been used to describe these shore-skirting bars, and their accompanying troughs, found inside the bars, have been referred to as <u>lows</u>. However, the word <u>ball</u> implies a roundish protuberance rather than an elongate ridge and <u>low</u> suggests a basin or valley, so that it seems advisable to use the more descriptive terms bar and trough for which there is ample precedent. In order to distinguish this type of bar from other types which include emergent bar beaches, a qualifying adjective is suggested so that they will be termed here <u>longshore bars</u> and <u>longshore troughs</u>.

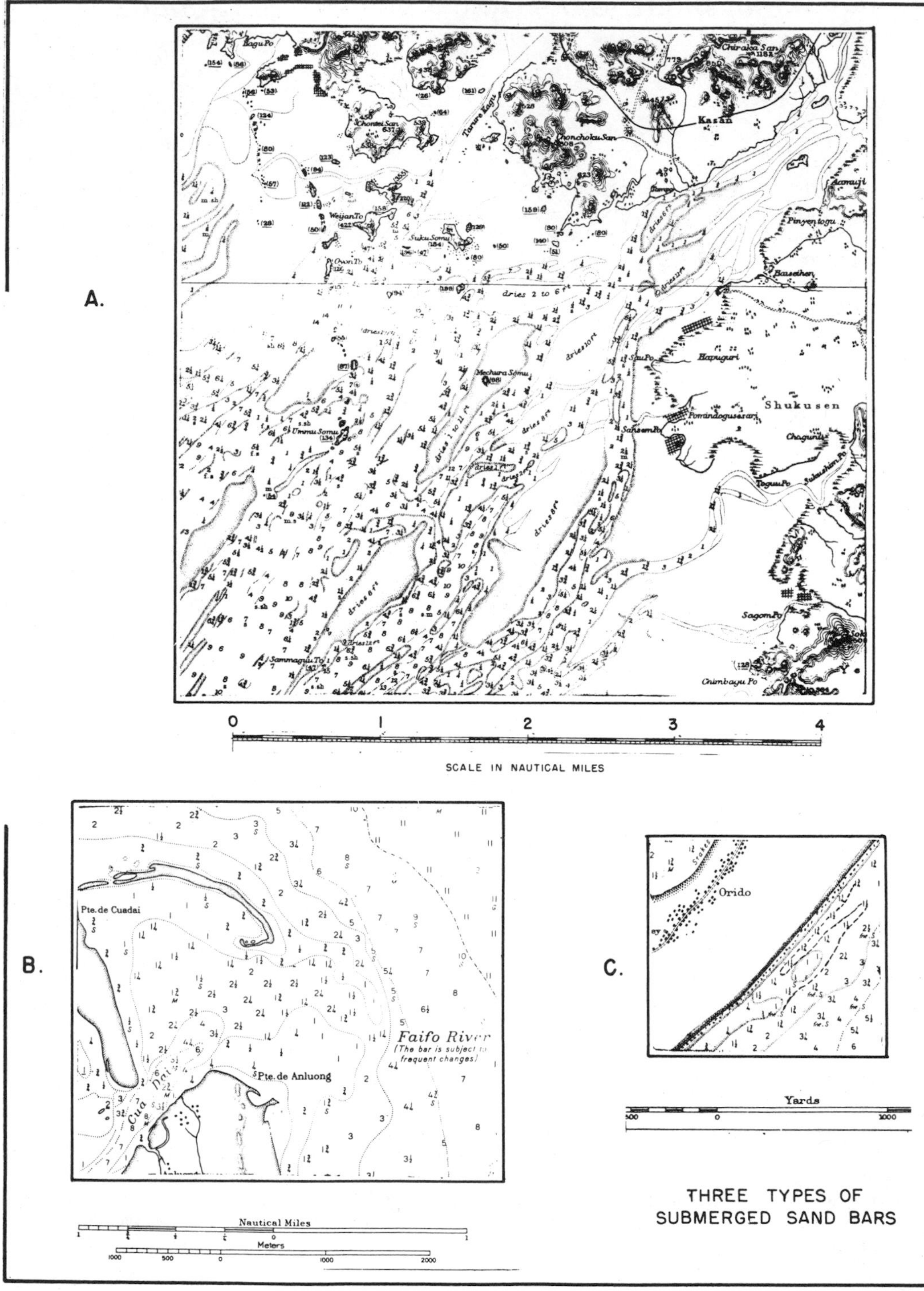
A.
SCALE IN NAUTICAL MILES
B.
Pte. de Cuadai
Faifo River
(The bar is subject to frequent changes)
Pte. de Anluong
Nautical Miles
Meters
C.
Orido
Yards
THREE TYPES OF SUBMERGED SAND BARS

FIGURE 1

Anyone accustomed to surf swimming along relatively straight sandy beaches can hardly have failed to encounter the longshore-bars and troughs while wading out into the breakers. During periods of high waves the longshore-bars can be detected from any high place which commands a view of the surf zone, because the waves break along these shoal ridges (figure 2) and usually re-form on the inside breaking again either on inner longshore-bars or on the low tide terrace, a feature which is very common near the low tide level. If the water is clear during periods of small waves, the bars can be seen as variations in the color shading in the water along the shore.

Longshore-bars have been described by various authors, and suggestions have been made regarding their origin. A good summary of the earlier literature will be found in Johnson (6) and in Evans (2). A considerable number of profiles of the longshore-bars were made by German investigators, notably Otto (10) who studied a series of bars over a period of five years. Measurements by the U. S. Coast and Geodetic Survey (15) provided other information. Profiles of the longshore-bars of Lake Michigan were made by Evans who attempted also to determine the variation in position of these bars during one summer.

The work of Keulegan (6) with the Beach Erosion Board during and following World War II produced valuable wave-tank observations and an analysis of the mechanics involved in bar formation. Extensive field studies of longshore-bars were initiated during the war by the Department of Engineering of the University of California (Berkeley). Most of this material is not yet published although numerous profiles from various parts of the west coast have appeared as mimeographed laboratory memoranda from the Berkeley group. Important investigations were also conducted by King and Williams (9) which included field and tank observations.

At Scripps Institution information on longshore-bars has come particularly from daily measurements made along the 1,000-foot Institution pier during 1937 and 1938 (14) Subsequent measurements along this pier were made at various intervals, and profiles have been obtained along seven other California piers, all of which revealed the existence and the changing positions of bars. The pier profiles, particularly those made daily in 1937 and 1938, provide the greatest source of information for the present report since they contain the only available information on the changing depths and positions of the bars and troughs which can be definitely related to short period changes in waves, currents, tides, and winds. In a sense it is unfortunate that this information should come from pier profiles since piers form a sufficient obstruction both to waves and currents to set up special conditions. On the other hand nothing has as yet been developed to take the place of pier soundings for making accurate profiles through large breakers. Self-propelled sea sleds (4) may form an exception of this statment, but they have not been very successful. The breaker zone is the locus for bar and

AIR VIEW OF WAVES BREAKING ON LONGSHORE — BAR
ALONG WEST COAST NORTH ISLAND, NEW ZEALAND
(PHOTOGRAPH BY WHITES AVIATION, LTD., NEW ZEALAND)
FIGURE 2

trough development. Measurements with Dukws, the amphibious vehicles developed during the war, are fairly satisfactory during small-surf periods, but these periods are not as significant as large-wave periods. No records are available in which profiles were repeated at short intervals during which a daily check on sea conditions had been made.

Adequacy of Pier Profiles as a Source of Information

Since much of the information for the present discussion comes from pier profiles, some justification seems required of the use of this evidence from locations which are under special influences. Evans (2, p. 479) observed that the bars along the east shore of Lake Michigan were disturbed and broken up in the vicinity of piers. The piers tend to deflect longshore currents seaward as rip currents which in turn develop gaps in the bars. However, rip currents are not particularly common along Scripps pier. A much more common rip exists somewhat north of this pier. Furthermore the longshore-bars appear to be well developed along the piers of southern California and, so far as can be told from somewhat fragmentary information, do not appear to be particularly different from the bars on either side. As will be shown subsequently, the changes of bars and troughs have a definite relationship to wave and current conditions. Therefore it seems likely that the changes in the pier profiles are a fairly reliable index of what is happening on either side. Finally, the soundings were taken half way between piles where the least effect would occur and, in the case of the Scripps pier, a boom extended the sounding wire 10 feet outside the rail, further decreasing the piling influence.

If at some time in the future information becomes available from the slopes which are entirely free from pier influence, it will no doubt be more satisfactory. However, an enormous amount of difficult work must be accomplished before sufficient information is available to replace that coming from the thousands of pier profiles.

Characteristics of Longshore-bars and Longshore-troughs

King and Williams (9) express the opinion that the feature referred to here as a longshore-bar is typical only of "tideless seas" such as the Mediterranean and Baltic, and that another type which they call ridge and runnel is found exposed by low tide on beaches having large tidal ranges. The present investigations in areas where appreciable tidal ranges operate make this distinction seem of no great importance, although it may well be that the longshore-bars in the areas of large tidal range are less continuous and cut by more channels, partly as a result of runoff from the exposed troughs during low tide.

The study of the hundreds of west coast profiles shows a relationship between the depths of longshore-bars and longshore-troughs. If as suggested by Passarge (11) the longshore-bars are built by sediment thrown shoreward by the breaking waves, there should be excavation of a trough on the outside and building of a ridge on the inside of the

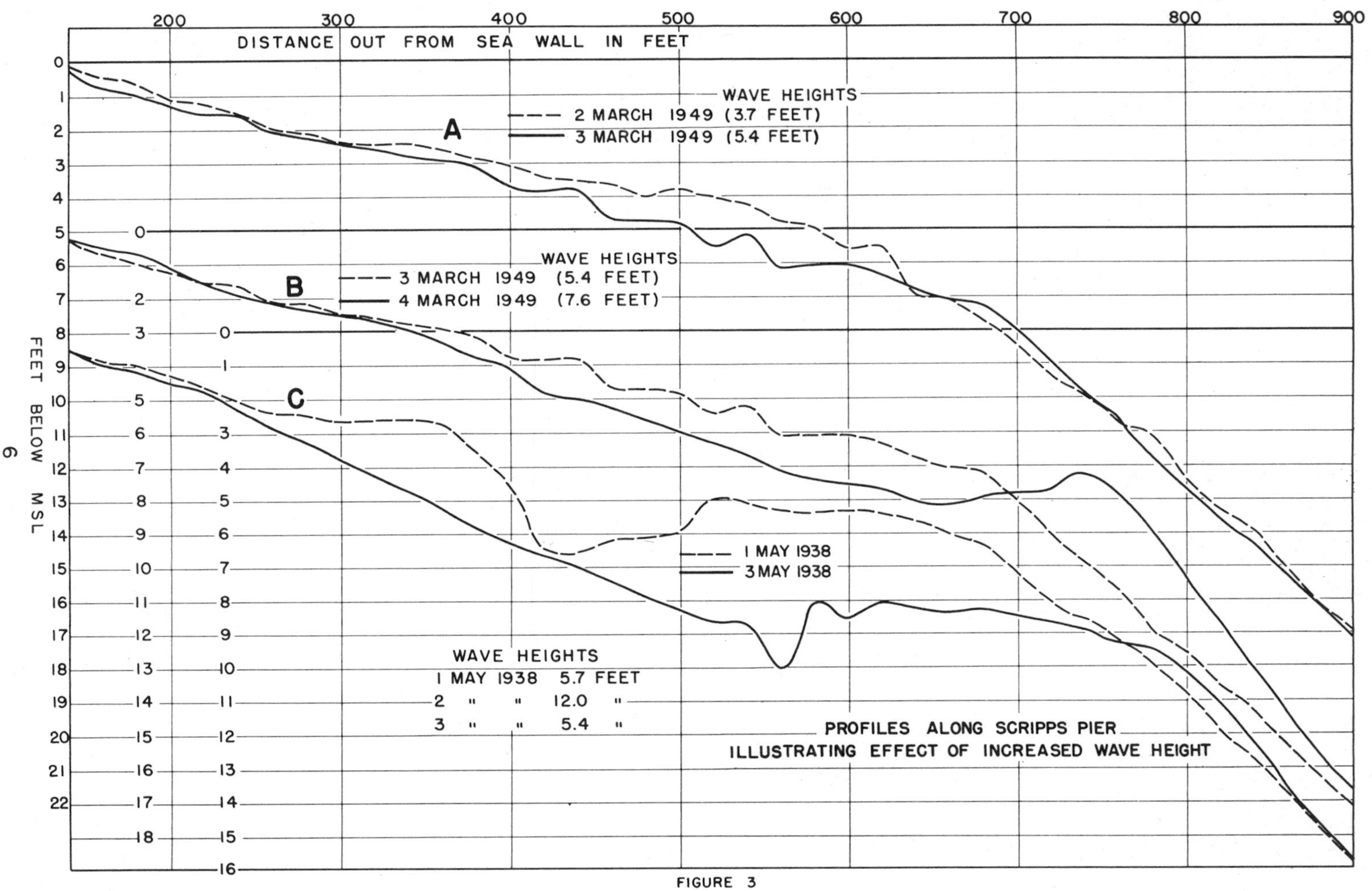
DISTANCE OUT FROM SEA WALL IN FEET
FEET BELOW MSL
WAVE HEIGHTS
2 MARCH 1949 (3.7 FEET)
3 MARCH 1949 (5.4 FEET)
A
WAVE HEIGHTS
3 MARCH 1949 (5.4 FEET)
4 MARCH 1949 (7.6 FEET)
B
C
1 MAY 1938
3 MAY 1938
WAVE HEIGHTS
1 MAY 1938 5.7 FEET
2 " " 12.0 "
3 " " 5.4 "
PROFILES ALONG SCRIPPS PIER
ILLUSTRATING EFFECT OF INCREASED WAVE HEIGHT
FIGURE 3

plunge point. A longshore-bar formed in this way should have a trough cut into the sand slope outside the bar, whereas the slope inside should remain essentially unchanged. Furthermore, the bars should form shoreward of the breaking waves. The daily profiles along the Scripps pier show that virtually all of the large longshore-bars form slightly seaward of the plunge points of the larger breakers occurring at the time the profile was measured. Furthermore, the trough is excavated on the inside and the bar is either left as an erosion remnant or is built up outside the trough (figures 3 and 18). This relationship is substantiated by the numerous cases where bars are found which have no concavity in the offshore slope outside and was suggested previously by Evans but without supporting evidence. It is also indicated in Keulegan's tank experiments.

Most profiles show more than one bar. In some places, as described by Evans (2) there are several parallel bars comparable in size and extending for many miles along the shore. Plane photos demonstrate such features along the shores of Texas, Louisiana, Lake Michigan, and Chesapeake Bay (8). The lines of breakers in plane photos showing the bars along other coasts indicate less continuous features and in many places suggest that only one significant bar exists. The pier profiles of southern California suggest that at least in this area one bar is much larger than any others which may exist. A count of the 1937-38 profiles at Scripps pier shows 209 with indications of one or more bars in addition to the large bar, whereas 90 profiles show only one bar. The remaining 54 profiles gave no indication of the existence of a bar. Evans found that in general the distance between the second and third longshore-bar is greater than that between the first and second, and that the bars grow successively deeper away from shore. This appears also to be true of most of the Oregon and Washington profiles reported by Isaacs (5). On the other hand no rule could be found relative to the plural longshore-bars measured along the Scripps pier. In a considerable number of cases the outer bar, which was the largest, stood higher than that directly inside.

Longshore-bars are very common off gently sloping sand beaches. So far as can be ascertained, steeply sloping beaches, that is, beaches where foreshore slopes are in excess of about 4°, have only very narrow insignificant bars. During the investigation of beach profiles in Oregon and Washington, Isaacs discovered that bars were not present in bays of small dimension. The report by King and Williams (9) of cusp-like bars in such localities is significant. The profiles suggest that longshore-bars and longshore-troughs do not exist outside of the zone where waves break during periods of high surf. However, so many of the sections establishing the existence of the bars terminate at depths near this outer limit that the statement may not be valid. Other types of submerged bars are known to exist at greater depth (16, chart V).

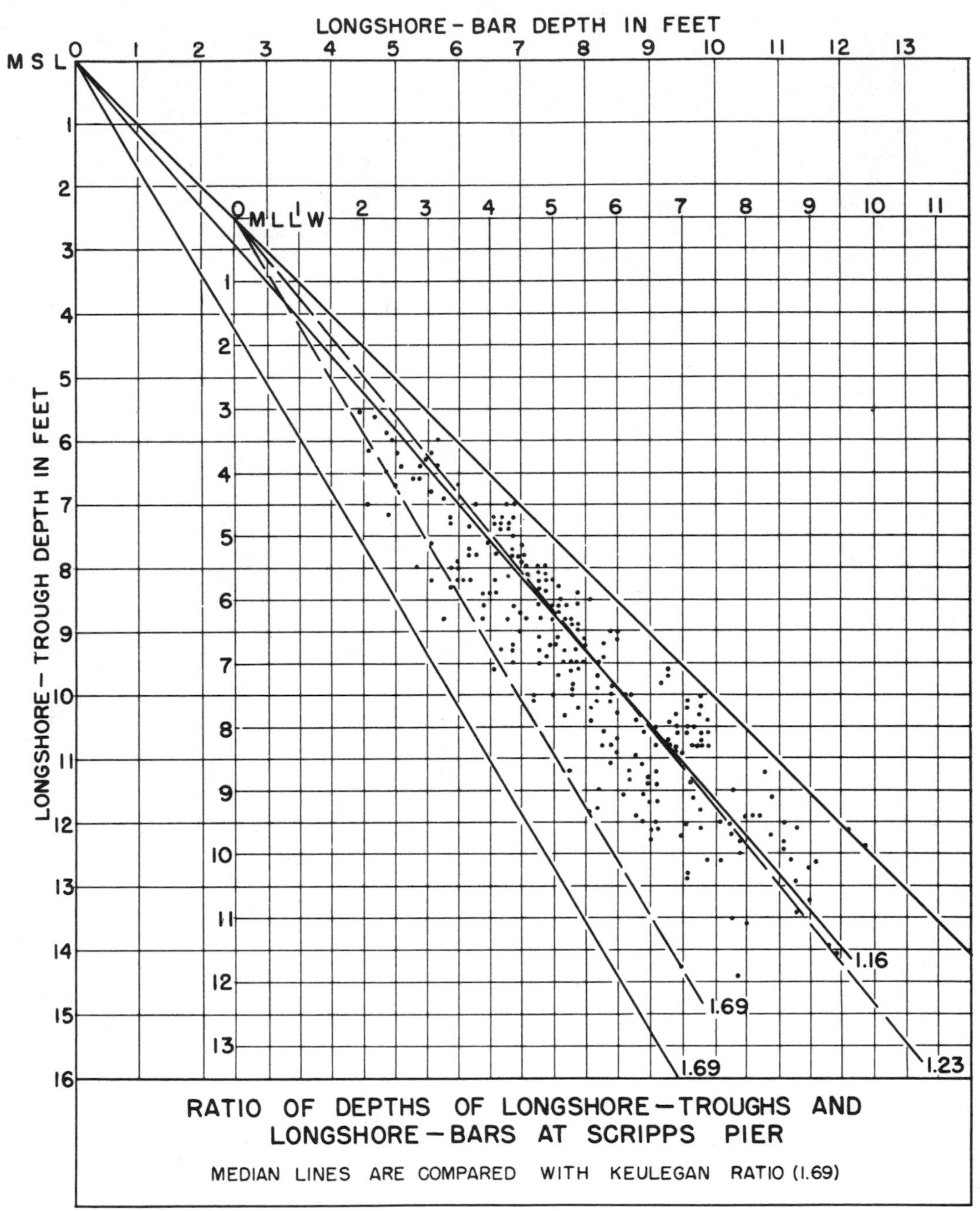

RATIO OF DEPTHS OF LONGSHORE—TROUGHS AND LONGSHORE—BARS AT SCRIPPS PIER

MEDIAN LINES ARE COMPARED WITH KEULEGAN RATIO (1.69)

FIGURE 4

8

Ratios of bar and trough depth are influenced by a number of variables of which the most important is the slight deepening which is found along the axis of the trough towards gaps in the bars. As a result largely of his tank experiments, Keulegan came to the conclusion that the depth of longshore-bars and longshore-troughs below still-water level showed a fairly constant relation such that the depth of the trough divided by the depth of the bar is approximately 1.69. A small number of profiles from Lake Michigan and the Baltic Sea seemed to verify his ratio. A test of the ratio was made from a much larger number of beach profiles using both mean sea level and mean lower low tide level instead of still-water level. The best reference level for this comparison is not easily determined. The rate of tidal fluctuation is small near the low and high tide marks, but it is also small between a low high and high low which are closely similar in height. The average height of the tide in the latter stage is close to mean tide (figure 14). One advantage of using mean lower low tide is that under some circumstances, particularly with large waves, the longshore-bars and longshore-troughs developed at the higher levels may be eliminated at low tide level whereas the low tide bars and troughs are likely to persist during high tides. A plot of these relations for 276 cases at Scripps pier (figure 4) shows that the mean relationship is 1.16 using mean sea level, or 1.23 using lower low tide. A much smaller number of sections along other California piers shows a relationship of 1.40 for mean sea level, or 1.63 for lower low tide, for the troughs to bars. A series of 116 profiles taken out from west coast beaches by the Engineering Department at the University of California, Berkeley, and by the Corps of Engineers, U. S. Army, gave a value of 1.39 using mean sea level, or 1.63 using lower low tide (figure 5). However, considering only the profiles reported by Isaacs (5) off Oregon and Washington, the ratio is close to that of Keulegan, being 1.60 using mean sea level, or 1.93 using lower low tide. This area is one of very large waves, 20-to 30-foot breakers being common during storms. A series of east coast profiles largely off Cape Cod (figure 6") indicate 1.34 using mean sea level, or 1.47 using mean low tide, for the average trough to bar ratio. All of these diagrams show a considerable spread in the ratio and in general indicate that the ratio decreases with depth. The ratios for mean sea level for the 4 groups average 1.3, or 1.5 for mean low water. The discrepancy with Keulegan's ratios may be explained because his sources of information were tank experiments which had bars and troughs of small amplitude.

Keulegan considered also that there was a common ratio between depths of the bar and what he called the bar base. He defined the latter as the line connecting the trough and the barless profile outside the bar. His experiments indicated that the depth of the bar had a ratio to the depth of the bar base, directly underneath the bar, of 0.58. This ratio was also tested from the beach profile. The large number of Scripps pier profiles shows the ratio for mean sea level to be 0.77. or 0.70 using lower low tide (figure 7). However, the profiles at other piers give 0.58 using mean sea level, or 0.46 using lower low tide. The west coast beach profiles give a

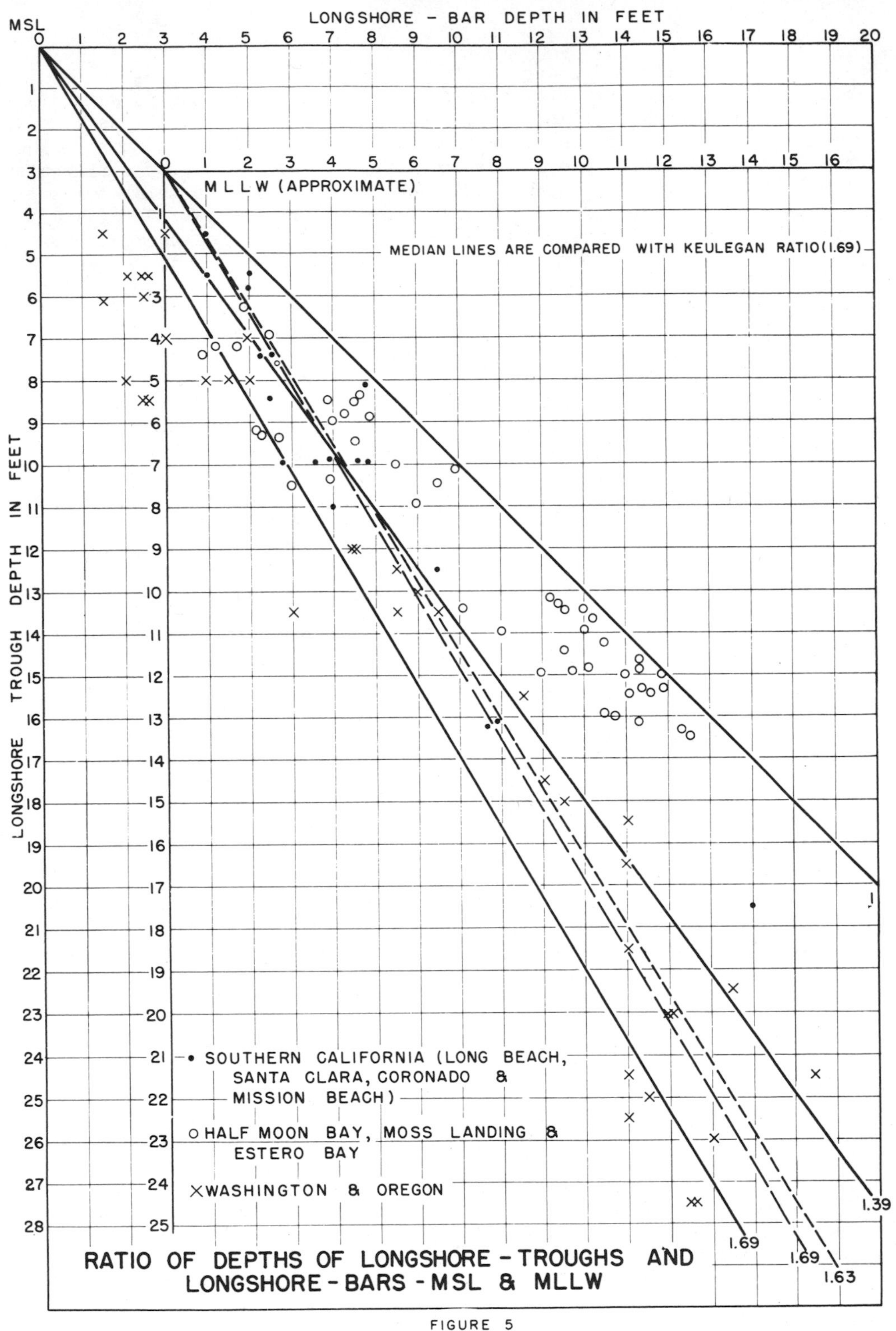
LONGSHORE - BAR DEPTH IN FEET
MSL
M L L W (APPROXIMATE)
MEDIAN LINES ARE COMPARED WITH KEULEGAN RATIO (1.69)
LONGSHORE TROUGH DEPTH IN FEET
• SOUTHERN CALIFORNIA (LONG BEACH, SANTA CLARA, CORONADO & MISSION BEACH)
○ HALF MOON BAY, MOSS LANDING & ESTERO BAY
× WASHINGTON & OREGON
1.39
1.69
1.69
1.63
RATIO OF DEPTHS OF LONGSHORE - TROUGHS AND LONGSHORE - BARS - MSL & MLLW

FIGURE 5

10

ratio of 0.63 using mean sea level, or 0.53 using lower low tide (figure 8), with distinctly higher ratios for the greater depths. The profiles off Oregon and Washington have a ratio of 0.57 using mean sea level, or 0.46 using lower low tide. The east coast sections largely off Cape Cod gave a ratio of 0.59 using mean sea level, or 0.53 using mean low tide (figure 9), also close to that of Keulegan. The average ratio for the 4 groups for mean sea level is 0.63 or 0.55 using lower low tide. Thus the Keulegan ratio for these parameters checks very well.

Relation of Bars and Troughs to Wave Height

Evans, Keulegan, and King and Williams all recognized the relationship between the wave size and bar depth. This relationship becomes still more evident from the study of the profiles taken along the California piers. The most substantial data were derived from the profiles along Scripps pier (figure 10). A plotting of bar depth against the wave heights which were measured daily shows a grouping of points around the median line but having a rather large spread. The same relation comes from the plot of trough depths against wave height (figure 11). The quartile lines in figures 10 and 11 drawn parallel to the median line indicate that more than 50% of the points lie within one foot of the median line. There can be little doubt but what deeper longshore-bars and longshore-troughs are found on the days of larger waves. The relationship was also tested by plotting the greatest wave height of the preceding 5 days against the bar depth (figure 12), but this does not group the points as close to the median line as did the plotting with accompanying wave heights.

A comparison of some of the daily profiles to changing wave heights indicates the nature of changes which take place (figures 3, 13, and 18). The complete record for the 1937-38 measurements of depth changes of both the longshore-bar and longshore-trough along with wave height and bar height above the trough is given in figure 14. The fact that the response to changing wave conditions is similar throughout the years leaves little doubt of the reality of this relationship. The bars accompanying relatively small waves are cut away by the large waves and new bars from outside. However, the building up of the new bars may lag so that the first effect of a day of large waves is likely to be a cutting away of the old bar and the excavation of a deep trough inside rather than the building of a new bar on the outside. The new bar often develops after the large waves have subsided. A comparison of the wave record over the year with the cross-sectional area of the bar is also instructive (figure 15). This cross-sectional area is derived by drawing a horizontal line out from the bottom of the trough to the point where the line intersects the outer slope and computing the area of the bar above the line. The bars clearly increase in size during the season of large waves, but the increase generally lags after a particular period of large breakers. Longshore-bar and longshore-trough depths show a relationship to maximum wave heights. Thus the bar crests off southern California are

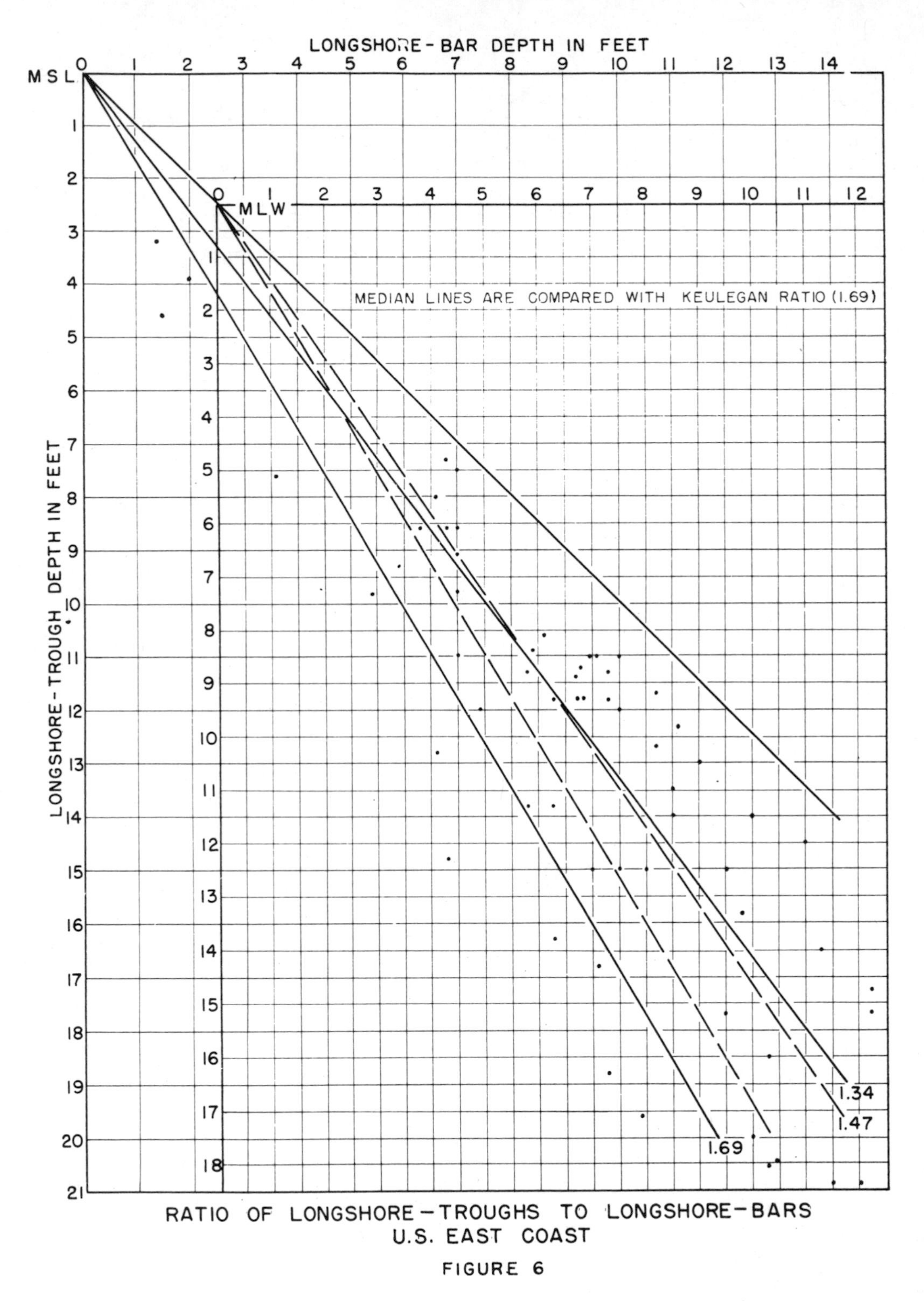

RATIO OF LONGSHORE—TROUGHS TO LONGSHORE—BARS
U.S. EAST COAST

FIGURE 6

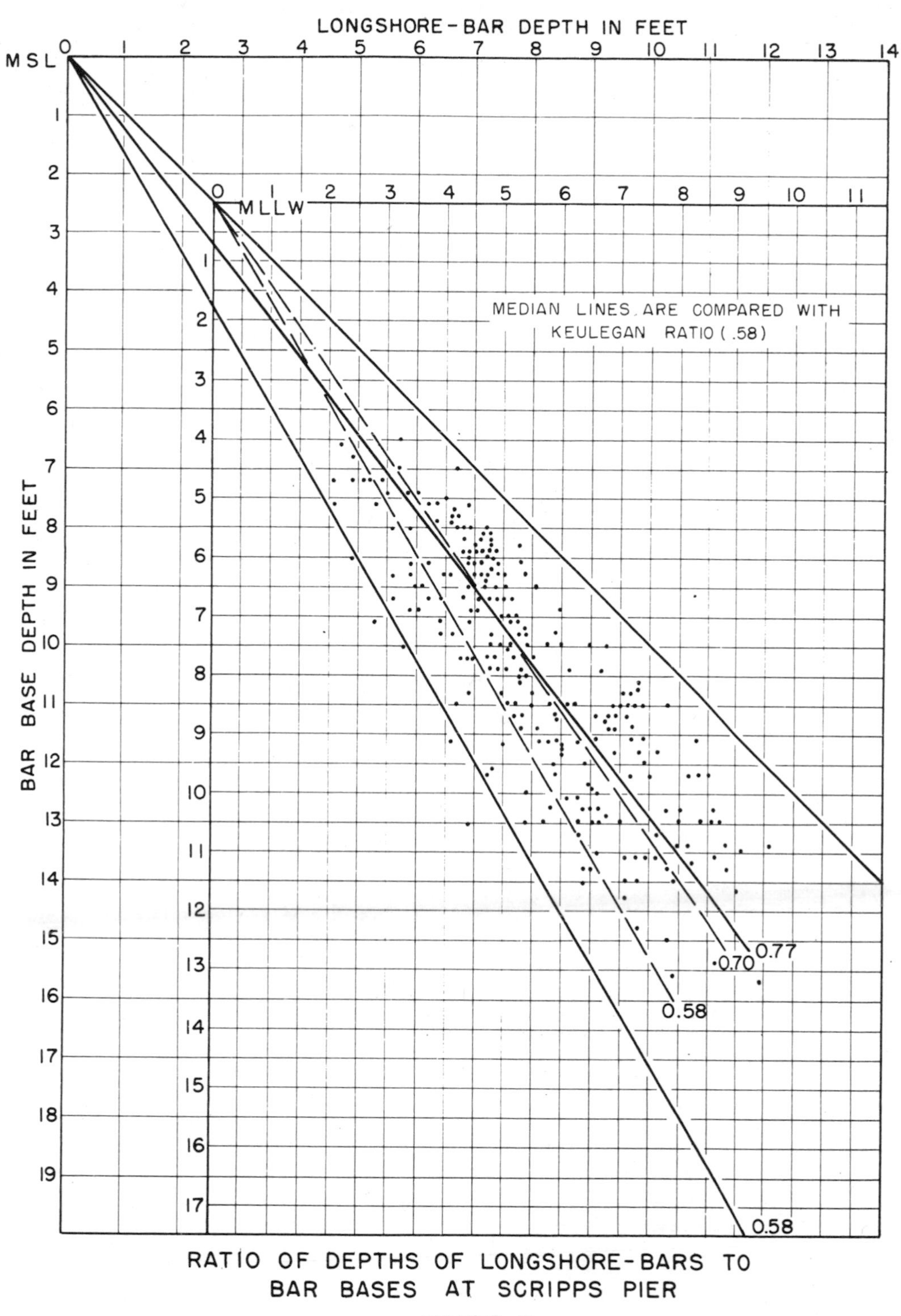

RATIO OF DEPTHS OF LONGSHORE-BARS TO BAR BASES AT SCRIPPS PIER

FIGURE 7

13

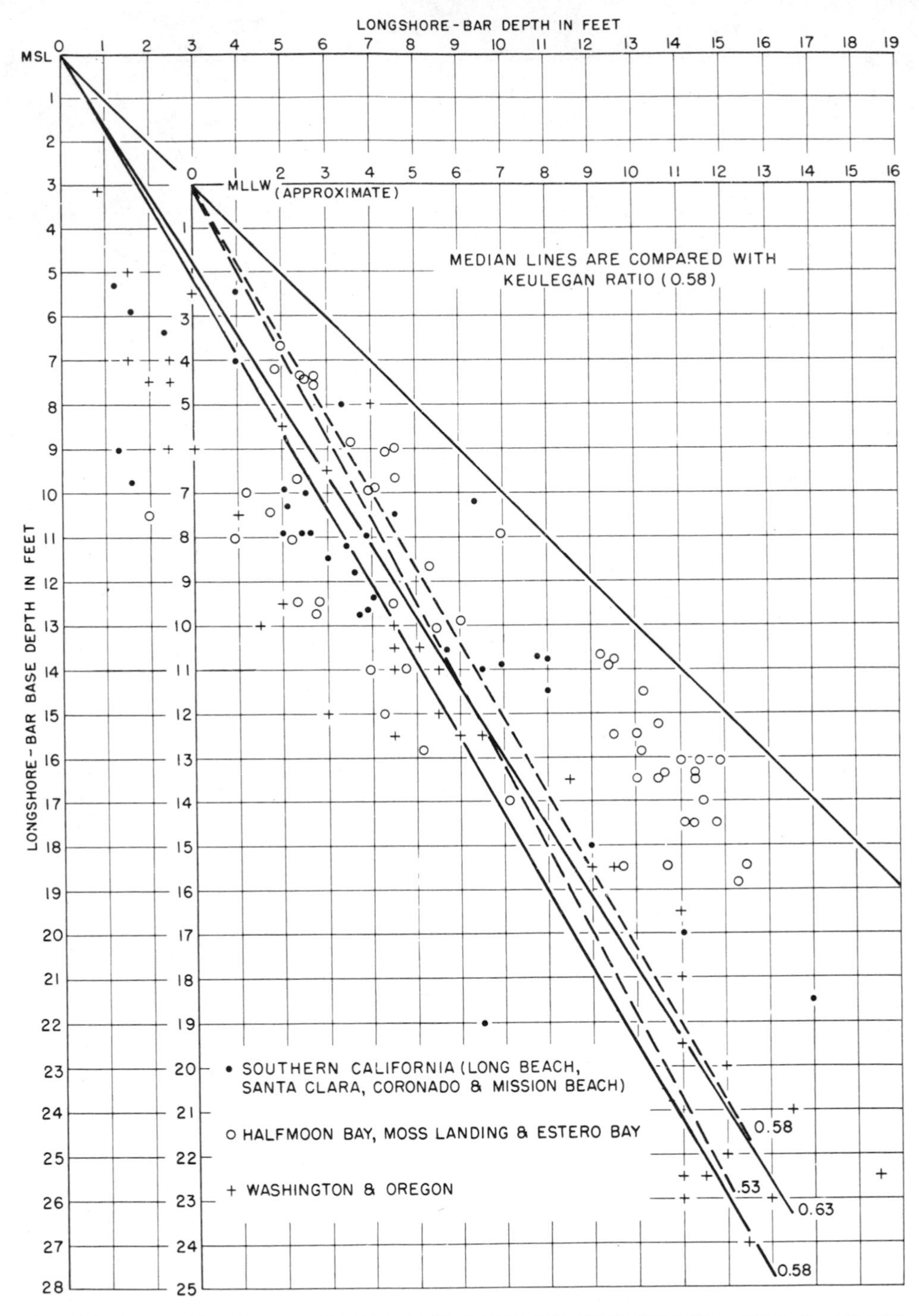

RATIO OF DEPTHS OF LONGSHORE BARS TO BAR BASES—MSL & MLLW

FIGURE 8

14

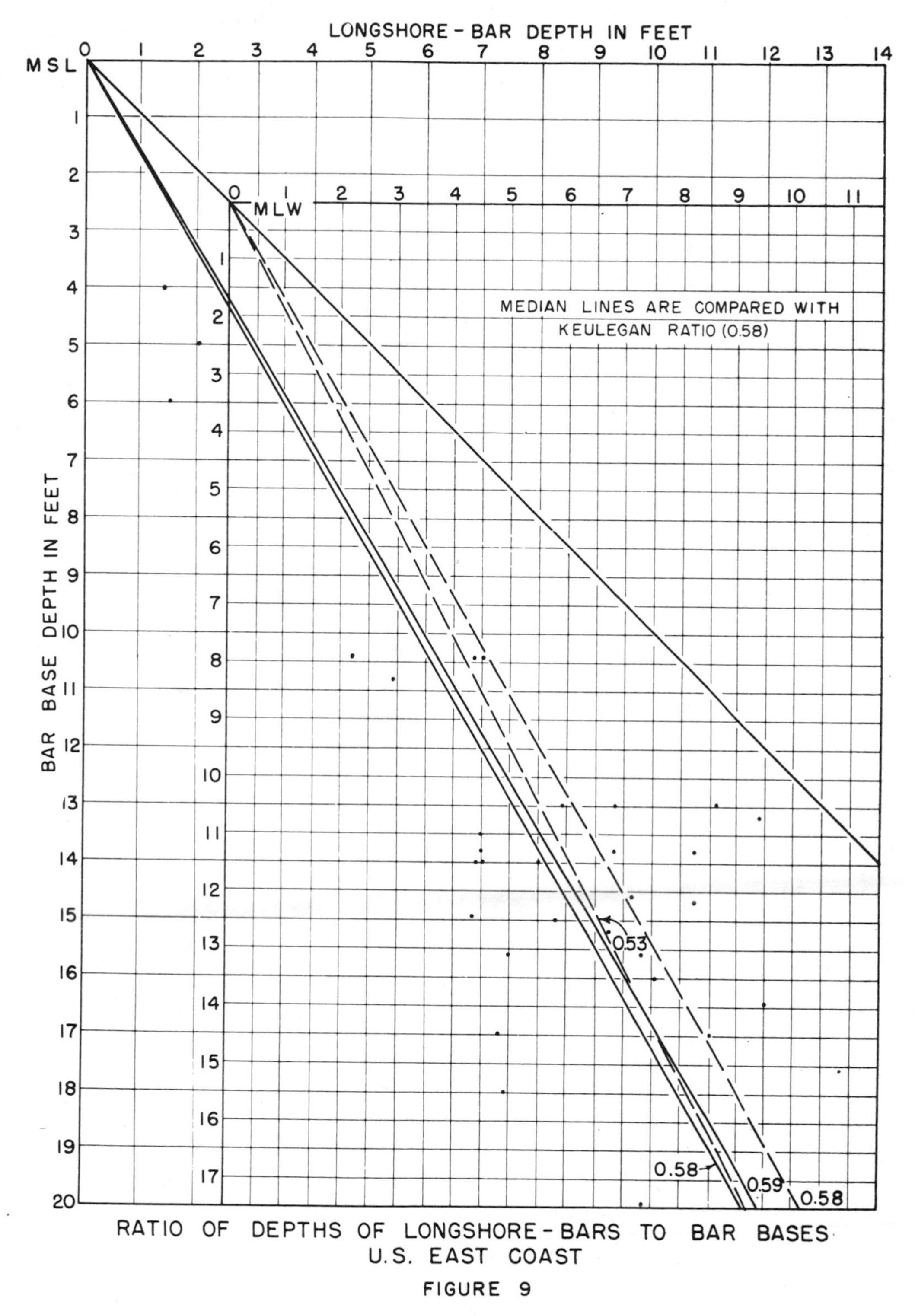

RATIO OF DEPTHS OF LONGSHORE-BARS TO BAR BASES
U.S. EAST COAST
FIGURE 9

15

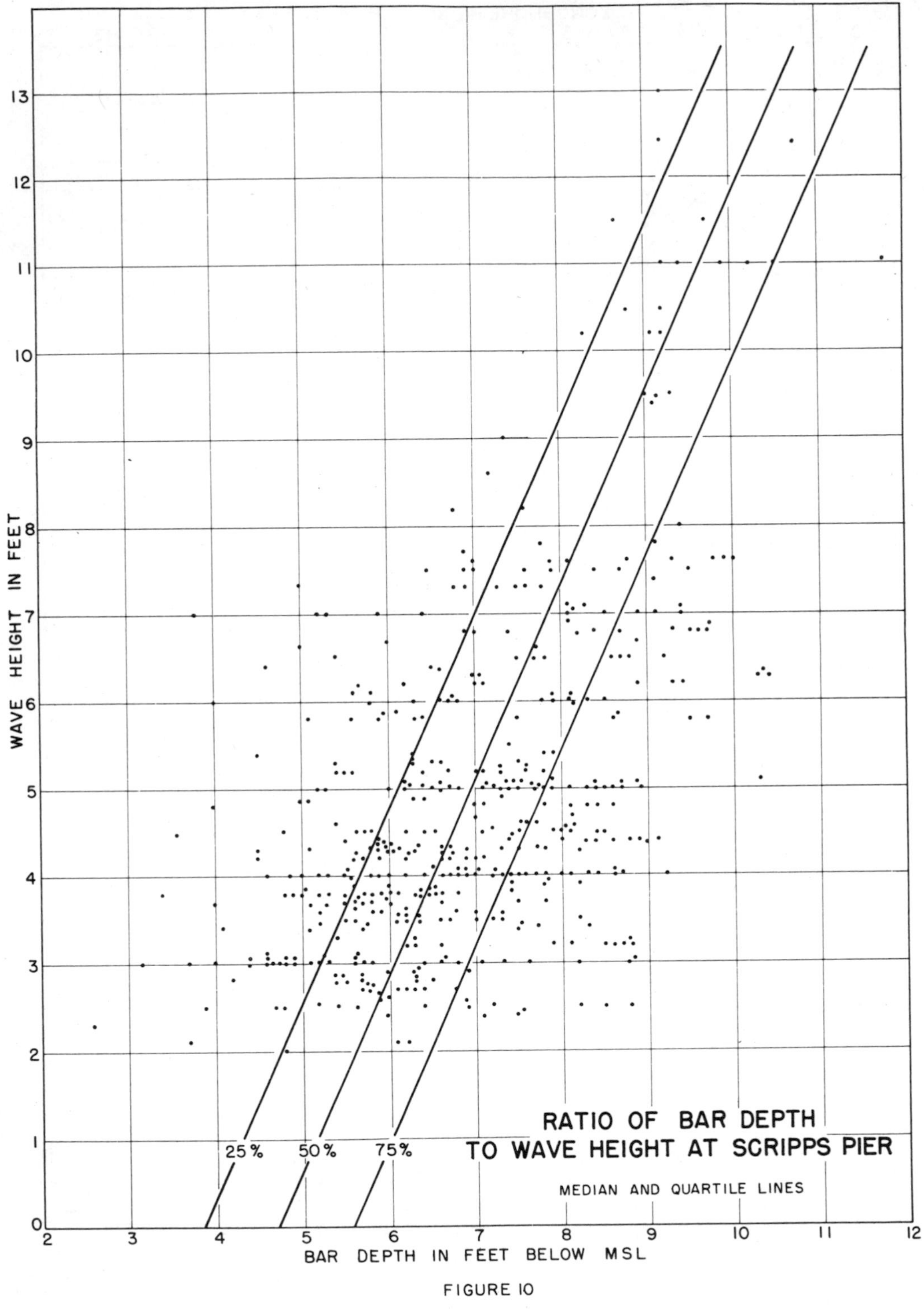
RATIO OF BAR DEPTH
TO WAVE HEIGHT AT SCRIPPS PIER
MEDIAN AND QUARTILE LINES
WAVE HEIGHT IN FEET
BAR DEPTH IN FEET BELOW MSL
25%
50%
75%
0
1
2
3
4
5
6
7
8
9
10
11
12
13
2
3
4
5
6
7
8
9
10
11
12

FIGURE 10

16

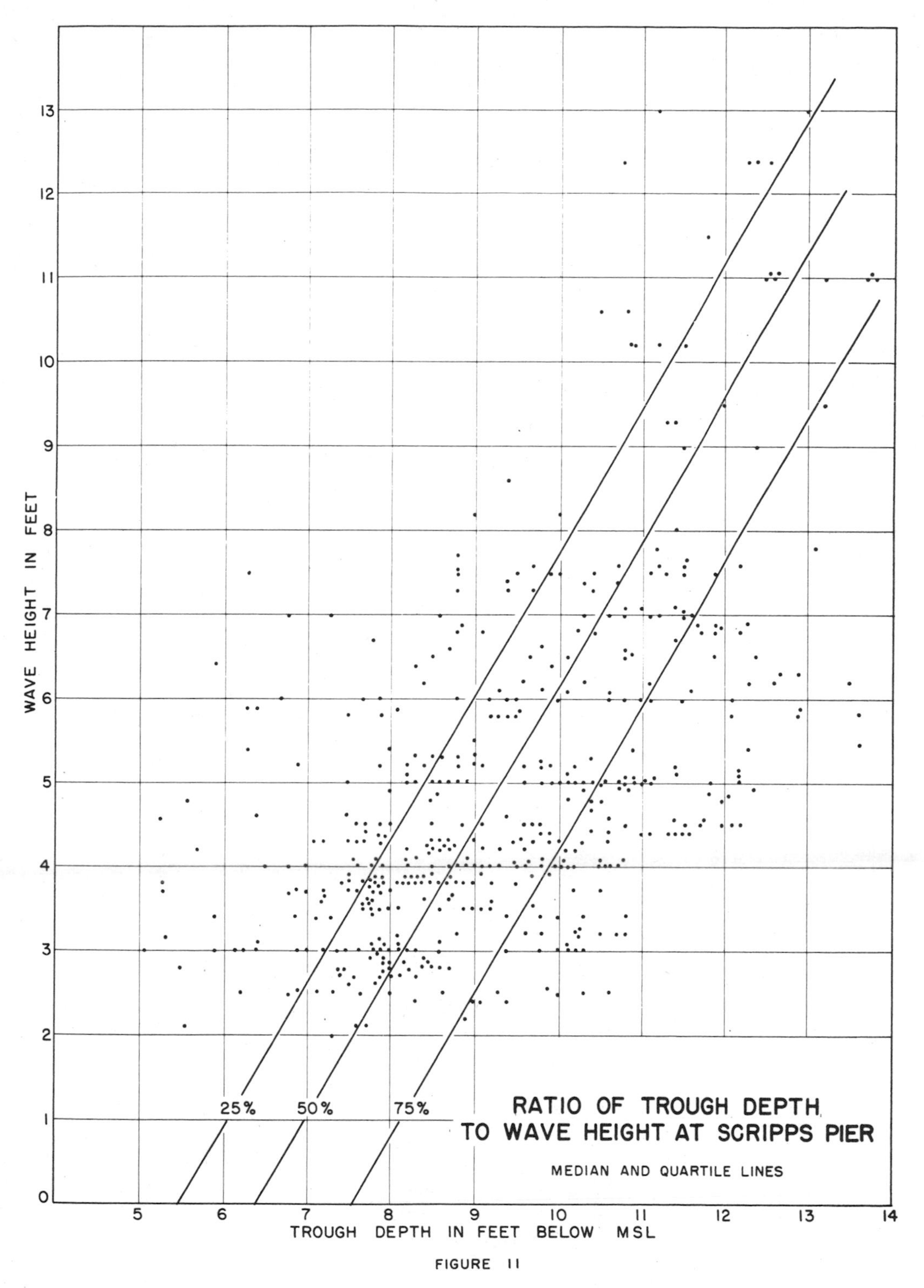

RATIO OF TROUGH DEPTH
TO WAVE HEIGHT AT SCRIPPS PIER
MEDIAN AND QUARTILE LINES
25%
50%
75%
WAVE HEIGHT IN FEET
TROUGH DEPTH IN FEET BELOW MSL
0
1
2
3
4
5
6
7
8
9
10
11
12
13
14

FIGURE 11

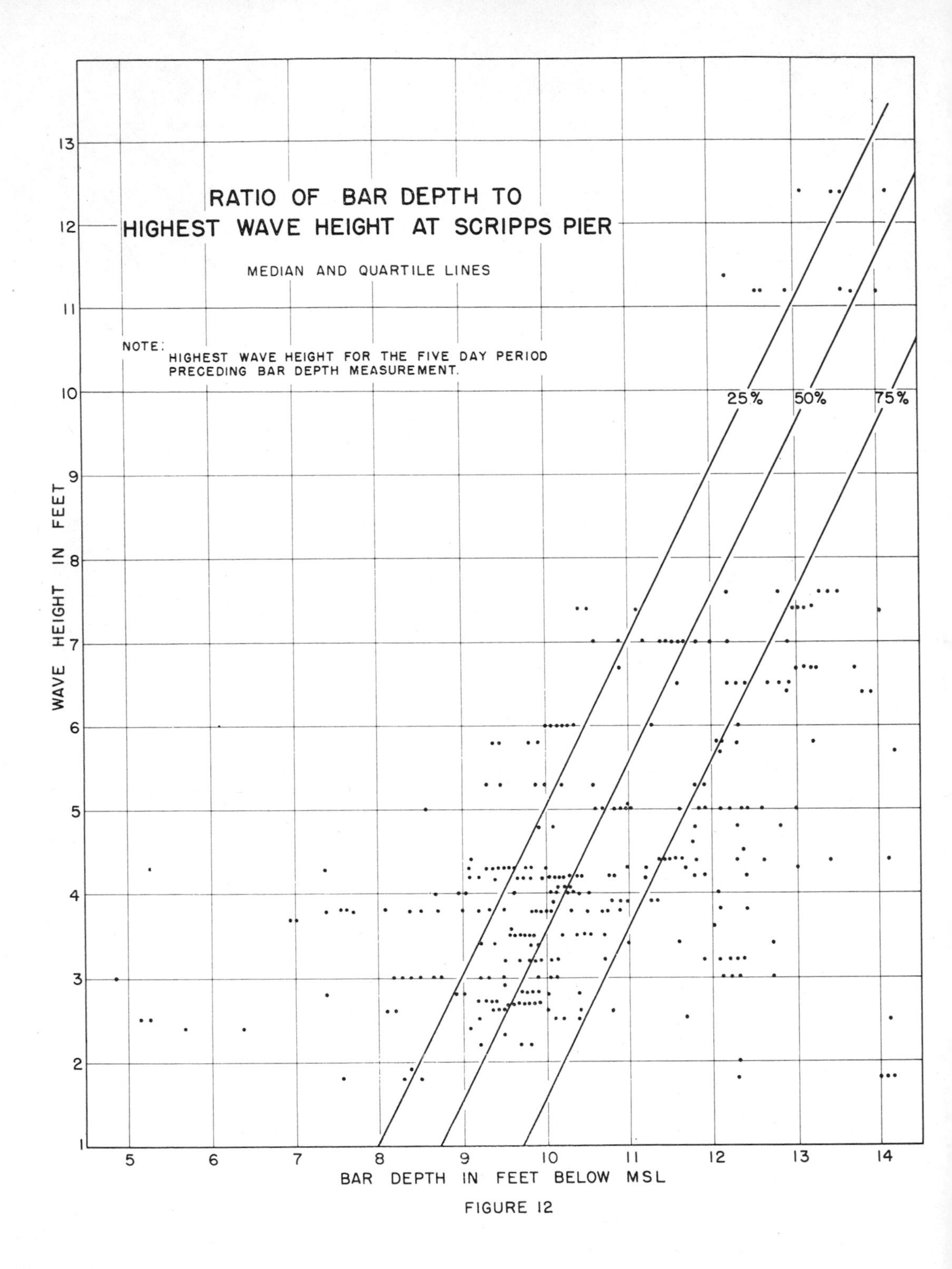
RATIO OF BAR DEPTH TO
HIGHEST WAVE HEIGHT AT SCRIPPS PIER
MEDIAN AND QUARTILE LINES
NOTE: HIGHEST WAVE HEIGHT FOR THE FIVE DAY PERIOD PRECEDING BAR DEPTH MEASUREMENT.
25%
50%
75%
WAVE HEIGHT IN FEET
BAR DEPTH IN FEET BELOW MSL
1
2
3
4
5
6
7
8
9
10
11
12
13
14

FIGURE 12

18

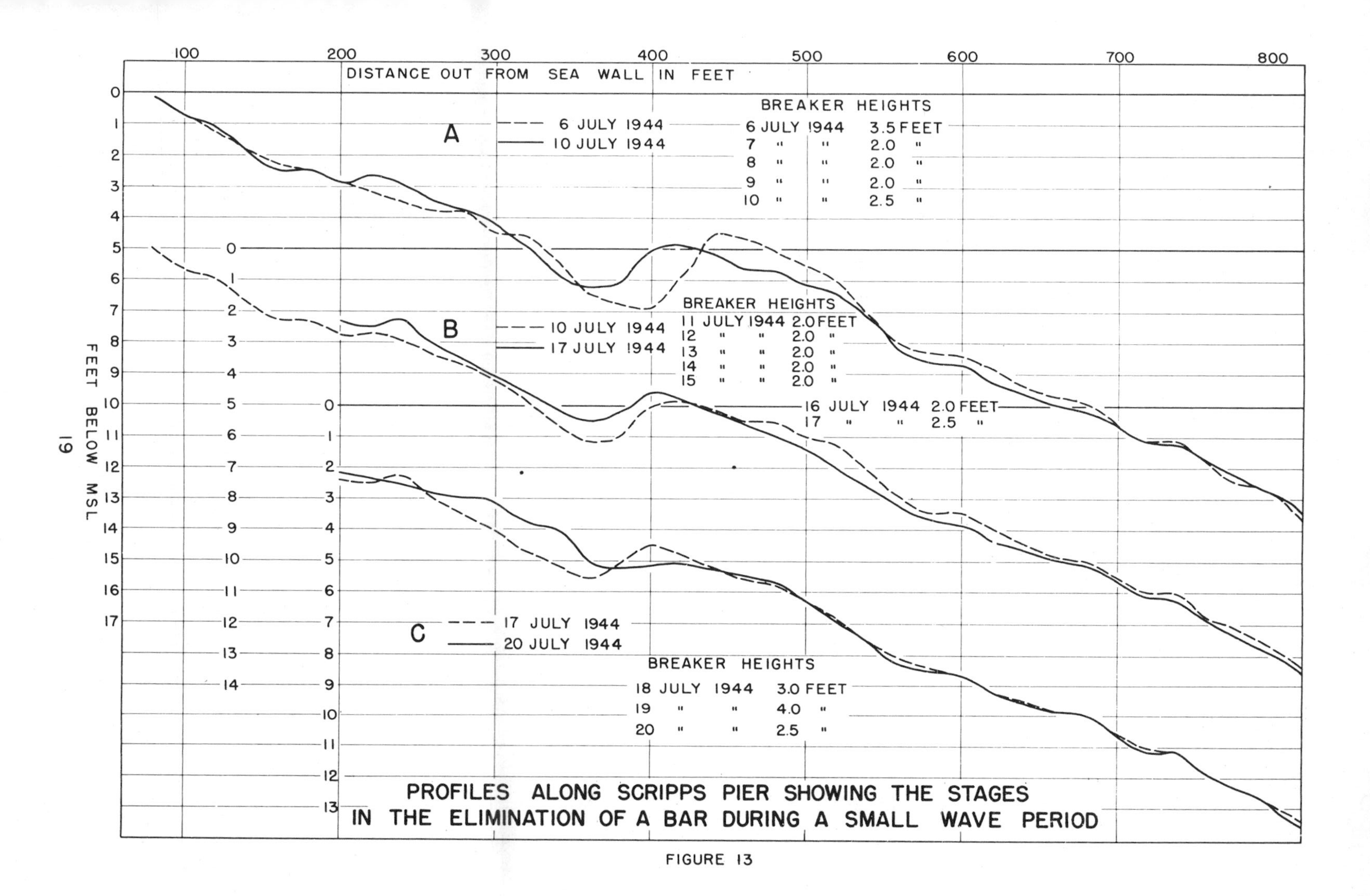

PROFILES ALONG SCRIPPS PIER SHOWING THE STAGES IN THE ELIMINATION OF A BAR DURING A SMALL WAVE PERIOD

FIGURE 13

almost all less than 10 feet below mean sea level. In this area breakers are rarely greater than 10 feet. Off Washington and Oregon 20-foot waves are fairly common and bars extend to depths of at least 18 feet. This difference can be seen by the groupings of points from southern California and from Washington and Oregon in figure 5.

A plotting of the bar depth at a series of piers in southern California was made in relation to the seasonal change (figure 16). All of the bars and troughs shoaled as the summer period of small waves advanced. A similar effect over a much longer period is shown by the depth of the bars and troughs along the Scripps pier during a 8-year period (figure 17).

Relation to Currents

The data do not permit as good a correlation between longshore-bar and longshore-trough development and currents as is possible with wave height. Currents were observed but not measured at several points along the Scripps pier during most of the period of daily sand measurement, and a Sverdrup-Dahl current meter was in use at the end of the pier during most of the time. It is unfortunate that there is not a complete record of currents in the zone of bar and trough development. However, out of 98 observations in this zone only 9 showed currents in a different direction from the currents observed at the pier end at the same time. There appear to be some striking relations between the area of the longshore-bar and the current direction (figure 15). South currents apparently increase the area of the bar and the strongest recorded south current was observed just before the bar grew to its largest area. North currents conversely appear to have the effect of decreasing the bar area. One case where large waves were accompanied by excessive fill took place in February, 1938, during a period of north currents (figure 18, A).

A profile obtained at Oceanside pier at a time of very large longshore currents inside the breakers indicated a very large bar and trough. Other evidence suggesting that strong currents are an important factor, at least in the formation of troughs, comes from the finding of deep troughs along the path of the longshore feeder currents to rip currents during times when the currents are particularly strong (13, pp. 350-352). When these channels are exposed by low tide, current ripples are found in them indicating flow along their axes (5). Isaacs reports that these ripples increase in magnitude along the troughs towards the rip channels which cross the bars. Near these cross channels he found giant ripples similar to those in tidal channels.

Relation to Tides

The longshore-bars and longshore-troughs are necessarily dependent on the tides since waves break farther out on the offshore slope during low tides than during high tides. The large range during spring tides

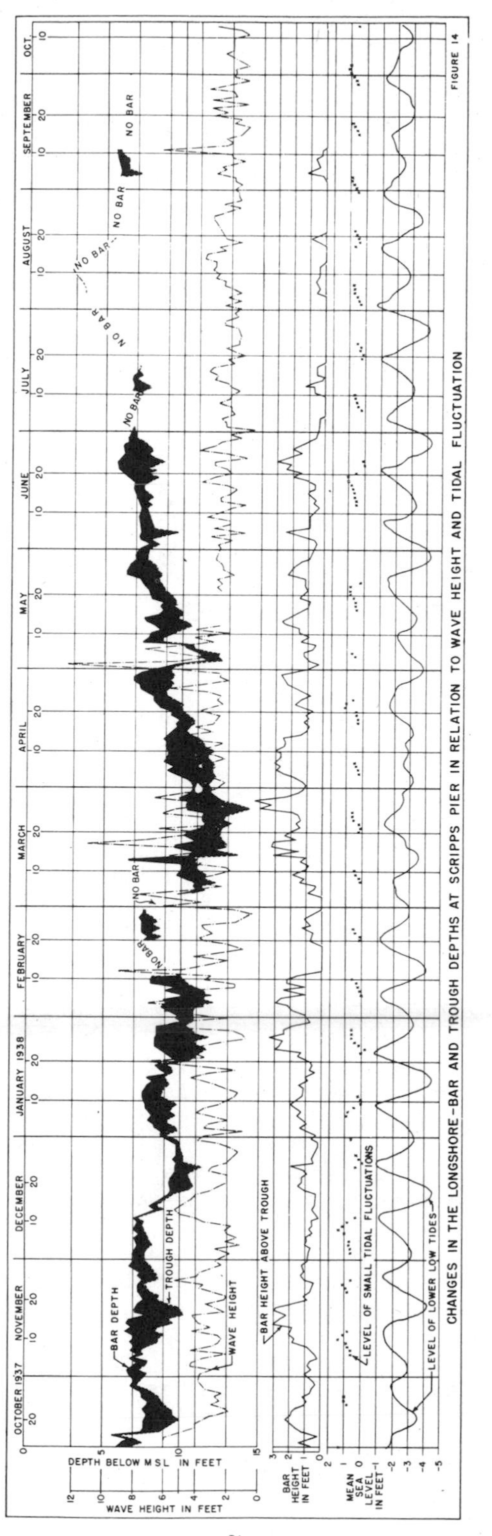

CHANGES IN THE LONGSHORE-BAR AND TROUGH DEPTHS AT SCRIPPS PIER IN RELATION TO WAVE HEIGHT AND TIDAL FLUCTUATION

FIGURE 14

cause a shifting of position of deposition and cutting. This was clearly shown during a 36-hour period of continuous profile measurement. The result of this shift should be a greater development of bars and troughs during or shortly after neap tides when the breaker line remains more constant in position. A similar effect comes from the small ranges between the contiguous low high and high low which accompany neap tides.

Examination of the bar development during 1937-38 shows that other influences such as wave height and current variation are more important than tide, but there is a rough correlation between the larger bars and neap tide periods (figure 14). The first of the records showed somewhat of an inverse ratio to tide stage because of the development of large waves at spring tide, but the more typical relationship is seen during most of the year when the bars are developed best during or shortly after neap tides. It will be noted in figure 14 that the growth of the bars is also somewhat dependent on the small fluctuations between high low and low high tides.

Origin of Longshore-bars and Longshore-troughs

Speculation concerning the origin of longshore-bars has been appearing in print for more than a century. De Beaumont (1) first recognized these bars as a natural result of the readjustment of submarine slopes to existing wave conditions. He considered that wave action removed material from the sea floor and built it up as a bar parallel to the shore. Johnson (6) supported this hypothesis as the early stage in the origin of the emergent offshore bars. The present investigation, however, supports Evans (2, p. 510) in the indication that longshore-bars are in equilibrium with wave conditions and that the migration of the bars is only a temporary or seasonal fluctuation and does not form the broad emergent beaches to which the name offshore bar is applied.

Russell (12) and Gilbert (3) concluded from their studies of the longshore-bars of the Great Lakes that the materials of the bars were derived from the beaches and carried along the shore until deposited where the "undertow" loses its force. Evans (2, p. 503) supporting Johnson gave the following arguments against the origin by longshore currents:

"1. Because of the acute angle at which the inner ball (longshore-bar) usually joins the shore

2. Because the ball (longshore-bar) does not receive any considerable amount of material from the locality where it joins the shore

3. Because of the failure of the ball (longshore-bar) to build above the water as does a spit or bar (bay bar)

4. Because of the lack of any considerable connection between the outer balls (longshore-bars) and the shore, although they are generally higher above the bottom and broader than ball No. 1 (innermost longshore-bar)

5. Because the line of the bottom of the lows (longshore-troughs) prolonged intersects the surface landward rather than seaward of the shore line."

The fifth point is significant since it indicates that erosion of the sea floor rather than simple deposition has taken place. However, it does not exclude longshore currents from an important part in the production of the longshore-bars. The other points depend on the assumption that longshore drift builds only structures which extend to the surface, such as spits, bay bars, and offshore bars. Actually there is no proof that this is the case.

One can agree better with Evans' suggestion (2, p. 508) that the longshore-troughs are the result of plunging breakers. However, his conclusion that the material set in suspension is carried out to form the bar may be only partially correct because the profiles along Scripps pier indicate that in many cases the material has come from the outside. Keulegan (7, figure 4) established through his experiments that the trough position is determined by the position of the plunging breaker. King and Williams (9, p. 80) considered that the bar forms at the "break point" as the result of seaward movement of material inside and landward on the outside.

The investigations at Scripps Institution do not offer any radically different hypothesis for the origin of the longshore-bars and longshore-troughs. The importance of the plunging of breakers is indicated by the close correspondence of the longshore-bar crest depths with the wave height. The waves steepen and start to break onto the bar and then plunge into the trough on the shoreward side of the bar. The position of the longshore-bar and longshore-trough is somewhat a matter of chance since given an even slope the position at which a series of newly developed large waves will break is dependent on the state of the tide when these waves begin their work. Many of the troughs may be initiated by the coincidence of several plunging waves along approximately the same line. Once developed the trough is likely to continue. Further indication of the importance of wave plunging in forming troughs and bars comes from the discovery that spilling waves such as accompany short-period storm waves with high local winds tend to obliterate longshore-bars. Under these conditions the longshore-bars and longshore-troughs at Scripps pier have been completely eliminated (figure 3, C) presumably because there was no breaker line. The inner longshore-bars and longshore-troughs are commonly destroyed by storm waves where the wind-driven waves break without plunging and no second set of breakers can be developed. This was nicely demonstrated by the effects of a northeast sorm which locally attained hurricane velocities along the outer side of Cape Cod, Massachusetts (figure 19).

Longshore currents have an important influence in the development of longshore-troughs. The slight net seaward transport which is observed along the bottom inside the breakers tends to concentrate the bottom water in the longitudinal depressions produced by the plunging waves, resulting in a strong flow parallel to the shore. These currents become feeders to the rip currents which break seaward through the bar at various points (13). The channels along these feeder currents may be exposed at low tide. Isaacs found that the rippled inner troughs partially exposed by low tide deepened progressively towards a point where they connected with a passage leading through the bar (a rip channel).

According to measurements made by Scripps Institution groups along the southern California coast, the rip currents become concentrated in the surface layers outside the breakers with the result that there is no very definite indication of a seaward net transport of water along the bottom. Since shoreward velocities under wave crests are somewhat in excess of the seaward velocity under wave troughs, sediment tends to move shoreward along the bottom unless the slope is so steep that the gravitative effect will counteract this movement up the slope. The shoreward movement is of course most effective near the breaker line in the relatively shallow water where the wave effect is most vigorous.

If the wave height and the water level should remain essentially the same, a profile of equilibrium will result because the shoreward movement of the sand will build the bar up only to the height where waves will break on top of it with sufficient force to carry away the sand which is being added. Similarly the trough will continue to deepen only to the depth at which the bottom current is capable of carrying away the sand contributed to it both from the landward migrating bar outside the breakers and from the seaward transport of material inside the breakers.

If the wave height changes, the equilibrium will be disturbed. If the waves are larger (assuming the same period) the breakers will plunge on top of the bar and thus cut it away (figure 18, B) and after a time a new bar will develop seaward of the old (figure 18, C). In the early stages of large waves the cutting effect is the most significant but the sediment which is largely carried out in the rip channels will tend to be redistributed along the outer slope and start moving in to build up the bars at depths which are compatible with the new wave height. The few instances where deposition took place along Scripps pier during large waves are not clearly understood, but they are apparently the result of the reversal of current which caused a blocking of the natural outlets of the troughs and thereby caused an accumulation of the sediment against the pier.

Lowered wave height will allow the bar to grow shoaler and to encroach onto the trough as in figure 13, A-C. If the feeder currents are no longer capable of maintaining the trough in its deep position because of the shoreward movement of the breaker zone, the trough

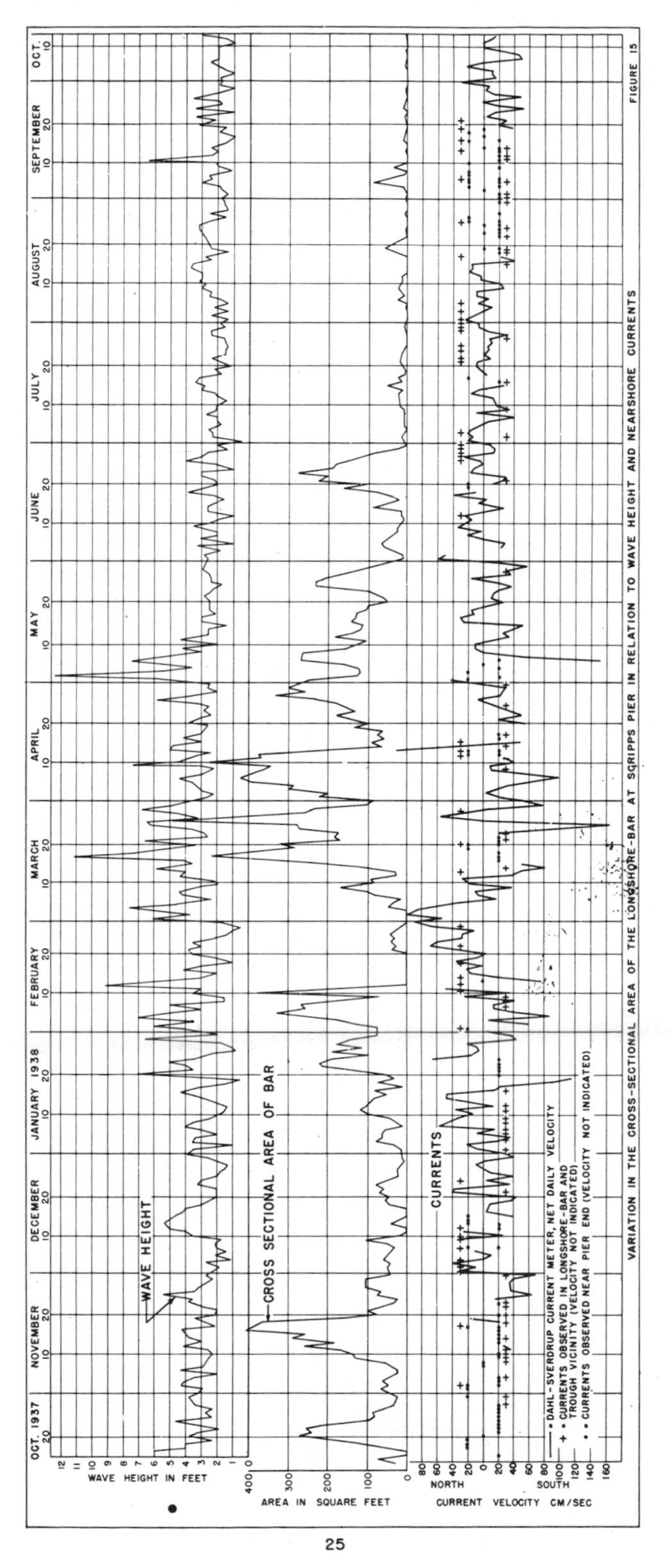

FIGURE 15

VARIATION IN THE CROSS-SECTIONAL AREA OF THE LONGSHORE-BAR AT SCRIPPS PIER IN RELATION TO WAVE HEIGHT AND NEARSHORE CURRENTS

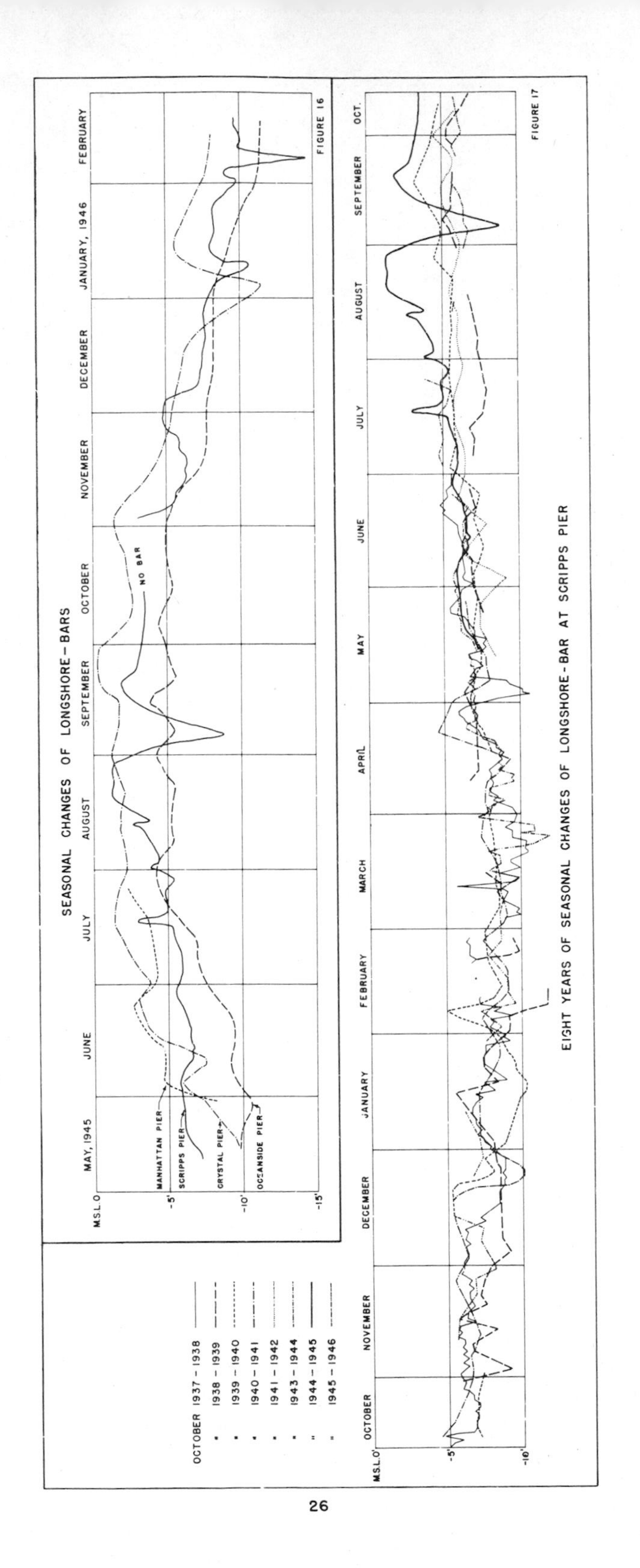

EIGHT YEARS OF SEASONAL CHANGES OF LONGSHORE-BAR AT SCRIPPS PIER

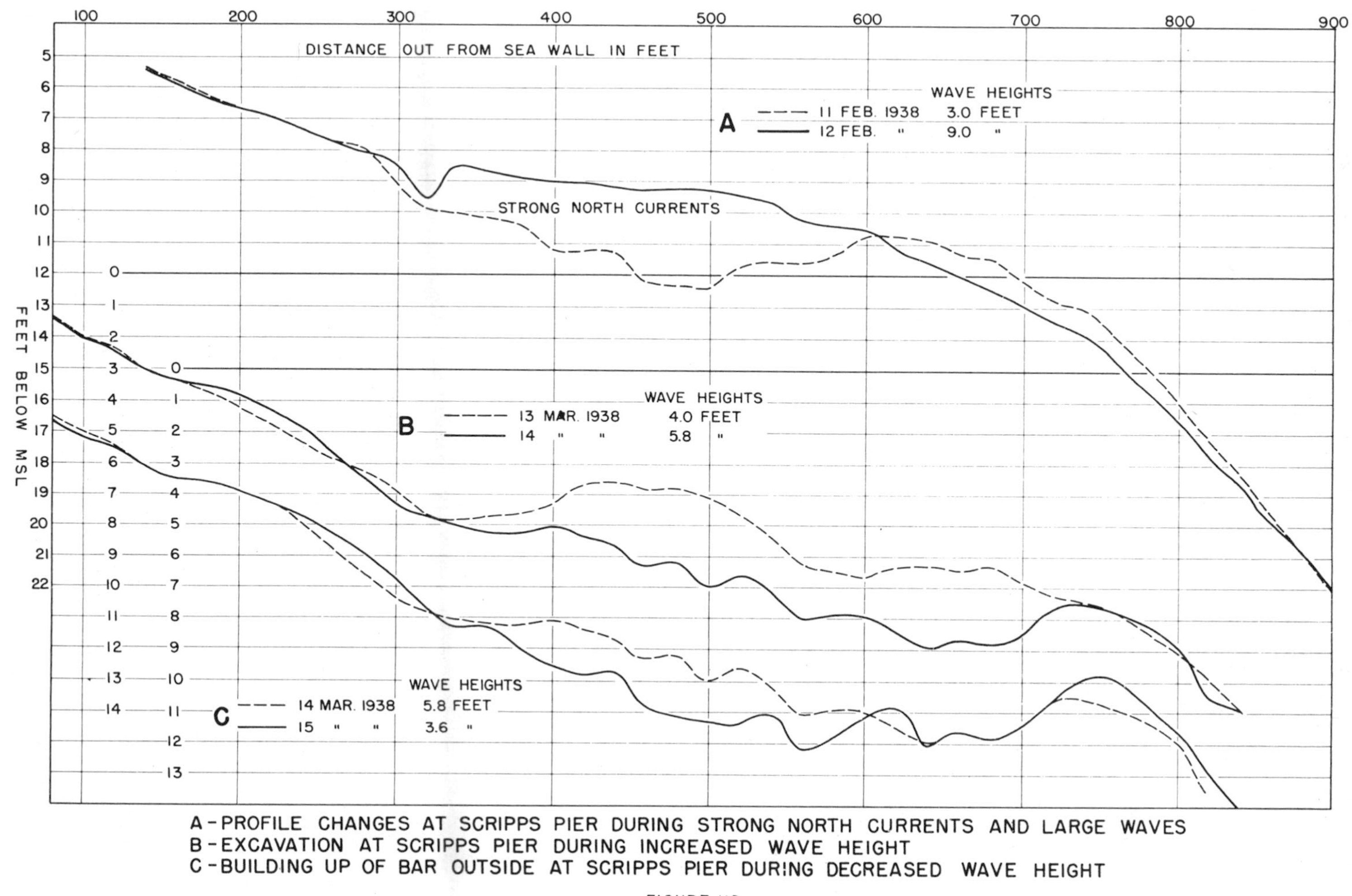

A - PROFILE CHANGES AT SCRIPPS PIER DURING STRONG NORTH CURRENTS AND LARGE WAVES
B - EXCAVATION AT SCRIPPS PIER DURING INCREASED WAVE HEIGHT
C - BUILDING UP OF BAR OUTSIDE AT SCRIPPS PIER DURING DECREASED WAVE HEIGHT

FIGURE 18

will fill and the slope become relatively even in the zone where the bar and trough had existed previously.

During periods of large waves the development of multiple longshore-bars is readily explained as the result of the re-forming of the waves over the troughs after breaking on the outer bar (figure 2). In shoaler water the wave breaks again with an exact repetition of the bar and trough forming process. Since the waves are smaller the topographic relief of the inner bars and troughs is less striking. Other cases of multiple bar development can be explained by a rising tide causing the waves to break at successively higher points along the slope. Wherever a good development of rip feeder currents is capable of carrying the sediment away from the plunge zone a trough will tend to develop. A falling tide, however, will tend to destroy the shoaler bars, as was shown in the tank experiments by King and Williams (9, p. 81). Longshore-bars and longshore-troughs developed by large waves may not be destroyed when the waves decrease and new bars and troughs are developed at shoaler depth. King and Williams developed two sets of bars in their tank experiments by decreasing the wave size. Isaacs found that during the small wave season he could anchor a wave meter on a deep outer bar off the Oregon coast and it remained undisturbed. These deep bars are the product of large winter waves.

The general absence of significant longshore-bars off steep beaches is readily explained by the fact that the waves break so close to the shore that there is no opportunity for the development of the longshore currents along the plunge zone. The association of well-developed cusps with these steep beaches is probably of some significance in this connection.

Summary and Conclusions

Slightly submerged sand bars extend parallel to the shore line off most gently sloping beaches. Hundreds of profiles of longshore-bars have been taken largely along the west coast. An analysis of these profiles has shown that there is considerable variation in the relation between depth of trough and depth of bar below mean sea level. The most representative relation shows that the troughs are 1.3 times as deep as the bars with reference to mean sea level, or 1.5 times as deep with reference to mean lower low water. The depth ratio between bar crest and bar base is 0.61 for mean sea level, or 0.55 for mean lower low water, which compares well with the 0.58 ratio indicated by Keulegan's experiments at the Beach Erosion Board laboratory. A source of difficulty in developing such ratios comes from the deepening of the longshore-troughs towards gaps in the longshore-bars.

The depth of the longshore-bars was found to be related to the height of waves and to the position of breakers. Plunging breakers excavate the bottom, producing the longshore-troughs. The material thus set into suspension is moved parallel to the shore by currents which turn seaward into rip channels forming gaps in the longshore-bars. Some of the bars and troughs are exposed by low tide. The

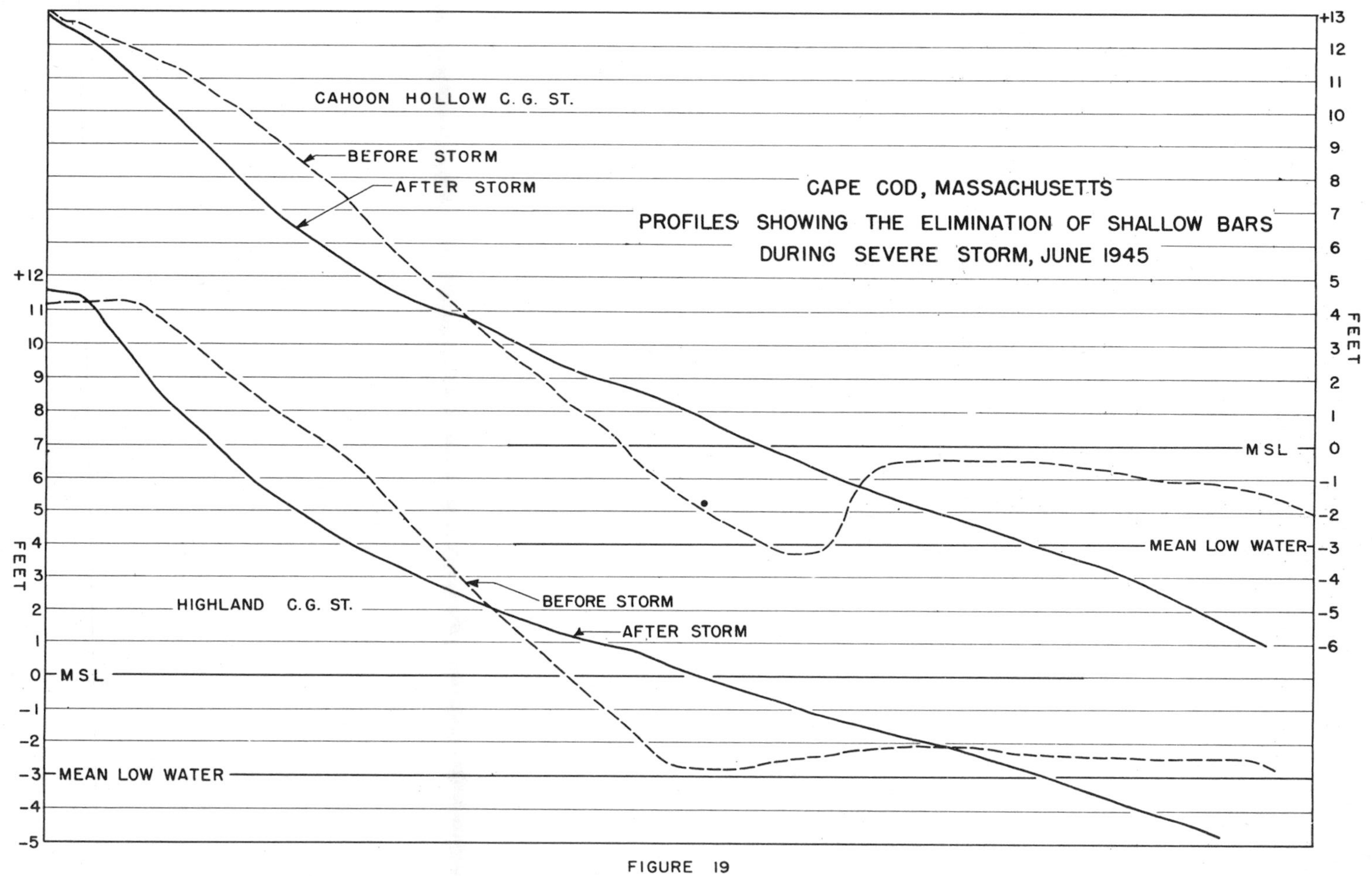
CAPE COD, MASSACHUSETTS
PROFILES SHOWING THE ELIMINATION OF SHALLOW BARS
DURING SEVERE STORM, JUNE 1945
CAHOON HOLLOW C. G. ST.
BEFORE STORM
AFTER STORM
HIGHLAND C. G. ST.
BEFORE STORM
AFTER STORM
MSL
MEAN LOW WATER
MSL
MEAN LOW WATER
FEET
FEET
+12
11
10
9
8
7
6
5
4
3
2
1
0
-1
-2
-3
-4
-5
+13
12
11
10
9
8
7
6
5
4
3
2
1
0
-1
-2
-3
-4
-5
-6

FIGURE 19

29

sand carried outside the bar is spread out over the slope by the expanding head of the rip current. Thence the shoreward drag under wave crests carries it back to build up the longshore-bars outside the troughs. The growth of these bars is limited by the depth at which the waves will plunge and prevent further sand encroachment.

Acknowledgments

The assistance received from U. S. Navy projects has been referred to previously. Expecial gratitude is expressed to E. C. LaFond who took many of the profiles along the piers, supervised the taking of a large part of the remainder, and was also a large contributor to determining the relations of the longshore-bars to the waves and currents. Much important information was contributed by John D. Isaacs from his extensive investigations of beaches along the west coast. Helpful suggestions were also provided by D. L. Inman and R. S. Arthur. Others who have taken an active part in the work include D. B. Sayner, Ruth Young, K. D. Langley, Ralph Gentoff, Mrs. Rosemarie Snow, and Mrs. Dorothy Williams.

REFERENCES

(1) Beaumont, E. de. 1845. "Lecons de Geologie Pratique." 557 pp. Paris.

(2) Evans, O. F. 1940. "The low and ball of the eastern shore of Lake Michigan." Jour. of Geol., Vol. XLVIII, pp. 476-511.

(3) Gilbert, G. K. 1890. "Lake Bonneville," U.S. Geol. Surv. Mono. I, p. 44.

(4) Isaacs, J. D. 1945. "A device for traversing the surf zone of ocean beaches.", Univ of Calif., Fluid Mech. Lab., Tech. Rpt. 90 (a).

(5) Isaacs, J. D. 1947. "Beach and surf conditions on beaches of the Oregon and Washington coast between August 27 and September 27, 1945." Univ. of Calif., Fluid Mech. Lab. Memo. HE-116-229.

(6) Johnson, D. W. 1919. "Shore processes and shore line development." 584 pp.

(7) Keulegan, G. H. 1948. "An experimental study of submarine sand bars." Beach Erosion Board, Office of the Chief of Engineers, Tech. Rpt. No. 3.

(8) Kindle, E. M. 1936. "Notes on shallow-water sand structures." Jour. Geol. Vol. XLIV, No. 7, pp 861-869.

(9) King, C. A. M., and Williams, W. W. 1949. "The formation of sand bars by wave action." Geog. Jour., Vol. CXIII, pp. 70-85.

(10) Otto, Theodor, 1911-12. "Der darss und zingst." Jahrb. d. Geog. Gesell. zu Greifswald, Band XIII, pp. 393-403.

(11) Passarge, Seigfred, 1912. "Physiologische morphologie," Mitt.d. Geog. Gesell. in Hamburg, pp. 195-198.

(12) Russell, I. C. 1885. "Geological survey of Lake Lahontan." U.S. Geol. Surv. Mono. II, pp. 92-93.

(13) Shepard, F. P., Emery, K. O. and LaFond, E. C. 1941. "Rip currents; a process of geological importance." Jour. Geol., Vol. 49, pp. 337-369.

(14) Shepard, F. P., LaFond, E. C., 1940. "Sand movements along the Scripps Institution pier, California." Am. Jour. Sci., Vol. 238, pp. 272-285.

(15) U. S. Coast and Geodetic Survey, 1890. "Cross sections of the ocean shore of Cape Cod." Appendix No. 13 for the Report for 1889.

(16) Veatch, A. C., and Smith, P. A. 1939. "Atlantic submarine valleys of the United States and Congo submarine valley." Geol. Soc. Am., Special Paper No. 7.

Editor's Comments on Paper 5

Continuing the approach established by Keulegan, E. D. McKee and T. S. Sterrett have investigated the effect of bottom slope, wave intensity, and sand supply upon longshore bars. Studied too were the internal primary structure of bars and beaches. In combination these two papers could well serve as an introduction for the beginner to wave-basin experimentation.

The authors' section on terminology on the second page of this article is particularly interesting. In this and other discussion within the paper McKee and Sterrett support the emergence of longshore bars. Upon closer examination, it will be found that their development of a barrier beach under these conditions does not differ too greatly from the possibility conceded by Evans some years earlier. As noted in the introduction to the paper preceding this, the topic of submarine bar emergence relative to barrier island origin will be dealt with more fully in a subsequent volume in this series.

At present Thomas S. Sterrett is with the Department of Geology of the El Paso Natural Gas Company, and Edwin D. McKee is with the U.S. Geological Survey in Denver. McKee received his A.B. degree from Cornell in 1928 and pursued graduate studies at the University of Arizona, University of California, and Yale University; he was awarded an honorary Sc.D. at Northern Arizona University in 1957. He has been a Grand Canyon National Park naturalist, Museum of Northern Arizona director, University of Arizona professor and department head, and both chief and research geologist in the Paleotectonic Map Section of the U.S. Geological Survey. In 1970 McKee was a visiting lecturer in Russia and Brazil.

5

LABORATORY EXPERIMENTS ON FORM AND STRUCTURE OF LONGSHORE BARS AND BEACHES[1]

EDWIN D. McKEE[2] AND THOMAS S. STERRETT[2]
Denver, Colorado

ABSTRACT

Beaches and bars have been formed during experiments conducted in a 46-foot wave tank at the Sedimentation Laboratory of the U. S. Geological Survey in Denver. By changing one variable factor at a time, elements responsible for major differences in primary structure and in shape of sand body have been determined. These elements are—differences in slope of sand floor (expressed in terms of water depth), intensity of wave action, and supply of sand. Stages in the growth of the bars and beaches were marked with dark layers of magnetite, and cross sections were preserved on masonite boards coated with liquid rubber, thus recording cross-stratification patterns and sand-body shapes.

Longshore bars are produced at the point of wave break. In very shallow water an emergent bar commonly forms; in somewhat deeper water a submarine bar is built; and in still deeper water no bar forms. Increase in intensity of waves tends to build a bar toward, and even onto, the beach. Weaker waves build bars upward to form barriers, with lagoons to shoreward. Abundant sand furnished on the seaward side of a growing bar simulates conditions caused by some longshore and rip currents, and causes gentle seaward-dipping beds to form. In contrast, a limited sand supply results in growth of bars that characteristically have shoreward-dipping strata of steeper angle.

Beach strata normally dip seaward at low angles from the crest to a point below water level. Offshore, the seaward extensions of these gently dipping beds include fore-set beds with relatively high angles which form a shoreface terrace. The sand body comprised of both sets of bedding builds outward if a large supply of sand is furnished. In shallow water, however, or at moderate depth where waves are strong, the period of beach growth is limited by the deposition of longshore bars which eliminate wave action as they grow into barriers and form lagoons. Under conditions in which no bar is built, growth of the beach and shoreface terrace is controlled by the amount of sand available; the proportion of top-set to fore-set beds is determined by the strength of waves.

INTRODUCTION

In order to determine variations in the form and structure of longshore bars and beaches a series of experiments was conducted in the Sedimentation Laboratory of the U. S. Geological Survey in Denver, during 1959. Various types of bars and beaches were reproduced. Differences in shape and in the cross-stratification of these experimentally formed sand bodies were recorded, establishing criteria useful in the interpretation of similar features preserved in ancient sedimentary rocks.

A metal tank, 46 feet long and 1½ feet wide, with plastic observation windows on one side and an electrically controlled wave generator at one end, was used for the experiments. Waves were formed by the back-and-forth movement of a perforated baffle for which the length of stroke and speed of movement was predetermined according to a table of wave calibration, and set at the beginning of each experiment. The floor of the flume sloped down toward the wave generator at an angle of 0°, 15′, 18″.

Sediment used for wave-tank experiments consisted of well sorted[3] and well rounded, screened sand of fine-grained quartz, showing the following size distribution

½ mm; coarse	7 per cent
¼–½ mm; medium	81 per cent
¼ mm; fine or smaller	12 per cent.

At the start of each experiment sand was distributed across the floor of the tank to form a seaward-sloping wedge that tapered to its termination approximately 30 feet from the beach area. At this time and during subsequent stages in the growth and evolution of the sand bodies, thin layers of magnetite were spread over the sand. The magnetite was introduced at regular intervals (15 minutes, 30 minutes, or 60 minutes, depending

[1] Read before the Association at Atlantic City, New Jersey, April 25, 1960. Manuscript received, April 26, 1960. Publication authorized by the Director, U. S. Geological Survey.

[2] U. S. Geological Survey.

[3] In experiments to be conducted later, poorly sorted sand will be used for comparison.

on the speed of growth) to serve as markers indicating steps in construction and also to record details of structure characteristic of the sand body.

Upon completion of each experiment a permanent record was prepared of the shape and structure formed during investigation. This record was made by draining water from the tank, cutting and removing a vertical slice of sand parallel to the wave direction, and then wedging a masonite board covered with liquid rubber against the vertical face. When the rubber hardened and a surface of sand had adhered to it, excess sand was removed from the board, leaving a sample cross section of the sand body. Additional records were made by tracing structure patterns on transparent paper through plastic windows of the tank, by making casts in liquid rubber of surface relief features such as ripple marks, and by keeping notes on the evolution of various structures.

Experiments were conducted under the direction of the senior author; details of tank operation were handled largely by the junior author.

TERMINOLOGY

In this series of experiments the term *beach* is applied to deposits of the *shore*, which is defined, following Johnson (1919, p. 160), as the zone over which the water line, or line of contact between land and sea, migrates. Extensions of beach deposits seaward of the low-water level have been considered a part of the beach by some authors, but are here referred to as deposits of the *shoreface* and include steeper dipping beds of the *shoreface terrace* (Johnson, 1919, p. 163). Sediments seaward of this terrace are referred as as offshore deposits.

The term *bar* is applied to the sand bodies formed offshore at the point of wave break and as a result of wave action. Depending on depth (variation in slope), amount of wave action, and availability of sand, these sand bodies may develop as (1) subaqueous masses that do not build up to water level, (2) masses that build up to but not above water level, or (3) masses that build to maximum wave height. All such sand bodies are termed *offshore bars* by Johnson (1919, p. 259), and he includes in the term such additional features as beaches and dunes where they are deposited on the emergent surfaces during a late stage.

Those bars that form only as subaqueous types have been called *balls* (in "lows and balls") by Evans (1940, p. 476), following earlier usage by various British geologists, but the inappropriateness of this term has subsequently been shown by Shepard (1950). The term *offshore bars* as used by Evans (1942, p. 846) was restricted to those sand ridges "lying approximately parallel to a shoreline and reaching above the water." Thus, he would consider two of the types of sand bodies that were formed in the present laboratory experiments as unrelated and genetically different.

Sand bodies of the types under discussion, where "submerged at least by high tides," are designated *longshore bars* by Shepard (1952, p. 1908). In contrast, any sand accumulations above high-tide level, added to such bodies, cause them to be termed *barriers*, according to Price (1951, p. 487–488), and *barrier beaches, barrier islands*, or *barrier spits*, depending upon physiographic features, according to Shepard (1952, p. 1905–1906).

Use of the term *longshore bar* in referring to various experimentally formed offshore sand ridges seems valid. In some of the higher sand bodies, especially those resulting from seaward sand feeding, however, low-angle structures comparable to those of a natural beach and in a similar position relative to sea level are formed, thus suggesting that these would be referred to in nature as *barrier beaches*. A further suggestion is made that the term *barrier* as used by Price, Shepard, and others does not represent a single structural or genetic unit as does a bar, but is a composite body composed of a longshore bar that forms its base or core, and a veneer or perhaps thicker covering of other sand bodies with different structures.

STAGES IN GROWTH OF LONGSHORE BARS AND BEACHES

Experiments in the series that forms the basis of this paper were begun, in each case, with a sand mass representing a shore cliff at one end of the tank and an even surface of sand sloping seaward from it toward the opposite end of the tank. In each experiment the degree of slope of the sand bottom was regulated by the height of water level, and thus determining the depth of water at any particular place. The strength of waves to be used was arbitrarily decided in advance and the wave generator was set for the appropriate speed.

After the wave generator was put in motion for an experiment, the variable factors that control sedimentation were not changed, and gradually the deposits became stabilized, terminating the experiment.[4] At that point no further change could be expected without altering one or more of several controls such as depth of water, degree of wave action, or amount of available sediment. Between the beginning of an experiment and the attainment of a stabilized condition, however, a continuous evolution of sedimentary forms and structures took place.

In each experiment, where water was sufficiently shallow for wave-formed orbits of water-particles to impinge on the sand floor, the initial structures to be formed consisted of scattered cusp-type ripple marks on the sand surface. These cusps, which are scattered, crescent-shaped bodies with steep slopes facing shoreward, then changed into parallel-type ripple marks, in series, as a uniform flow movement was brought about by wave motion (Scott, 1945). The ripple marks migrated shoreward, rapidly or slowly according to intensity of wave action, by the normal process of continuous erosion on the seaward side of each crest and deposition to lee (Fig. 1). All ripple marks were asymmetrical because of the forward movement of the waves, but their size varied greatly with changing conditions on the sea floor.

In the evolution of the sea floor, change from an even, ripple-marked surface (Fig. 2-*A*) to a low mound with rippled surface (Fig. 2-*B*), signifies the forming of an embryo offshore bar. This change takes place at approximately the point of wave break, as shown by Keulegan (1948, p. 2), and is caused by increased turbulence which tends to disrupt normal ripple-mark production and to pile up sand into a mound. The running time required to start growth of a submarine bar in the laboratory ranged from 4 to 21 hours, depending on each of the several variable factors concerned. As the bar grows vertically, ripple marks disappear from its surface because turbulence increases and waves of translation carry shoreward the disrupted sand from the ripple marks. The bar then continues to increase in size through the systematic addition of sloping sand layers to shoreward, which may be classed as fore-set beds, or in some bars, as top-set beds.

Vertical and lateral growth of a bar through the accretion of sand in the form of dipping beds is dependent upon an adequate source of sand. In nature this source, according to Davis (1909, p. 708), is the previously developed sand floor to seaward, but according to Gilbert (1890, p. 40), it is shore drift continually being introduced by longshore currents.[5] In the laboratory, sand obtained from the seaward floor alone, being relatively scant, caused growth largely on the lee or shoreward side of the initial mound or bar. This was because removal of sand to seaward with resulting increase of water depth in that direction resulted in shoreward migration of the point of wave break. In contrast, where new sand was constantly being fed artificially[6] to seaward of the bar, as though by a combination of longshore and rip currents, a shoal was produced, resulting in the seaward migration of the point of wave break and the accretion of sand to the seaward side of the bar (Figs. 4-*B*, 4-*D*).

In every experiment a series of changes in sand form and in structure evolves in the shore area at the same time that bars are forming to seaward. Beach deposits with strata dipping seaward at low angles form above water level and may be considered as top-set beds; the continuation of the low-angle beds beneath water level and the fore-set beds offshore beyond, dipping at moderately high angles, forms a shoreface terrace (Johnson, 1919, p. 163, 259). The evolution of these related shore and offshore deposits is controlled early in their history by the type and growth of longshore bars that form to seaward.

[4] In experiments to be undertaken, rising water level and lowering water level, simulating conditions of transgression and regression, will be introduced.

[5] Validity of the concept of bar growth by the work of longshore currents has been strongly questioned on several grounds by Evans (1942, p. 848); but more recently, Shepard (1950, p. 23) has shown that the combination of shore drift and rip currents feed much sediment to the seaward sides of longshore bars.

[6] Measured quantities were hand fed at regular, timed intervals.

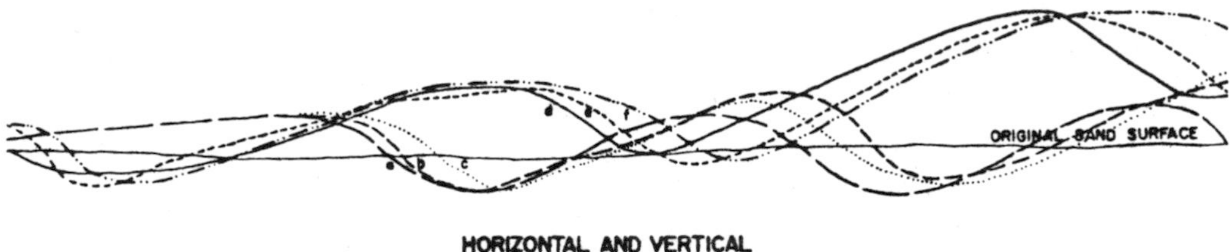

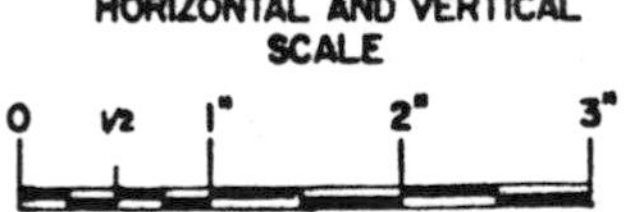

FIG. 1.—Migration of sand by wave-formed ripples from offshore toward beach. Tracing was made on window of tank during early stages of experiment on longshore bar growth. a, b, c = ripple mark development and migration (early stage); d, e, f = same (late stage). a, b, c = 3-minute time intervals; d, e, f = 5-minute time intervals.

VARIABLE FACTORS IN WAVE-TANK EXPERIMENTS

In the present series of experiments the growth of longshore bars and beaches took place under controlled conditions. Three variable factors proved to be of major significance—differences in slope of sand floor (expressed in terms of water depth), force or intensity of wave action, and abundance of available sediment. One factor was varied at a time, and after each change the sand body was allowed to evolve until it became completely stabilized with no further growth or change in shape. The time involved ranged from 8 hours where water was shallow and wave action weak, to 70 hours where water was deep and wave action strong.

Three depths of water measured 16 feet from the beach were used—0.4 feet, 0.6 feet, and 0.8 feet; these depths will be referred to hereafter as shallow, moderate, and deep, respectively. Actually they represent differences in slope of the sand floor, and these differences are responsible for major variations in the types of beaches and bars produced.

Intensity of wave action was controlled largely by wave frequency, which ranged from 12 to 48

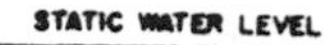

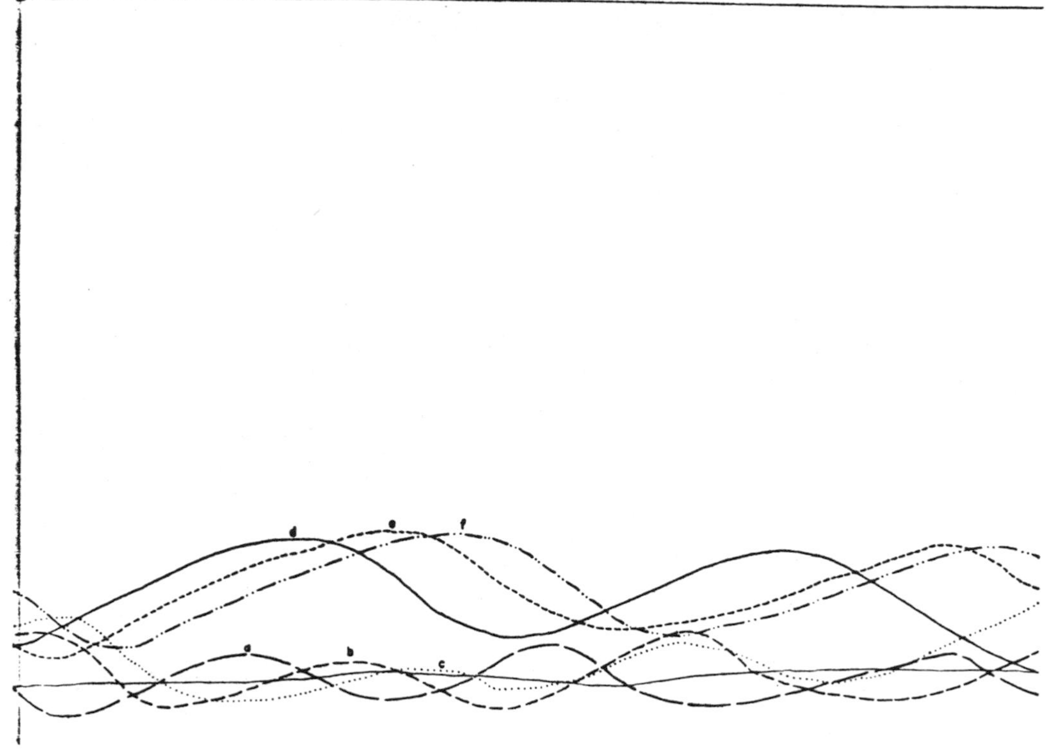

waves per minute, and this frequency was measured in terms of transmission speeds. For convenience in discussion, therefore, wave action is hereafter referred to as weak when it has a frequency of 16 waves or less per minute; as moderate when 20 to 28 waves per minute; as strong when 32 waves or more per minute.

Sources of sand were of two types (1) sediment derived from normal, sloping floor and from erosion of the shore, and no new sand introduced, and (2) sand fed artificially at constant rate into an area seaward of growing submarine bar or at base of shoreface terrace when no bar formed, conditions intended to simulate those resulting from introduction of sand by a combination of longshore and rip currents.

SHAPE AND STRUCTURE OF LONGSHORE BARS

Experiments 1 and 2—Water depth, shallow (gentle floor slope). *Wave action*, moderate. *Sand*, none added. Figures 2-*C*, 4-*C*.

Longshore bars form readily under these conditions. They start near the point of wave break, which is relatively far from shore, and build both upward and shoreward (Fig. 2-*C*). The resulting structure in cross section is a series of fore-set beds dipping shoreward at 20°–30° (Table I). For some reason, not full understood, the cross-strata normally are concave downward. Forward growth of the bar commonly buries the shoreward, ripple-marked surface on which it advances; vertical growth soon reaches water level and above, attaining a height equal to that of the maximum wave height, thus forming a lagoon to shoreward and terminating deposition in that direction. Sand bars formed under these conditions are relatively high in relation to their width; those formed experimentally have ratios of 1 to 11 (Table II; Fig. 4-*C*).

Experiments 3, 4, and 5—Water depth, shallow (gentle floor slope). *Wave action*, strong. *Sand*, none added. Figures 2-*E*, 3-*A*.

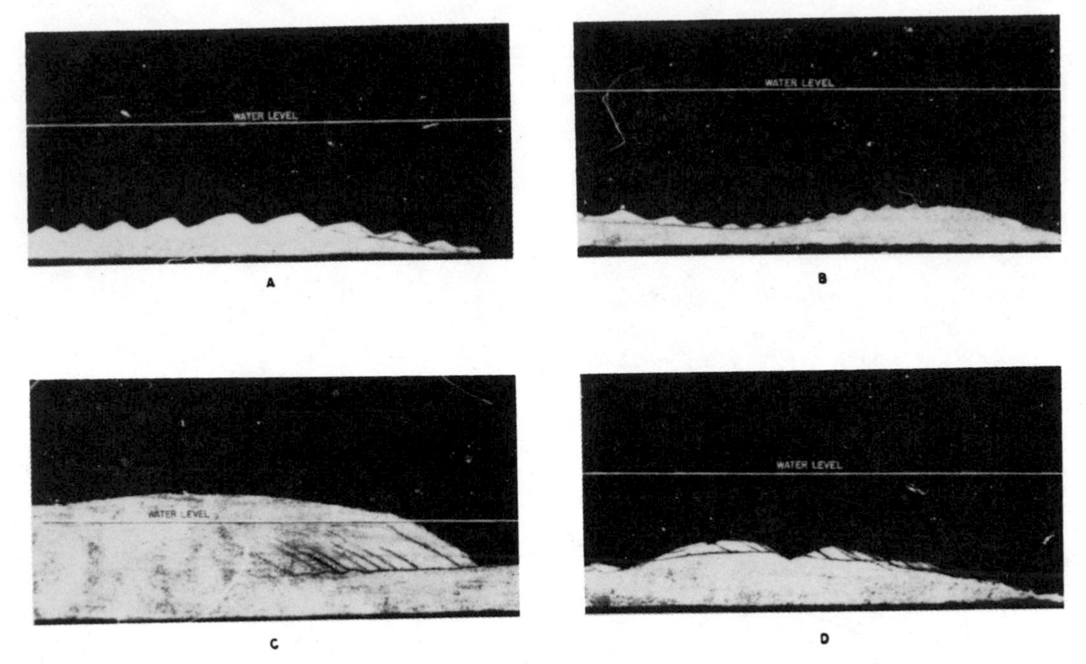

FIG. 2.—Cross sections of longshore bars formed in laboratory. Shore to right.

A, *B*—Low mounds formed during early stages of bar growth on flat, ripple-marked sand floor.
C—Bar formed with shallow water and moderate wave action; sand derived from seaward floor only.
D—Bar formed with moderate water depth and weak wave action; sand derived from seaward floor only.
E—Bar formed with shallow water and strong wave action; sand derived from seaward floor only.
F—Bar formed with moderate water depth and moderate wave action; derived from seaward floor only.
G—Bar formed with deep water and strong wave action; sand derived from seaward floor only.

Where shallow depth is maintained but intensity of wave action is increased to "strong," a longshore bar begins to form farther from shore than in the preceding experiment—a result of the point of wave break being in deeper water. The bar builds upward and shoreward, but mostly shoreward because of stronger wave action than before. Commonly, the upper surface of the bar barely attains water level, but the bar builds shoreward all the way to the beach and even buries the lower part of the foreshore. The resulting structure in cross section consists of a series of relatively long fore-set beds, dipping shoreward at 10°–20°, and concave downward (Figs. 2-*E*, 3-*A*; Table I). Its ratio of height to width as determined in the laboratory is about 1 to 28 (Table II).

Experiment 6—Water depth, moderate (moderate floor slope). *Wave action*, weak. *Sand*, none added. Figure 2-*D*.

Under these conditions a longshore bar will accumulate only after considerable time, and it will remain as a submarine mound, rising but a short distance above the original sand floor. Intensity of wave action apparently is too slight to allow further accumulation and the bar moves back and forth locally by alternate erosion and deposition along its margins. Its structure consists of shoreward-dipping strata with angles of 28°–30° (Fig. 2-*D*). Its ratio of height to width is 1 to 9.

Experiments 7 and 8—Water depth, moderate (moderate floor slope). *Wave action*, moderate. *Sand*, none added. Figures 4-*A*, 2-*F*.

At moderate depth as in the preceding experiment, but with stronger wave action (moderate strength), longshore bars will grow to water level and advance shoreward, much like those formed in shallow water when wave action is strong. Thus, an effect similar to that caused by increasing wave action in shallow water is obtained at greater depth where a submarine bar builds upward and shoreward because of an increase in

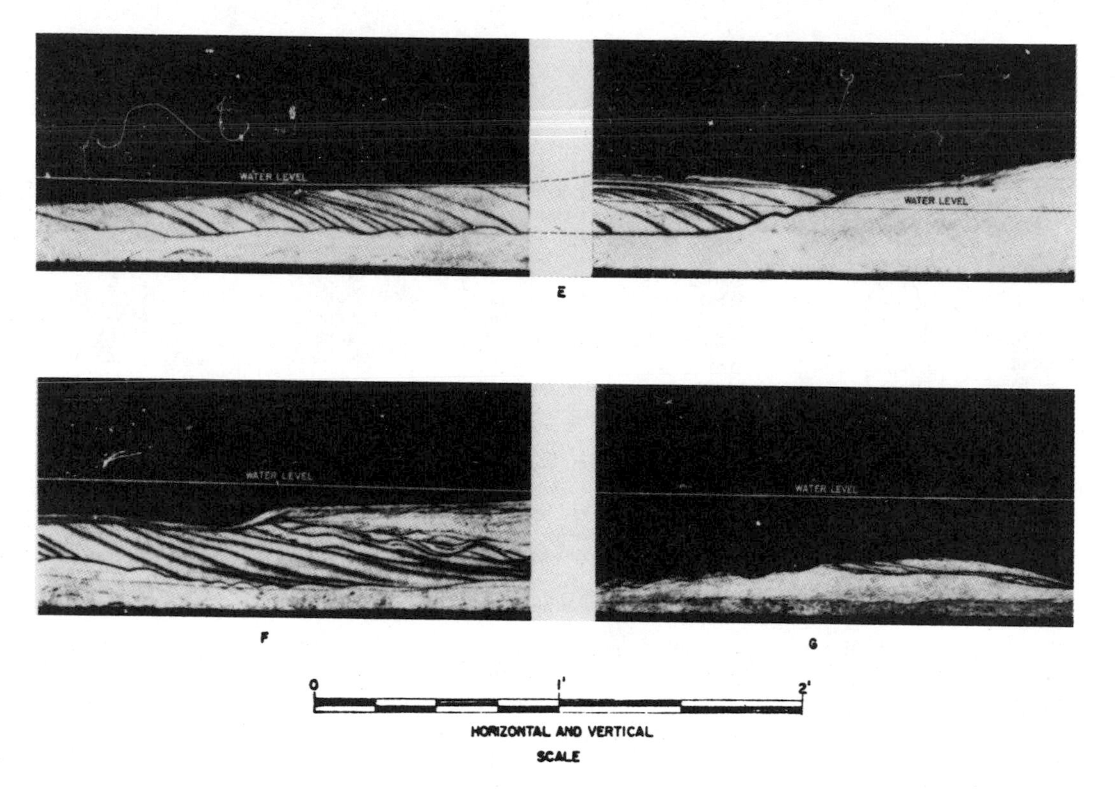

wave action (Figs. 4-*A*, 2-*F*). Such bars advance to and over the lower part of the beach; their structure includes fore-set beds that are concave downward and that dip 23°–30° shoreward. The height-to-width ratio of this type of bar as formed in the laboratory is 1 to 21.

Experiments 9 and 10—Water depth, deep (steep floor slope). *Wave action,* weak to moderate. *Sand,* none added. No figure.

In deep water with weak to moderate wave action no bar forms, regardless of the time involved.

Experiment 11—Water depth, deep (steep floor slope). *Wave action,* strong. *Sand,* none added. Figure 2-*G*.

Under these conditions submarine bars commonly form. In structure and shape (Fig. 2-*G*) they resemble those formed in water of moderate depth with weak wave action, and like them, never mature into any more than low mounds well below water level. Fore-set beds dip 20° shoreward and the ratio of height to width in the sand body is 1 to 12 (Tables I and II).

Experiment 12—Water depth, shallow (gentle floor slope). *Wave action,* moderate. *Sand,* artificial seaward feeding. Figure 4-*B*.

With new sand constantly being furnished artificially on the seaward side of a growing submarine bar, conditions are attained which simulate natural bar growth from the introduction of sand by longshore and rip currents. In shallow water, with wave action moderate, sand-feeding to seaward results in building a bar upward and slightly seaward until it reaches water level where it becomes stabilized (Fig. 4-*B*). Cross sections normal to the shore show an original core or nucleus composed of relatively steep, shoreward-dipping beds like those of preceding experiments. Superimposed on this core are low-angle beds dipping gently to seaward (Table I). The overall shape of the bar is represented by a height-to-width ratio of 1:10.

TABLE I. ANGLE OF DIP OF CROSS-STRATA IN BARS[a]

Exp. No.	Environment			Seaward-dipping Strata (degrees)	Shoreward-dipping Strata (degrees)
	Water Depth	Intensity of Wave Action	Sand Available		
1	Shallow	Moderate	Bottom only	None	28–30
2	Shallow	Moderate	Bottom only	None	20
3	Shallow	Strong	Bottom only	Horizontal	10–15
4	Shallow	Strong	Bottom only	None	14–20
5	Shallow	Strong	Bottom only	None	15–20
6	Moderate	Weak	Bottom only	None	28–30
7	Moderate	Moderate	Bottom only	None	23–30
8	Moderate	Moderate	Bottom only	None	25–30
9	Deep	Weak	Bottom only	No bar	No bar
10	Deep	Moderate	Bottom only	—	—
11	Deep	Strong	Bottom only	None	20
12	Shallow	Moderate	Seaward feeding	6–7	16–25
13	Shallow	Strong	Seaward feeding	4–6[b]	17–25
14	Deep	Moderate	Seaward feeding	No bar	No bar
15	Deep	Strong	Seaward feeding	—	—

[a] In fifteen experiments completed, not every possible combination of the three variables could be included, but data for experiments omitted probably can be estimated reasonably well on the basis of the combinations selected.
[b] Terrace front—30°.

Experiment 13—Water depth, shallow (gentle floor slope). *Wave action,* strong. *Sand,* artificial seaward feeding. Figure 4-*D*.

Under these conditions, the resulting structure is relatively complicated (Fig. 4-*D*). The waves move some sand forward over the original submarine core to form shoreward-dipping strata; the remaining sand accumulates as nearly horizontal beds on top, or is deposited to seaward and dips gently in that direction. Thus, the bar grows larger in three directions until it reaches sea level and can build farther only in a seaward direction. On the seaward side of the bar the beds dip 4°–6°; on the shoreward side they dip 17°–25°. Height-to-width ratio is 1:19.

Experiments 14 and 15—Water depth, deep (steep floor slope). *Wave action,* moderate to strong. *Sand,* artificial seaward feeding. No figure.

TABLE II. HEIGHT-TO-WIDTH RATIOS OF LONGSHORE BARS AS DETERMINED IN LABORATORY EXPERIMENTS

Exp. No.	Environment			Dimensions		Ratio H:W
	Water Depth	Intensity of Wave Action	Sand Available	Height (feet)	Width (feet)	
1	Shallow	Moderate	Bottom only	0.35	3.8	1:11
2	Shallow	Moderate	Bottom only	0.4	4.6	1:12
3	Shallow	Strong	Bottom only	0.25	7.0	1:28
4	Shallow	Strong	Bottom only	0.25	7.0	1:28
5	Shallow	Strong	Bottom only	0.25	7.0	1:28
6	Moderate	Weak	Bottom only	0.2	1.8	1:9[a]
7	Moderate	Moderate	Bottom only	0.42	8.8	1:21
8	Moderate	Moderate	Bottom only	—	—	—
9	Deep	Weak	Bottom only	None	None	None
10	Deep	Moderate	Bottom only	—	—	—
11	Deep	Strong	Bottom only	0.2	2.3	1:12[a]
12	Shallow	Moderate	Seaward feeding	0.6	6.2	1:10
13	Shallow	Strong	Seaward feeding	0.5	9.5	1:19
14	Deep	Moderate	Seaward feeding	None	None	None
15	Deep	Strong	Seaward feeding	—	—	—

[a] Submarine bar.

TABLE III. ANGLE OF DIP OF TOP-SET BEDS ON BEACHES AND FORE-SET BEDS ON SHOREFACE TERRACES

Exp. No.	*Environment*			*Top-set Beds Beach (degrees)*	*Fore-set Beds Shoreface Terrace (degrees)*
	Water Depth	*Intensity of Wave Action*	*Sand Available*		
4	Shallow	Strong	Bottom and shore	7–9	27–28
6	Moderate	Weak	Bottom and shore	9–10	29–30
7	Moderate	Moderate	Bottom and shore	9–11	27–30
9	Deep	Weak	Bottom and shore	8–12	29–31
10	Deep	Moderate	Bottom and shore	11–13	28–31
15	Deep	Strong	Seaward feeding	12–15	30

In deep water, even with strong wave action and with considerable offshore sand-feeding, no bar accumulates. All sand moves steadily into the beach area.

SHAPE AND STRUCTURE OF BEACHES AND ADJACENT SHOREFACE TERRACES

Experiment 4—Water depth, shallow (gentle floor slope). *Wave action,* strong. *Sand,* none added. Figure 3-*B*.

Beaches in shallow water areas (gently sloping floors) normally have only a short period of growth because longshore bars build up rapidly to seaward and change those factors responsible for beach accumulation. Where wave action is weak the bars form barriers, thus eliminating wave action to shoreward and causing lagoons to be formed; where wave action is strong, the bars build shoreward and sometimes partly bury the beach. In laboratory experiments shallow-water beaches washed by strong wave action are built up largely as top-set beds with gentle seaward dips of 7°–9°. Shorter but steeper dipping (27°–28°) fore-set deposits are formed in the shoreface area to seaward, but they cannot be considered a part of the beach proper (Fig. 3-*B;* Table III), for they are below low water level.

Experiment 6—Water depth, moderate (moderate floor slope). *Wave action,* weak. *Sand,* none added. Figures 3-*E*, 3-*C*.

Under these conditions a complex structure pattern is formed. It consists of relatively long, high-angle fore-set beds below water level and low-angle top-set beds partly above and partly below (Fig. 3-*E*). The beach grows seaward relatively far because no longshore bar of sufficient height to eliminate wave action is formed. Stabilization is attained, however, when sand advances shoreward in the form of migrating ripple marks and mingles with the sand of fore-set beds that are building seaward. This mingling results in a zone where opposing forces produce alternations in deposition and erosion, and it results in complex interfingering structures (Fig. 3-*C*). In the laboratory such composite sand bodies have height-to-width ratios of approximately 1 to 12 (Table IV).

TABLE IV. HEIGHT-TO-WIDTH RATIOS OF SAND BODIES COMPOSED OF BEACH AND SHOREFACE TERRACE DEPOSITS COMBINED

Exp. No.	*Environment*			*Dimensions*		*Ratio H:W*
	Water Depth	*Intensity of Wave Action*	*Sand Available*	*Height (feet)*	*Width[a] (feet)*	
4	Shallow	Strong	Bottom and shore	0.09	1.5	1:17
6	Moderate	Weak	Bottom and shore	0.30	3.5	1:12
7	Moderate	Moderate	Bottom and shore	0.28	3.0	1:11
9	Deep	Weak	Bottom and shore	0.46	4.0	1:9
10	Deep	Moderate	Bottom and shore	0.57	6.0	1:10
15	Deep	Strong	Seaward feeding	0.45	4.5	1:10

[a] Measured horizontally from base of terrace fore-sets to approximate shoreward pinchout.

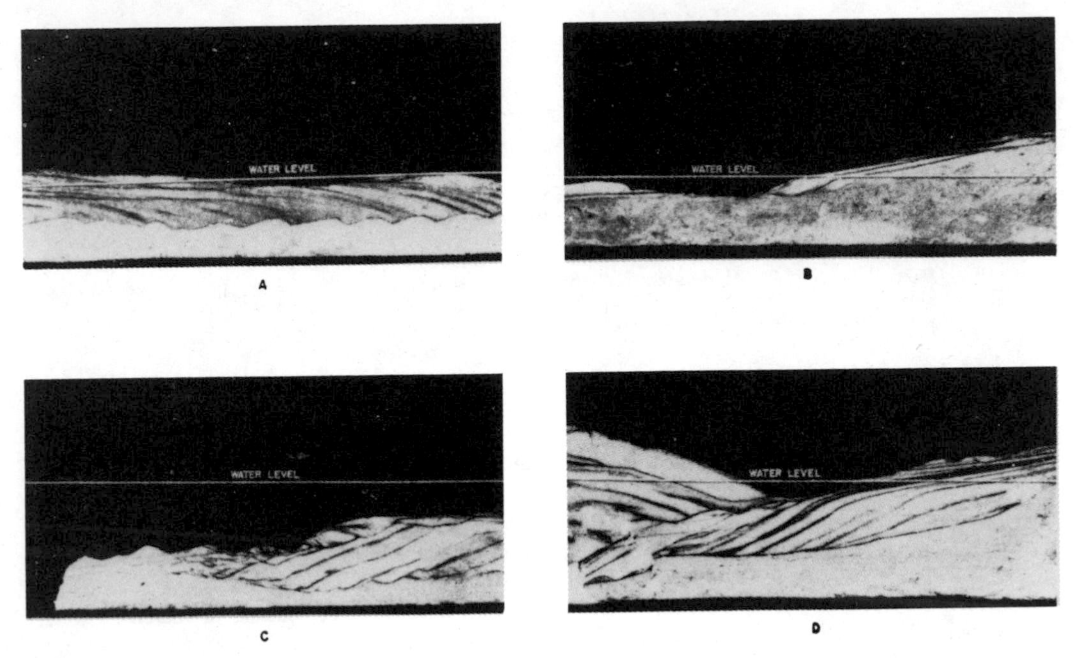

FIG. 3.—Cross sections of beaches, shoreface terraces, and a longshore bar formed in laboratory. Shore to right.

A—Bar formed with shallow water and strong wave action; sand derived from seaward floor only.
B—Beach formed with shallow water and strong wave action; sand derived from shore.
C—Shoreface deposits formed with moderate depth and weak wave action; sand derived from sea floor.
D—Beach (with fore-set beds of bar to left) formed with moderate depth and moderate wave action; sand derived from shore.
E—Beach and shoreface deposits formed with moderate water depth and weak wave action; sand derived from shore.
F—Beach and shoreface deposits formed with deep water and weak wave action; sand derived from shore.

Experiment 7—*Water depth*, moderate (moderate floor slope). *Wave action*, moderate. *Sand*, none added. Figure 3-*D*.

Beaches formed in the laboratory under these conditions contain structure patterns resembling, in most respects, those formed by weaker waves in the preceding experiment. Because of greater wave strength, however, top-set beds here build to a greater thickness and at a higher angle (9°–11°). Further, the ultimate seaward growth of these associated beach and shoreface deposits is limited by advancing sand of a longshore bar that approaches and overrides them (Fig. 3-*D*). The height-to-width ratio of the beach and shoreface deposits upon stabilization is about 1 to 11 (Table IV).

Experiment 9—*Water depth*, deep (steep floor slope). *Wave action*, weak. *Sand*, none added. Figure 3-*F*.

In this environment the process of forming gently dipping top-set beds on the beach proper and steeper dipping fore-set beds in shoreface areas (Fig. 3-*F*), is similar to that described for other environments. Only a relatively thin series of top-set beds is formed because the waves are weak and both top-set and associated fore-set beds dip at lower angles than those formed under conditions of stronger wave action. Because longshore bars, which eliminate wave action near shore, do not form under conditions of deep water, beach extension seaward continues longer than otherwise and is terminated only when waves can

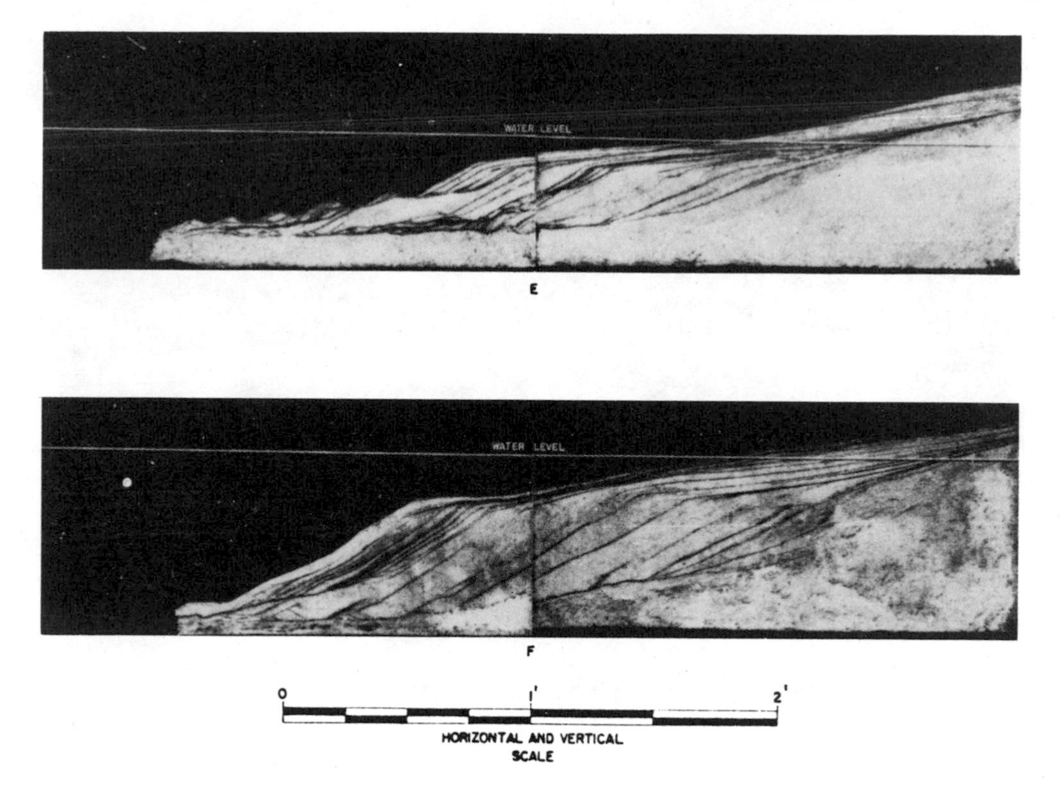

no longer obtain a supply of sand from the shore. The ratio of height to width in this type of sand body is 1 to 9 (Table IV).

Experiment 10—Water depth, deep (steep floor slope). *Wave action*, moderate. *Sand*, none added. Figure 4-*E*.

A final experiment in beach construction with no sand introduced offshore was in deep water (steeply sloping floor) with moderate wave action. The resulting sand body shows, in cross section normal to the strand (Fig. 4-*E*), a very thick series of top-set beds. Both top-set and fore-set beds have relatively steep dips—11°-13°, and 28°-31°, respectively. The height-to-width ratio obtained in the laboratory was 1 to 10.

Experiment 15—Water depth, deep (steep floor slope). *Wave action*, strong. *Sand*, artificial seaward feeding. Figure 4-*F*.

Continuous feeding of sediment at base of the shoreface terrace, intended to simulate conditions of sand introduction by longshore and rip currents, results in a complex form and structure pattern (Fig. 4-*F*). Experiments with deep water (steeply sloping floor) and with strong wave action result in a conflict of forces in which advancing waves move sand shoreward to meet other sand carried seaward by backwash. Top-set beds on the beach, which are comprised of shore-derived sand, dip seaward at 12°-15° and form a landward-thickening wedge. To seaward, meanwhile, newly introduced sand advances landward forming a complicated ripple and scour-and-fill structure. This deposit builds up as rapidly as the sand supply permits. Much of the new sand, moreover, is carried farther shoreward, ultimately contributing to construction of fore-set beds as the

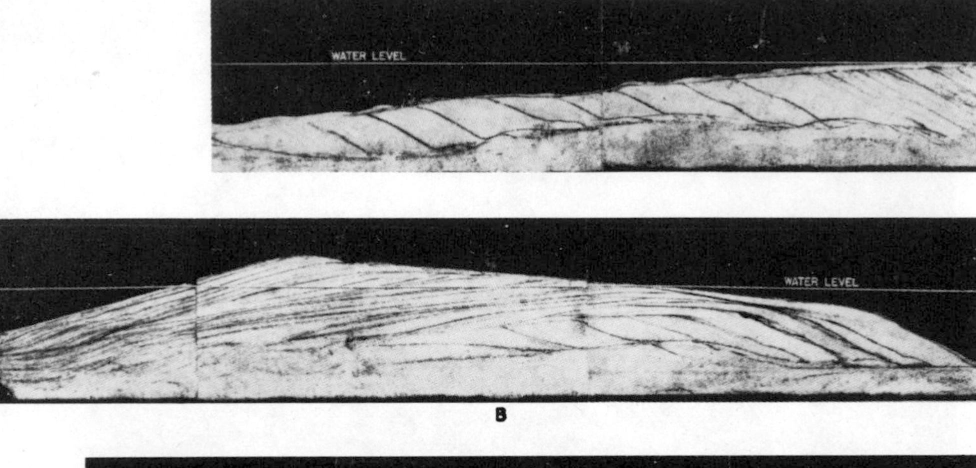

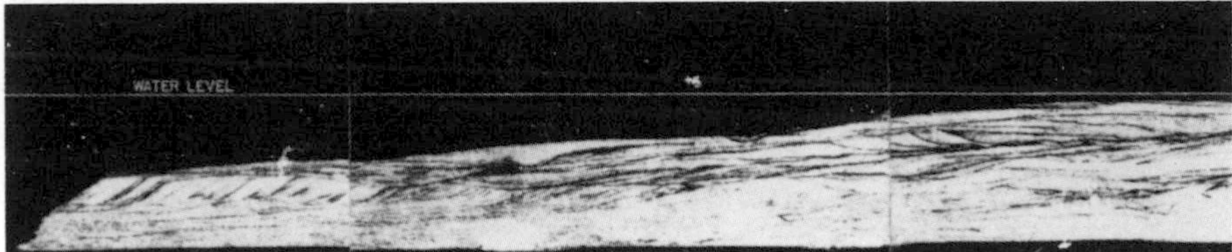

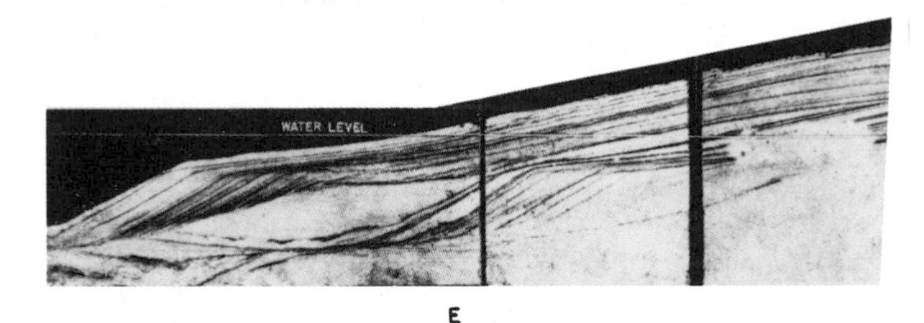

FIG. 4.—Cross sections of longshore bars, shoreface deposits, and beaches formed in laboratory. Shore to right
A—Bar formed with moderate depth and moderate wave action; sand derived from seaward floor only.
B—Bar formed with shallow water and moderate wave action; sand fed to seaward.
C—Bar formed with shallow water and moderate wave action; sand derived from seaward floor only; illustrates shape but not structure.
D—Bar formed with shallow water and strong wave action; sand fed to seaward.
E—Beach and shoreface deposits formed with deep water and moderate wave action; sand derived from shore.
F—Beach and shoreface deposits formed with deep water and strong wave action; sand fed to seaward.

backwash redeposits this sand in the form of a shoreface terrace.

The wedge shape of top-set beds seems to result from the continuous compression of the seaward parts of these beds. This compression results from the force of waves which break near shore because of deep water (steeply sloping floor) and which exert greatest pressure in the area referred to. Intensity of the force increases as the fore-set beds build out from shore, causing the point of wave break to be progressively farther to seaward.

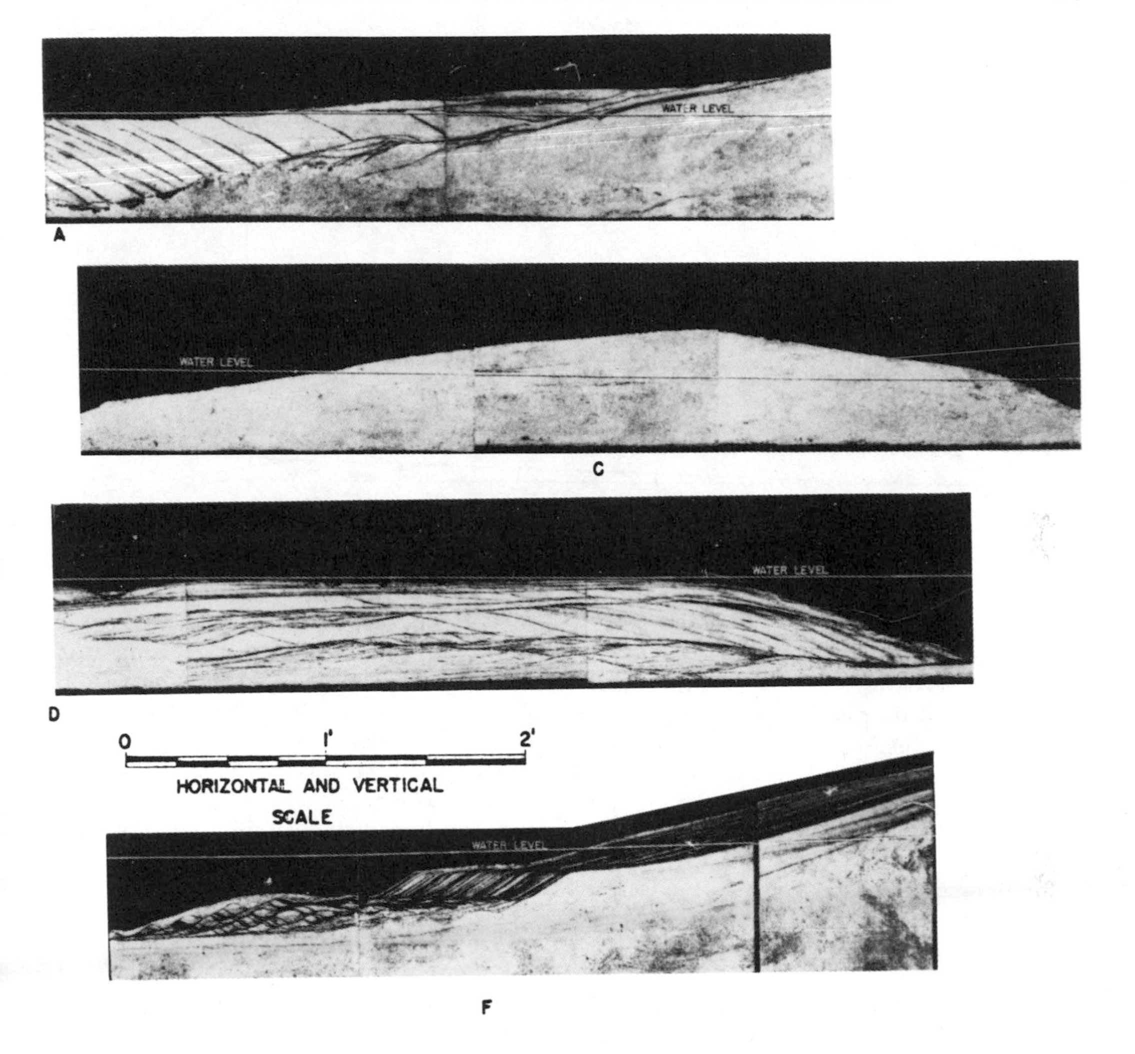

STRUCTURES IN SOME MODERN BEACHES AND BARS

A comparison of foreshore beach structures formed during laboratory experiments with those formed on modern beaches at several localities shows a close similarity. In both actual and experimental beaches the strata dip gently seaward with long, even slopes; in both, the structure pattern as shown in cross section is due largely to slight differences in the dip of strata in adjacent sets (Thompson, 1937, p. 732). Only in degree of dip is any noticeable difference detected; those strata formed in the experimental laboratory are, in general, somewhat steeper than those of natural beaches. Studies of foreshore beach strata on the Gulf Coast of Texas, for example, show a range from 2° to 5°, and on the coast of southern California between 7° and 10° (McKee, 1957, p. 1708–1715). By way of comparison, foreshore beach strata deposited in laboratory experiments dip from 7° to 15° (Table III).

No data are available, insofar as the writers know, concerning structures on modern shoreface terraces. In the laboratory, as shown in Table III,

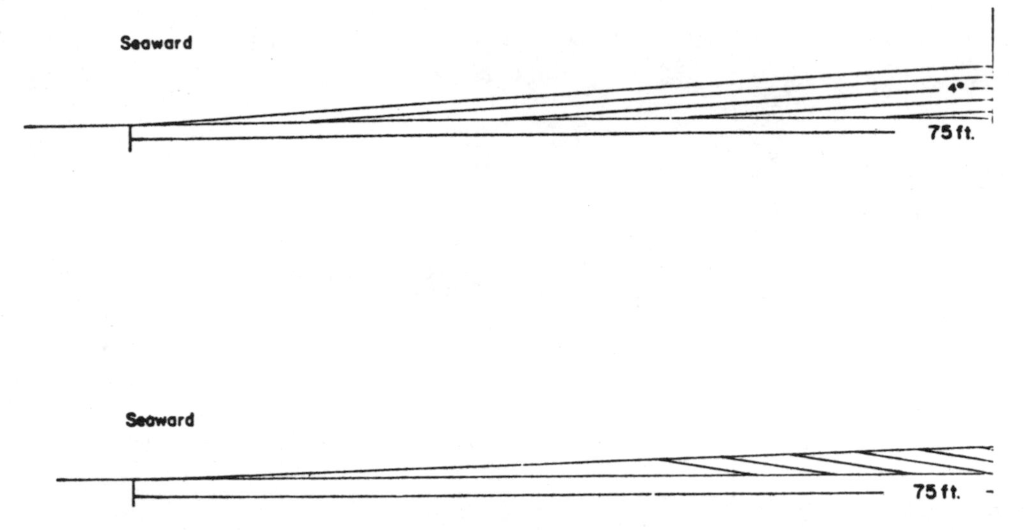

strata on the front of this terrace dip 27° to 31° seaward, forming in cross section a pattern which in most respects is not unlike that made by fore-set beds of a delta. Differences in beach strata formed under similar conditions, suggest that these dips probably are somewhat greater than those formed in nature. Confirmation of this comparison, however, must await field investigations.

Stratification of some longshore bars formed in the laboratory may be compared with that in bars examined by the senior author and C. Teichert in 1959 off the north coast of Bimini in the Bahama Islands. The Bimini bars lie roughly parallel to the shore; they stand above low-tide level and are barely awash during high tide. Cross sections trenched normal to the shoreline during low tide showed that large parts of the bars were formed mainly of strata dipping shoreward at 16° to 20°, but in some places, seaward of the bar crest, upper strata dipped seaward at 4° to 5° (Fig. 5). Thus, the structures in these modern bars resemble those formed in the experimental laboratory under conditions of shallow water (gently sloping floor), artificial sand feeding to seaward, and moderate wave action (Fig. 4-*B*).

Longshore bars, other than the type just discussed that builds to the surface and advances shoreward, have been described by numerous geologists, but no information concerning their structure seems to be available. Submarine bars that have accumulated on the relatively steep slopes off southern California (Fig. 6) have been examined by Shepard (1950), and those that have accumulated in Lake Michigan have been studied by Evans (1940) and others.

Bars that have built above high water level in the shallow waters of the Texas coast and that have compound structures resulting from the accretion of beach and dune deposits, are discussed by Shepard and Moore (1955) and by Fisk (1959). On the basis of sample data from many borings, most of the sediment underlying Padre Island, for example, is considered (Fisk, 1959, Fig. 13; Table 2) to be shoreface sand; overlying the shoreface sand on the present barrier is beach sand, and above it dune sand, indicating a seaward advance of these facies (Fig. 7). Unfortunately, data describing stratification in this sequence of sands are meager.[7]

The present general lack of information concerning primary structures both in modern shoreface deposits and in longshore bars makes impossible any detailed or systematic comparison

[7] The volume "Recent Sediments, Northwest Gulf of Mexico," edited by Shepard, Phleger, and van Andel (1960, Am. Assoc. Petroleum Geologists, Tulsa, Okla.), appeared as this paper *was* in press.

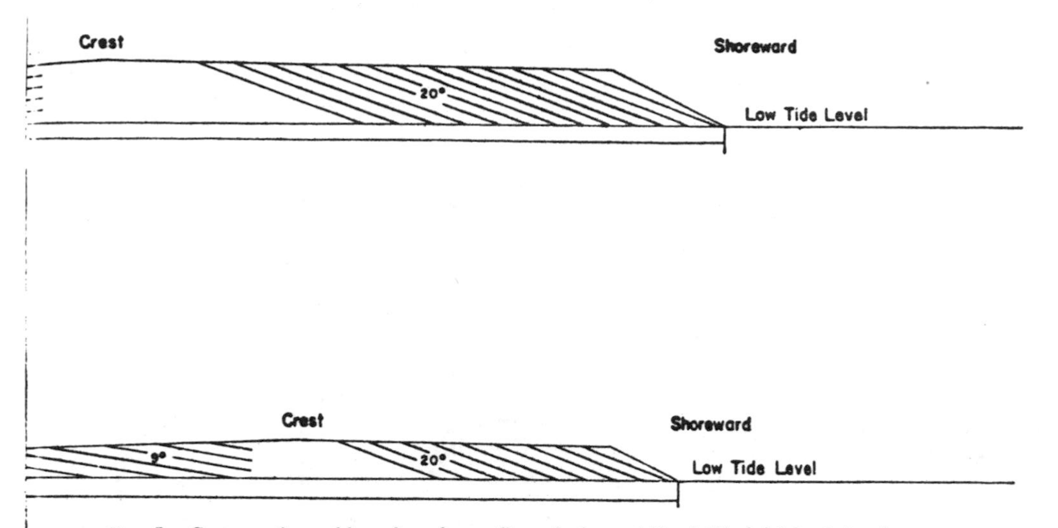

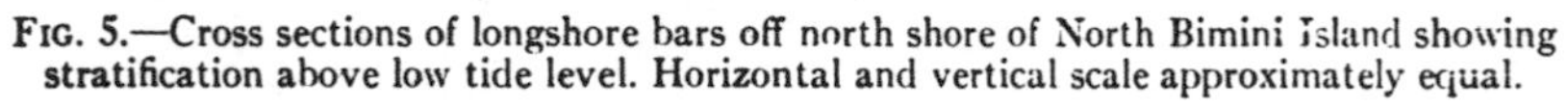

FIG. 5.—Cross sections of longshore bars off north shore of North Bimini Island showing stratification above low tide level. Horizontal and vertical scale approximately equal.

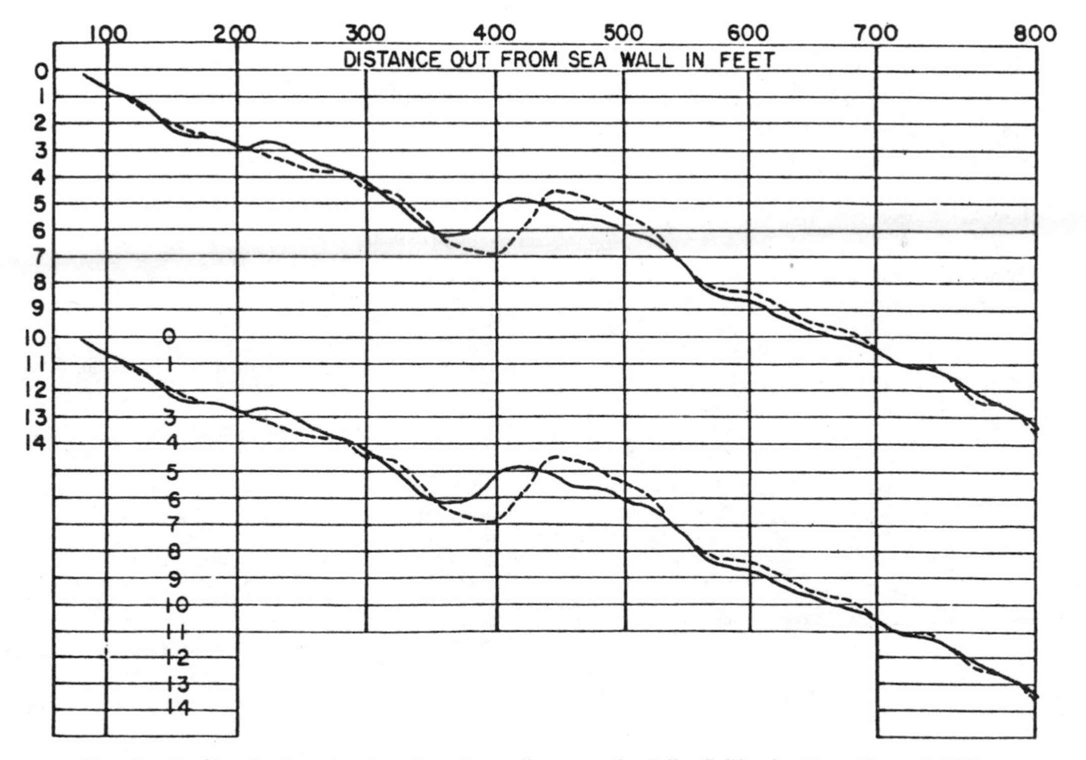

FIG. 6.—Profiles of submarine, longshore bars off coast at La Jolla, California. From Shepard, 1950.

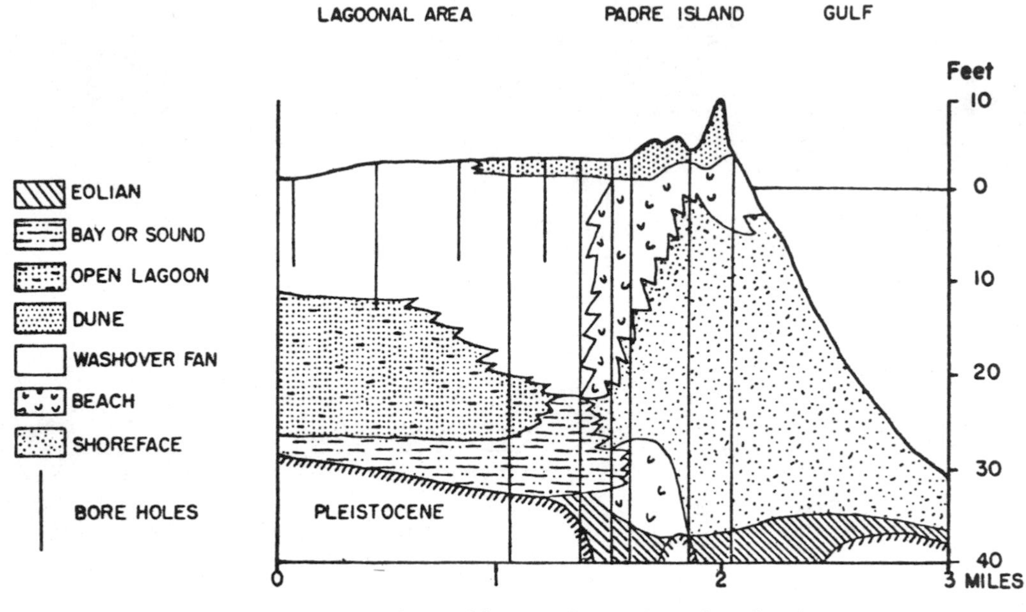

FIG. 7.—Cross section of post-Pleistocene shore and nearshore deposits at Padre Island, Gulf Coast of Texas. From Fisk, 1959.

between them and their supposed counterparts formed in the laboratory. Thus, the suggestion is clear that an important field for future investigation lies in a precise recording of details concerning structure patterns that are representative of these sand environments. The obvious next step, then, is the interpretation of ancient sand bodies through comparison with the modern and with the laboratory examples.

REFERENCES

Davis, W. M., 1909, Geographical essays: Boston, p. 708.

Evans, O. F., 1940, The low and ball of the eastern shore of Lake Michigan: Jour. Geology, v. 48, no. 5, p. 476–511.

——— 1942, The origin of spits, bars, and related structures: Jour. Geology, v. 50, no. 7, p. 846–865.

Fisk, H. N., 1959, Padre Island and the Laguna Madre flats, coastal south Texas: 2nd Coastal Geog. Conf., Coastal Studies Inst., Louisiana State Univ., Baton Rouge, p. 103–151.

Gilbert, G. K., 1890, Lake Bonneville: U. S. Geol. Survey Mon. 1, 438 p.

Johnson, D. W., 1919, Shore processes and shore-line development: New York, John Wiley and Sons, Inc., 584 p.

Keulegan, G. H., 1948, An experimental study of submarine sand bars: U. S. Beach Erosion Board Tech. Rept. 3, 40 p.

McKee, E. D., 1957, Primary structures in some Recent sediments: Am. Assoc. Petroleum Geologists Bull., v. 41, no. 8, p. 1704–1747.

Price, W. A., 1951, Barrier island, not "offshore bar": Science, v. 113, p. 487–488.

Scott, Theodore, 1945, Sand movement by waves: U. S. Beach Erosion Board Tech. Mem. 48, 37 p.

Shepard, F. P., 1950, Longshore-bars and longshore-troughs: U. S. Beach Erosion Board Tech. Mem. 15, 32 p.

——— 1952, Revised nomenclature for depositional coastal features: Am. Assoc. Petroleum Geologists Bull., v. 36, no. 10, p. 1902–1912.

——— and Moore, D. G., 1955, Central Texas coast sedimentation: characteristics of sedimentary environment, recent history, and diagenesis: Am. Assoc. Petroleum Geologists Bull., v. 39, no. 8, p. 1463–1593.

Thompson, W. O., 1937, Original structures of beaches, bars, and dunes: Geol. Soc. America Bull., v. 48, no. 6, p. 723–751.

Editor's Comments on Paper 6

One of the first of a flurry of spit investigation reports that came out in the 1960s was this paper by Clarence Kidson, whose research interests include fluvial denudation chronology, movements of sediment in the littoral zone, and coastline evolution during the Quaternary. Dr. Kidson has been a professor of physical geography since 1964, and Head of the Department of Geography since 1968, at the University College of Wales, Aberystwyth, Wales. Prior to that, from 1954 to 1964, he was Head of the Coastal Research Section of the Nature Conservancy in London. Kidson was born in Yorkshire, England, in 1919, and educated at King's College, University of London. He received his B.A. in 1947, his Ph.D. in 1959, and was a lecturer at University College of the South-West, Exeter, from 1947 to 1954.

In this paper, Kidson takes up the problems encountered where longshore transport forming spits and bars is affected by local tides and currents. Of particular interest are twin spits opposite each other across an embayment. Did these grow from opposing shorelines toward one another through counterdrifting; or are they, in reality, two segments of a breached spit or baymouth bar? These, and the possible variations between them, are the subject matter of an interesting series of investigations at four sites along the British coast.

6

Sonderdruck aus
Zeitschrift für Geomorphologie
Annals of Geomorphology — Annales de Géomorphologie

Herausgegeben von H. Mortensen, Göttingen

in Gemeinschaft mit J. P. Bakker, Amsterdam / A. Cailleux, Paris / D. L. Linton, Sheffield / N. Nielsen, København / R. J. Russell, Baton Rouge / H. Spreitzer, Wien / F. E. Zeuner, London

Neue Folge Band 7, Heft 1 (1963), S. 1—22

The growth of sand and shingle spits across estuaries

by

C. KIDSON, The Nature Conservancy, Wareham

With 11 figures and 4 photos on plates

Introduction

The drift of beach material alongshore is complicated wherever rivers discharge land water into the sea and wherever inlets permit the diurnal ebb and flood of the tide to cut across the direction of longshore drift. There is ample evidence that such interruptions in the ideally smooth curves of the shoreline are not barriers to continued transport of beach material alongshore. BRUUN & GERRITSON (1959) have shown that coastal inlets are naturally by-passed by sand, and KIDSON, CARR & SMITH (1958 and 1959) have demonstrated that even exceptionally strong tidal currents into and out of estuaries do not prevent the passage of shingle. However, only a part of the beach material succeeds in crossing such obstacles. A large and varying proportion is deposited in the mouth of the inlet to form spits and bars. Where the discharge from the estuary is negligible, the deposited material succeeds in closing the gap, as for example at Slapton Ley (WORTH [1904]) and Loe Pool (TOY [1934]) in the South West peninsula of England [1]). In all other cases there is a continual battle between river and tide on the one hand, and the tendency to the deposition of both fluvial sediments and material transported alongshore, on the other. In major estuaries like the Mersey and the Humber there is, under present conditions, not the remotest possibility that a spit or bar could close the inlet. Between this extreme and that of the completely enclosed lagoon, there is every possible gradation and variation. The complexity is increased in those cases where the opposing forces are so nicely balanced that the spit is breached at intervals either by the sea or by the river which it has succeeded in diverting.

In such naturally complex situations, it is not surprising that some confusion, and lack of agreement in interpretation, frequently arises. One of the

1) All locations referred to in the text are shown either on the general site map (Fig. 1) or on the detailed area maps (Figs. 2—10).

still unresolved problems concerns the way in which spits, which appear to grow from opposite banks of an inlet, have evolved. Have they, in fact, grown in opposite directions, or is their form relict in nature, and does it stem from the breaching of one spit, thus giving the appearance only, of growth from more than one direction? The present paper is intended to review opinions currently

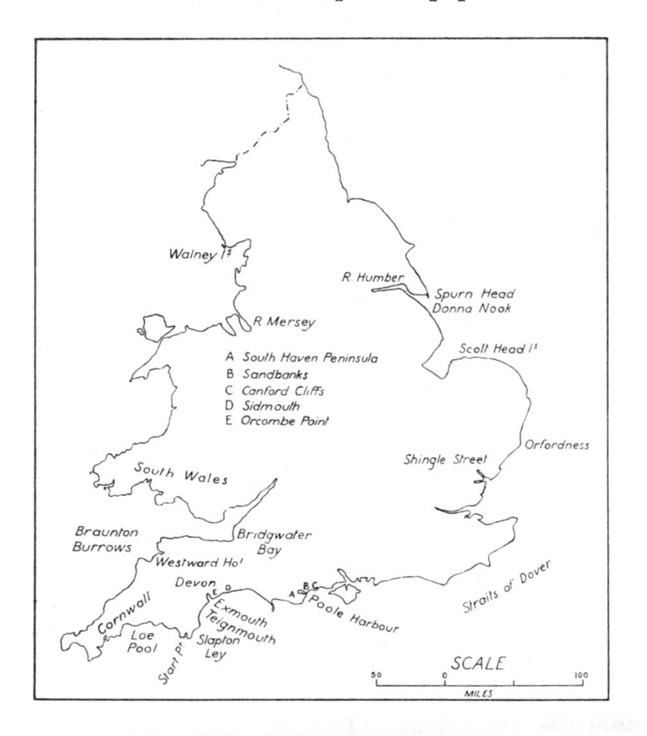

Fig. 1. Location Map

held and to put forward experience gained from detailed work on British coasts which seems to call for a modification of current views. While the examples are all from the British Isles the results are of general application.

Present views on spit growth

ROBINSON (1955) has argued strongly in favour of uni-directional growth followed by breaching. STEERS (1948, p. 415) considered that such a process "seems to be the most reasonable explanation" for the depositional features on either flank of the Humber entrance, Spurn Head on the northern shore and Donna Nook on the southern. On the face of it one could scarcely have a more obvious explanation even though the Humber example appears to be a very extreme case. In most areas growth is clearly dominantly from one direction and in many localities there is a history of repeated breaching.

There are, however, difficulties in accepting this simple explanation as holding true universally. There are good grounds for believing that in certain cases it is unacceptable. STEERS (1948, p. 89) referring to the recurves at the northern and southern ends of Walney Island, Lancashire, suggests that a local counter drift may be responsible for their contrary growth. He compares Walney Island, in this respect, with Scolt Head Island, Norfolk (STEERS [1948], pp. 358—361). Where spits in adjacent estuaries appear to grow in opposite directions, it is clearly not possible to invoke the breaching hypothesis. GRESSWELL (1957, p. 85) cites the examples of the spits across the mouths of the Rivers Exe and Teign in Devon. Here, within 5 miles of each other, the Denn at Teignmouth, and Dawlish Warren, opposite Exmouth, point in diametrically opposed directions. Gresswell suggests: "Perhaps a small headland between the two is the dividing line, and waves coming from E. S. E. drive material away from the headland in both directions".

If growth in opposing directions can take place in adjacent estuaries, there would seem to be reasonable grounds for expecting it to happen sometimes within the confines of a single inlet. WARD (1922, p. 105) suggested that the double spits across many of the harbour entrances on the south coast of England have, in fact, grown towards each other.

It is the purpose of this paper to put forward the results of recent detailed work on spits on the east, south and west coasts of England as a contribution to the solution of this difficult but interesting problem.

One of the underlying difficulties which the exponents of the "Breach hypothesis" appear to have to contend with is an inability to concede the existence or the importance of longshore drifting of beach materials in more than one direction, either along adjacent stretches of shoreline or under differing conditions on a single beach. Thus, ROBINSON (1955, p. 35) is unwilling to accept the opinions of WARD (1922, p. 97) and STEERS (1948, p. 289) that a local counter drift may be responsible for the building of the Sandbanks spit in the opposite direction to South Haven peninsula at the entrance to Poole Harbour. He appears to misinterpret the reply of the Rev. G. H. WEST (1885, pp. 427—428), to a British Association questionnaire. This reply, which, insofar as it is concerned with beach drifting, is concerned with shingle only, is confined to the area well to the east of the harbour mouth. West states that "gravel which has come from the cliffs (Canford Cliffs to Christchurch) distinctly travels east". ROBINSON transformed this opinion into a statement refuting the views of Ward and Steers, on what is predominantly a sand spit, as follows: "These views are at variance with the findings of the British Association (1885) Committee appointed to investigate various aspects of erosion along the south coast. At Poole.... the dominant direction of beach drifting along the whole length of the shoreline adjacent to the Harbour entrance is stated to be to the east or north-east, depending on the local trend of the coastline. No hint of the existence of any counter drift is given in the reports".

KING (1959, p. 376) discussed Robinson's paper and concluded "This hypothesis, put forward by Robinson, seems more satisfactory than other theories which suggest that this type of spit is due to local counter drift resulting from irregularities along the shore. It allows the drift of beach material to be continuously in one direction which agrees with observations and is supported by cartographic evidence".

It is the view of the present writer that while double spits resulting from breaching undoubtedly exist, the hypothesis is only partially applicable, that

1*

beach drifting is rarely, if ever, uni-directional and that any theory of spit evolution must give great weight to counter drifting.

Theoretically it ought to be possible to resolve many of the difficulties in interpretation in model experiments in wave tanks. In practice, however, so many independent variables, known and unknown, are involved, that this expedient is impracticable (INMAN & NASU [1956]). Only two methods of examining the problem are really possible. Firstly map evidence can be used to trace developments over the last century or so. For pre-19th century information, maps must always be treated with great reserve (CARR [1962]). Secondly, field observations and experiments on the growth of spits and the movements of their constituent materials provide evidence of the highest value. Such field trials suffer, however, to some extent, from the limitation that they are, of necessity, relatively short term. An ideal solution is the combination of both these lines of approach.

Growth of spits across a new inlet: Bridgwater Bay, Somerset

Only rarely is it possible to trace the growth of spits across an inlet from a new beginning. Such an opportunity arises, however, on the coast of Somerset, where a major large scale breach, initiated in 1783 (LOCKE [1792]), cut off a considerable area of the alluvial lands at the mouth of the River Parrett and created Stert Island. The breach widened between 1783 and 1840 (as shown by map evidence). The subsequent growth of the mainland spit at Stert Point (which later incorporated Fenning Island), and of the spits on Stert Island, are very relevant to the present study. Figures 2, 3 and 4, trace the growth of the spits across the gap, from its inception to the present day.

Fig. 4. illustrates the 20th century developments and shows clearly that the width of the inlet has been reduced, by the growth of spits from both flanks, to less than a third of its maximum width. The air photograph (Plate I) gives a vivid picture of the present position. The cartographic evidence is, by itself, proof of independent growth in opposite directions, since reliable maps exist for the whole period of development. There is, however, strong supporting evidence, not only in the alignment of the shingle recurves at Stert Point and of the spit at the southern end of the island, but also from direct observations of the movement of beach material. At Stert Point itself, observations carried out over a period of a year, with shingle markers, show conclusively that movement and spit building is from west to east. Table I summarises part of these experiments and shows a steady eastward movement from the point of injection of the marked material to the end of the shingle spit some 1200 feet away.

WARREN (1956) showed that beach material produced by erosion on Stert Island drifts south, that is, in a direction opposite to that at Stert Point, and is incorporated in the spit on the southern end of the island. Repeated surveys at annual intervals, carried out by the Nature Conservancy, have also shown that, at the northern end of the island, the great mass of residual shingle is steadily overwhelming the Spartina immediately to the south of it. Every indication is, then, of movement towards the south. Direct observations support in a very striking way, the conclusion arrived at from strong cartographic evidence, that the Stert Point and Stert Island spits are growing towards each other. This is not the appearance of growth resulting from the breaching of a single feature but incontrovertible proof of real growth from opposite directions resulting from local counter drifting.

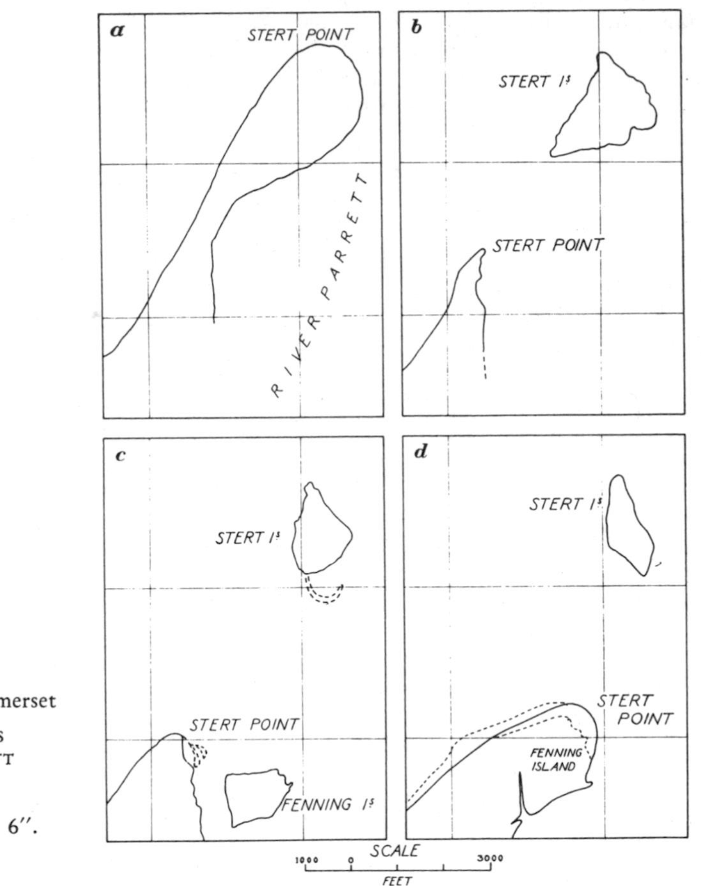

Fig. 2. Bridgwater Bay, Somerset

a) 1782 DAY and MASTERS
b) 1819 CHARLES CHILCOTT
c) 1853 Admiralty Chart (ALLDRIDGE)
d) 1902 Ordnance Survey 6″. 1st Revision

Evolution of opposing spits across a bayhead estuary where no history of breaching exists: The Taw-Torridge Estuary, Devon

Where long straight beaches, such as those on parts of the east coast of England, are studied, it is easy to visualise the steady progression of beach material in one direction. Even where occasional headlands supervene and divide a coastline into broad sweeping arcs, as along the south coast from Start Point to the Straits of Dover, it may be possible to reconcile unidirectional longshore drift and the existence of spits which appear to be built in opposite directions. Where, however, the coastline is crenulate and headlands, acting as gigantic groynes, occur every few miles, it is clear that the effects of local irregularities in coastal outline will assume greater significance than any theoretical consideration such as the angle between the direction of approach of the dominant waves and a generalised coastal alignment. The Atlantic coasts of Devon and Cornwall show

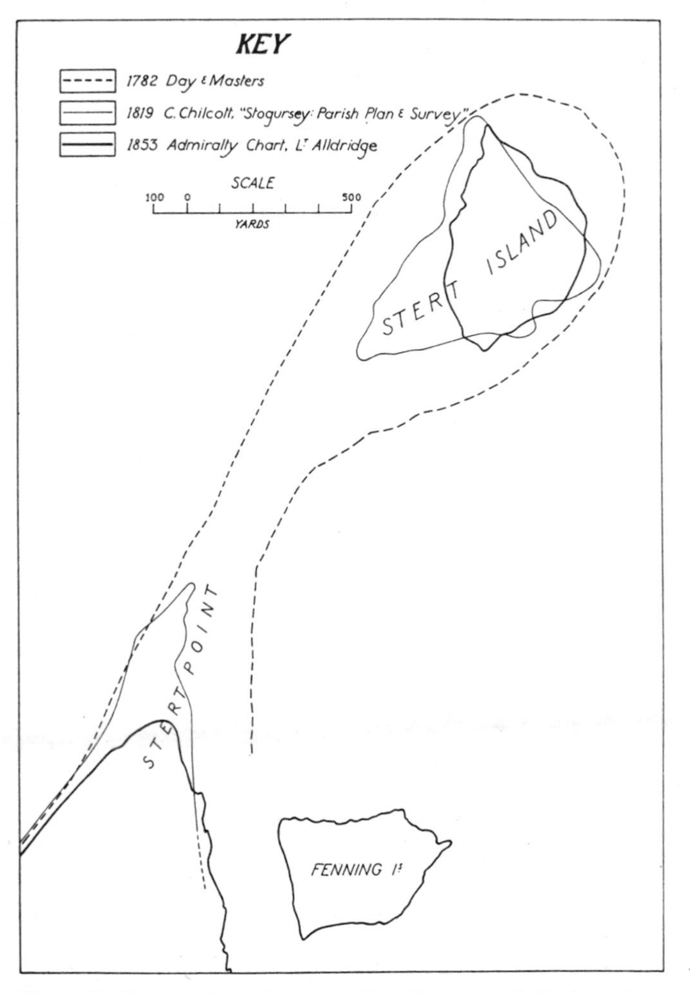

Fig. 3. Bridgwater Bay, Somerset. Development of the breach

a succession of small bays and estuaries, separated by headlands and promontories, where the longshore movement of beach material must be largely local, and where no one direction can be regarded as dominant [1]).

WARD (1922, p. 58) suggested many years ago that "a headland of sufficiently marked projection not only mechanically opposes the longshore drift of

---

[1]) This must not be taken to imply that headlands are a complete barrier to sand transport (see below). For example TRASK (1955) has demonstrated movement round headlands in southern California, in depths of up to 30 feet.

beach material but may interrupt the transporting currents which determine its direction of motion and generate currents moving landward along each flank from its outer end". More recently DAVIES (1959), in discussing bay head beaches

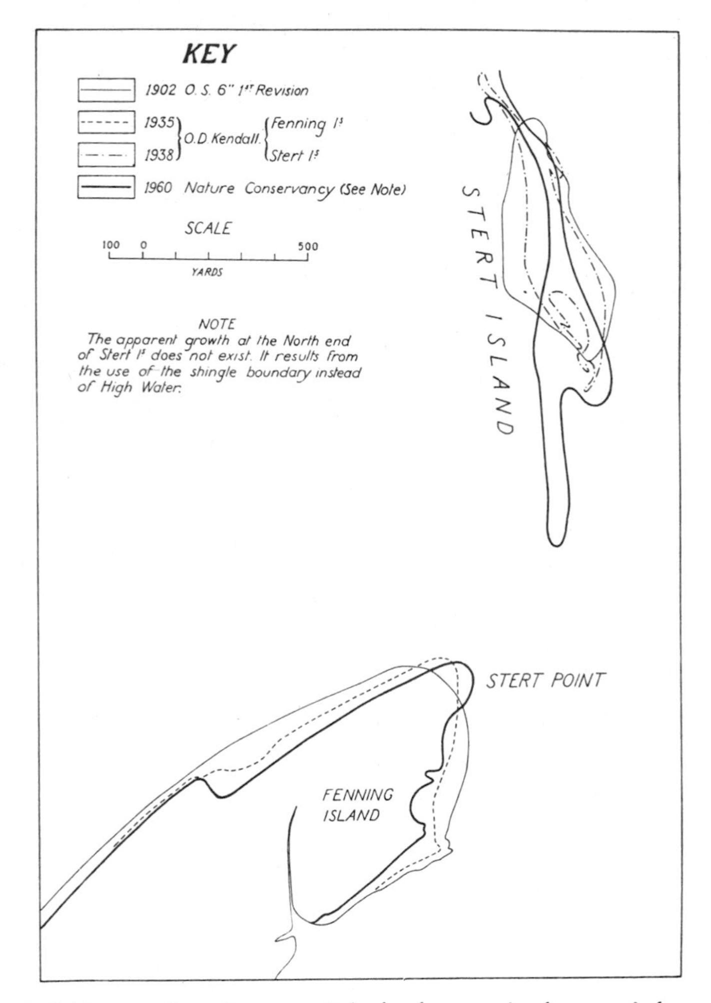

Fig. 4. Bridgwater Bay, Somerset. Spit development in the twentieth century

wrote "Lack of concordance between beach curve and refraction curve virtually always implies that the former is sharper than the latter. When this is so, waves will approach the beach obliquely, thus causing beach drifting and longshore currents from either end of the beach toward the centre". While the differing uses of the term "currents" by these two authors are confusing, it is not necessary in this context to discuss the relative functions of wave and current in longshore transport. The important point is that if their views on longshore movement are valid assessments of shoreline conditions, spits and bars must grow in opposing

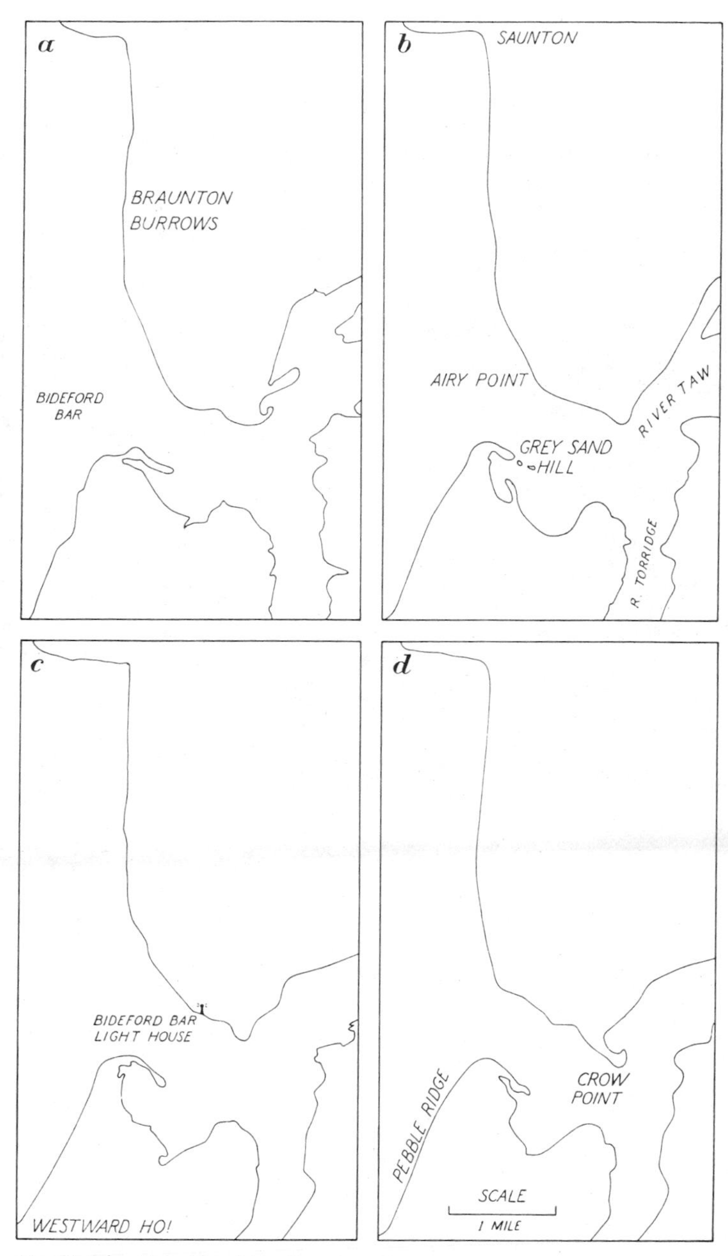

Fig. 5. The Taw-Torridge Estuary, Devon

a) 1808 MURDOCH MACKENZIE
b) 1850 (circa) Ordnance Survey 1″.
1st Edition (revised) incorporating DENHAM's 1832 coastline
c) 1855 Admiralty Chart (ALLDRIDGE)
d) 1903 Ordnance Survey 6″. 1st Revision

Plate I.
Oblique aire-photograph of Stert Point and Stert Island. J. K. Joseph (No. OB2. 25. vi. 1955) Crown Copyright Reserved

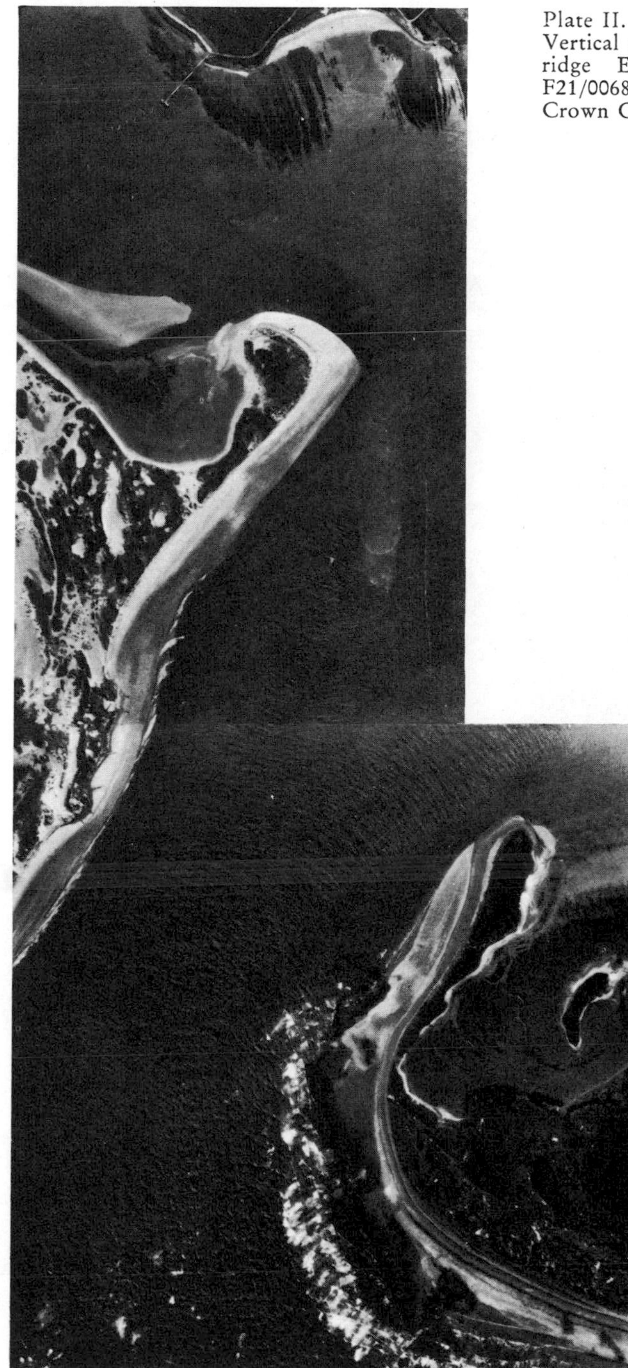

Plate II.
Vertical air-photograph of the Taw-Torridge Estuary. (58/RAF/2776. Prints F21/0068 and F22/0047 30. iv. 1959) Crown Copyright Reserved

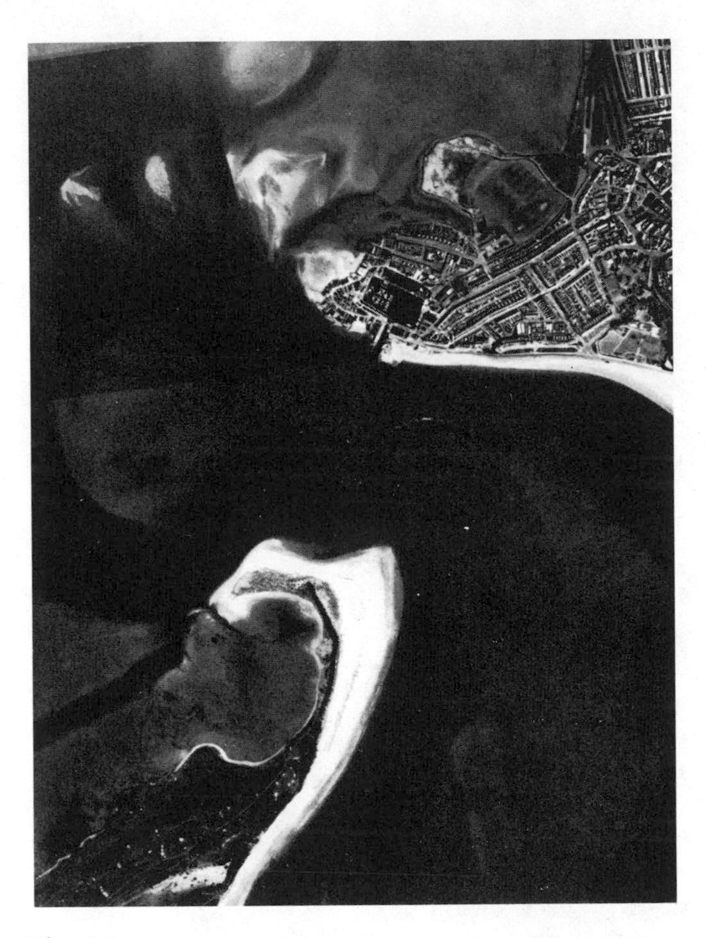

Plate III.
Vertical air-photograph of the Exe Estuary. (58/RAF/2776. F22/ 0025 abd 0026. 30. iv. 1959) Crown Copyright Reserved

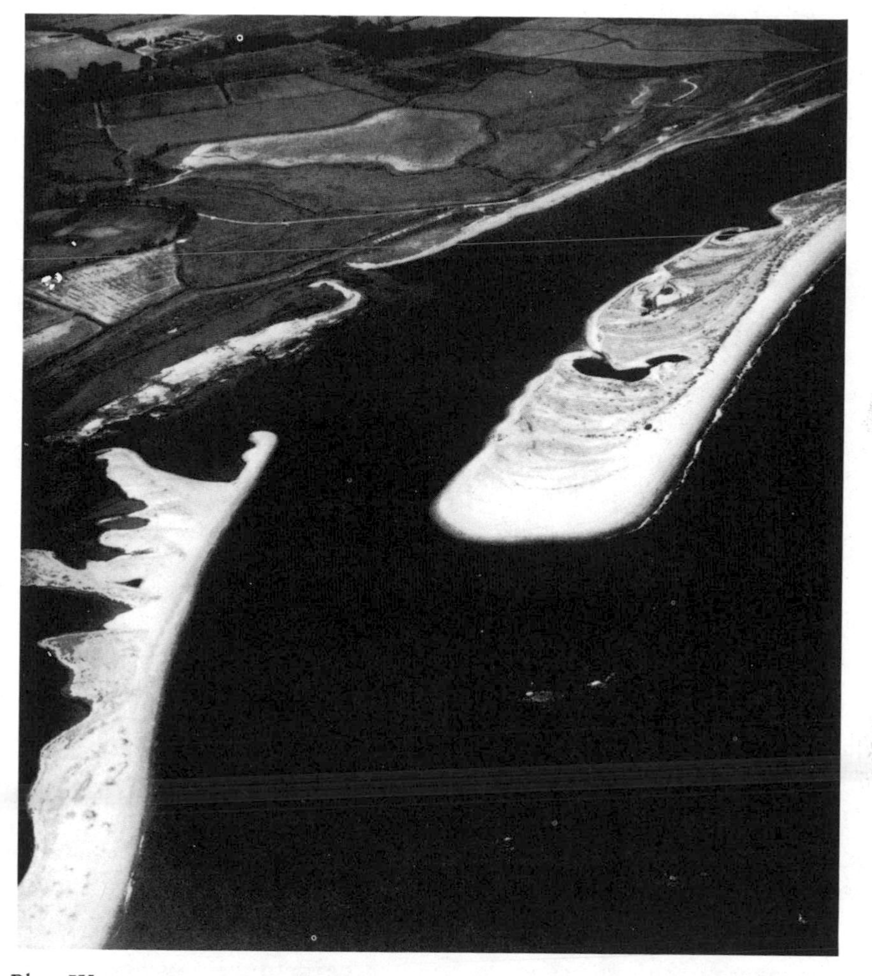

Plate IV.
Oblique air-photograph of the entrance to the River Ore. J. K. Joseph (KJ. 52. 16. vi. 1959) Crown Copyright Reserved

directions from opposite flanks of a bay. There are many mid bay bars which support such a conclusion. For example, KING (1959, p. 374) quotes the case of those in Dingle Bay, Eire. Bay head spits will clearly be built from both directions under similar conditions, where an estuary, entering at the head of a bay, prevents the formation of a continuous bay head beach.

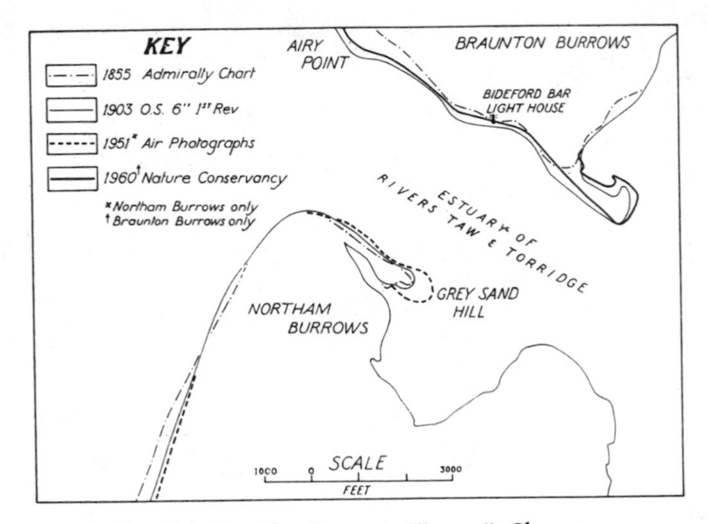

Fig. 6. The Taw-Torridge Estuary. "Recent" Changes

The great sand dune system of Braunton Burrows and the Westward Ho Pebble Ridge in Devon, on the northern and southern flanks respectively of the Taw-Torridge estuary, appear to have been built independently, from opposite directions, in just such a bay head situation. No record exists which suggests that the "Breach hypothesis" needs to be invoked to interpret these features. Every map, since that of Saxton of 1575, shows a situation essentially similar to that of to-day. The relative stability of the area is clearly shown in Figure 5, which gives outlines of the two spits at intervals during the 19th century, and in Figure 6 which gives superimposed outlines of the areas of greatest change in the very mouth of the estuary in the hundred years to the present day. This very stability and a history free from serious breaches in either feature, in itself suggests that there is no need to look for any hypothetical explanation of their origin. The interpretation which naturally springs to mind is that spits have grown from either flank of the bay and have formed the base for wind blown sand to create the dunes which now exist. Such evidence as there is on the drift of beach material supports these conclusions.

In an examination of the beach sands of this coast, STUART & SIMPSON (1937) showed that constituent minerals decreased in frequency in shore sands with increasing distance from their origin. Thus minerals derived from the granite masses of Cornwall diminished northwards, suggesting a tendency to northerly drift. Certain other minerals occur only in the north of the area and increase in frequency northwards. Some of these, for example, enstatite, hypersthene, clino-enstatite and clino-zoisite are not of Cornish or Devonian provenance. Most of them are present in the dune sands of South Wales. The inference is that they originated in the north of the area, perhaps in glacial times, but that their decreasing frequency in the shore sands southwards in present day beaches indicates a drift tendency in that direction matching the northerly drift suggested by the granite minerals. The presence of these far travelled minerals demonstrates that headlands are by no means complete barriers to the drift of beach materials, particularly of the finer fractions.

The recurved ends of the two spits, Crow Point on the north of the river outlet, and Grey Sand Hill on the south, as shown in Figure 6 and Plate II, clearly stream down drift. The groynes at Westward Ho indicate drift at the top of the beach, from the south, and those close to Bideford Bar lighthouse, on Braunton Burrows, show a drift from the opposite direction. STUART & HOOKWAY (1954) showed at Westward Ho, that marked pebbles on the pebble ridge moved steadily northward and that the products of the erosion of the southern end of the spit resulted in the building of new ridges near the northern end. STUART (1959) showed that pebbles from the Westward Ho Pebble Ridge are not carried across the estuary but form a recurved ridge towards Grey Sand Hill.

While there is no direct evidence from long term drift experiments, on the Braunton side of the estuary, all the available information points to a drift to the south. STUART (1959) observed "The shift of detritus is from Airy Point to Crow, and I have myself seen very large quantities moved from Airy Point to Crow in three days".

There appears to be a marked tendency to accretion in the Airy Point region (see Figure 6) which is matched by a tendency to erosion in the north. This is a long term trend and in recent years there has been some erosion between Airy Point and Crow Point. This has been ascribed locally to the effects of sand and gravel extraction in the estuary, at present being carried on at the rate of 80,000 to 100,000 tons per annum. A large part of this is won from the Crow Point area. The extraction of gravel complicates the picture but does not appear to be a major influence (see STUART [1959]). The picture is also complicated by the fact that here, as in most areas of dunes, much beach material is removed by wind (see KIDSON & CARR [1960]).

In this large estuary, then, (see Plate II) two major spits, with blown sand dunes superimposed, appear to have developed from opposing directions and with no suggestion that there is any need to account for their position, form or history in terms of breaching of a single spit.

Spit evolution related to breaching: The Exe estuary

The spits at the mouth of the River Exe, in Devon, Dawlish Warren and Exmouth Point (Plate III) give an example of repeated breaching of one spit which would appear to be completely unrelated in any direct way, to the growth of the second spit on the other side of the estuary. Figure 7 gives the outlines of the two features at intervals throughout the last century. In 1809 the breaching of Warren Point is clearly shown but by 1839 the spit was fully recovered. In

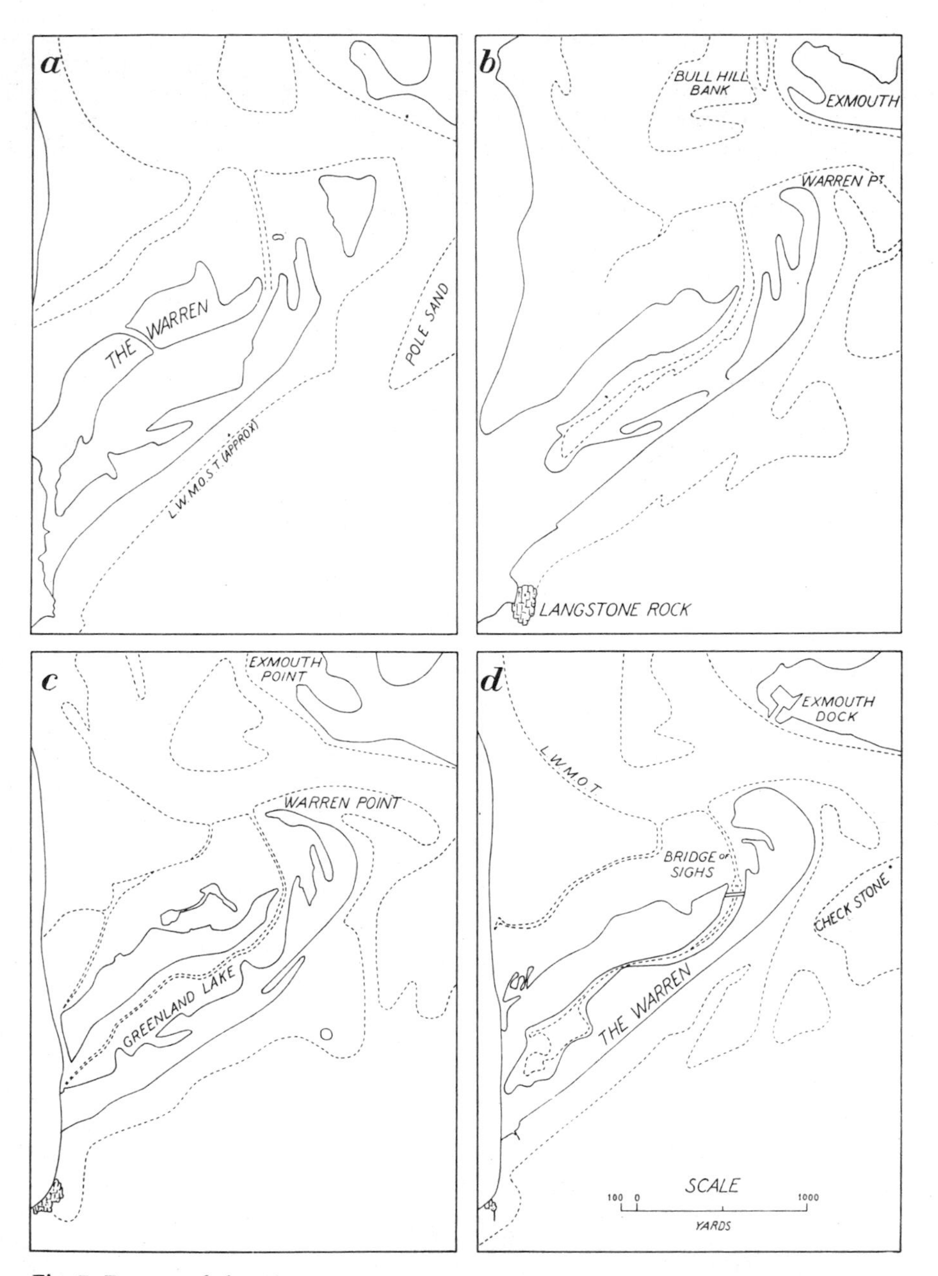

Fig. 7. Estuary of the River Exe. Devon

a) 1809 Ordnance Survey 1″. 1st Edition
b) 1839 JAMES M. RENDELL
c) 1851 Admiralty Chart (SHERINGHAM)
d) 1903/4 Ordnance Survey. 6″. 1st Revision

the Admiralty Chart of 1851, the whole of the distal point of Dawlish Warren had been removed by erosion but by 1903 the spit extended further towards Exmouth than ever before. KIDSON (1950) has shown that this periodic breaching and destruction of the eastern end of the spit is a regular occurrence. Recovery appears in the past to have been equally certain. The reality of the process of breaching and destruction followed by recovery is shown in Figure 8 in which recent outlines are superimposed on that taken from the 1851 Admiralty Chart.

During the whole of the period covered by reliable maps and charts, changes in the Exmouth spit even including man made changes such as the excavation of

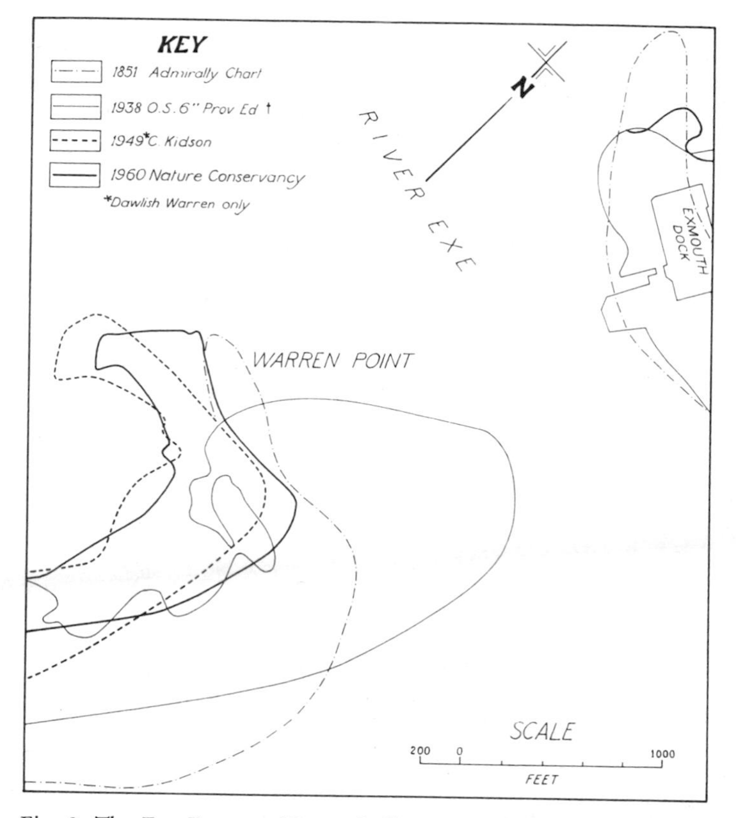

Fig. 8. The Exe Estuary. "Recent" Changes

the harbour, have been minor by comparison with those of Dawlish Warren. There is no cartographic evidence of dramatic increases in the size of the Exmouth spit to match the sudden change on the Warren. The channel of the River Exe, which between the two spits is never shallower than 20 to 25 feet even at low water springs, has remained in its present position throughout these seemingly cyclic fluctuations. There is quite clearly no question of the development of Exmouth Point following the attachment of a severed part of the Dawlish Warren spit to the eastern shore.

Spit evolution and breaching: The entrance to the River Ore

The spits on either side of the entrance to the River Ore in Suffolk (Plate IV) combine many of the growth features of those already discussed. STEERS (1926) has traced the evolution of Orford Ness by the drift of material from the north, and has also (1948, p. 389) indicated the part played by the breaching of the distal point of the spit (North Weir Point) in the genesis of the feature at Shingle Street as follows, "Here (i. e. at North Weir Point) the spit

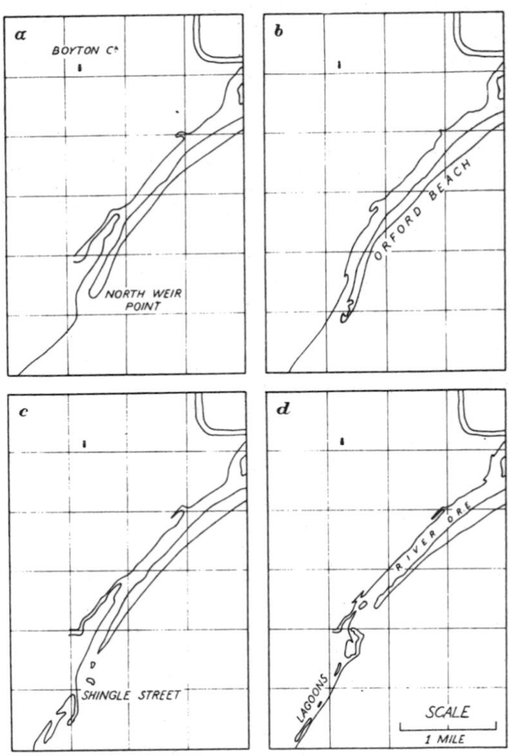

Fig. 9.
The Entrance to the River Ore, Suffolk
a) 1847 Admiralty Chart (BULLOCK)
b) 1880 Ordnance Survey 6″. 1st Edition
c) 1902 Ordnance Survey 6″. 1st Revision
d) 1945 Air Photographs

is thin and unstable, and in the great storm of 1897, a mile or more was cut off This shingle was piled up in quantities at Shingle Street, the effect of the storm being to add to that already existing there . . .".

The history of spit development in the century to 1945 is shown in Figure 9. Between 1847 and 1880 North Weir Point was built southward. This movement was reversed by 1895 as shown by the Admiralty chart of that year (Fig. 10) and the cycle of southerly growth was, finally, terminated by the storm breach of 1897. Recession is clearly indicated by the Ordnance Survey map of 1902. On the same map, the shingle cut off from North Weir Point can be seen tied to the mainland at Shingle Street and already forming into small spits growing in both directions. The outline derived from air photographs of 1945 shows firstly that North Weir Point in very much further north than it was in 1902. The same map clearly shows also the growth of the Shingle Street spit in a northerly direction, that is, in the opposite direction to Orfordness. The contrast between the 1945 Shingle Street coastline and that of 1902 demonstrates that on this part of the coastline a northerly movement is dominant. This cartographic evi-

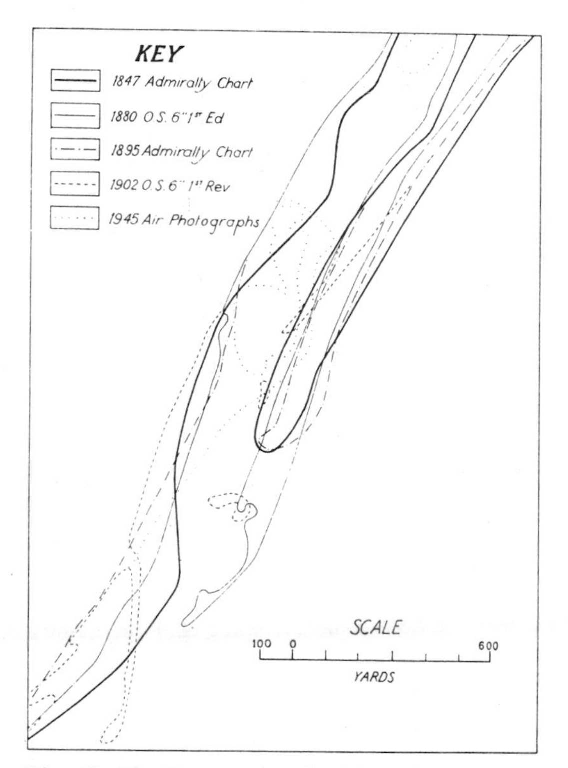

Fig. 10. The Entrance to the River Ore. Changes 1847—1945

dence is shown superimposed in Figure 10 and in this form indicates clearly that breaching is a periodic occurrence at North Weir Point and that much of the material for spit building at Shingle Street is derived in this way. The detailed processes of the growth of spits in opposite directions are shown, for the period 1945 to 1960, by the superimposed outlines given in Figure 11. This shows that the growth of North Weir Point, is accompanied by erosion on the flank of the spit on the opposite shore. The implication here, which is supported by the history of North Weir Point is that, if the "cycle" develops to its 'conclusion', as it did in the period 1880—1895, the growth of the major spit is accompanied by the destruction of Shingle Street spit, a relationship between the growth of the distal point of the major spit and shoreline recession on the opposite bank referred to by GUILCHER (1954).

Even in an example such as this, in which the growth of one spit is directly linked to and controlled by the evolution of a major spit and in which the breaching of the major spit is almost a pre-requisite of the growth of the minor, counter drifting cannot be dismissed. The Shingle Street spit does not have the appearance of having grown to the north, it does so grow. The cartographic

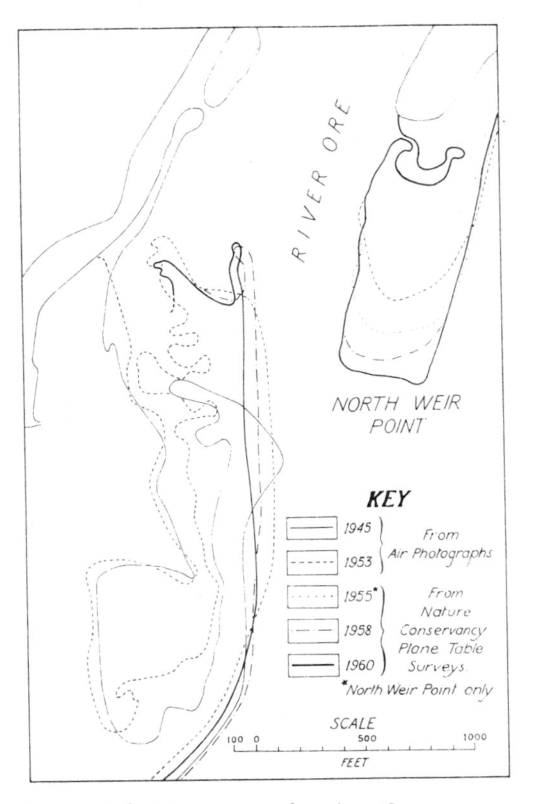

Fig. 11. The Entrance to the River Ore.
"Recent changes"

evidence by itself is overwhelming in this context, and here it is backed up by incontrovertible evidence from drift experiments. Cobb (1957) records the northerly movement of beach material on the Suffolk coast at Shingle Street, where the dominant direction, as reflected in the massive Orfordness shingle spit, would appear to be southerly. Kidson, Carr & Smith (1958), using radio-active tracers, showed that northerly drift persisted for long periods at Orford-ness. Yet that spit is clearly built from materials drifted from the opposite direction. It has been demonstrated that similar counter drifting exists elsewhere. Steers (1960) at Scolt Head Island, recorded a drift to the west under the influence of waves approaching from the north-east and an easterly drift under the influence of waves from a north-westerly quarter. Kidson (1960) has shown the same process to be operative on the Bridgwater Bay coast.

This experimental work shows that on many parts of the coast, material drifts in both directions. Usually one direction is dominant and contrary movement is cancelled out when the waves approach again from the dominant direction. However, on irregular coastlines and particularly in inlets and estuaries, some material is moved by counter drift into, as it were, sheltered positions, where it remains untouched when drift is resumed in the dominant direction. It is here suggested, that the Shingle Street spit is one such "sheltered" locality and that many of the spits which grow in similar situations can be accounted for in a similar way.

A traverse round the British coasts will show that spits growing in opposing directions are much more common than might be supposed. Frequently islands lying close inshore, such as Scolt Head Island and Walney Island, which have their opposite ends washed by the ebb and flood of the tides of neighbouring estuaries, give proof positive that building can take place in opposite directions. Occasionally major spits in adjacent estuaries, for example, the Exe and the Teign, give the lie to any dogmatic insistence on unidirectional drift. Spits which grow towards each other from the opposite shores of inlets or estuaries seem to be as common, and perhaps more common, than single barriers. In many cases the complementary spit passes almost unnoticed because it is often dominated by a much larger structure. How many geographers, who speak familiarly of Spurn Head, have ever heard of Donna Nook? Frequently the minor spit is deprived of some of its material by the presence of a major "partner" which partially oversteps it, for example Dawlish Warren and Exmouth Point. Sometimes the second spit is a transient but recurring feature which is eroded out of existence by the growth of its opposite number, only to be reborn when the larger spit oversteps its critical limit. Orfordness and Shingle Street spits illustrate this type.

Counter Drift

The detailed configuration of the coast is clearly of importance in the growth of double spits. The flood tide aided rather than opposed by wave action helps to build depositional features in the very throat of an estuary. This accounts for relatively minor details such as recurved ridges, for example Grey Sand Hill and Crow Point in the Taw-Torridge estuary, rather than for major features such as Braunton Burrows and the Westward Ho Pebble ridge. The important point is that on almost all coasts drift in more than one direction takes place. It is important everywhere and sometimes vital, but it is often overlooked. Only rarely is it possible to refer to detailed drift studies carried out on a scientific basis for a sufficiently long period to cover a wide range of conditions. Frequently the drift on any coast is determined by reference only to the accumulation in the lee of comparatively short groynes, i. e. the resultant of the drift at the top of the beach. Both laboratory and beach experiments show, however, that there are two belts of maximum movement and deposition, one at the top of the beach and the other just below the break point of the waves. This has been well shown with shingle marked with radioactive tracers (KIDSON, CARR & SMITH [1958]) and with sand marked with fluorescent dyes (ZENKOVITCH [1958]). Many casual estimations of the direction and amount of beach drifting thus often take cognisance of only one part of the process. In this connection, it is important to note that the material cast up on the top of the beach can only be cast there by waves. Currents of any kind can only operate on the lower part of the beach and they are able to carry the smaller rather than the larger fraction of beach material.

Table I

Stert Point, Shingle Drift experiment
Eastward travel from point of injection

Date	Mean distance in feet	Maximum distance in feet
28th March 1956	Injection	
24th April 1956	145	374
23rd May 1956	190	385
18th June 1956	280	1140
18th July 1956	844	1096
15th August 1956	973	1114

It has sometimes been observed that larger material can move in the opposite direction to the smaller, at one and the same time. Marsh quoted by Steers (1926, p. 325) wrote "with waves acting against the ebbing tide at Orford Ness the author has seen shingle being thrown southwards by the breaking waves, and sand in suspension being drifted northwards by the ebb".

Table II

Shingle Drift Experiment. Somerset coast Stolford Farm site.
Movements from 24th July to 19th August 1955

Marker Size	Movement in Feet (mean position of recovered markers)		
	High Tide Zone	Intermediate Zone	Low Tide Zone
Large (500 cc)	131 east	160 east	41 east
Small (250 cc)	53 east	15 west	47 west

Even with shingle, there is differential movement varying with relative size and position on the beach. Table II gives details of part of shingle drift experiments carried out on the coast at Bridgwater Bay, Somerset. Some 200 markers, the sizes of which conformed to those of the beach shingle, were laid down.

The Table shows the movement of the arithmetical mean of the positions of all the markers recovered. The percentage of recoveries was so high that burial could be discounted as having a negligible effect. The large markers all moved toward the east, further in the high tide and intermediate zone of the beach than in the low tide area. The small markers moved eastward in the high tide zone but not nearly as far as the large markers. In the intermediate and low tide zones, the net effect was a displacement in the opposite direction.

The evidence from shingle drift experiments shows that beach material moves in more than one direction. Even with large beach material, however, the experimental evidence is limited. The problems of marking and tracing sand

and silt, under littoral conditions, have not yet been overcome. The quantitative data is still restricted so that dogmatic assertions that the drift of beach material is continuously in one direction are completely unjustified.

As an illustration of the complexity of the problem, the literature on Chesil Bank, in Dorset, is revealing. COODE (1853) suggested that the material of the beach came from the west and moved to the south east. PRESTWICH (1875) argued that the pebbles came from the Portland raised beach and moved in the opposite direction. STRAHAN (1898) agreed with COODE but CORNISH (1898) came to the conclusion that material came from the west and from Portland, to the south east and that it moved in both directions. RICHARDSON's (1902) experiments seemed to show movement eastwards but, since the markers used differed materially from the beach shingle, they cannot be wholly accepted. The argument still rages. LEWIS (1938) suggested that the grading of the shingle "seems conclusive proof that there is no permanent drift in either direction". ARKELL (1947) agreed with Cornish that waves from differing directions move pebbles in opposite paths. One can but agree with STEERS (1948, p. 277) that a final answer must await the outcome of detailed experimental work on the beach. This serves to show that the direction of longshore drift is not easily understood and that "counter drift" cannot be lightly dismissed. Its significance in the evolution of spits appears to have been much under-estimated in the past.

Conclusion

The four examples considered in some detail in this paper serve to show that opposing spits which really grow towards each other are not confined to sheltered situations. It might be argued that the Exe and Teign estuaries, for example, lying at the western end of Lyme Bay, are sheltered by Start Point from the direct influence of waves of very long fetch. Braunton Burrows and the Westward Ho pebble ridge, however, could scarcely face these Atlantic waves more squarely. The spits at Stert Point and Stert Island are only approached by waves attenuated after a long passage over mud flats. North Weir Point and Shingle Street, on the other hand, are pounded by waves which break directly upon them, since depths immediately offshore can exceed 20 feet. This twinning of spits can thus take place in a variety of situations.

Breaching of a spit sometimes provides the material from which another, pointing in the opposite direction, develops. It would seem, however, that counter drift is responsible for re-sorting the material and refashioning it into a new spit. Growth in the opposite direction is real and the breaching is only a part of the process. In other cases spits grow in opposing directions under the influence of counter drift and breaching plays no part in the process. Counter drift appears to exist on most coasts. Its significance is largely a function of the detail of the coastal outline. The growth of double spits across the mouths of estuaries must always owe something to counter drifting. Only occasionally is breaching important. Only rarely is it the dominant factor.

The author wishes to acknowledge the facilities provided by the Hydrographic Department of the Admiralty, and for the use of a number of Charts and Original Roughs, in the compilation of this paper.

Maps referred to in the text.	*Appendix*
A. Bridgwater Bay	
1. 1782 The County of Somerset. Day and Masters.	K. 37. 6. 8. TAB (British Museum)
2. 1819 Stogursey. Parish Plan and Survey. CHARLES CHILCOTT.	Somerset County Records Office. Taunton.
3. 1853 Original rough of the Port of Bridgwater. ALLDRIDGE.	Admiralty Chart (BI./L 9785).
4. 1902 Ordnance Survey 6″ 1st Revision.	(Published 1904).
5. 1935/38 O. D. KENDALL	"The Coast of Somerset" (I) Proceedings of the Bristol Naturalists Society. 8(2) 1936 p. 301. "The Coast of Somerset" (II) ibid. 8(4) 1938 p. 503.
6. 1960 The Coast of Somerset from Lilstock to Stert Island. The Nature Conservancy.	Unpublished survey 1956—7 revised annually to 1960. Scale 1/1000.
B. Taw-Torridge Estuary	
7. 1808 Part of the North Coast of Cornwall and Devonshire in Bristol Channel. MURDOCH MACKENZIE Junr.	Chart 643 (Admiralty Hydrographic Office). Surveyed 1771. Corrected 1808. Graeme Spence.
8. 1832 A chart of N. E. coast of Devonshire between Hartland Point and Combe Martin including the Bar and Ports of Barnstaple and Bideford ... Lt. DENHAM.	Admiralty Chart 2127(11) originally published privately.
9. 1850 (circa) Ordnance Survey 1st Edition Revised (incorporating DENHAM's 1832 coastline).	
10. 1855 Barnstaple and Bideford. G. M. ALLDRIDGE.	Admiralty Chart. D. 1824.
11. 1903 Ordnance Survey. 6″ 1st Revision.	
12. 1960 Braunton Burrows. The Nature Conservancy.	Unpublished survey 1957—1960. Scale 1/2500.
13. 1809 Ordnance Survey 1″. 1st Edition.	
C. Exe Estuary	
14. 1839 The Mouth of the Exe. JAMES M. RENDELL.	Many copies e. g. 454. i. 5(2) (British Museum). Estate survey in the possession of the Earl of Devon.
15. 1851 Exmouth Harbour. Capt. W. L. SHERINGHAM.	Admiralty Chart 2290.
16. 1903/4 Ordnance Survey. 6″ 1st Revision.	
17. 1938 Ordnance Survey. 6″ Provisional Edition.	
18. 1949 Dawlish Warren. C. KIDSON.	Trans. Inst. Brit. Geog. 16. 1952 (1950) p. 74.
19. 1960 Dawlish Warren. The Nature Conservancy.	Unpublished Surveys 1955—1960. Scale 1/2500 and 1/1000.
D. Entrance to the River Ore	
20. 1847 Orford Haven. Capt. F. BULLOCK.	Admiralty Chart B 2720 (Original Rough L 7288).
21. 1881 Ordnance Survey 6″. 1st Edition.	
22. 1895 River Ore and Approaches. Comm. G. PIRIE.	Admiralty Chart B 5833.
23. 1902 Ordnance Survey 6″. 1st Revision.	
24. 1960 Orfordness and Shingle Street. The Nature Conservancy.	Unpublished surveys 1953—1960. Scale 1/1000.

2*

Zusammenfassung

Gezeiten-Mündungen und andere Meeresbuchten, sind keine vollständigen Hindernisse für die Wanderung von Strandmaterial längs der Küste. Durchtransport kommt vor, aber viel Material wird abgelagert in Form von Landzungen und Barren, die dahin tendieren, die Flußmündungen zu verstopfen. Kleine Flüsse werden gelegentlich zu Lagunen zurückgestaut. In allen anderen Fällen besteht ein beständiges Ringen zwischen Fluß und Gezeiten einerseits und der Ablagerungstendenz andererseits. Häufig sind Landzungen an beiden Ufern eines Ästuars entwickelt. Früher wurde angenommen, daß in solchen Fällen das Wachstum nur von einer Seite her Platz gegriffen habe und daß das Wachsen auch in die andere Richtung nur scheinbar sei. Neue Untersuchungen an Landzungen der West-, Ost- und Südküsten der Britischen Inseln haben jedoch gezeigt, daß Landzungen in benachbarten Ästuaren in gegenläufige Richtungen und sogar quer zur gleichen Bucht wachsen können. Das Vorhandensein einer „Gegenströmung" in und nahe bei Gezeiten-Mündungen wurde in Experimenten nachgewiesen, und es wurde wahrscheinlich gemacht, daß in vielen Fällen die Gegenströmung für das Wachstum der zweiten Landzunge verantwortlich ist. Beispiele für ein solches Wachsen werden beigebracht sowohl aus Gebieten, in denen Durchbrüche niemals stattgefunden haben, als auch an Stellen, wo, ungeachtet einer Fülle wiederholter Durchbrüche, diese offensichtlich in keinem Verhältnis zum Wachstum der zweiten Landzunge stehen. Zusätzlich werden Beispiele vorgelegt von Ästuaren, wo das Material für die zweite Zunge aus der Durchbrechung der ersten herrührt, wo aber das Wachstum der Zungen in gegenläufigen Richtungen kartographisch und experimentell vorgeführt werden kann. Endlich wird dargelegt, daß die Verfrachtung von Strandmaterial an jeder beliebigen Küste nur selten in einer Richtung erfolgt, und daß in vielen Buchten „Gegenströmungen" besonders bezeichnend sind.

Résumé

Les estuaires et autres baies marines ne sont pas un empêchement total à la migration des matériaux du rivage le long de la côte. Ceux-ci traversent, mais une grande partie se dépose sous forme de flèches et de barres, qui tendent à obturer les embouchures des fleuves. De petites fleuves sont à l'occasion barrés et forment des lagunes. Dans tous les autres cas, il y a lutte permanente entre fleuve et marées d'une part et tendance au dépôt d'autre part. Souvent des flèches sont rattachées aux deux rives d'un estuaire. On admettait auparavant, dans de tels cas, que la croissance ne s'était effectuée que dans une seule direction, que la deuxième flèche résultait d'une percée dans la crête, et que la croissance dans l'autre direction n'était donc qu'une apparence. Cependant de nouvelles recherches sur les flèches des côtes W, E, et S des Iles Britanniques ont montré que des flèches pouvaient croitre suivant des directions opposées dans des estuaires voisins, et par suite obliquement pour la même baie. L'expérience démontre l'existence d'un "contre-courant" dans et près des estuaires; dans de nombreux cas, ce contre-courant serait responsable de la croissance de la deuxième flèche. Des exemples d'une telle croissance sont donnés tant dans les régions où il n'y a jamais eu de percées, que là où, nonobstant un remplissage de percées répétées, celles-ci n'ont visiblement aucun rapport avec la croissance de la deuxième flèche. En plus sont étudiés des exemples d'estuaires où le materiau de la deuxième flèche provint de la rupture de la première, mais où la croissance des flèches dans des directions opposées peut être démontré par l'expérience et la carte. Finalement il est suggéré que le transport du materiau du rivage pour n'importe quelle côte n'a lieu que rarement dans *une* direction; et que dans de nombreuses baies les contre-courante sont particulierement importante.

References

1. Arkell, W. J.: The Geology of the Country around Weymouth, Swanage, Corfe and Lulworth. Mem. Geol. Surv. 1947. 340.
2. Bruun, Per, & Gerritson, F.: Natural by-passing of sand at coastal inlets. Journal of the Waterways and Harbours Division. Proceedings of the Am. Soc. of Civil Engineers, **85**, 75—107, 1959.
3. Coode, Sir J.: Description of the Chesil Bank ... Proc. Inst. Civil Engineers, **12**, 520, 1853.
4. Cornish, V.: Chesil Bank. Geogr. Journ. **11**, 628, 1898.
5. Cobb, R. T.: Shingle Street: Suffolk. Annual Report of the Field Studies Council. **1956/7**, 31—42.
6. Carr, A. P.: Cartographic record and historical accuracy. Geography **47**, 135—144. 1962.
7. Davies, J. L.: Wave refraction and the evolution of shoreline curves. Geographical Studies 5, 7—8, 1959.
8. Guilcher, A.: Morphologie littorale et soumarin. Paris, 1954, 64.
9. Gresswell, R. Kay: Beaches and Coastlines. London. 1957, 85.
10. Inman, D. L., & Nasu, Noriyuki: Orbital velocity associated with wave action near the breaker zone. Beach Erosion Board, Technical Memorandum. **79**, 2 (1956).
11. Kidson, C.: Dawlish Warren: A study of the evolution of the sand spits across the mouth of the River Exe in Devon. Trans. Inst. Brit. Geog. **16**, 69—80, 1952 (1950).
12. Kidson, C., Carr, A. P., & Smith, D. B.: Further experiments using radioactive tracers to detect the movement of shingle over the sea bed and alongshore. Geogr. Journ. **124**, 210—18, 1958.
13. Kidson, C., & Carr, A. P.: The movement of shingle over the sea bed close inshore. Geogr. Journ. **125**, 380—9, 1959.
14. Kidson, C.: The Shingle Complexes of Bridgwater Bay. Trans. Inst. Brit. Geog. **28**, 78, 1960.
15. Kidson, C., & Carr, A. P.: Dune Reclamation at Braunton Burrows, Devon.. Journ. Inst. Chartered Surveyors **93**, 298, 1960.
16. King, C. A. M.: Coasts and Beaches. London, 1959, 376.
17. Locke, R.: An essay on the subject of New Taxing Eighteen Parishes in Somersetshire (under the new name of Huntspill level) with the repairs of sea wall in Huntspill Parish, two years since cast down by the sea. Letter to George Templer Esq: 1792. — Original in the possession of W. J. Greener, Huntspill Court, Bridgwater, Somerset. — Copy at Somerset River Board, West Quay Road, Bridgwater.
18. Lewis, W. V.: The Evolution of Shoreline Curves. Proc. Geol. Assoc. **49**, 116, 1938.
19. Prestwich, J.: Phenomena of the Quarternary period in the Isle of Portland and around Weymouth. Quart. Journ. Geol. Soc. **31**, 29—53, 1875.
20. Richardson, N. M.: An experiment on the movements of a load of brickbats deposited on Chesil beach. Proc. Dorset. Nat. Hist. and Arch. Field Club. **23**, 103, 1902.
21. Robinson, A. H. W.: The Harbour Entrances of Poole, Christchurch and Pagham. Geogr. Journ. **121**, 33—50, **1955.**
22. Steers, J. A.: Orford Ness. Proc. Geol. Assoc. **37**, 306—325, 1926.
23. Steers, J. A.: Coastline of England and Wales. Cambridge 1948, 415.
24. Steers, J. A.: (Ed). Scolt Head Island. 1960, 36.
25. Stuart, A., & Simpson, B.: The Shore Sands of Cornwall and Devon from Land's End to the Taw-Torridge Estuary. Trans. Roy. Geol. Soc. Cornwall, **17**, 13—40, 1937.
26. Stuart, A., & Hookway, R. J. S.: Coastal Erosion at Westward Ho, N. Devon. Report to the Coast protection Committee of Devon County Council, 1954, 13.

27. STUART, A.: Report on the winning of sand and gravel from the Taw-Torridge Estuary with special reference to the effect of the use of grabs in the channel on the state of Instow Beach. 1959, 5.
28. STRAHAN, A.: The geology of the Isle of Purbeck and Weymouth. Mem. Geol. Surv. 1898. pp. 203—5.
29. TOY, H. S.: The Loe Bar. Geogr. Journ. **83**, 40, 1934.
30. TRASK, P. D.: Movement of Sand around southern Californian Promontories. Beach Erosion Board. Technical Memorandum, **76**, 31, 1955.
31. WEST, G. H.: in Report of the Committee on the Rate of Erosion on the sea-coasts of England and Wales. British Assoc. 55th Report, 427—428, 1885.
32. WORTH, R. H.: Hallsands and Start Bay. Report and Trans. Devon Assoc. **36**, 302, 1904.
33. WARD, E. M.: English Coastal Evolution. London, 1922, 105.
34. WARREN, A.: The coast of Bridgwater Bay. 1956. Unpublished Thesis. Fitzwilliam House, Cambridge.
35. ZENKOVITCH, V. P.: Emploi des luminophores pour l'etude du mouvement des alluvions sablonneuses. Bulletin d'information C. O. E. C. X. Anné No. 5. Mai. 1958, 243—253.

Editor's Comments on Paper 7

Our first historical study of spit and bar growth is this by A. P. Carr. Covering the period from 1945 to 1962, the paper traces changes in North Wier Point, Shingle Street, and the mouth of the River Ore in Suffolk, England. As stated by Carr, the study revealed evidence contradictory to some previously held ideas on spit development.

Since the River Ore investigation continued after the publication of the 1965 paper, the author has graciously provided an up-to-date commentary on his latest observations for publication in this volume. This additional information will be found here immediately following the original article. Points b and d in this new section, dealing with relationships between the spit and bar, touch upon a most interesting concept in spit development that will be discussed in a later paper by F. J. Meistrell.

Alan P. Carr studied geography (specializing in biogeography), geology, and economics at University College, London. After teaching for a few years, he joined the Physiographic Section of the government-sponsored Nature Conservancy. In 1964 he was put in charge of the section, which was then involved in both coastal research and in providing geomorphic advice for the Conservancy's regional staff. In 1971 Carr moved to the Unit of Coastal Sedimentation to be in charge of its foreshores program. Now 42, Mr. Carr reports to the editor that he has a wife and three recalcitrant children; the latter two facts revealing that many of us connected with this volume have more than professional interests in common.

Shingle Spit and River Mouth: Short Term Dynamics

A. P. CARR, B.SC.

(*Senior Scientific Officer, Physiographical Section, Nature Conservancy*)

Revised MS. received 20 August 1964

Introduction

ALTHOUGH CONSIDERABLE attention has been paid to the growth of sand and shingle spits, most of the studies have been qualitative rather than quantitative, and often of a somewhat general nature. The need for more detailed work has been widely recognized. Thus J. A. Steers[1] said '. . . far more attention must be given to local examples. A number of the mistakes that have been made in the past about coastal features are probably due to the neglect of observations in specific small areas'. This paper examines changes in one such area, the mouth of the River Ore, Suffolk, during the period 1945 to 1962, although some reference to changes in 1963 is also included. The data obtained help to shed light on the means by which earlier changes took place there and suggest a number of features in spit development which may not have been fully recognized elsewhere.

North Weir Point is the distal end of the shingle structure which extends from Aldeburgh in the north in a southerly and south-westerly direction for almost ten miles (Fig. 1, inset). The northern section between Aldeburgh and Orfordness consists of an eroding beach which has receded continuously landward for several centuries at least. This beach is backed by reclaimed salt marsh which separates the shingle coast-line from the River Ore. The river itself runs roughly parallel to the shore and has the appearance of being diverted from its original exit near Slaughden. Salt marsh is again present landward of the cuspate foreland of the Ness itself, but is scarcely evident on the seaward side of the river south of that point.

South of Orfordness point, with its complex pattern of shingle ridges, the spit narrows to reach its minimum width of under fifty yards at high water mark, about three-quarters of a mile north of the present river mouth (Survey Point 17 on Figure 2, which shows the various survey points referred to in this paper). Over the whole of this stretch as far as the distal end of the spit the structure consists of a series of sub-parallel ridges on the riverward side of which recurves may be present. Very rapid growth of an initially tenuous distal point, or its breaching, may result in an extreme form of these, with the recurves separated by tidal pools such as those shown immediately north of Survey Points 27 and 29 in Figure 2. Figure 2 also indicates the ridge structure to the north of North Weir Point. The subsequent infilling of one of the pools is apparent from Figure 3.

While the width of the spit varies, both its composition and height show a remarkable uniformity. The material itself is almost exclusively flint, probably derived from offshore at times when sea-level was lower than at present, or from the glacial gravels of the cliffs to the north. It ranges from a maximum long diameter of about three inches down to one-sixth of an inch, through the whole range of sizes included under the term 'pebble' in the Wentworth scale.

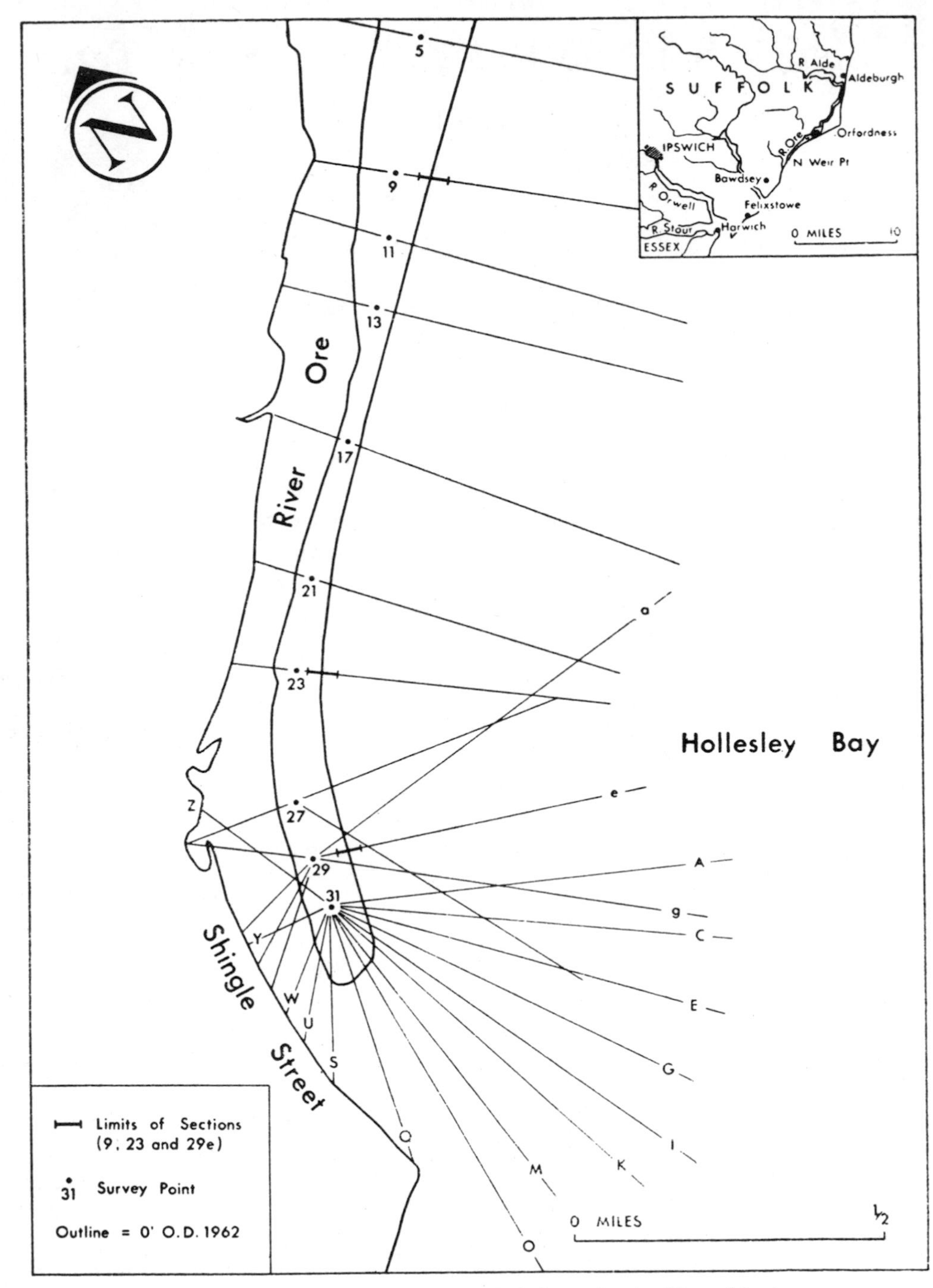

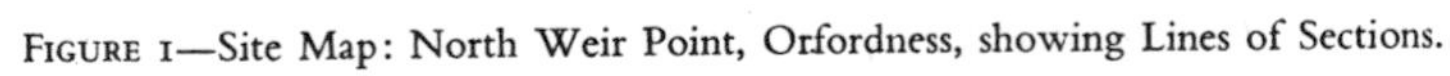
FIGURE 1—Site Map: North Weir Point, Orfordness, showing Lines of Sections.

Individual ridges or constituent parts of a ridge may show a predominance of a particular grading of material. That part of the spit below low water mark shows a similar size gradation, but farther offshore mud and sand can also be found.[2] The ridges on the river side reach a maximum height of just above 6 feet O.D. Newlyn, while the beach crest varies between about 11 feet and 16 feet O.D., dependent on the orientation of the spit at a particular place and also upon the conditions under which the ridge was built and later modified. The intervening swales may be as much as five feet lower than the ridges, but are generally only about one foot lower.

The ridge structure at Shingle Street on the other side of the estuary to the south of North Weir Point is in many ways comparable, although generally lower, far more ephemeral in character, and with the recurves facing the reverse direction. Between the two lies the River Ore. On either side of the river's main channel are the North and South Shingle Banks, areas of considerable size reaching roughly −3 feet O.D. Since low water springs is at approximately −5 feet O.D. and high water springs +4 feet O.D. some proportion of the two banks is exposed for part of each tidal cycle. The North Shingle Bank is separated from North Weir Point by a narrow swash channel, while the South Bank may or may not be attached to the shore there, depending upon conditions prevailing over a given period. Until 1963 the South Shingle Bank had greater relief than the North, there being a 'spine' which ran along it normal to the shore and which continued on to form the river bar. The latter curves round to the north and is characterized by a gentle slope facing the river side but a very steep seaward slope. The location of these various features is shown in Figures I, 4, 5 and 6 and in parts of Figure 7.

Wave attack may be expected from any direction from north-north-east to south-south-west, the range being again in

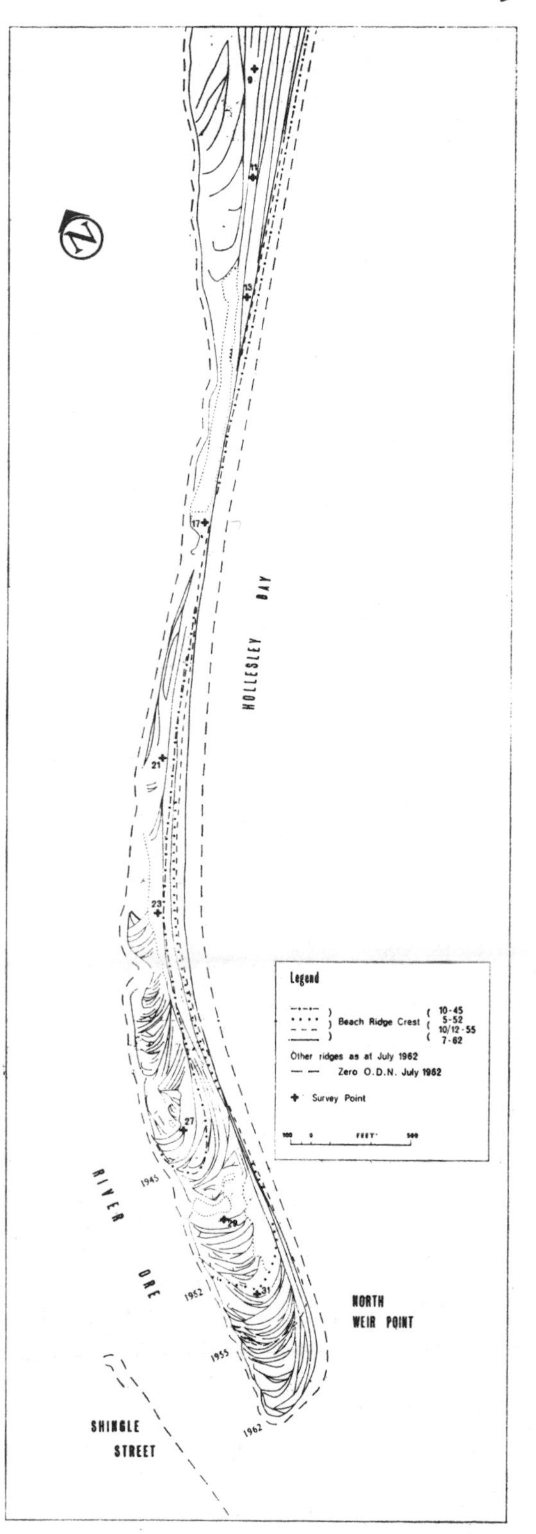

FIGURE 2—North Weir Point and adjacent area: shingle ridge structure and development.

part dependent upon the orientation of the shore at a particular point. The direction of maximum fetch is from the north-east although winds from this quarter are comparatively rare.[3] For most of the length of the spit, apart from the distal point, the beach shelves steeply to about −18 feet to −20 feet O.D. so that the waves are but little attenuated and break close in to the shore. This is not true in the area of the shingle bank where, except for the bar, the shore-line shelves much more gently (Figs. 4, 5 and 6). On the North Shingle Bank, for example, the north-east face rises gradually from −18 feet O.D. to about −4 feet O.D. over roughly a quarter of a mile. The bar has, however, the same abrupt drop reminiscent of much of the length of the spit, as is shown in Figures 4, 5 and along Line 31k of Figure 7, although of course the relative heights are different.

Work in the area during 1957 and 1959,[4] using shingle marked with radioactive tracers, showed that movement in the offshore zone was almost non-existent and that the material there could be discounted as a source of supply to the beach itself, at least under present-day conditions. The 1957 experiment also indicated that it was possible for marked shingle to travel alongshore both northward and southward, and for it to cross the estuary of the River Ore, with its tidal currents of up to 7 knots. The subsequent experiments at the mouth of the river in 1959 suggested that if material reached the North Shingle Bank it was unable to cross the estuary to the Shingle Street side. The shingle merely progressed slowly towards the river channel and contributed towards its infilling. The route taken by material in crossing the estuary was therefore thought likely to be farther seaward and linked in some manner to the river bar and the South Shingle Bank. As will be shown later additional information is now available which helps to confirm these ideas.

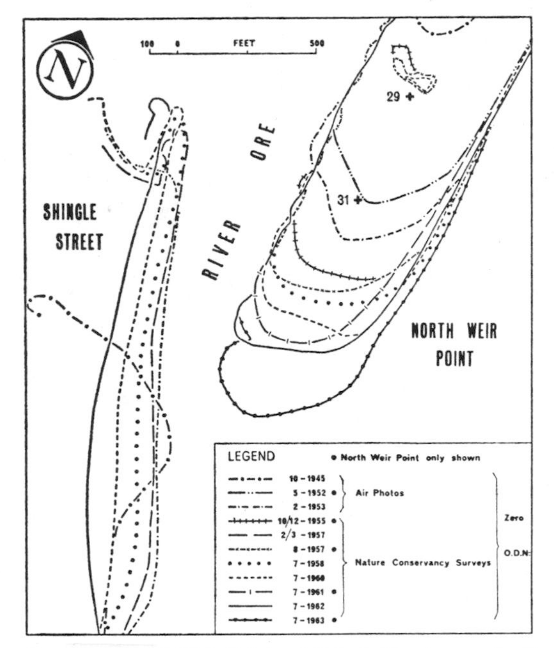

FIGURE 3—North Weir Point and Shingle Street: growth of the distal point and associated landward recession (1945–63).

The present study is based mainly on data obtained by the Physiographical Section of the Nature Conservancy between 1955 and 1962, supplemented by aerial photography taken from 1945 to 1953. While occasional reference is made to earlier maps and charts, this is done mainly to see to what extent the present-day changes were reflected in the past. Figure 9 gives an indication as to how the changes over the period 1945–63 are related to those over the last century and a half. Ground survey, based on a theodolite traverse, was commenced in 1955. Detail was mapped on a scale of 1/1000, using a telescopic alidade incorporating Beaman scales. Throughout the subsequent years plane table revision and repetitive levelling have been used. Offshore detail has been obtained by an MS26A Kelvin Hughes inboard echo-sounder incorporated into the Section's survey launch. To fix the position of the boat during the echo-sounding runs, sextant angles were read on to beacons erected onshore, some 1400 readings being taken to provide control during the 1962 survey. The lines of section are shown on Figure 1.

This paper falls broadly under three headings: the nature of the accretion at the distal end of the shingle spit at North Weir Point, together with the related recent erosion of the Shingle Street shore-line on the other side of the River Ore; the changes on the shingle banks and offshore; and the alteration in the orientation of the spit itself.

The Relationship between the Distal Point and Shingle Street

Figure 3 indicates the detailed changes of the outline of North Weir Point between 1945 and 1962. It gives an indication of the average extent of prolongation over this period, roughly seventy-five feet per year, and also the variation in outline which may occur in the distal point itself. Growth is erratic; there was so little between the annual surveys of 1956 and 1957 that it was impossible to show both outlines on Figure 3 while far more than an average growth occurred between July 1962 and July 1963. Not only is the build-up far from uniform—it is clearly still less uniform over a longer period, as is indicated in Figure 9—but also the final appearance of a shingle ridge bears little relationship to the outline when it is first created. This is apparent in Figure 2 in which the 1945, 1952 and 1955 beach ridge crests and distal points are superimposed on the ridge structure existing in 1962. In addition, as the distal point has grown farther towards the south, the associated beach ridges immediately north of the tip have become greater in height subsequent to their original development. Thus both ridges and recurves may be refashioned or even removed by subsequent wave attack, and in fact repeated annual surveys since 1955 have shown three occasions on which recurves at the distal point have been eliminated between successive surveys and others built in their place. The maximum recession followed by rebuilding, that between the July 1958 and March 1959 surveys, was of the order of 100 feet—that is, greater than the annual rate of growth, but the length of the spit at the time of each survey was virtually identical.

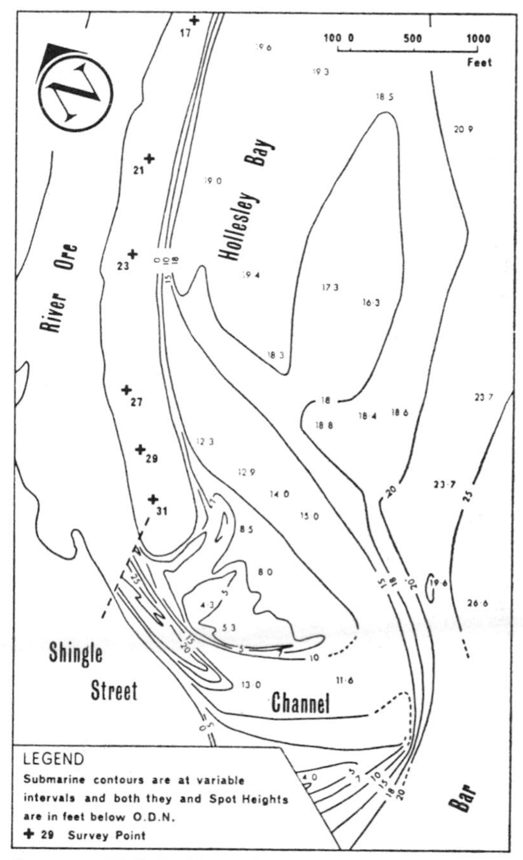

FIGURE 4—Hollesley Bay and the mouth of the River Ore: hydrographic chart, July 1958.

The major phases of destruction of North Weir Point during the last century and a half, over which reliable evidence is available, differ in that the actual river channel at the mouth was fundamentally altered, but this alteration appears to have been subsequent to the initial breach rather than a part of it. This aspect will be referred to again in another context.

Figure 3 also depicts the relationship of North Weir Point since 1945 to the initial northerly growth of the Shingle Street shore-line, and to its subsequent landward recession caused by the development of the main spit towards the south.[5] While this precise relationship is liable to vary it can again be demonstrated that there appears to be a minimum width of the River Ore and that this has remained almost constant, at least since about 1800. Where the river estuary is wider than this, as, for example, following a major destructive phase of the spit, the Shingle Street shore-line is unprotected by North Weir Point, and to some extent by the offshore shingle banks also. It is therefore subject to wave attack from the same range of directions and with the same intensity as that on North Weir Point and is then characterized by a complex pattern of lagoons, in part the legacy of the former river course, and recurves. This is distinct from the present phase in which the straight Shingle Street shore-line is sheltered by North Weir and in which its recession landward is the effect of tidal currents at the river mouth. These may work in concert with the small refracted waves which can still reach the shore there and which produce the recurves on North Weir Point. This stage is obliterating the evidence of the previous phase of development (Fig. 3). Only when the critical minimum width of the river is reached does landward recession of this sort occur. It appears to be an essential corollary to the southward progression of the main spit.

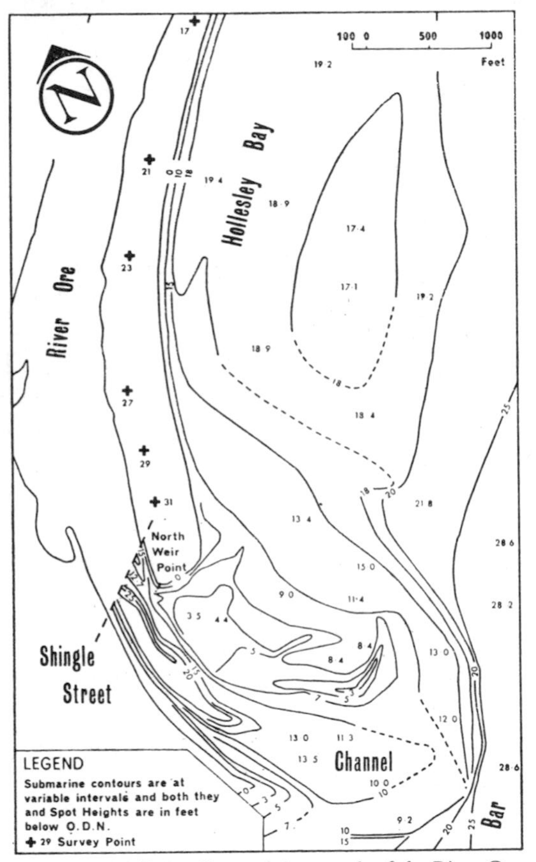

FIGURE 5—Hollesley Bay and the mouth of the River Ore: hydrographic chart, July 1962.

Almost continuous recession of the seaward face of the beach is evident on the Ness itself from the position of the lighthouse as far north as Aldeburgh. The mechanism and cause of landward recession is different as between the two examples mentioned (the relationship of Shingle Street to North Weir Point, and that of Orfordness Point to Aldeburgh). They do, however, have a comparable orientation and it might be argued that erosion in both is a response to the direction of dominant wave approach from the north-east. This is unlikely, since in the area immediately north of North Weir Point, which possesses a similar directional trend, accretion occurs. Elsewhere even this apparent similarity in orientation is absent and only minor changes of the foreshore take place in the long run. There is little evidence of serious erosion on the length of beach from Orfordness Point south as far as Survey Point 5 of Figure 1.

Rebuilding of the spit following major phases of destruction occurs on remarkably parallel lines. Figure 9 shows that, while development is erratic in its extent, during rebuilding the spit has the same orientation as its precursors, and is even formed in the same relative position.[6]

Successive distal points are not produced landward of each other. While the end of the present spit, which has been growing southward since about 1921, initially infilled part of the River Ore and thus appeared to be developing landward of earlier phases,[7] subsequent erosion by the river changed the relationship. The effect of the river in this respect is indicated by the truncated recurves just south of Survey Point 23 on Figure 2.

Changes in the Shingle Banks and Offshore

Any tendency for growth of the distal end of the spit *landward* is prevented because of the effect upon deposition of the tidal race at the mouth of the estuary and because of the depth of the river channel. Since the latter is as much as 28 feet to 30 feet below O.D., vast quantities of material are necessary to raise the level up to high water mark. Farther seaward the channel is shallower (Fig. 7, Line 31k, contrasted with 31q) so that the amount of material required is less and extensions of the spit are correspondingly more rapid. Where shingle is deposited depends on wave type and approach, and on tide height at the time of transport. This was reflected in the distribution of marked material in the tracer experiments in 1957. Since even at high tide waves of eighteen inches or over break on the seaward face of the Shingle Banks their destructive force is largely dissipated there. This in itself means that accretion is likely to be greater on the shallow North and South Banks immediately to landward, and is one of the factors which helps to explain the development of the spit parallel to the river channel. The difference in the rate of growth is brought out in Figures 4 and 5 which show the topography below zero feet O.D. for the years 1958 and 1962. (It should be noted that, in order to depict the structure of the shingle banks more precisely, the contours are drawn at variable intervals.) Figure 6 is derived from the two hydrographic charts and shows isopleths of accretion and erosion over the four-year period. Thus in this figure, although the various topographical features remain readily recognizable, it is their displacement *horizontally* which is shown, although the measurements are in terms of the amount of accretion and erosion recorded *vertically* between 1958 and 1962.

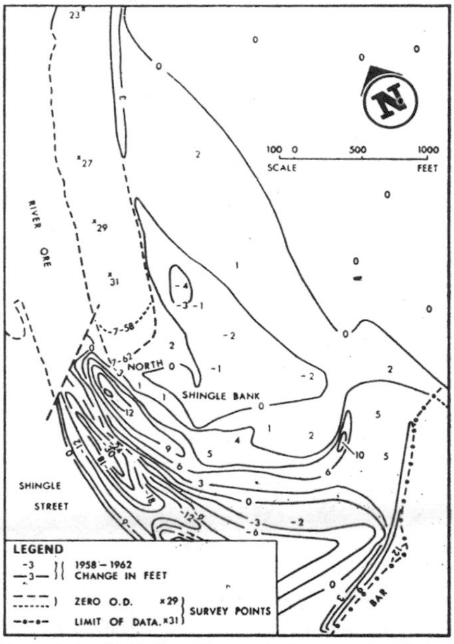

FIGURE 6—Hollesley Bay and the mouth of the River Ore: offshore change, (1958–62).

Figure 6 brings out clearly a number of points:

1. It shows the accretion on the channel side of the North Shingle Bank, and the consequent erosion of the Shingle Street shoreline as the river attempts to maintain its minimum width and depth. This has already been considered. The progress of spit, North Shingle Bank and river channel toward the south, and their effect on the outline, orientation and development of the river bar and the South Shingle Bank is shown in greater detail in Figure 7, where representative sections are drawn. These also give an indication of the variation from year to

year in the amount of change that occurs, both on the Point and offshore. Only rarely has the annual change at the Point been at variance with the longer-term trend in its direction and this has never been so offshore. Apart from the greater consistency in the movement of both the North Shingle Bank and of the seaward edge of the bar towards the south compared with that of the distal point, growth offshore has been at an appreciably faster rate over the period; so that with time the channel and bar have become progressively farther from the distal end of the spit. Even the period between the 1962 and 1963 surveys, with its long duration of easterly weather, did not reverse this trend. It is not at present clear whether this is always the case or

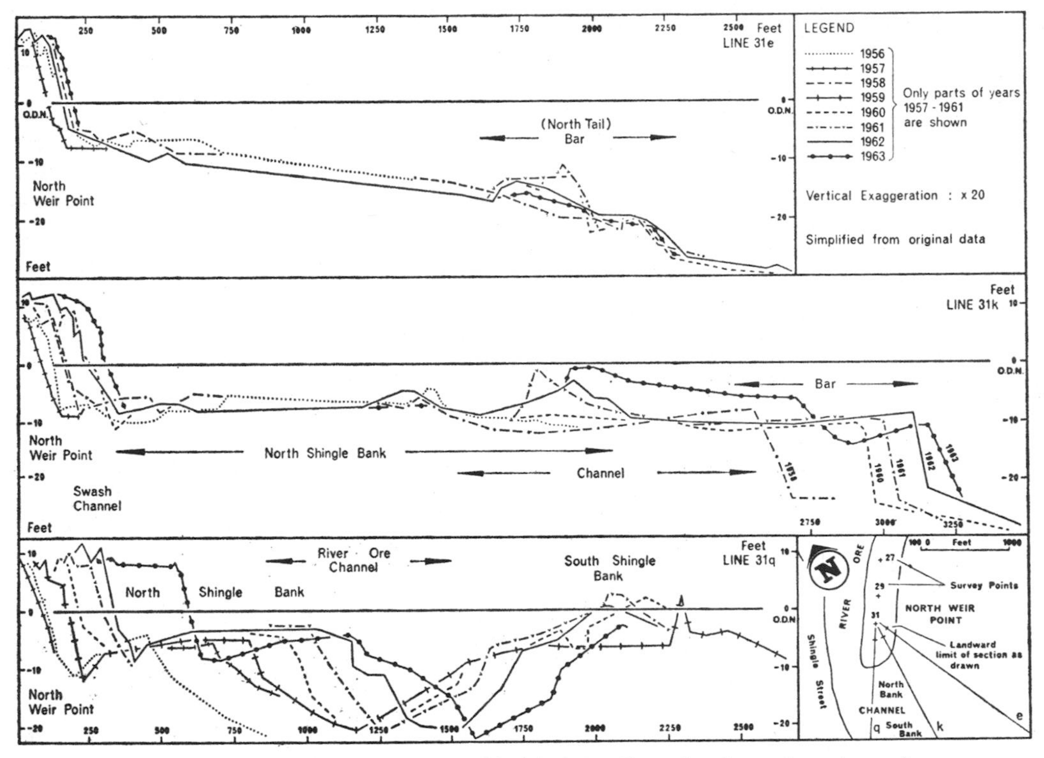

FIGURE 7—North Weir Point and the estuary of the River Ore: lines of section to show changes between 1956 and 1963.

whether it represents only a particular phase in the sequence of development. Long-term cartographic evidence is not very helpful in elucidating this question.

2. Directly seaward of Survey Point 23 there is a narrow zone of accretion parallel with the spit and this will be considered later in more detail.

3. To the south-east of Survey Point 23 the stability of the sea-bed is well shown. This is also true of the area to the north of it. The *offshore* tracer experiments referred to above, in which negligible movement was detected, were carried out here. There is some evidence to suggest that well to the north of the area described limited movement does occur, but this is probably associated with the actual Ness, and is likely to be parallel with the offshore banks, rather than onshore.

4. Offshore of Survey Point 27 and south-eastward of it, there is a broad area of accretion,

although this has been only of the order of one to two feet in height over the four-year period. To the north it is bounded by the stable area previously referred to, while on the south it runs into the bar, which in turn is joined to the South Shingle Bank. It seems probable that during the 1957 experiments marked material travelled along the seaward edge of this zone of accretion to reach the South Shingle Bank and the Shingle Street shore. Between this zone of accretion and the North Shingle Bank there is a smaller area of erosion with losses of up to four feet in height over the period 1958–62. The latter is a response to the displacement of the North Shingle Bank southward in association with the growth of the spit. If this erosion did not occur the bank would become progressively greater in extent, as the spit continued its development.

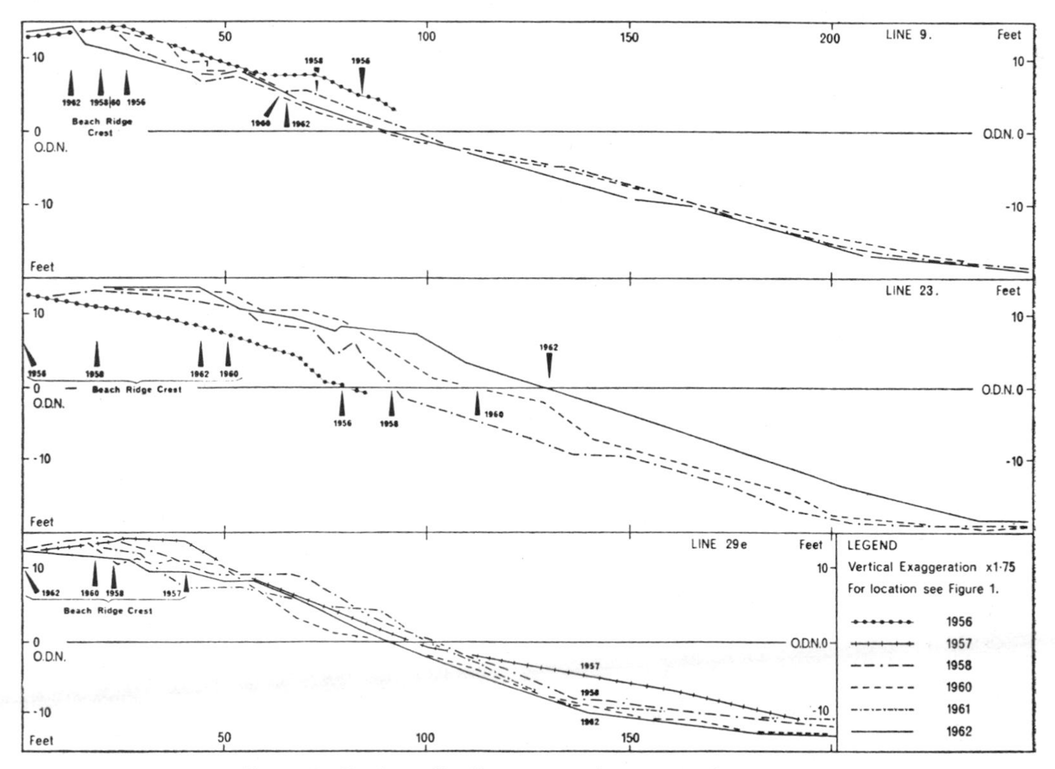

FIGURE 8—Beach profiles, lines 9, 23, and 29e, seaward, 1956–62.

Subsequent Changes in Spit Orientation and Beach Ridge Structure

Changes in orientation of the spit north of the distal point may occur as the structure extends its length southward. Part of Figure 2 indicates the progressive widening of the spit and the reduction in the concavity of the shoreline as prolongation occurs. The concavity (near Survey Point 23) was initially produced about 1921. At that time the spit finally retreated to a position close to Survey Point 23 and subsequent regrowth of the distal point had a trend at an angle to the residual part of the structure. During the period from 1921 successive distal points have, as in earlier times, been developed in a virtually straight line towards the south but modifications to the orientation of the spit north of its southward-progressing tip have taken place. Figure 8 shows the limited erosion which has occurred on the seaward face of Line 9, and is typical of that part of the beach. It also indicates the almost continuous build-up on Line 23, amounting nearly to a doubling of the width of the spit above high water mark over a period of 17 years. Farther south the tendency to erosion of the seaward face of the spit nearer the point is shown by Line 29e. South of Point 29 again erosion turns to accretion at the distal point itself (Fig. 7, Lines 31k and 31q). The net effect of this differential erosion and accretion is to straighten the spit as its length is extended. A comparable change in the orientation of the spit can also be demonstrated between 1872 and 1903. During that time the spit did not build continuously southward. Indeed, over the latter part of the period retreat northward appears to have been the case. In fact this impression is in part incorrect, since the partial breaching and subsequent shortening between 1880 and 1895 did not initially eliminate the southern portion but only reduced its height to about that of low water mark. G. Pirie's chart of 1895 still shows the shingle banks well to the south of the exposed part of North Weir Point spit, and clearly related to the maximum extent of that spit's growth some years earlier.

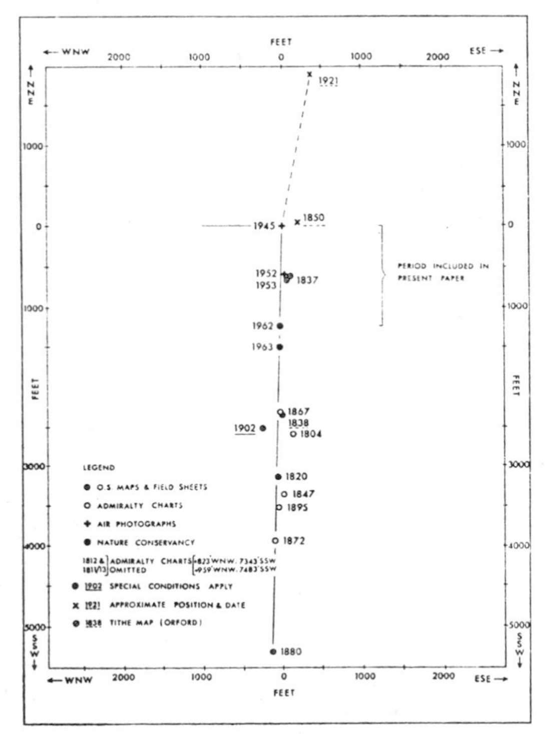

FIGURE 9—North Weir Point: the distal point 1804–1963, showing its variation in length but consistency in orientation. Note the anomaly in the positions for 1837 and 1838.

It is suggested that the straightening of the spit as it is lengthened is the result of the southward migration of the river channel and the shingle banks. This in turn has the effect of increasing the range of directions from which wave attack on the concave part of the spit (near Survey Point 23) can operate. For example, waves approaching this part of the coastline from a south-easterly direction no longer break on the North and South Shingle Banks, since these have become relatively displaced toward the south. As a result the whole wave energy is retained until they break on the beach instead of the near-shore zone. Thus the forces acting both there and farther south change with time. In this case slow but progressive accretion occurs with increasing influence of waves from a southerly quarter.

The change in the width of the spit as a result of reorientation is of considerable importance. It has already been mentioned in another context that the shingle recurves ultimately fashioned may well be the result of a series of numerous periods of wave attack and of consequent modifications to the original structure, and that this is also true of the straight beach ridges themselves, which run parallel to the foreshore and frequently extend over a considerable length. The near-doubling of the width of the spit, especially north of, but to some extent south of Survey Point 23, has been achieved by the addition of new ridges in front of those existing previously. They gradually merge into earlier ones at each end, and as they become vegetated in a similar fashion little evidence of their independent development remains. That beach ridges can develop in this way is at variance with some of the ideas previously advocated and with the suppositions based upon them.[8] It means that continuous and apparently contemporaneous ridges can be made up of two or more components built at widely differing times. It also suggests that it is not necessarily true that the material available for accretion at the distal point is provided by the net erosion of the area *immediately* updrift even when the orientation of that beach is comparable to other eroding sites near by. Rates of movement of beach material on this stretch of coast are sufficiently rapid for there to be no need to look for a source of material immediately adjacent to the end of the spit to provide shingle for accretion at the distal point. The net movement of beach material is from the north and most of that used in the building of the spit has probably been re-sorted many times since first becoming subject to shore processes.

The consistency in height of an individual ridge when first formed is very striking. It was this that prompted Steers to suggest the rolling *onshore* bodily of new ridges from below low water mark.[9] The experimental work already alluded to and the hydrographic data in Figures 4, 5 and 6 have shown that this does not happen in this area. This is not to say, however, that shingle already present as part of the spit below low water mark is not pushed up above it by constructive waves normal to the shore, but the amount of material available is so limited as to be only sufficient to produce minor ephemeral ridges below the beach ridge. To the extent that onshore movement appears to occur elsewhere (as, for example, in the landward progression of barrier beaches in the Warren Point area of Dawlish Warren, Devon) it is probably an indication that the dynamics of the beach zone extend farther seaward than initially conceived, either through the nature of wave attack, or because of the depth of the sea bed offshore.

Apart from the recurves, the individual ridge crest on the Orford shingle spit frequently shows a difference in height along its length of as little as three to four inches over a mile. Since the origin of the ridge lies, not from offshore, but in the longshore drift of material, it is to be expected that this slight variation in height should be progressive either from north to south or vice versa, dependent upon the direction of oblique wave approach at the time of construction. Here too a further point should be made. Much has been written about 'constructive' and 'destructive' waves, but there are all stages between these extremes. Most 'constructive' waves comb down beach material to some extent. Further, the incidence of attack may result in the same waves having a predominantly constructive role on part of the foreshore and a mainly destructive one at another. Observations have shown this to be the case at Orford (for example, north and south of Survey Point 23 (Figs. 1 and 2)), and elsewhere (such as in the shingle 'complexes' at Bridgwater Bay, Somerset).[10]

Conclusion

This paper has described the results of routine recording at North Weir Point mainly over the period 1955–62 and indicated the mechanism through which the changes are achieved: that is, the building up of the distal end of the spit, the infilling of the river channel, the consequent erosion of the Shingle Street shoreline and the related changes on the shingle banks and immediately offshore. It has also briefly considered how far certain of these developments are typical of the past. The zones of accretion and erosion offshore, north of the North Shingle Bank, probably represent the route taken by beach material in its path across the estuary, a route the existence of which was inferred in earlier experimental work.[11]

It has also been pointed out that neither an individual beach ridge, nor a recurve, is necessarily of uniform origin or date throughout, and that the changes in orientation of the spit and the doubling in its width at the most concave point, rather than the severe erosion which might be expected there, are at variance with such earlier ideas on spit development as those of D. W. Johnson. These ideas stemmed in part from a lack of understanding of the nature of movement of beach material and of the rate at which that movement took place. The stability of the sea bed in the offshore area from north of Survey Point 5 to the latitude of Survey Point 23 is in agreement with the lack of movement found in this area in tracing experiments, using marked shingle, during 1957 and 1959.

There are two particular aspects on which further information is needed. First, since the river channel and the river bar appear under present conditions to move southward at a faster rate than does the distal point of the shingle spit, what bearing does this have on the ultimate rate of growth and the eventual breaching of the spit itself? Does the spit sow the seeds of its own destruction, or is this differential growth a feature typical of only one phase in the whole sequence of development? Even the rapid growth of North Weir Point spit from July 1962 to July 1963, consequent upon the long period of easterly weather during the winter months, failed to reverse the more rapid progression of the Banks and bar towards the south as compared with North Weir Point itself. Secondly, how does the breaching of the spit occur, and what are its direct and indirect effects? The solution of these problems would appear to be largely dependent upon the conditions occurring in future years.

NOTES

[1] J. A. STEERS, *The Coastline of England and Wales*, 2nd. ed. (1964), 62.

[2] In work carried out by the Physiographical Section of the Nature Conservancy below low water mark at Chesil Bank, Dorset, during 1960 and 1961, the same features were also present, but there grading is less perfect below low water mark.

[3] See, for example, Table II in C. KIDSON, 'Movement of beach materials on the east coast of England', *E. Midland Geogr.*, No. 16 (1961), 14.

[4] C. KIDSON, A. P. CARR and D. B. SMITH. 'Further experiments using radioactive methods to detect the movement of shingle over the sea bed and alongshore', *geogr. J.*, 124 (1958), 210–18 and C. KIDSON and A. P. CARR, The movement of shingle over the sea bed close inshore', *geogr. J.*, 125 (1959), 380–9.

[5] Another example of this type of recession is given in: A. GUILCHER, *Morphologie littorale et sous-marine* (Paris, 1954), 87.

[6] The most useful of the cartographic sources which provide evidence of long-term changes are:

Date of Survey	*Title*	*Surveyor*	*Sheet or Ref. No.*
1804	Orford Ness to Whitaker Beacon	G. Spence (Hydrographic Office)	664
1812	Hollesley Bay	Capt. F. W. Austen (Hydrographic Office)	789*
1811–13	A revision of Spence's chart apparently partially based on Austen's survey	(Hydrographic Office)	722*
1820	Two inches to one mile field sheets	E. B. Metcalf (Ordnance Survey)	OSD 315c
1837	Two inches to one mile field sheets	J. L. Hall	38, North-east
1838–43	Tithe Maps of parishes in the Orford area		
1847	Naze of Orfordness	Capt. F. Bullock (Hydrographic Office)	L7288
1867	R. Ore and Alde including Orfordness	Comm. E. K. Calver (Hydrographic Office)	B1446
1872–73	Approaches to Harwich	Comm. J. Parsons (Hydrographic Office)	A3671
1879–82	Six-inch map, 1st edition	(Ordnance Survey)	Various
1895	River Ore and approaches	Capt. G. Pirie (Hydrographic Office)	B5833
1902–03	Six-inch map, 2nd Edition	(Ordnance Survey)	Various

In addition some of the relevant six-inch sheets underwent partial revision mainly in 1925–28 and 1953.)

1945	16 Oct. 1945	1/10000	Air photographs (R.A.F.)	Various
1952	21 Mar. 1952	1/5000		
1953	19 Feb. 1953	1/5000		

(Other sorties on various scales available covering the period September 1955–59)

* Hydrographic Office, Admiralty charts Nos. 789 and 722 both depict a North Weir Point farther south than that known at any other date. A. H. W. Robinson, in his *Marine Cartography in Britain* (1962), 128, has already called into question Capt. F. W. Austen's survey in other respects, but there is sufficient other evidence to suggest that the position given for the spit is substantially correct.

The 1902 six-inch Ordnance Survey Map indicates an attenuated distal point deflected landward towards the river. The orientation of the remainder of the spit is quite typical.

Since the position for the distal point in 1921 given on Figure 9 is only approximate, distances are measured from the 1945 air photograph plot, rather than the most northerly position of the spit during the period 1804–1962. J. A. Steers (see Note 9), also gave a site comparable to the one indicated for the spit's maximum recession.

[7] See 1945 outline from air photographs in Figure 10 of C. Kidson 'The growth of sand and shingle spits across estuaries', *Z. Geomorph.*, 7 (1963), 1–22.

[8] For example, D. W. Johnson, *Shore Processes and Shoreline Development* (New York, 1919), 405–7; 'The shoreline advances seaward throughout a considerable portion of its extent simultaneously. . . . ', 408.

[9] J. A. Steers, 'Orford Ness: a study in coastal physiography'. *Proc. Geol. Ass.*, 37 (1926), 313 and 322.

[10] C. Kidson, 'The shingle complexes of Bridgwater Bay', *Trans. Inst. Br. Geogr.*, 28 (1960), 75–87 and C. Kidson and A. P. Carr, 'Beach drift experiments at Bridgwater Bay, Somerset', *Proc. Bristol Nat. Soc.*, 30, (1961) 163–80. The bodily movement of the 'complexes' at Bridgwater Bay is caused through the variation in the efficacy of wave attack on the differently orientated parts of each structure. The erosion of the western face provides material for accretion immediately to the east (downdrift), not as in the case of the North Weir Point spit. However, it is only because movement of beach material at Bridgwater Bay is so slow and because each complex is self-contained that this adjacent source of supply has to be invoked at all.

[11] C. Kidson, A. P. Carr and D. B. Smith, op. cit.

Author's Commentary
Changes in the Ore Estuary up to 1970

The paper reprinted above incorporated data obtained up to 1962–1963. Since that time a number of the questions posed in it have been answered although the fundamental one—that of how the shingle spit finally breaks up—remains in the realms of conjecture.

Topographic and hydrographic surveys continued on an annual basis until 1970. From 1964 onwards the survey lines were extended to cover a greater area offshore, especially to the south of the bar. Continuous wave and tide records were obtained for 1967–1968 and anemometer records from 1966 onwards. More comprehensive tidal current data were collected in 1967, 1968, and 1969. Bottom samples were obtained from time to time.

The information shows:

a) The spit has always been rebuilt in the same relative position following periods of destruction. This was originally suggested by cartographic and historical data for the period from 1804 A.D. and has been confirmed by peat deposits (5390 ± 110 B.P.) found at a depth of 6.4 to 7.9 m in line of the spit. Since there is a deeper river channel to landward and a corresponding drop at the edge of the bar to seaward, it implied that the relative positions of bar, spit, and river must have remained constant throughout the period that the spit has been of its present order of length. Progressive landward recession appears impossible even although recorded elsewhere.

b) Onshore windrun and significant wave height and frequency do not correlate well, at least with the recent growth of the spit. This suggests that the initial disposition of the offshore banks and bar is more important in the accretion of the spit than other factors.

c) Calculations over the past 12 years suggest that since about 1967 accretion in the estuary has become dominant over erosion, possibly being as much as 6 to 7 times as great. Until 1967, even with the sometimes rapid growth of the spit the two processes had been finely balanced. The dramatic and sudden change in the hydraulic efficiency of the estuary had not been anticipated.

d) The fact that the bar had become further and further from the distal end of the spit was suggested in the previous paper as a possible cause of the latter's final break-up. However, subsequent data has shown a reversal of this relationship and it is now believed to be merely part of the mechanism of the development of the spit. As spit and bar diverge, tidal current velocities fall, friction of the bed increases and deposition, at least of the finer material subject to tidal currents, becomes more feasible between them.

e) Since 1963 there have been many other changes in detail of the various features at the mouth of the River Ore yet some—such as the way in which the deepest part of the river has remained opposite survey point 29—have shown a remarkable constancy.

Over the whole period during which research has been carried out the spit has continued in its southerly progression. Nevertheless, light has been shed on possible causes of recession and these, together with a more detailed evaluation of post-1963 changes, are published elsewhere.

Editor's Comments on Paper 8

The development of bars at the mouths of rivers is dependent upon all the variables existing in the river and at the shore. Among others, this includes river flow, sediment discharge, wave action, tidal currents, vegetation, and coagulation (flocculation) of suspended matter. V. N. Mikhailov, utilizing laboratory models and field observations, attempts to evaluate the relative significance of these factors. Beyond this, he then considers the regulatory interaction between bar development and channel flow.

Dr. Mikhailov was born in 1932. He graduated from the Geographic Faculty of Moscow University in 1954, then did postgraduate work at the university. Since 1958 he has been with the State Oceanographic Institute in Moscow. Mikhailov's specialty is river hydrology, particularly the hydrology of deltas and river mouths.

The following article is reproduced by the permission of UNESCO.

8 Hydrology and formation of river-mouth bars

V. N. Mikhailov

INTRODUCTION

Due to the effect of river sediment material discharge and action of the sea the river-mouth (or delta-channel) bars (shoals), formed by sediment deposits, are commonly the most variable delta areas. Seaward advance of a delta and extension of its river channels occur principally in bar areas. As a result of shoaling, instability of bottom relief and intensive sediment accumulation within river channels, the delta-bar areas become a serious obstacle to navigation and, therefore, necessitate a comprehensive investigation of bay-mouth bars.

This report deals with some problems associated with the hydrological régime and the formation of bars in the mouths of plain rivers.

MAJOR ACTIVE FACTORS

Bars are characteristic of the greater majority of rivers entering seas and lakes. Aquaintance with cartographic, geological and geomorphological data, as well as with sea- and lake-bottom topography, shows that river-mouth bars are formed in various physiographical and hydrological environments.

The formation of a bar, its size and shape depend on the intensity of sea and freshwater interaction processes taking place at the mouth of a river.

Factors such as sea currents, wave action, tidal currents, sea-level rise, etc., can in a number of instances prevent the formation of a large delta. Nevertheless, bars are usually formed even under similar environmental factors and the bar features can be traced at almost any river mouth.

Many factors affect the formation of bars. There is still no clear evidence of their relative significance. Despite the fact that under varying environmental conditions the influence of one factor or another may differ, it is necessary to point out that entirely different rivers reveal many common features in the process of bar formation. Therefore, the existence of certain general regularities common to the process of bar formation can be expected in any environment.

The presence of these general regularities is explained by the fact that the main factor predetermining the formation of a bar is the river water and the sediment discharge. To form a bar in the mouth of a river, it is sufficient to have a water flow that can transfer river sediments. The action of the river-water flow and the sediment discharge builds up a background upon which the effect of other factors, deforming and shaping an already formed bar, is displayed.

Wave action in the process of bar formation has a complex and contradictory effect. It may sweep away sediment formations created by the river action, and build up new ones. Observations carried out at a number of river mouths have shown that whilst sweeping out bar formations, wave action broadens them and increases water depths over the bars. Under the influence of storms bars move towards the land. Wave action washes away the shoreline and a portion of the delta banks and bars, transports the washed-out material and accumulates it at some other place, thus forming shoals and sand-bars. The wave action changes the configuration of an initial bar formed by river action. In a number of instances the action of waves can block up the mouth of a river with banks.

A similar destructive and shaping effect on bars can produce tidal and ebb-flood currents varying in their direction. Like the wave action, these cause migration and redeposition of bar sediments at bar areas thus causing changes in water depth above the bar and sedimentation deposits within navigable channels.

Under similar environmental conditions the balance of the sediment deposit at the bar area is of great importance. If the transportation power of the longshore sediment flow caused by the action of waves and sea currents exceeds the volume of sedimentation discharged by the river, the bar is destroyed. Otherwise, it grows intensively. Evidently, therefore, the adjustment of the river sediment discharge may be viewed as a radical step to control sediment deposit at bar areas.

The growth of vegetation on the top of the sediment formations built up by a river flow enables them to consolidate. While hampering the further movement of sediments carried by river flow, the vegetation cover facilitates the bar development, promotes the bifurcation of the river channel and contributes to the growth of the delta. In some deltas the vegetation cover grows so actively that it seems to have a decision effect on the delta formation.

There is no doubt that such passive factors as coastline configuration, sea-bottom relief underlying the bar, geological structure, etc., play no less an important role in the formation of bars.

The author has tried to investigate the mechanism of bar formation in the case when a river flow is the sole or decisive factor. Among the factors determining bay-bar formation of a river-mouth bar, the river current is the most prominent one; a similar approach to the problem permits the study of the process which constitutes the background upon which other factors display their effect. In the long run it will help to estimate more precisely the role of other factors and, above all, the role of the wave action and tidal currents.

THE ROLE OF RIVER FACTORS IN THE FORMATION OF RIVER-MOUTH BARS

Laboratory experiments

The investigation of regularities in the formation of a river-mouth bar can be greatly assisted by model testing. Such tests have been carried out by the author at the Laboratory of Channel Processes, U.S.S.R. Academy of Sciences.

At the outset the task was to investigate the velocity structure of a freshwater stream discharged into a reservoir and to determine its role in the process of bar formation. The experiments have shown that there is no regularity in the distribution of stream velocities in the river flow: with the widening of the flow velocities decrease both along and across the stream. Decrease of velocities along the stream starts from the mouth entrance and develops in a relation near to the exponential, while across the stream it proves to follow a parabolic curve (close to the mouth) and a curve approaching the Gaussian probability curve (away from the mouth). The application of hydraulic methods allows an estimation of the field velocity of the stream to be made (Mikhailov, 1959).

Material transported by the model stream, deposited in full conformity with the stream velocity structure: the decrease of the velocity along the stream resulted in the formation of a bar in front of the mouth, and a non-uniform distribution of velocities across the stream led to a formation of sand shoals. The formation of bars and sand shoals developed simultaneously and

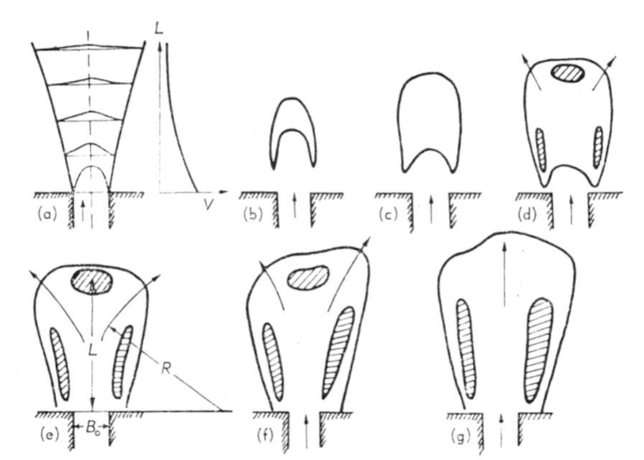

FIG. 1. Diagram of bar and island formation process in the mouth of a flow (plotted according to laboratory experiments data): (a) to (g) show the successive stages of the process.

there appeared a "finger" of river sediments deposited in the reservoir (Fig. 1a, b and c). The intensity with which the bar grows is proportional to the amount of sediment discharge and is inversely proportional to the angle of the shore slope in the reservoir.

The object of further experiments was to reveal how a relief formed at the mouth of a stream influences the structure of a river-water flow. Bottom drifts deposited in the upper areas of the stream brought about a rise of the channel bottom and the appearance of longitudinal and transverse slopes of water surface above the bar. The presence of a transverse slope led to the smoothing of the flow when it passed over the bar, the widening of the bar and a subsequent bifurcation of the flow and the formation of an island in the middle (Fig. 1d and e).

Bars in river mouths

Field observations of river mouths carried out by the State Institute of Oceanography in the Soviet Union and analysis of data on foreign rivers show that at places where the river discharge is a decisive factor, the process of bar formation at the mouth of natural rivers is similar to that obtained under laboratory conditions. Prominent examples of the development of the bottom relief in the mouth of rivers formed at open shores as a result of the breaking-through of beach barriers are the Vistula river mouth and the mouth of the Sulak river falling into the Caspian Sea.

The process of bar formation at the mouth of other rivers which have sufficiently large water and sediment discharge is similar to the above. The mouth of the Danube, the Mississippi, the Amu-Darya, the Kuban and other rivers (Bates, 1953; Mikhailov, 1958, 1962; Samoilov, 1952) were also formed in a similar way.

The bars at the mouth of similar rivers are composed of river-sand deposits and only at normal water stages are they covered with silt. There is a distinct relationship between the flow velocities of river current and the composition of sediment particles deposited over the bar (larger particles accumulate at places where current velocities are greater). The discharge of a river at its various stages plays an important role in the process of bar formation: bars are formed principally at a high-water stage. All the above-mentioned phenomena support the assumption that river factors occurring at the mouth of a majority of rivers play a principal role. Therefore, river-mouth bars should be viewed primarily as formations built up by river action. This observation allows two major conclusions to be drawn: (a) the channel régime of the water flow should influence the channel régime of the bar; (b) the hydraulic methods of computing channel deformations may be applicable to the bar.

The role of the river-sediment discharge

The river-bed load plays an active part in the formation of river-mouth bars. A bar composed of a bed load is similar in shape to rapids with gentle upstream and steep downstream slopes, and, as in rapids, the shoaling of bar areas takes place during the flood stage. As bars are formed mainly during the flood stage and as during this period the velocities of currents passing over the bar are relatively high, the suspended sediment material is not deposited over the bar but is swept farther seaward. This is proved by the field measurements of current velocities taken at river mouths. During the normal stage suspended sediments can deposit over the bar area and higher in the upstream direction.

The sand-banks shifting from place to place within the river channels can destroy the bar by creeping over it. Having studied the river-channel depth régime

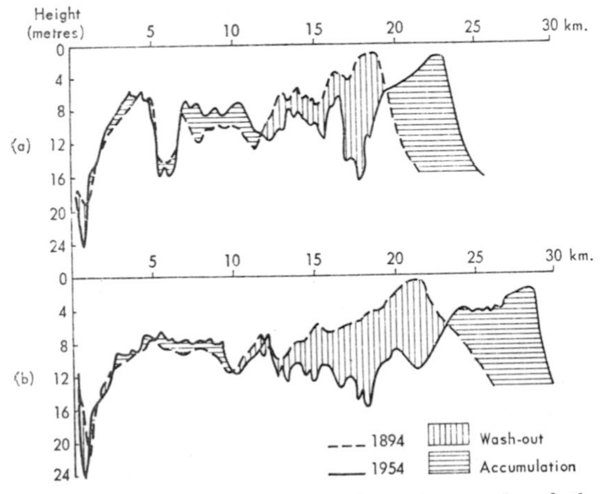

FIG. 2. Advance of bars seaward at the mouths of the (a) Ochakov and (b) Old Istanbul channels in the Danube delta.

and the expected fluctuations of the river discharge, it is possible to forecast the water-depth régime over the bar.

With the accumulation of sediments at the mouth of a stream, the bar moves seaward and the stream channels elongate (Fig. 2). The bar displacement of the three heterogeneous zones of channel-shaping activity is involved with: (a) the accumulation and rise of the channel bottom; (b) the scouring of the bar upstream slope; and (c) the accumulation of the sediment material on the bar's seaward slope.

Morphometric characteristics of the river-mouth bar

It is well known that there exists a relationship between the morphometry of a river channel and the dynamics of its stream. The author has made an attempt to apply hydraulic methods when computing the morphometric characteristics of a river-mouth bar (Mikhailov, 1958). It was found that the increase of the flow velocities and the water discharge led to the increase of both the minimum attainable radius of the river-stream curvature in the bar area and the distance from the mouth to the ridge of the bar. This relationship was produced theoretically:

$$\frac{L}{B_0} \sim A \left(\frac{h_0}{B_0}\right)^{1/3}$$

where L stands for the distance to bar ridge or island; h_0 and B_0 are the depth and width of the channel at the mouth entrance; A is the ratio coefficient which depends but very little on the sediment composition.

The development of the formula required the use of: (a) parameters for longitudinal and transerve flow equilibrium over the bar

$$J_{\text{long.}} = \frac{V^2}{C^2 h} \text{ and } J_{\text{trans.}} = \frac{V^2}{gR}$$

where J is the longitudinal and transverse slopes of the water surface over the bar, V and h are average velocity and stream depth, C is the Shezi coefficient, R is radius of stream curvature; (b) the approximate relationship between $J_{\text{long.}}$ and $J_{\text{trans.}}$

$$\frac{J_{\text{long.}}}{J_{\text{trans.}}} \sim k \frac{B_0}{L};$$

(c) the approximative relationship between distance to the bar L, channel width in the mouth B_0 and radius of stream curvature R:

$$L \sim \sqrt{RB_0}.$$

On the basis of laboratory data and cartographic materials of a number of river mouths (Fig. 3) an empirical relationship approaching the theoretical ($A \sim 12$, exponent 0.37) was found which proves the close correlation between the dynamics of the river flow and its bottom relief in the mouth. Thus obtained the empirically deduced relationship shows that streams with greater values of the ratio width to depth form

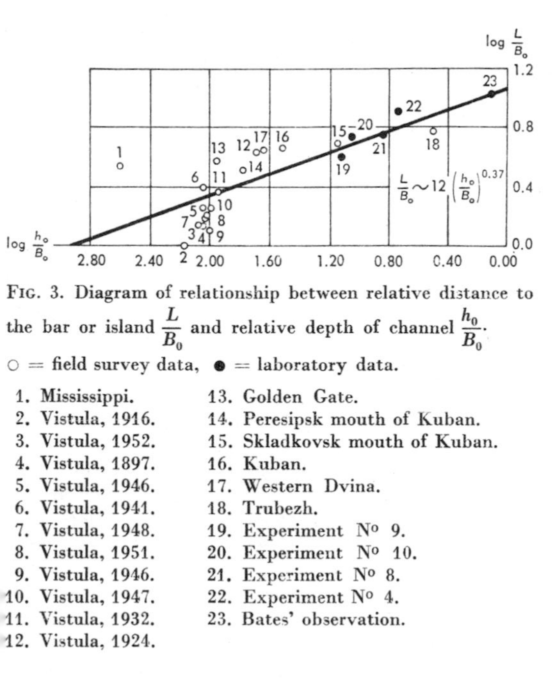

FIG. 3. Diagram of relationship between relative distance to the bar or island $\frac{L}{B_0}$ and relative depth of channel $\frac{h_0}{B_0}$.
○ = field survey data, ● = laboratory data.

1. Mississippi.
2. Vistula, 1916.
3. Vistula, 1952.
4. Vistula, 1897.
5. Vistula, 1946.
6. Vistula, 1941.
7. Vistula, 1948.
8. Vistula, 1951.
9. Vistula, 1946.
10. Vistula, 1947.
11. Vistula, 1932.
12. Vistula, 1924.
13. Golden Gate.
14. Peresipsk mouth of Kuban.
15. Skladkovsk mouth of Kuban.
16. Kuban.
17. Western Dvina.
18. Trubezh.
19. Experiment N° 9.
20. Experiment N° 10.
21. Experiment N° 8.
22. Experiment N° 4.
23. Bates' observation.

the bar and the island nearer to the mouth than those with smaller values of the $\frac{B_0}{h_0}$ ratio. Therefore, the shapes of channels in river mouths are geometrically dissimilar and depend on the size of a river. This relationship may be utilized for approximate computations of channel deformations in river mouths with a predominant influence of river discharge and in the construction of river-mouth models.

The radius of the stream curvature over the bar is also associated with hydraulic characteristics of the flow (Mikhailov, 1958). For natural river mouths the following relationship is valid: $R \sim kQ^{0.5}$.

The author used the above relationships and the data obtained by Leopold and Maddock (1953) for rivers for the equation $h \sim h_1 Q^{0.4}$, $B \sim k_2 Q^{0.5}$, where B and h are the depth and the width of a channel, Q is channel-shaping volume of water.

The periodic processes in deltas

Laboratory experiments have shown that the processes of bar formation are of a periodic nature. An intensive widening of a bar which results in a bifurcation of the river flow and the formation of an island is further replaced by the straightening and the narrowing of the flow and, finally, by the washing-out of the island (Fig. 1f and g). In these environmental conditions a new bar is formed farther seaward in the same direction. Later, the bar widens again and forms a new island, etc. If an island so formed is somehow protected from the washing-out, two channels each with a bar in its mouth are developed on each side of the island. Under laboratory conditions the island was consolidated by growing a grass cover over its surface (the experiment was temporarily discontinued).

Similar periodic processes of widening and narrowing the flow passing over the bar were recorded in the mouth of natural rivers (the Danube, the Vistula and others). It is rather difficult to observe these processes—frequent and precise measurements being required—which, however, are indispensable for forecasting the navigable conditions of the water depth over the bar.

If the island built up in a bar area is not washed out, two short, symmetrical channels are usually formed. The subsequent process of delta development follows along the lines of periodic formation of new islands and the growth of a number of channels. During this process the symmetry of the initial division of the flow passing over the bar is usually rapidly disturbed. The principal cause of this phenomenon is probably the inability to maintain for a long period of time a constant volume of water through each newly formed channel; this should inevitably lead to the straightening of the greater-capacity channel. It is due to the fact that the minimum permissible radius of flow curvature in the greater-capacity channel is also larger. This circumstance may explain the present state of the Keil channel mouth in the Danube river delta. Out-of-date maps show that the initial delta channels bifurcated symmetrically. Later, one of the channels received a greater volume of water flow and straightened, while the other, through having lost its water flow, turned its course at a more obtuse angle with respect to the base channel.

As a result of this activity the traces of symmetric bifurcation of the water flow into delta channels gradually diminish beginning from the delta apex towards its more recent marginal areas.

CONCLUSIONS

As a result of analysis and extrapolation of field observations and specially planned laboratory experiments one may reach the conclusion that the formation of bars and river-mouth islands is characteristic of the majority of river mouths, located in various physiographic environments, and that their formation is predetermined principally by the dynamics of the river flow and the volume of the river-sediment discharge. Other factors (wave action, tidal currents, growth of vegetation, coagulation of suspended sediments, etc.) deform or shape the bar formed by the river flow, and facilitate or retard its development.

During the first phase of the bar-forming activity in an environment where the effect of river factors is predominant, the velocity structure of a river stream is of a decisive significance, as the bar and the river-mouth sand shoals are formed in conformity with this structure. At a later stage the newly-created forms of relief reveal their own effect on the river stream by widening the stream and even by bifurcating it into separate channels. At this phase of bar development it is possible to speak of the interaction processes taking place between the river stream and the forms of relief created by the latter. If bar-consolidating factors are not involved, the process of bar formation develops along the lines of periodic widening and narrowing of the river flow.

The position of an already formed bar or island in the mouth of a flow is predetermined by the velocity structure of the river stream entering the sea. By utilizing hydraulic methods, relationships were derived which facilitated the calculation of the distance to the bar and the radius of the flow curvature. The testing of these relationships based on the data of field surveys undertaken in the mouth of a number of rivers and laboratory experiments confirm their validity.

Résumé

Hydrologie et formation des barres dans l'embouchure des fleuves (V.N. Mikhailov)

L'analyse et l'extrapolation des résultats obtenus au moyen d'observations sur le terrain et d'expériences de laboratoire spécialement conçues à cette fin permettent de conclure que la formation de barres et d'îlots dans l'embouchure des fleuves est une caractéristique de la plupart des embouchures situées dans divers milieux géomorphologiques et que cette formation est prédéterminée principalement par la dynamique du fleuve et par le volume des sédiments que celui-ci dépose. D'autres facteurs (action des vagues, courants de marée, végétation, coagulation de sédiments en suspension, etc.) déforment et façonnent la barre formée par l'écoulement des eaux du fleuve et en facilitent ou en retardent l'évolution.

Au cours de la première phase de la formation de la barre dans un milieu où l'effet des facteurs fluviaux prédomine, la distribution des vitesses d'un cours d'eau revêt une importance capitale, car la barre et les bancs de sable de l'embouchure se constituent conformément à cette distribution. Par la suite, les formes de relief nouvellement créées font sentir à leur tour leurs effets sur le fleuve en l'élargissant, voire en le scindant en plusieurs bras distincts. A ce stade de la formation de la barre, on peut dire qu'une interaction se produit entre le fleuve et les formes de relief qu'il a créées. Si aucun facteur n'intervient pour renforcer la barre, la formation de celle-ci a pour effet d'élargir et de rétrécir périodiquement le cours d'eau.

La position d'une barre ou d'un îlot déjà formés dans l'embouchure d'un fleuve est prédéterminée par la distribution des vitesses du cours d'eau qui pénètre dans la mer. Au moyen de méthodes hydrauliques, l'auteur a obtenu des relations permettant de calculer la distance de la barre et le rayon de courbure du cours d'eau. L'exactitude de ces relations a été confirmée par les résultats d'enquêtes effectuées sur le terrain dans un certain nombre d'embouchures, ainsi que par des expériences de laboratoire.

Discussion

V. N. NAGARAJA. Dr. Mikhailov's paper is interesting, bringing out briefly the mechanism of the formation, development, stabilization, etc., of bars in estuaries. The author's findings are based on his observations of the river mouths of the Danube, Vistula and the Sulak in the U.S.S.R., all discharging into seas with insignificant tidal action.

When, on the other hand, strong tides and waves occur at the river mouth, the shoal-forming processes undergo great changes. A nearby example is the River Hooghly joining the Bay of Bengal. But for the great tidal activity the river would have been dead long ago. There are many other rivers which are similarly maintained by tidal action.

Since all important rivers are used for various purposes—navigation, industrial establishments, etc.—new forces come into play. Navigation channels, jetties and other structures, dredging and spoil disposal, etc., exert great influence on the river-mouth bars. Littoral currents also play an important part.

Bibliography/Bibliographie

BATES, Ch. C. 1953. Rational theory of delta formation. *Bull. Amer. Assoc. Petrol. Geol.*, vol. V, no. 9, p. 37.

LEOPOLD, L. B.; MADDOCK, T. 1953. *The hydraulic geometry of stream channels and some physiographic implications.* (United States Depart. Inter. Geolog. Survey, Prof. paper 252.)

MIKHAILOV, V. N. 1958. *Laws of island ("oseredkov") formation in the mouth of rivers.* (Scientific reports, Geological and geographical series. no, 1.)

——. 1959. *Hydrodynamics of rivers entering a basin.* (Publications of the State Institute of Oceanography, no. 45.)

——. 1962. *Processes of bed formation in a single-channel stream flowing into a tideless sea.* (Publications of the Oceanographic Commission of the U.S.S.R. Academy of Sciences, vol. X, no. 3.)

SAMOILOV, I. V. 1952. *The mouths of rivers.* Moscow. (Coll. Geography.)

Editor's Comments on Paper 9

In the added commentary to his earlier paper A. P. Carr noted the relationship that appears to exist between offshore banks and bars and the spits that develop landward of them. In a significant, but as yet unpublished, manuscript F. J. Meistrell points out the evolution of a spit upon the shoaling bar, or platform, that is built out ahead of the spit. The alternating cyclic growth of the spit and platform, and slope changes on the spit, were developed in a wave-basin model and then subjected to statistical analysis, as is described in his report. For this insight into spit development the Meistrell concept surely deserves a place in the spit and bar literature.

Frank Joseph Meistrell received the B.S. degree from Columbia University in 1965, and the M.Sc. from the University of Alberta at Edmonton, Canada, in 1966. Besides individual course work at the Université de Paris, the Sorbonne in Paris, and the University of Tennessee, he has undertaken studies toward the doctorate at the University of Natal, Durban, South Africa, and taught for a while at Hunter College in New York City. Since 1966 Meistrell has been employed in commercial geological exploration for copper, lead, and zinc in British Columbia, Canada, and in eastern Turkey; gold in south Africa; and oil and zinc in the southeastern United States. He is now 32, and resides in Florida.

UNIVERSITY OF ALBERTA

THE SPIT-PLATFORM CONCEPT: LABORATORY OBSERVATION OF SPIT DEVELOPMENT

A THESIS

SUBMITTED TO THE FACULTY OF GRADUATE STUDIES

IN PARTIAL FULFILMENT OF THE REQUIREMENTS FOR THE DEGREE

OF MASTER OF SCIENCE

DEPARTMENT OF GEOLOGY

by

FRANK J. MEISTRELL, B. S.

EDMONTON, ALBERTA

April, 1966

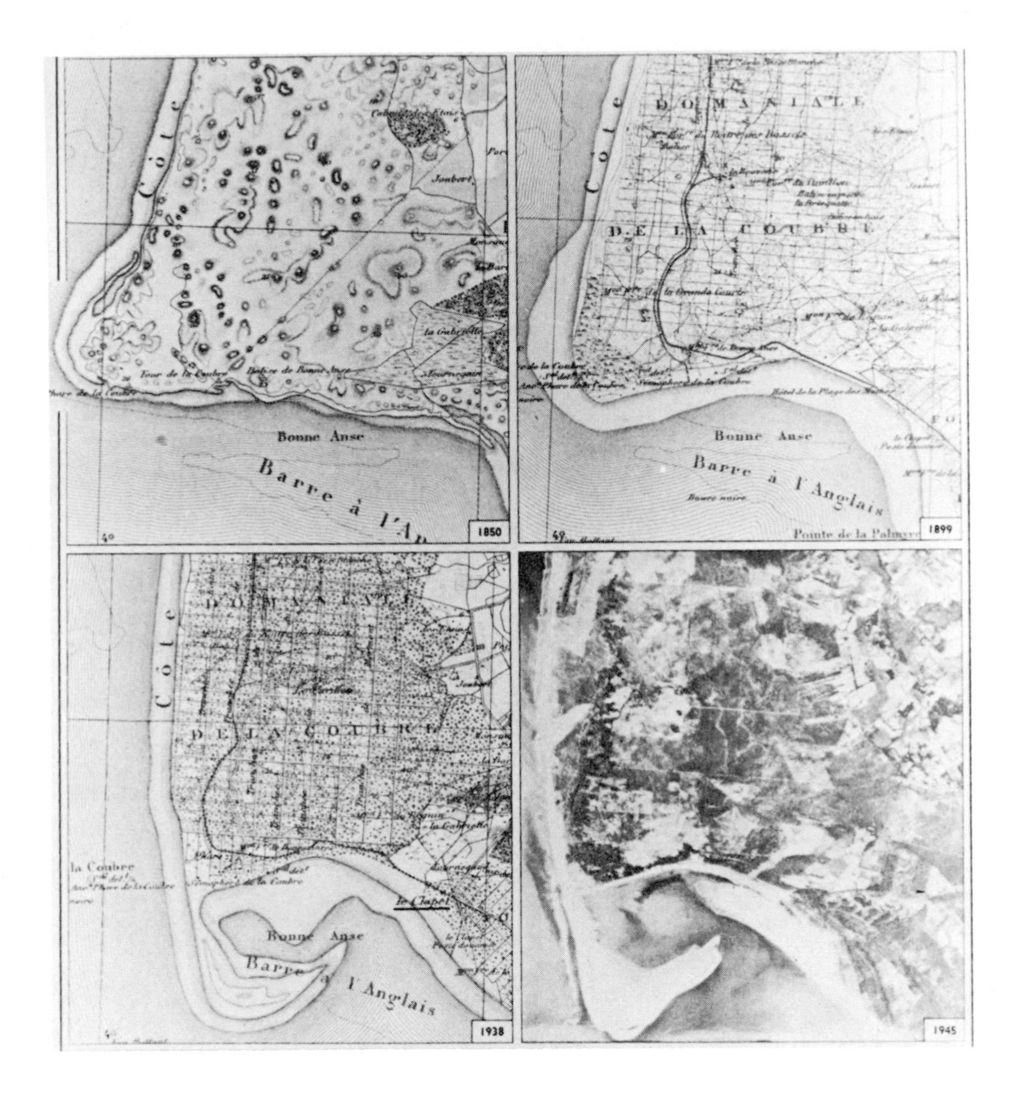

FRONTISPIECE

Development of a spit at La Coubre, France, during the period 1850 - 1945. The experimental model was a counterpart of this type of natural situation.

La Coubre, France is located on the west coast of France on the Bay of Biscay on the north edge of the mouth of the Garonne River.

Scale: 1:80,000 (Relief Form Atlas, 1956).

I do not know what I may appear to the world, but to myself I seem to have been only a boy playing on the sea-shore, and diverting myself in now and then finding a smoother pebble or a prettier shell than ordinary, whilst the great ocean of truth lay all undiscovered before me.

Sir Isaac Newton (1642-1727)

ABSTRACT

Spits and their sequential development were studied by a series of 27 wave tank experiments.

Waves with known parameters were directed obliquely against an erodible headland, inducing shore drift. In all cases a spit and platform developed at the down drift end of the headland. Similar features are present along modern coasts.

Slopes of the resultant beaches and length and width of the spit-platform structure were measured. Statistical methods were employed in finding the relationship between beach slope and wave parameters.

A spit-platform concept was derived from experimental results and statistically substantiated.

1. The platform is an embankment elevated above the shelf, but below mean low water level.

The spit is a ridge on the upper surface of the platform, partially emergent above mean high water.

The spit-platform structure is a large scale primary sedimentary structure formed principally by beach drifting.

Development of a spit is dependent on the presence of a platform

2. Growth of spit and platform is inversely related and occurs in alternating cycles.

Slope of headland beach is a function of grain size of beach material, wave steepness, mass transport, wave energy, wave height and wave length.

Slope of spit-platform beach is a function of grain size of beach material, wave energy, depth ratio, wave height and wave length.

3. With time, spit-platform beach slope increases until it equals headland beach slope. This change advances progressively along the spit-platform structure extending the headland beach.

ACKNOWLEDGEMENTS

The author wishes to express his gratitude to each of the following individuals, without whose kind assistance this work would not have been possible.

The thesis was supervised by A. J. Broscoe, Assistant Professor, Department of Geology, University of Alberta. He gave willingly of his time and energy, and the author is most appreciative for his many helpful suggestions and criticisms.

Professors Blench, Verschuren and Peterson of the Department of Civil Engineering assisted by providing the experimental apparatus and the facilities of the Hydraulics Laboratory of the University and by furnishing helpful advice in several disucssions. Mr. G. Schook assisted the author in measuring the waves generated in the tank. The technical staff of the Hydraulics Laboratory were most helpful in both the building of, and the continued maintenance of, the equipment.

Professor K.W. Smillie of the Department of Computing Science, and Dr. B. Mellon of the Research Council of Alberta, assisted the author in developing the statistical analysis of the data collected. All the mathematical calculations required were performed by the IBM 7040 computer of the Computing Science Department.

A word of appreciation is due the graduate students of the Departments of Geology, Civil Engineering and Computing Science who gave many useful suggestions throughout the course of the research. Mr. A. Johnson and Mr. J. Robinson of the Department of Geology were particularly helpful.

The initial phases of this work were conducted under the guidance of Professor Rhodes W. Fairbridge, Department of Geology, Columbia University. Under his sponsorship the author spent the 1964 field session studying spits on Long Island, New York. Mr. M. Schwartz of that department loaned his wave tank apparatus for the preliminary experiments. This work was done with the collaboration of Mr. Eric Clausen.

Professor G. D. Williams, Department of Geology, University of Alberta, loaned the author the photograph of the lake spit in Saskatchewan, Canada.

Mr. Frank Dimitrov, of the Department of Geology, University of Alberta, aided in the preparation of the plates, and took some of the photographs presented. Miss L. Summach typed the manuscript.

The author's wife, Diane, gave unfailing encouragement throught the course of these endeavors.

The author, in thanking all of the above for their kind and generous assistance, assumes full responsibility for the final manuscript.

TABLE OF CONTENTS

LIST OF ILLUSTRATIONS

Figures

PLATE

CHAPTER ONE

INTRODUCTION

Introduction

Since the early days of recorded history, man has been concerned with coastal features. Descriptions of the coast of the Mediterranean and Black Sea in "The Periplus of Scylax" and the Journal of Captain Pytheas existed as early as 400 B.C. In part, the requirements of coastal engineers were responsible in this century for the attempts at correlation between the descriptive aspects of coastal geomorphology and the dynamical aspects of geology and oceanography.

Previous Work

Many authors have written on the constructive agents involved in the development of coastal features. Amongst these are Gulliver (1899), Johnson (1919), Twenhofel (1932), Evans (1939), Thornbury (1954), and King (1959). Many authors, such as Shephard (1963) and Bascom (1964), have contributed to the understanding of waves in their natural environment. Wiegel (1964) and others have dealt with the problems of the interrelationships between coastal structures and oceanographic parameters.

The interrelationships between the various components which make up a coast, and the properties of the different types of waves are extremely complex. The use of models as a means of producing a workable facsimile of a coastal phenomenon, by establishing a situation with known parameters, has been expounded by Bagnold (1946), Bruun (1954), Bruun and Kamel (1964), Einstein (1948), Hubbert (1937), H.W. Johnson (1949), Kemp (1961), King (1959), Krumbein (1944), Reynolds (1933), Saville (1950), and Silvester (1960). As Thornbury (1954, p. 446,7) has summarized, spits have been thought to be the result of shore drift, or waves of major storms, or beach drifting.

Definitions

Although there are many types of coastal features, perhaps the most fundamental is the beach. Much research, both in the field and in the laboratory, has been conducted on the various component parts of the beach and the structures associated with the beach.

The definition of a beach used in this paper is essentially the same as given by Bascom (1964). Specifically, it is that profile which has its upper portion as land and its lower portion in water, and along which there is movement of material under surf action. The nomenclature of beach features as defined by Shepard (1963, p. 168) is also used in this report. The caption of Figure 1 includes definitions of some fundamental shoreline terminology.

A beach is the progenitor of a great many shoreline structures found in nature. A spit, it is suggested, is simply an extension of the beach. Certainly spit formation is closely related to the marine geologic processes operative upon the associated beach. It was decided that a fuller understanding of the complex nature of beaches might be provided by a study of the formation of spits. This study might also furnish some additional information on shoreline structures such as tombolos and baymouth bars. It is suggested, in fact, that the spit is the progenitor of these features. An interesting aspect of this birth sequence is that, with time, all of these structures eventually become part of, and indistinguishable from, the beach from which they originated.

Scope of the Research

The purpose of this research was firstly to study the development of spits in a wave tank under different known wave conditions, and secondly to correlate this spit development with the known wave parameters in order to obtain some indication as to their relationship.

The geomorphic setting was one of an erodible headland under oblique wave attack, the result of which was shore drift, and the subsequent formation of a spit at the down-drift end of the headland (Frontispiece). An experimental matrix of 27 different wave conditions was designed. Each of the 27 tests began with an uneroded headland.

Bruun (1962) evolved the theory that beach profiles change with variation in water depth. Schwartz (1965) substantiated this theory with both experimental and beach studies. The unifying dynamical link between headland-spit structure and different wave conditions might best be expressed in the slope of the resultant beach. Therefore, six separate slope measurements were taken, three on the headland beach and three on the spit beach, at uniform time intervals during the experiment.

Studies were made of hydrographic charts of coastal areas which had spits and related structures. Field work on spits was also conducted.

The basic terminology used is seen in Figure 1.

1. The shelf area is the general level of the submarine bottom surface surrounding a headland.

2. The platform is the structure, an embankment in character, adjacent to, and connected with the headland which is elevated above the shelf, but below the mean low water level.

3. The spit is the structure, a ridge in character, located on the upper surface of the platform and partially emergent above the mean high water level.

4. The spit-platform structure, connected at one end with the land, terminating in open water, elongate in character, straight to curved in plan, is a large scale primary sedimentary structure composed of a platform and a spit.

Figure 1. Terminology

Shelf area: the general level of the submarine bottom surface surrounding a headland.

Platform: the structure, an embankment in character, adjacent to, and connected with the headland which is elevated above the shelf, but below the mean low water level.

Spit: the structure, a ridge in character, located on the upper surface of the platform and partially emergent above the mean high water level.

Spit-Platform: connected at one end with the land, terminating in open water, elongate in character, straight to curved in plan, is a large scale primary sedimentary structure composed of a platform and a spit.

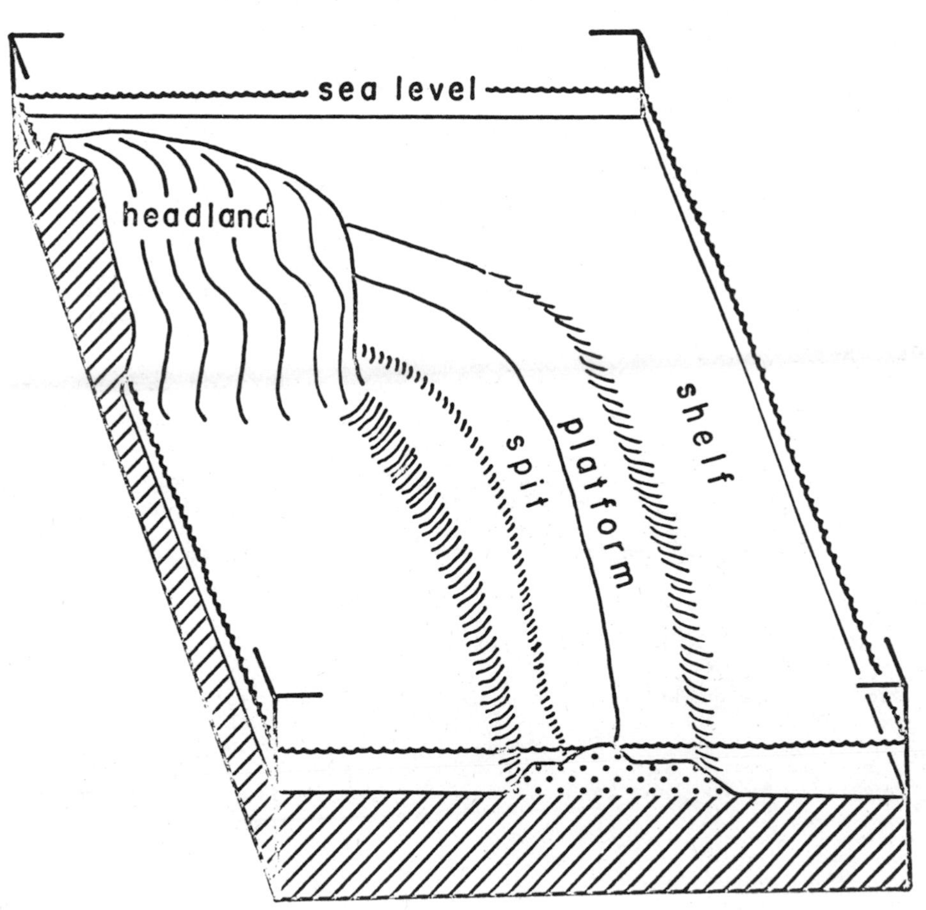

CHAPTER TWO

APPARATUS AND PROCEDURES

Introduction

The apparatus and procedures associated with the research can be broadly considered in three groups. Basically, several different wave types were created by a wave generator in a medium sized wave tank. These waves attacked a headland structure built in the tank on a sand base at the end opposite to the wave generator. Data on the beach slopes and spits formed were recorded for each case. These data and the various wave parameters were then statistically analyzed using an IBM computer, in order to determine any interrelationships.

Apparatus

General Statement

The apparatus used in the course of the research may be divided into three main groups. This grouping corresponds to the different phases of the work. The first group consists of the wave tank and associated equipment; the second group is the sand and sand structures built in the wave tank; the third group is the IBM 7040 computer and related equipment.

Wave Tank and Associated Equipment

The wave tank and wave generator (Plate 1-A, Figure 2) used was similar to the types described in the literature by Nortov and Levehenk (1963), Samarin (1962), Inman and Bowen (1963) and Schwartz (1965). The tank was constructed of 1/4 inch thick steel plates and was ten feet long, six feet wide, and two feet deep.

The wave generator was of the swinging shield type (Plate 1-A, Figure 2) as described by Bascom (1964) and Schwartz (1965). The driving force was a 3/4 horse-

A. PLAN VIEW

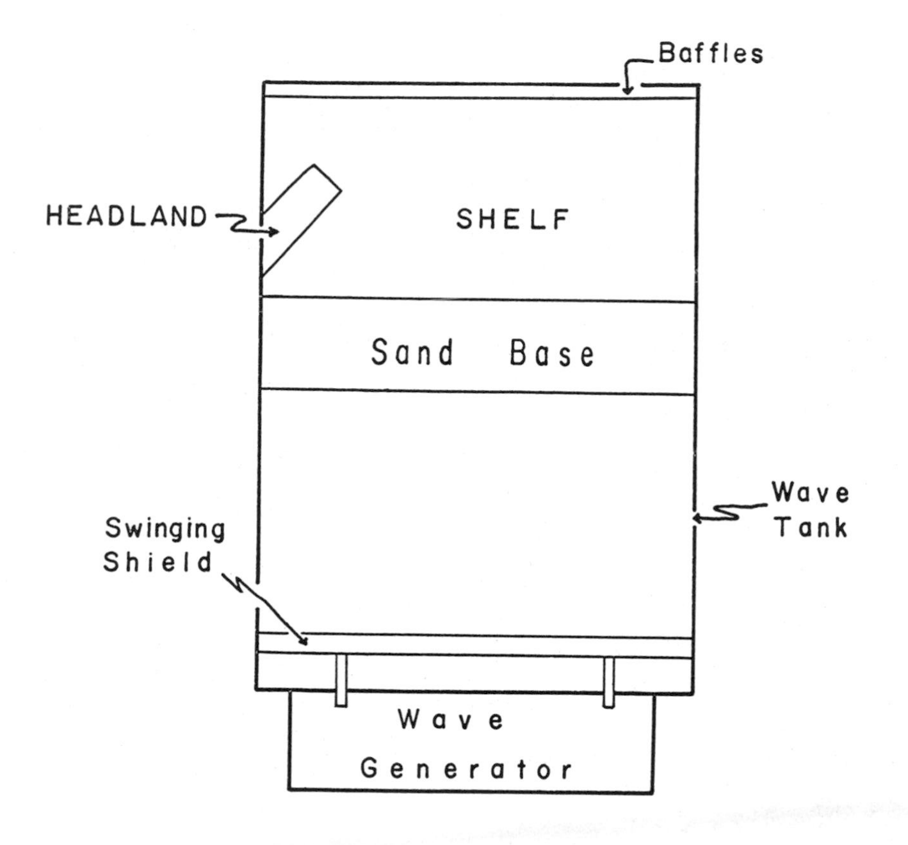

B. CROSS SECTION (long side of tank)

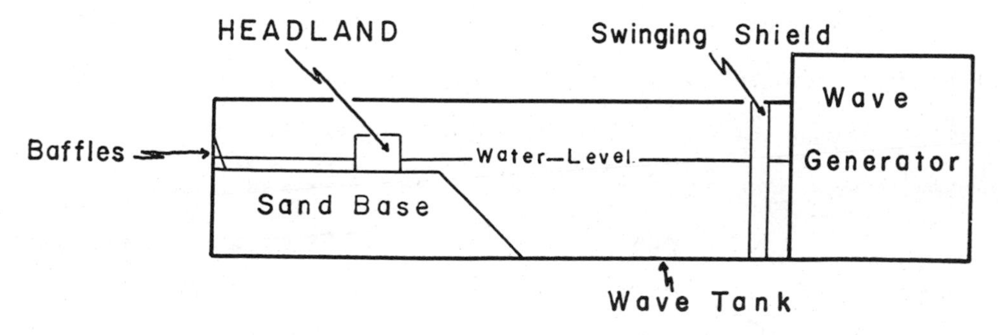

Figure 2. Diagramatic sketch of apparatus.

power PH Induction motor, which oscillated the shield by means of eccentrically set rocker arms connected to the motor by belts and adjustable pulleys. This mechanical arrangement enabled the author to both vary the length and the frequency of the stroke of the swinging shield. The shield was faced with a plywood sheet one-half inch thick, six feet wide, and two feet high, hinged to the bottom of the tank. This size facing of the shield allowed just enough space at either end so as to permit free movement along the side of the tank.

A Mosely Autograf X-Y Recorder was used to accurately measure the 27 wave types created in the wave basin. A pressure transducer of the standard type made in the Hydraulics Laboratory was coupled to the recorder. The recorder was thus able to plot a continuous graph of the wave. This graphical representation of the wave was accurate to about 0.2% on the ordinate axis and 0.1% on the abscissa axis.

Sand and Sand Structures

The general sand structure created in the wave tank for each experiment consisted of two parts (Figure 2). The first part was a sand base upon which the second part, the headland structure, was built. The sand base was built at the end of the tank opposite the wave generator. The front edge of the sand base was parallel to the swinging shield. The base was one foot high and was level for 3.7' from the back of the tank and then sloped to the bottom at an angle of 26°. The bottom edge of the sand base was 6' from the back of the tank. The upper level surface of the base is called the shelf area (Figure 1).

A box was constructed so as to form a mold for the headland structure. The resulting headland structure had dimensions of 1.5 feet in length, 0.5 feet in width and an initial height of 0.5 feet. The shape of the headland was rectangular. In map view, the headland slanted towards the back of the tank at an angle of 67° with respect to the side of the tank and was 3 feet from the back of the tank. Use of the mold insured that the headland had a constant position, shape and angle for each experimental test.

Baffles of expanded aluminum mesh and 1/4" mesh screening were placed along the entire back of the wave tank opposite the wave generator. The mesh had the effect of damping waves which otherwise would be reflected from the rear wall of the tank, creating undesirable complex wave parameters. The accumulation of water on the shelf area as a result of the passage of waves over the shelf necessitated the construction of an outflow channel. A trench was dug behind the baffles and then extended along the side of the tank, opposite the headland structure, to the edge of the shelf. The excess water on the shelf returned through this channel to the deep basin part of the tank.

Several types of sand were used in the different phases of experimentation (Figure 3, A-D). The size distribution of the sand were initially tested (Figure 3, B,C). Similar spit-platform structures were formed in all cases tested. The sand used in the early experiments conducted at Columbia University (Figure 3-D) also gave similar spit-platform structures.

The instrument used for measuring the slope of the beach was a three-dimensional mobile pointer. The pointer could be moved in the two horizontal axis, X and Y, and the vertical axis, Z. The X axis was parallel to the ends of the wave tank, the Y axis was parallel to the long side of the tank, and the Z axis was perpendicular to the bottom of the tank. The pointer was on the end of a measuring rod which was movable in the Z axis. The measuring rod moved on a bar in the Y axis. The bar moved in the X axis. The bar was attached to an open square frame six feet on each side which was securely fastened to the tank. The position of the pointer on the X and Y axis was recorded to the nearest 0.01 feet, and on the Z axis to the nearest 0.001 feet. The entire instrument was made of aluminum. Movement was effected by hand operated gears.

Figure 3: Histogram of Sand Size Distribution

A - Sand used in main experiments at University of Alberta.

B,C - Sand also tested at University of Alberta.

D - Sand used in experiments at Columbia University.

E - Sand from upper berm, mid-spit, Eaton's Neck Spit, Long Island, New York.

F - Sand from distal end of Eaton's Neck Spit, Long Island, New York.

G - Sand from upper berm, mid-spit, Lloyd's Neck Spit, Long Island, New York.

H - Sand from distal end of Lloyd's Neck Spit, Long Island, New York.

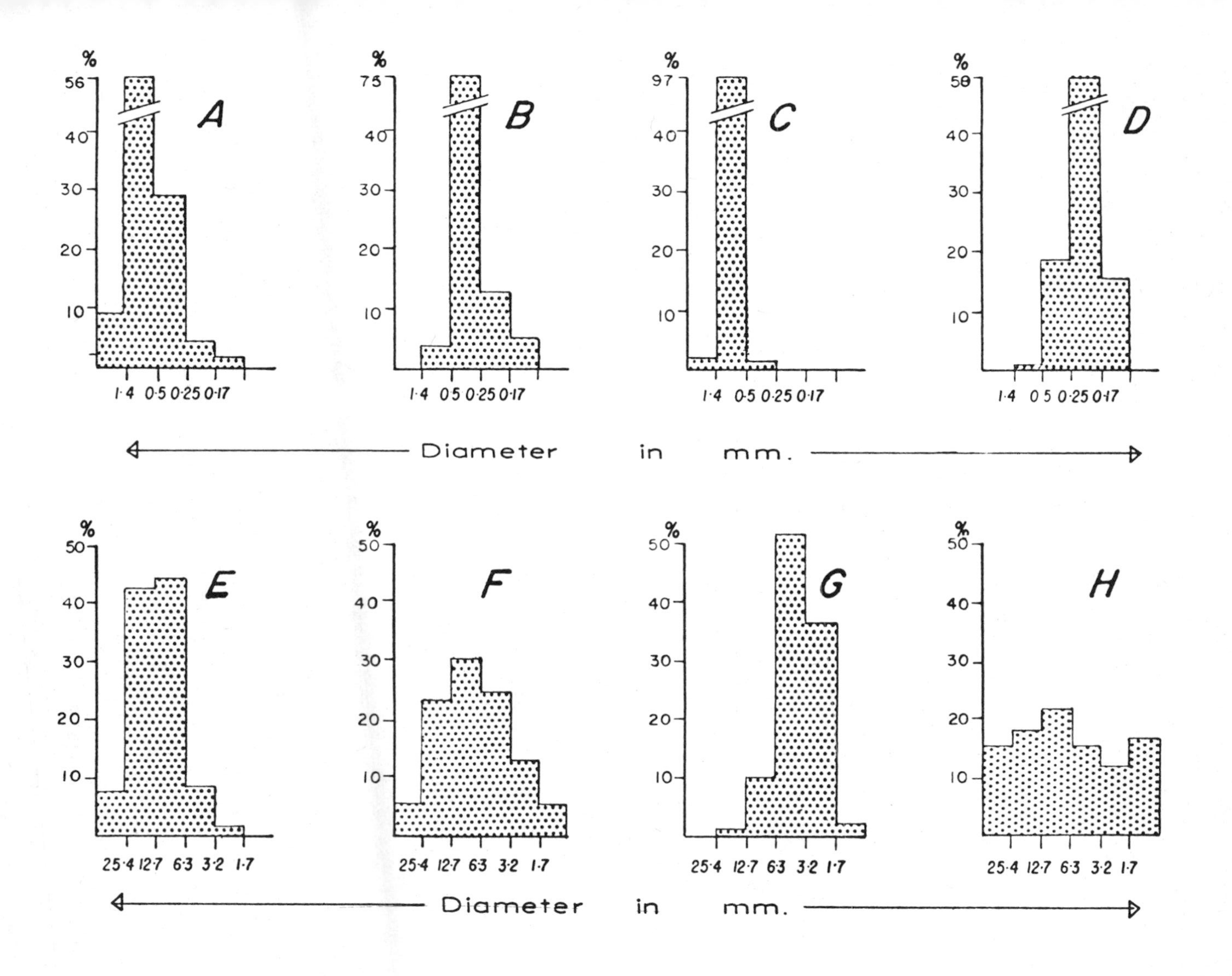

%
56
40
30
20
10
A
1·4 0·5 0·25 0·17
%
75
40
30
20
10
B
1·4 0·5 0·25 0·17
%
97
40
30
20
10
C
1·4 0·5 0·25 0·17
%
58
40
30
20
10
D
1·4 0·5 0·25 0·17
Diameter in mm.
%
50
40
30
20
10
E
25·4 12·7 6·3 3·2 1·7
%
50
40
30
20
10
F
25·4 12·7 6·3 3·2 1·7
%
50
40
30
20
10
G
25·4 12·7 6·3 3·2 1·7
%
50
40
30
20
10
H
25·4 12·7 6·3 3·2 1·7
Diameter in mm.

Computer

The computer used for the various mathematical calculations undertaken was an IBM 7040. Also used were the associated machines such as key punch, card sorter, reproducer and printer. The author wrote the program which was used in the computation of the beach slope and horizontal beach width from the X, Y, Z, co-ordinate data. The programs used for statistical analysis were obtained from the Library of the Computing Science Department, University of Alberta.

Procedures

General Statement

The procedural aspects of this research are divided into three main groups. The first group consists of the generation of the different cases of waves; the second group consists of the measurement of the slope of the beach, of the headland and spit-platform structure; the third group deals with the analysis of the data collected.

Different Wave Conditions

An experimental test matrix was designed which consisted of 27 different wave cases (Figure 4). This matrix was based on the three parameters that could be controlled in the tank. These were: (1) the length of the stroke of the swinging shield, i.e. the initial amplitude of the wave form, (2) the frequency of the movement of the shield, i.e. the period of the wave, (3) the depth of the water in the tank, i.e. the water depth on the shelf area. In establishing the test matrix, high, medium and low values were selected for each of the controlled parameters of wave amplitude (height), wave frequency (period) and water depth. This produced the matrix of 27 wave cases which were used to study the development of spits.

Prior to starting the experimental tests, the wave cases were produced in the tank and were recorded by the Autograf Recorder. It is important to note that the wave amplitude did not have three constant values associated with the three constant stroke

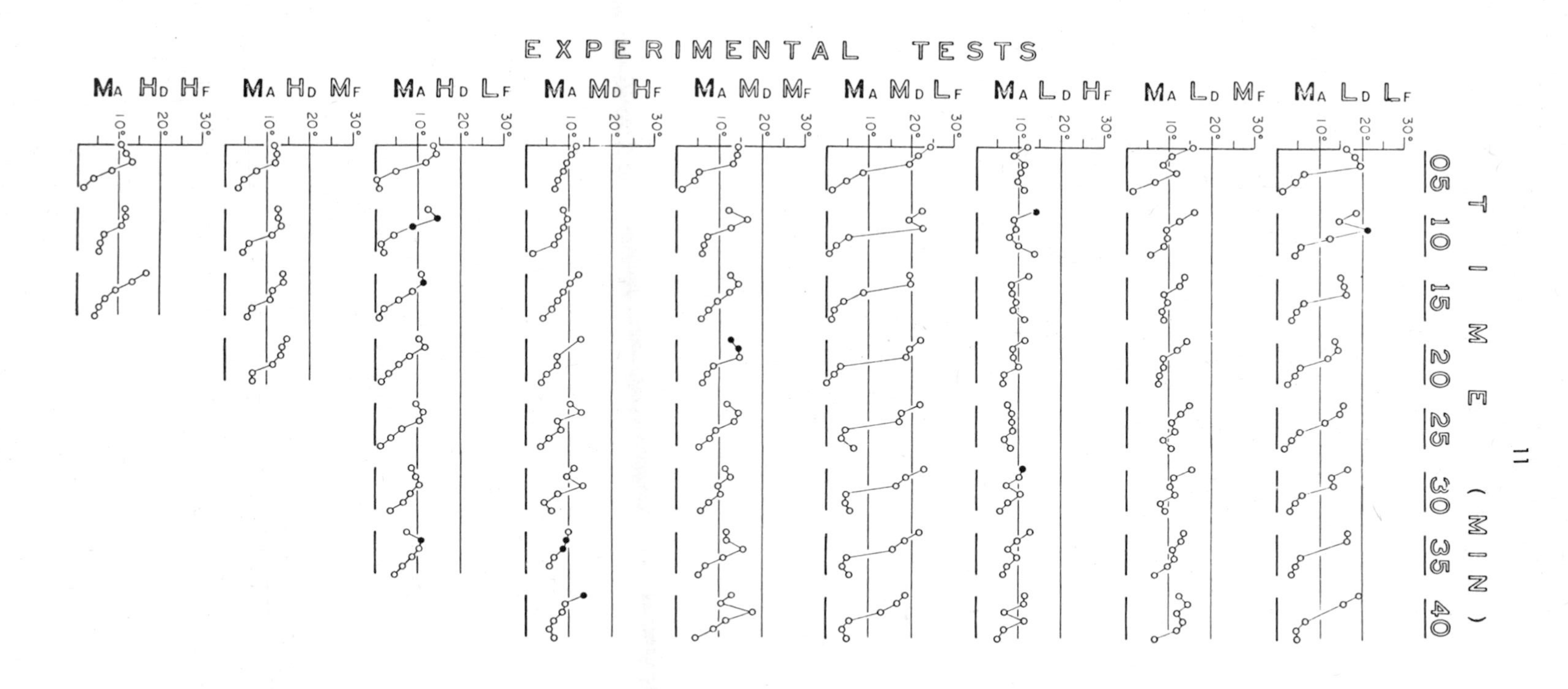

Figure 4. Experimental Test Matrix

The experimental tests were designed in a 3x3x3 matrix. Each one of the parameters had a high (H), medium (M), and low (L) value.

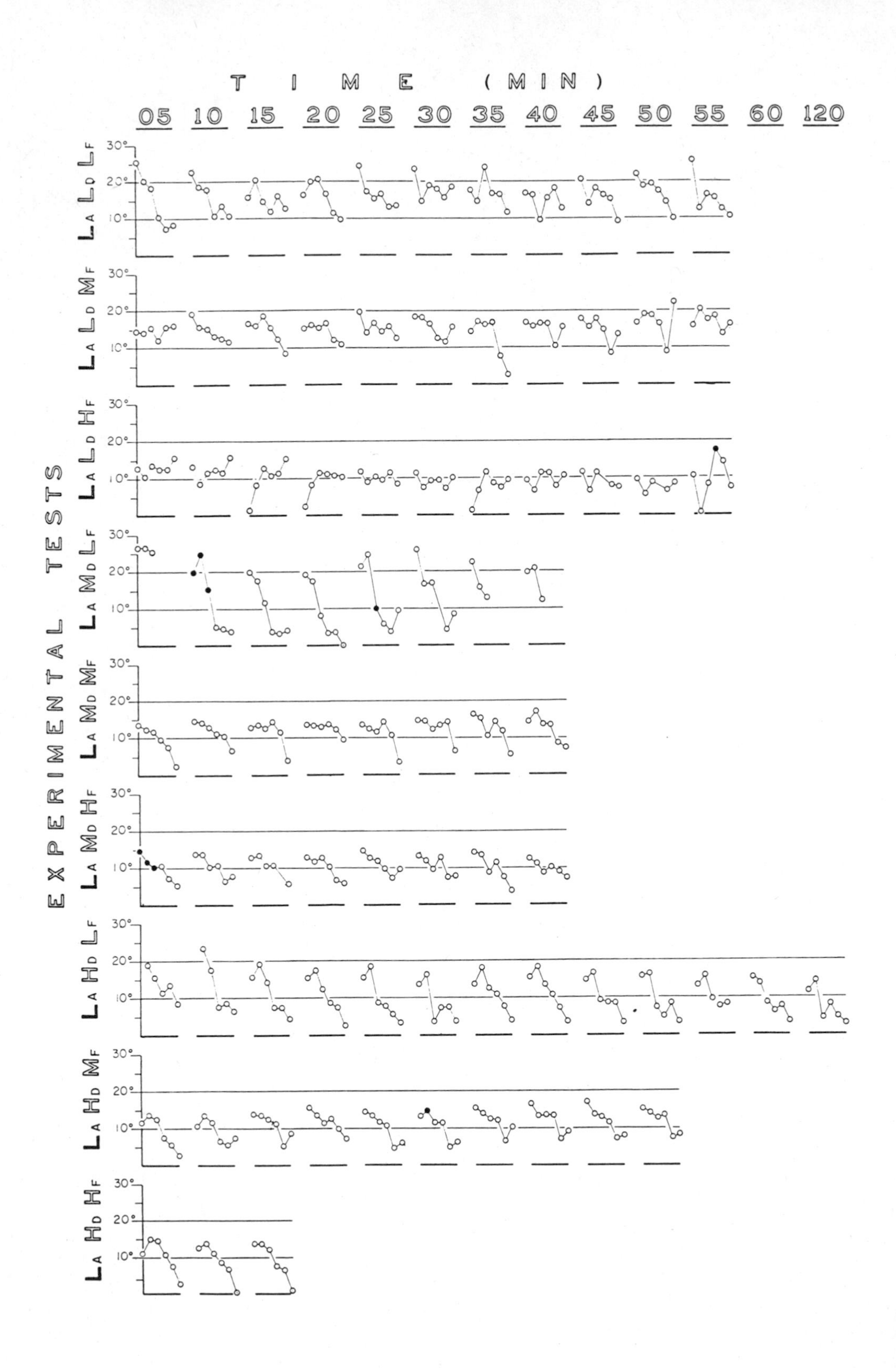
T I M E (MIN)
05 10 15 20 25 30 35 40 45 50 55 60 120
EXPERIMENTAL TESTS
LA LD LF
LA LD MF
LA LD HF
LA MD LF
LA MD MF
LA MD HF
LA HD LF
LA HD MF
LA HD HF
30°
20°
10°

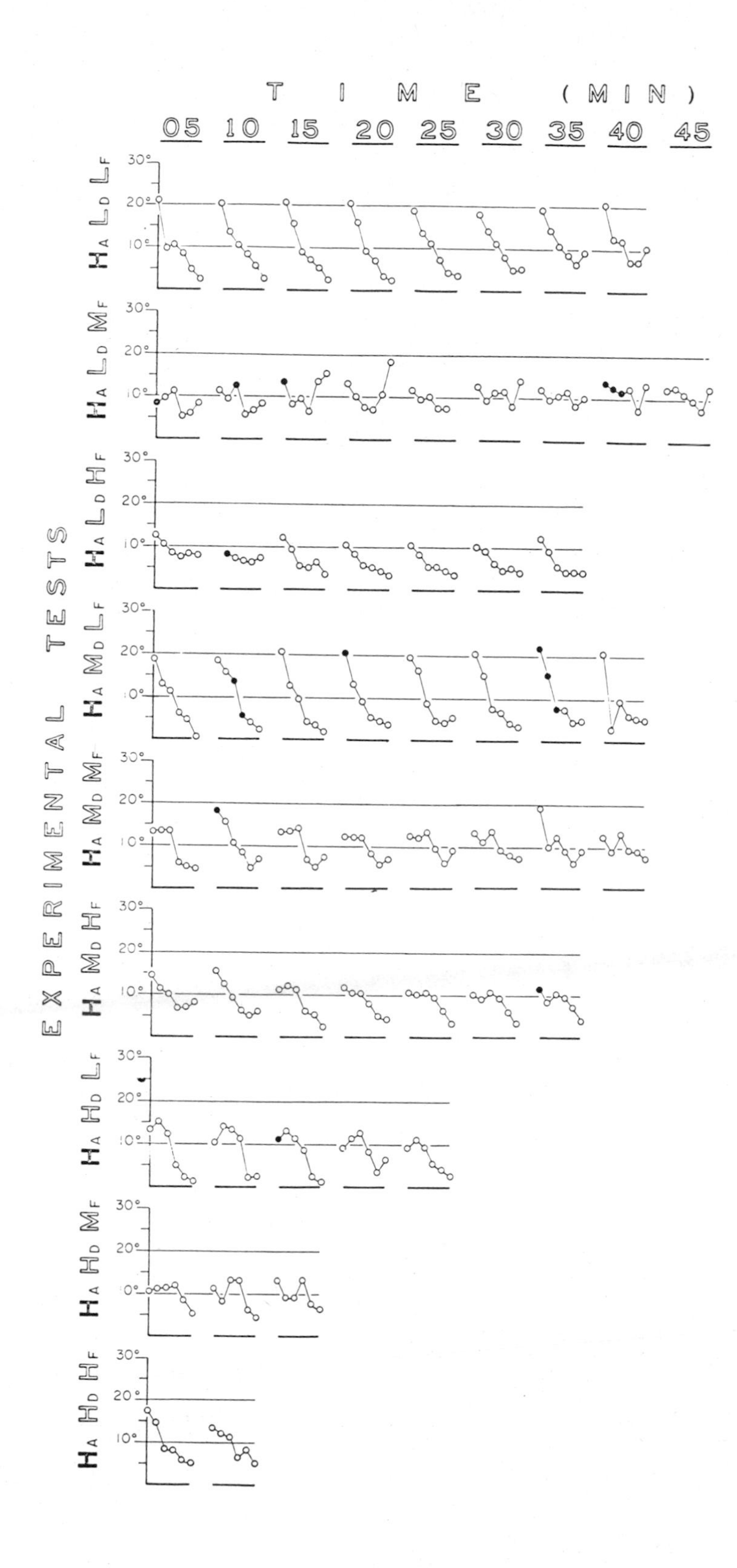
T I M E (MIN)
05
10
15
20
25
30
35
40
45
EXPERIMENTAL TESTS
30°
20°
10°

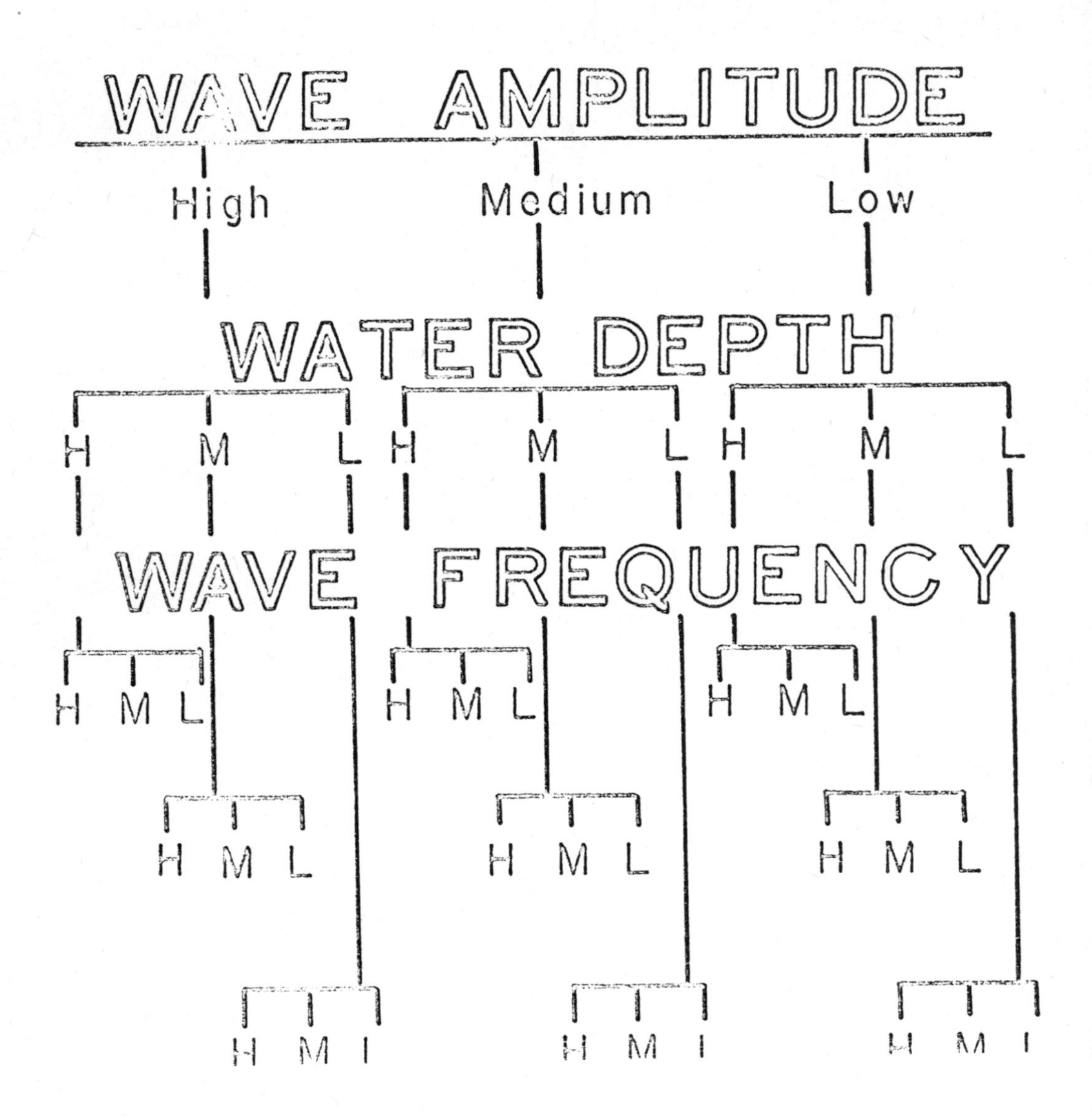
WAVE AMPLITUDE
High
Medium
Low
WATER DEPTH
H M L H M L H M L
WAVE FREQUENCY
H M L
H M L
H M L
H M L
H M L
H M L
H M L
H M L
H M L

settings of the swinging shield. Rather, the amplitude was modified by the interaction with the changes of the frequency and the depth of the water. The various values for wave amplitude (height), and wave steepness are detailed in Appendix One, Part A.

Each of the 27 experimental tests was conducted in the same manner. The three wave parameters were set. The mold for the headland structure was placed in position and filled with moist sand which was then tamped down in a standard manner. The mold was removed. The waves were generated for five minute intervals.

Measurement of Beach Slopes

At the end of each five minute interval wave action ceased and data needed to make six slope measurements on the headland and spit-platform structure were recorded. The actual measurements recorded were the X, Y, Z coordinates of 12 points (6 pairs) from which the slope values were computed. These coordinate measurements were taken at pre-determined locations, as described in the following paragraph. The length and width of the spit and the platform were also recorded. This sequence of wave action and measurement was repeated until the spit-platform structure either showed a slow steady growth for at least twenty minutes or reached the back of the tank. When either of these conditions had occurred, the experimental test was complete. The headland and spit-platform material was removed and the shelf area was levelled. The next experimental test was then begun.

In order to determine the mean slope of each beach profile, a point was chosen at the top of the profile. An imaginary line, perpendicular to the shoreline and in the plane of the beach slope, was extended from the upper point to the bottom of the profile, thus locating the bottom point. The X, Y, Z coordinates of these two points were recorded. By simple trigonmetric methods, the slope of the beach profile and the true horizontal width of the beach was calculated. These calculations were done by the computer. The length and the width of both the spit and the platform were measured with a scale to 0.01 foot accuracy.

The location of the six beach slope measurements and the measurements of the length and width of the spit-platform are shown in Figure 5. Positions One, Two, and Three were located on the headland, dividing it into approximately three equal parts. Position One was placed 3" from the side of the tank so as to avoid any undesired effects from any boundary condition. Position Two was approximately midway on the headland structure and position Three was at the end. Positions Four, Five and Six were located on the Spit-Platform structure, dividing it into approximately three equal parts. Position Four was adjacent to the headland but on the spit ridge. Position Five was midway between position Four and Position Six which was the end of the spit. The three headland positions (1,2,3) remained essentially constant, while the spit-platform positions (4,5,6) were variable, dependent on the growth of the structure.

The slope of the beach at any one of the six positions was not a true linear feature. Rather the slopes were either convex or concave in profile. The slope measurements derived were therefore mean slopes of the beach profile (Appendix One, Part B). The headland beach slope measurements were taken from the top of the berm crest to the crest of the shore face (Shepard, 1963, p. 168). The spit-platform beach slope measurements were taken from the top of the outermost spit ridge (berm crest) to the upper outermost edge of the platform (crest of the shore face).

The length of both the spit and platform were measured in a straight line from the terminus of the headland to the end of the individual structure, regardless of any curvature that might have developed. The width of the spit and platform was taken perpendicular to the length line and at the respective structure's widest portion. Since the spits tended in many cases to be compound in morphology, the width was determined for only the outermost ridge.

Analysis of Data Recorded

The data from the 27 experimental tests were analyzed in two groups. These two groups were beach slope measurements and length and width of the spit-platform

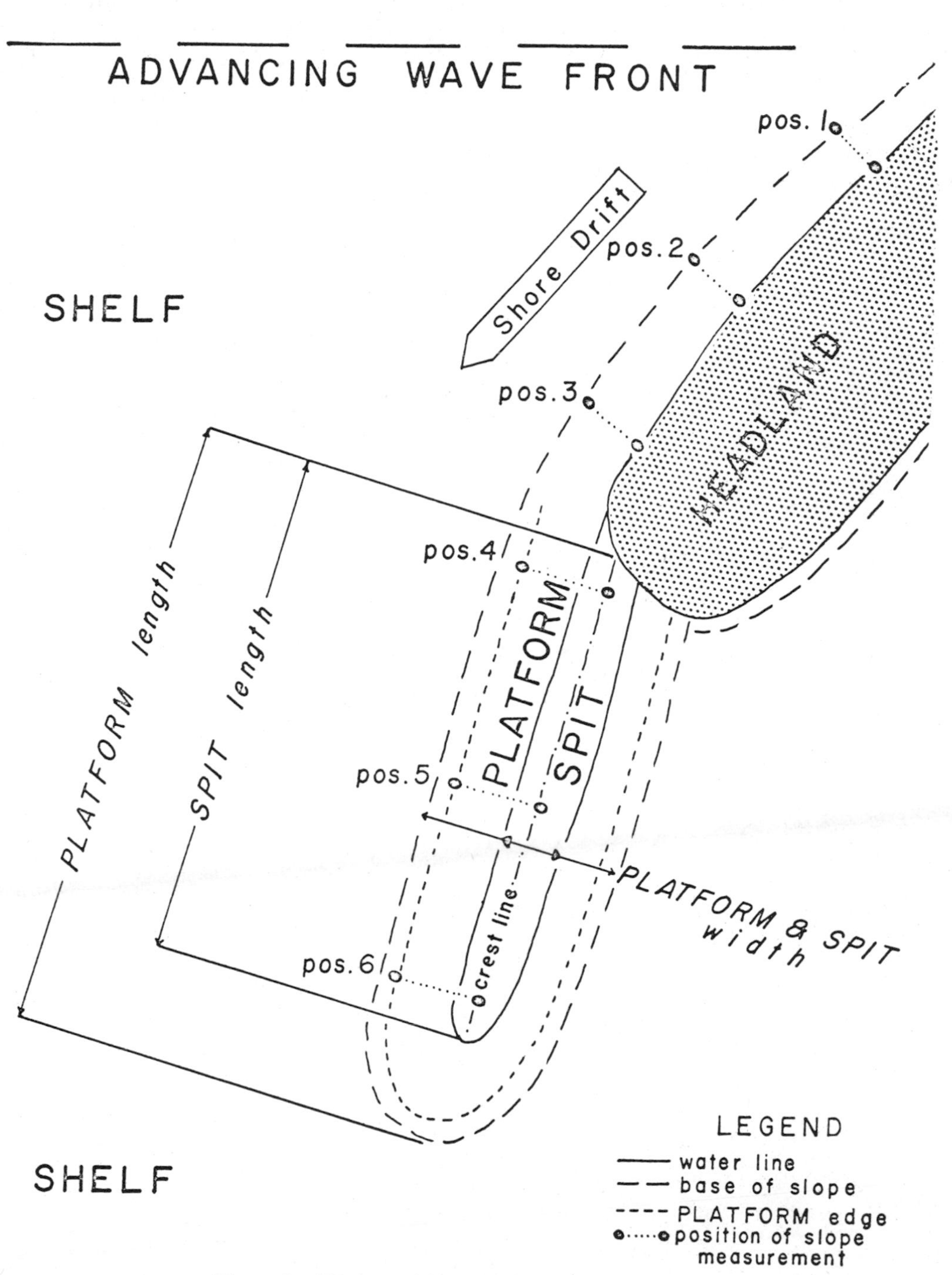

Figure 5. Positions of Measurements Recorded.

structure. Each group was analyzed with respect to the wave conditions. The slope measurements were analyzed using the statistical techniques of analysis of variance and multiple regression. The length and width measurements of the spit-platform structure were considered by comparison of volumetric changes with time.

Errors

Two sources of error existed with respect to the data collected, experimental error and operator bias. The experimental error was the result of the difficulty of placing the tip of the pointer exactly on the sand surface. In practice this tip was usually below the upper sand surface. It was determined that this error probably affected the value of the angle of the slope plus or minus one degree. The operator bias error was due to the difficulty in locating the second point exactly normal down the beach face. This error was judged to affect the value of the angle of the slope plus or minus one degree. The angle of the slope of the beach profile therefore has an estimated accuracy of plus or minus two degrees.

CHAPTER THREE

THE SCALE FACTOR

Introduction

In order for model experiments in a wave tank to have any geologic significance, it is essential that there be a correlation between the results obtained from the wave tank and known natural conditions. This correlation is accomplished by use of a scale factor. Initially the scale factor was established by means of comparison of the dimensions of waves in the tank with the dimensions of waves as found in the ocean. Then, as a check, the same scale factor was applied to the dimensions of the model spit and water depth in the tank and correlated with similar parameters in nature. If the scale factor does indeed correlate the model results with those found in the geologic situation, then one is justified in tentatively extending the conclusions based on the model study to the natural geologic situation.

The Scale Factor

Bridgman (1922) and others have shown through dimensional analysis that a scale factor is a dimensionless number which equates a measurable quantity in one system with a similar measurable quantity in another system, given that the two separate systems are of the same type. As Rector (1954) and others have pointed out, the results obtained from models are not actual complete reproductions of larger scale natural processes, but rather they similate, with reasonable accuracy, the principles which appear to be involved in a given situation.

For the purposes of this paper, it is defined thatthe fundamental measurement of the wave parameters is length. The measurements recorded on the spit-platform and headland are in terms of length and the dimensionless unit of slope (Figure 5). The depth of the water is also a unit of length. It is seen then, that the basic unit of measurement in this model experiment is the unit of length. The scale factor should then be a dimensionless number which approximately equates the length measurements in the tank to the length measurements in the corresponding natural situation.

Blench (1965, oral communications) suggested that it was only necessary to correlate the range of tank waves with the range of ocean waves in order to derive a scale factor. This was done by scaling the length and height of the tank waves so as to be comparable with the length and height of similar ocean waves. The scaling of the tank wave heights and lengths was done directly by multiplying by the scale factor. By using the formula $L = 5.121^2$ (Shepard, 1963, p. 80, where L = deep water wave length and T = deep water wave period) the periods for the ocean waves were determined. Direct observations and measurements of the waves generated in the tank indicated a wide range of wave types. The tank waves correspond to the swell and sea classes of ocean waves as shown in the ocean wave spectrum presented by Bascom (1964, p. 9). These are the most common types of waves on a coastal shoreline. The scale factor obtained for the model experiments conducted in the tank used in this reseárch was: unit length (Tank Waves) x 200 = unit length (Ocean Waves). The different tank and scaled wave parameters are presented in Figure 6.

In order that this scale factor be considered valid, it must be applied to the other length quantities in the tank. This serves as a check. The depth of the water on the shelf area, when scaled, was 18 feet (low), 23 feet (medium), and 32 feet (high). These depths are within the range found in nature in near-shore areas.

Comparison of sizes of material furnished a further check. The mean grain size used in the experiment (Figure 3, A) is about 1/10th the mean grain size of material found on two spits located in Long Island, New York - (Figure 3, E-H). Material on a shingle spit reported by Carr (1965) in England is about 200 times as large as the material used in the experiment. The use of very fine material in the tank would result in the introduction of the factor of cohesion, thereby destroying the

WAVE PARAMETERS

	EXPERIMENT						NATURE Scale Factor: EXPERIMENT x 200					
	PERIOD (sec.)	LENGTH (ft.)	HEIGHT (ft.)	STEEPNESS	VELOCITY (ft./sec.)	DEPTH RATIO	PERIOD (sec.)	LENGTH (ft.)	HEIGHT (ft.)	STEEPNESS	VELOCITY (ft./sec.)	DEPTH RATIO
	T	L	H	S	C	D	T	L	H	S	C	D
				H/L	L/T	h/L				H/L	L/T	h/L
HIGH	*0.65*	*2.16*	*.0614*	.0284	3.32	.546	9.2	432	12.28	.0284	47	.546
MEDIUM	*0.79*	*3.19*	*.0366*	.0114	4.04	.378	11.2	638	7.32	.0115	52	.378
LOW	*1.65*	13.93	*.0195*	.0068	8.45	.089	23.3	2789	3.90	.0014	121	.089
notes	SLANTED NUMBERS indicate measured values UNDERLINED NUMBERS indicate mean values											

Figure 6. Wave Parameters.

H = deep water wave height, L = deep water wave length,
T = deep water wave period, h = depth of deep water.

observed dynamic similarity between tank and actual beaches.

In some of the preliminary experimental tests conducted at Columbia University, the effect of scaling the surface tension of the water was investigated by adding a wetting agent to the water. The wetting agent used was Triton CF-32, a non-ionic surfactant, mixed in a concentration on the order of 50 parts per million. It was found, as Schwartz (1965) has reported for similar experiments, that the wetting agent produced no noticable effects with regard to beach slope development, or movement of sand particles. The wetting agent did, however, reduce the ability of the waves to form breakers.

Natural spits range in length from tens of miles (Sandy Hook, New York) to tens of feet (Lake Spit-Canada, Plate 1,F). Grain size of material on these spits ranges from sand to cobble. The scale factor presented here, derived from wave length and height, provides satisfactory correlation with the common coastal situations found in nature. It would be unreasonable, in the light of known ranges of size of natural materials and existing spits to expect any more precise agreements between model and natural examples than that presented in the preceeding paragraphs.

CHAPTER FOUR

MORPHOLOGY OF THE SPIT-PLATFORM

Introduction

The geomorphic setting created in the tank was an erodible headland under oblique wave attack. The result of the wave attack was shore drifting, and the subsequent formation of a spit at the down-drift end of the headland (Frontispiece and Plate 1-C). The spit thus formed is but the partially emergent part of a larger submarine stru ture, the Platform (Figure 1).

Headland

The headland was eroded by wave attack and material was moved by shore drifting. This movement was principally beach drifting in nature. A fine discussion of shore processes is presented by Strahler (1963). Prior to initial wave attack, the wet sand composing the margin of the headland slumped into the water and assumed a slope which was the angle of repose of the sand in quiet water. Under wave attack a beach profile developed. The headland beach profile developed in these tank experiments is of the same type as reported in detail by Sivakov (1963) for his beach profile experiments. The profiles formed in the present experiments had all the characteristic sections of a beach as indicated by Shepard (1963, p. 168). These were a foreshore section, with component parts of berm crest, beach face and low tide terrace, and an offshore section, with component parts of shore face, longshore trough and longshore bar. It was noted that the longshore trough coincided with the plunge point of the incoming breaking waves. In some experimental tests the longshore trough and bar did not form.

The actual beach profile was not a planar surface. Rather it was at times concave upward or convex upward or both in part. Therefore, the slopes measured on the headland beach (Positions 1,2,3) were mean values from the berm crest to the upper edge of the shore face. Slumping at the front face of the headland occurred from time to time in each experiment. This was due to the undercutting action of the waves and the unconsolidated nature of the materials forming the headland face. Rounding of the headland face was done at irregular intervals in order to reduce the occurrence of slumping. When slumping had occurred and the beach profile had not as yet readjusted at the end of a five minute interval, slope measurements for the affected positions were not recorded. This accounts for the missing points in some of the experimental tests (Appendix One, Part B). The headland and a developed spit-platform are seen in Plate 1-C. This is typical of the structures that were built in all 27 tests.

The equivalent example in nature of a headland eroding and supplying sediments to a spit is seen in Plate 1-D. This is the headland at Lloyd's Neck, New York. A spit is to be found at either end of this headland. A similar spit is to be found nearby on Eaton's Neck. A longitudinal view from the spit crest of the Eaton's Neck Spit is seen in Plate 1-E. A small lake spit in Prince Albert National Park, Canada, is seen in Plate 1-F. This is similar to a spit found in a Michigan Lake as reported by Wulf (1963).

The Spit-Platform

Due to wave attack, sediments were eroded from the headland and moved, principally by beach drifting, down the headland beach to the deeper water of the surrounding shelf area. A study by Boldyrev and Nevesskiy (1964) of headland erosion and subsequent deposition on the Chushka spit (in the area of the Kerch Strait between the Sea of Azor and the Black Sea) indicates results similar to those observed in this model study. In the model, the sediment upon reaching the deeper water at the end of the headland beach was deposited in a successive series of fore-set beds. This physical extension of the beach is the beginning of the platform structure. The platform has two distinct characteristics which differentiate it, morphologically, from the beach. The platform tends to build in length normal to the predominant wave front direction, and it has a markedly less steep profile than the headland beach slope (Appendix One, Part B).

As the platform continues to increase in length, three major characteristics become evident: (1) the depth of water above the platform remains constant, (2) the structure is basically composed of fore-set and top-set beds, and (3) a spit ridge forms on the top of the platform.

The depth of water above the platform remains constant. The height of the platform above the general level of the shelf is directly proportional to the depth of the water on the shelf. In the early experimental tests conducted at Columbia University, the depth of water on the shelf area was reduced to 3 mm., no platform formed, and the spit formed directly on the shelf. Another test was conducted with several deep troughs placed on the shelf in the area of platform growth. The platform maintained a constant depth of water above in spite of the irregular shelf topography.

With continued growth of the platform, a series of top-set beds were formed, which were composed of the fine fraction of the material being supplied by the erosion of the headland. The coarse fraction of the eroded material continued to be moved over the top of the platform to its outer edges where it was deposited to form the new fore-set beds. The formation of a platform occurred in all 27 experimental tests.

For each experiment, at a point in time of the development of the platform, a spit started to form on the top of the platform. The time interval between the initial formation of the platform and the subsequent formation of the spit varied from experiment to experiment. However, in all experimental tests, a spit did develop. The spit grows first as a subaqueous mound and then develops as an emergent ridge above the level of wave action. This emergent ridge, attached at one end to the headland, is the structure commonly considered as a spit. The spit is composed of the coarse fraction of the eroded material from the headland. It has a steep slope on the seaward side and a less steep slope on the landward side. The lateral growth of the individual spit ridge is principally due to overwash. This has been shown in the natural case by Yasso (1964). However, spit complexes may also grow laterally by the formation of new ridges to the seaward thus truncating and bypassing the older ridges. This process, in large part, gives rise to the many geomorphological terms applied to spits. Continued growth of the spit in the tank produced examples of all the geomorphic structures associated with spits (i.e. - beach ridges, simple, compound, complex, and recurved spits).

Several important relationships between the spit and the platform are evident from observations of the experimental tests. The platform always develops in advance of the spit. The spit never develops in any location other than on top of the platform. The typical cross-section of this relationship is seen in Plate 1-B. This cross-section is one of many that were taken for the entire length of the spit-platform structure in an experimental test. All cross-sections were essentially identical. The platform continues to grow while the spit is developing. The spit can be a very temporary structure subject to the various wave conditions, whereas the platform is a stable structure. It should be understood, however, that both the spit and the platform are together only temporary structures with regard to overall coastal development.

In order to study the role of the spit-platform structure with regard to coastal development, several experiments were conducted for periods as long as 48 hours. As is well known, the development of a spit acts to straighten a coastline. This was the observed case in the tank. The front of the headland beach and spit, with time, tended to form a long straight line with a very small acute angle between the direction of the beach front and direction of the wave fronts. During this protracted experiment, the spit-platform structure, as described, was found only at the distal end of the long straight coastal shoreline. The remaining portion became a beach in profile. Specifically, it was observed that, as the spit-platform structure continued to grow in length, the spit-platform beach slope adjacent to the headland increased in steepness until it approximated the headland beach slope. This change advanced progressively along the spit-platform structure, creating additional headland beach while the spit-platform beach was to be found only at the distal end of the lengthening spit-platform structure. Thus the spit-platform structure is the constructive agent for the eventual extension of the beach.

In order that there to be any validity to model studies, the results obtained must have counterparts in the natural environment. An extensive study of the available hydrographic charts and topographic maps show that, in all cases known to the writer, spits are built on submarine platforms. Four representative cases are presented in Figure 7, A-D. Sidney Spit (Figure 7-A) is a long narrow spit extending northward from Sidney Island near Vancouver Island. It rests on a platform outlined by the dotted line. This platform is widest to the east. The headland contour line indicates a height of 285 ft. Figure 7-B does not have a spit emergent at the present time. However, it seems apparent that the area enclosed by the 12 foot contour line will soon become an emergent spit ridge resting on a platform. Point Frances Island (Figure 7-C) is also found in the Vancouver area in the Bellingham Bay area. The spit on the northern end of the island is recurved in nature. The platform is clearly visible. It is interesting to note that the platform has extended by forming a second lobe while the spit has not as yet grown onto the newly formed platform. A tombolo forming between Point Frances and the mainland can be seen. Sandy Point on the north end of Block Island (Figure 7-D) is a simple small spit built on a large platform. In each of these cases (Figure 7-A, B, C, D) a headland is supplying sediments to form spits resting on platforms. One other case is worth mentioning. A Pleistocene spit-platform can be seen on the topographic map of the Calumet City Quadrangle in the area south of Chicago (United States Geologic Survey, 1960, topographic map, Calumet City Quadrangle). It begins at about the location of the Glenwood-Dryer Interchange.

The Spit-Platform Concept - Morphology

From the observations of the models in the tank experiments, and a study of the hydrographic charts and topographic maps, the spit-platform concept may be stated, in part, as follows:

1. Morphologically, a spit is a partially emergent ridge which forms on the top of the larger submarine embankment, the platform. Both are formed principally due to beach drifting of material from the headland beach.

2. The initial prerequisite for the development of a spit is the prior existence of a platform, except for the rare case when the depth of water on the shelf is so shallow so as to not require the formation of a platform.

3. The development of the spit-platform structure is a constructive phase in the extension of coastal beaches.

Figure 7. Selected Hydrographic Charts of Spits

A and C - CHS CHART 3449 Race Rocks to Turn Pt. - depth in fathoms
Scale 1:87,560.

B - USCG chart 265, Nantucket Island, depth in feet
Scale 1:52,176.

D - USCG chart 269, Block Island, depth in feet
Scale 1:14,763.

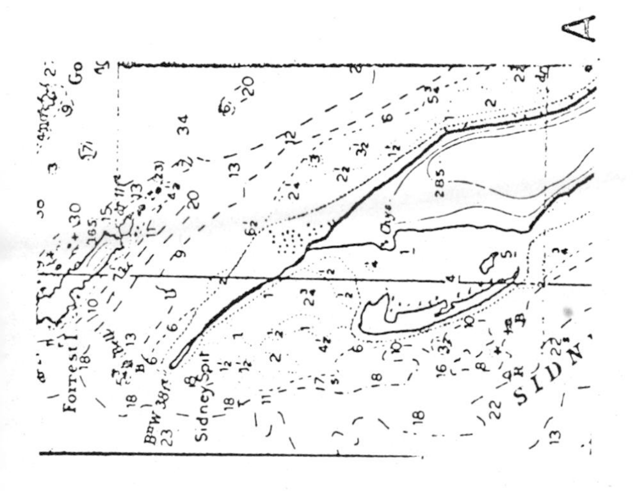

Forrest I
Sidney Spit
SIDN

A

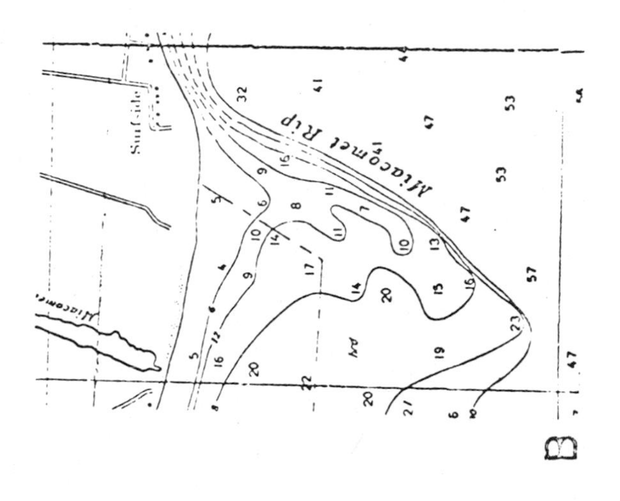

Miacomet Rip
Surfside

B

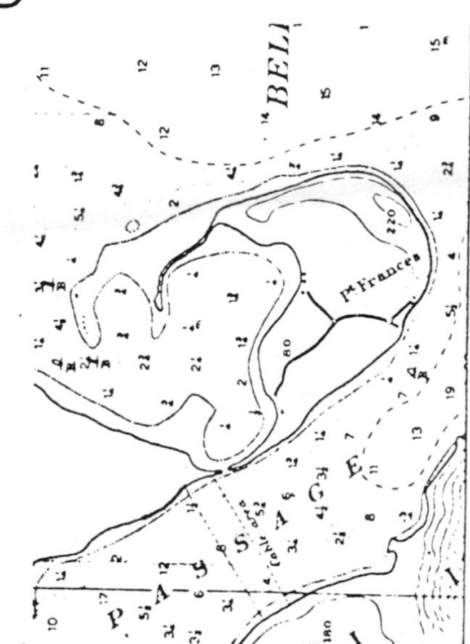

Pt Frances
PASSAGE
BELL

C

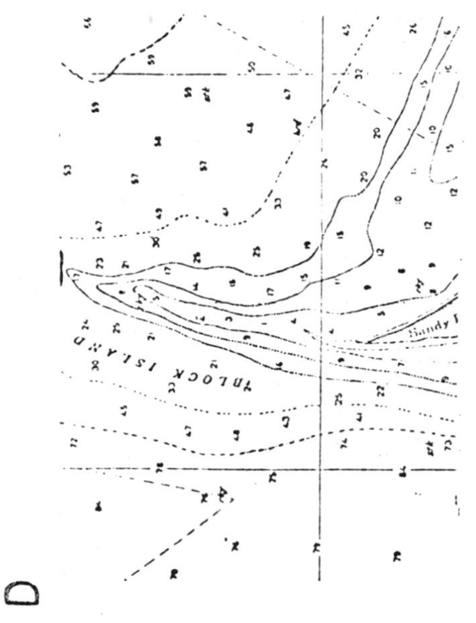

BLOCK ISLAND

D

CHAPTER FIVE

DYNAMICS OF SPIT-PLATFORM DEVELOPMENT

Introduction

The dynamics of shore processes, ocean waves, and their interrelationships are at present not clearly understood. The primary purpose of the research undertaken was to study the sequential development of spits under different wave conditions. The secondary purpose was to attempt to gain an understanding of the dynamic interrelationships between the wave parameters and spit-platform development. This was attempted by two methods. The first method was to statistically analyze the data (beach slopes and wave parameters) in order to develop an equation which expresses any relationships present. In a situation of the same general type as the research undertaken here, Harrison, et al. (1965) employed similar statistical techniques in order to develop an equation for the beach - ocean - atmosphere system at Virginia Beach, Virginia. The second method was to analyze the data on the growth of the spit-platform structure in order to study the dynamic relationship between spit and platform growth.

Beach Slope Equations

The dynamical link between the wave parameters and the structures produced in the tank was investigated by measurement of the slope of the beaches developed. Many authors have shown that beach profiles are a reflection of given wave conditions. However, the nature of this interrelationship is only generally understood. At present it is thought that the beach slope is predominantly a function of the grain size of the beach material and secondarily a result of the wave energy (Shepard, 1963).

In order to develop an equation showing the relationship between beach slope and wave parameters, an analysis of variance was initially performed on the variables.

The computer program used was the Analysis of Variance for Factorial Design.

The variables used in the analysis of variance were the wave parameters, time, position and beach slope. These were annotated as follows:

INDEPENDENT VARIABLES

A = amplitude (wave height)

D = depth of water on the shelf

F = frequency (wave period)

T = Time

P = Position

DEPENDENT VARIABLE

y = slope value

The independent variables A, D, and F each had three levels, namely, a high, medium, and low (Figure 4). The variable P was divided into two parts, i.e. headland position (1,2,3) and spit-platform position (4,5,6). The slope value (y) was an averaged value of the three slope values for each of the headland and spit-platform positions (1,2,3-4,5,6). It should be noted that the independent variables did not have numerical values associated with them in this analysis. Rather, they were arbitrarily designated. The dependent variable however, had a numerical value.

The statistical technique of analysis of variance indicates the magnitude that each independent variable and each combination of independent variables contributes towards the total variance of the dependent variable. The combinations of the independent variables were taken two at a time. The technique also indicates the contribution of the 1st, 2nd, and 3rd order component of each variable and each combination of variables. From this array of possible variables, the analysis of variance method indicated fifteen 1st and 2nd order terms that were of significance, at the 5% level, in explaining the variance of beach slope values.

The next step in developing a beach slope equation was to employ multiple regression techniques. Basically, multiple regression statistically finds the best fit surface for a set of points. The multiple regression analysis gives three facts concerning the equation of the surface. The first is that it computes the coefficients for each term in the equation. Second, it computes the total percentage of variance of the dependent variable, beach slope, explained by the equation and also the percentage explained by each separate term of the equation. Third, it computes the standard deviation of the estimated beach slope value (computed from the equation) and the actual observed beach slope value. This deviation is an indication of how closely the equation predicts the actual observed value of a given beach slope.

The independent variables first used for the multiple regression equation were those shown to be significant from the analysis of variance. It was evident from the analysis of variance that variable P (position) was of major importance. Therefore, the data were divided into two separate groups; namely, headland beach slope values and spit-platform beach slope values. This result is readily apparent from an examination of the data in Appendix One, Part B. Arbitrary high, medium and low values of amplitude, depth, and frequency were used in the analysis of variance method. The measured values of these variables were used in the multiple regression analysis.

The entire data set was analyzed first. This indicated that the terms with T (time) were of no significance. This was verified by computing the equation for each time group separately. The terms with T were therefore discarded. The remaining terms to be used in the two regression equations (headland beach and spit-platform beach) were then solely in terms of amplitude, depth, and frequency. These terms were arrived at on the basis of mathematical significance. It therefore became necessary to give physical meaning to the remaining terms.

Mathematically it can be shown that the following equivalences are valid:

1. L (wave length) varies as F^2
2. E (wave energy) varies as A^2F^2

3. S (steepness) varies as A/F^2

4. q (mass transport) varies as A^2/F

5. d (depth ratio) varies as D/F^2

These relationships are derived from the wave equations presented by Shepard (1963) · hers. The wave length (L) is the legnth of the wave in deep water. The wave energy (E) is the energy per unit surface area of the wave. The steepness of the wave is the ratio of the height of the wave to the length of the wave. The mass transport (q) is the discharge in terms of volume transported forward per unit of wave crest length per unit time. The depth ratio (d) controls the character of the wave form: deep water wave (d > 1/4L); intermediate water wave (1/4L > d > 1/20L); shallov water wave (1/20 L > d). These five terms, expressed as values of A, F, and D, were used in a new regression equation, in as much as they represented physcial characters of the waves.

Once the equation for the tank data had been derived, a new set of data for the nature of ocean situation was generated applying the scale factor to the tank data. Multiple regression analysis was then applied to this new natural data set using the same variables as were used for the tank data. The results obtained were similar.

Through the statistical techniques of analysis of variance and multiple regression the following relationships were established. These equations express in a general way the effect of the different wave parameters on the slope of the resultant beach. The detailed equations are found in Appendix One, Part C.

1. Headland Beach

$$y = f(S, q, E, A, L) \quad (1)$$

slope = f(steepness, mass transport, energy, wave height, wave length)

Percentage of variance explained: Tank - 51% Nature - 45%

Standard deviation of variance: Tank - 2.2° Nature - 2.4°

2. Spit-Platform Beach

$$y = f(E, d, A, L) \quad (2)$$

Slope = f(energy, depth ratio, wave height, wave length).

Percentage of variance explained: Tank - 49% Nature 38%

Standard deviation of variance: Tank - 2.4° Nature 2.4°

For complex natural phenomena, the percentage of variance explained by the above equations is reasonable in light of the variation inherent in the complex interrelationships encountered. There is, however, a check on the equation. This check is the experimental error. The standard deviation of the variance is determined by a comparison of the observed value of the beach slope with the value as computed by the equation. If the standard deviation is approximately equal to the experimental error, then the data have been used to their maximum level of significance. It is seen that the standard deviation of the two equations does closely approach the experimental error of two degrees.

As many investigators have realized, grain size has an effect on the slope of the beach. Only one type of sand was used for these experiments (Figure 3-A), which were not designed to investigate the effect of particle size on the beach slope. Therefore, it is suggested that the function of a grain size term (s) in the final equation relating natural physical parameters to beach slopes would be that of a factor by which the terms of equations 1 and 2 would be multiplied. Therefore, the new equation relating the slope of the beach to wave parameters should be:

1. Headland Beach

$$y = g(s)\ f(S, q, E, A, L) \tag{3}$$

2. Spit-Platform Beach

$$y = g(s)\ f(E, d, A, L) \tag{4}$$

Dynamics of Sediment Distribution

The length and width of the spit and platform were recorded for each experiment (Figure 5). The source of the sediment being deposited on the spit and platform was the headland eroded under wave attack. The measurements made possible the study of the sediment distribution between spit and platform. This study provided an understanding of another aspect of the dynamics of spit-platform development. It was observed that the spit and platform structures were receiving sediments at the same time, and the sediments were being deposited in different locations on the respective structures. It was necessary to unify these diverse data in order to study the spit-platform growth in all tests. It was decided, therefore, to study the volumetric change with time of both the spit and the platform.

The height of the platform, as previously discussed was a function of the depth of the water on the shelf. Therefore, the data were divided into three groups based on water depth. Volume is a function of the parameters of length, width and height. It was observed in the tank for each experiment that the height of the platform remained constant, and the height and width of the outermost spit ridge remained constant. Since the platform maintained a constant height, the approximate volume of the platform would then be a function of its length and width (base area). The approximate volume of the spit, since it had a constant width and height, would be a function of its length. In order to study the dynamics of sediment distribution between spit and platform, it was necessary to compare volumetric changes with time.

A graph was prepared of the change in volume between five minute intervals plotted against total time. This graph is based on the data for the nine medium water depth experiments (Figure 4). The individual volumes of spit and platform were computed for each five minute interval in each of the nine tests. The computed volumes at the end of any one time span for all nine tests were averaged. This mean volume was then subtracted from the next mean volume (in time) and this difference (change in volume or Δvolume) was then plotted against time.

The results are shown graphically in Figure 8. It is seen that when the rate of growth of the platform declines, the spit grows uniformly, while when the platform grows uniformly, the rate of growth of the spit declines. Based on observations of the model in the tank, the growth relationship between spit and platform may be generally

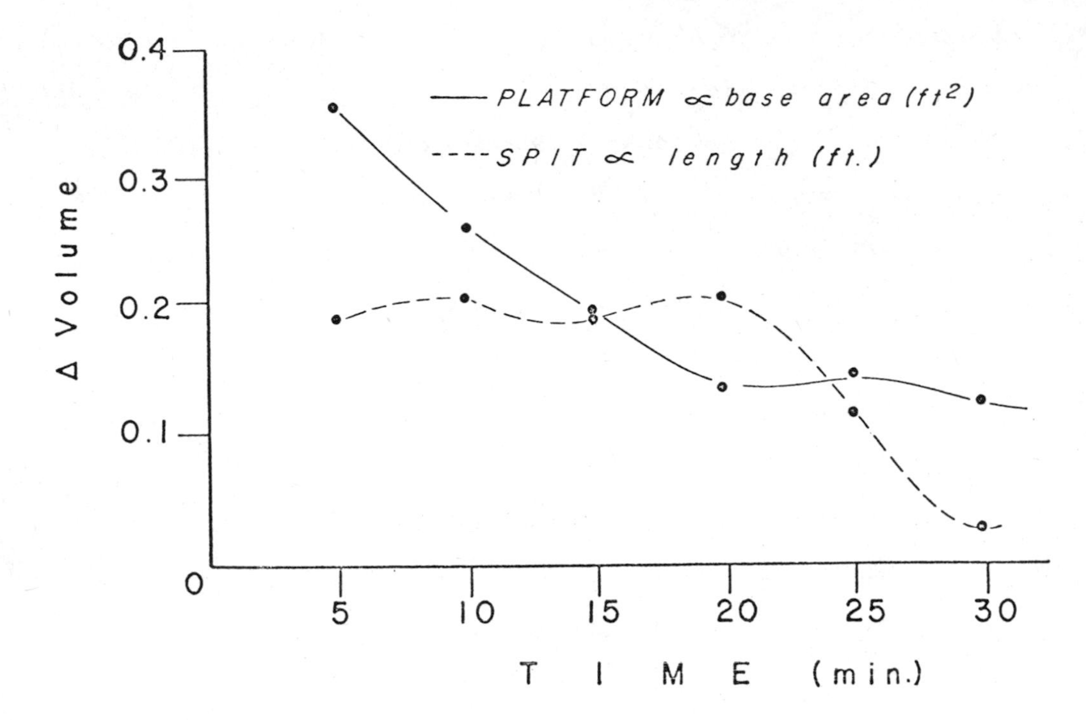

Figure 8. Growth of Spit-Platform with Time

Sediment eroded from headland is source of sediment for spit-platform. Data averaged from all Medium water depth experimental tests. The data from the high and low experimental test groups produced similar graphs.

stated as an inverse one. Thus as one grows faster the other grows slower. Similar results were obtained from plots of the deep water and shallow water cases.

The cyclical relationship shown in Figure 8 is readily observable in the tank. The platform grows in advance of the spit, and then the spit grows onto the newly formed platform. Because of the complex morphology of the spit-platform structure, the significance of the period of the cycle cannot be determined from the data recorded. Because the volumes computed for the spit and platform are only approximations, the graph is regarded only as a general verification of the observed sequential development of the spit-platform structure in all experimental tests.

The Spit-Platform Concept - Dynamics

From observations of the model in the tank experiments, and a statistical analysis of the data recorded for each experiment, the spit-platform concept may be stated, in part, as follows:

1. The slope of the headland beach is a function of grain size of beach material, wave steepness, mass transport, wave energy, wave height, and wave length as follows:

$$y = g(s)f(S, q, E, A, L) \tag{3}$$

2. The slope of the spit-platform beach is a function of grain size, wave energy, depth ratio, wave height and wave length as follows:

$$y = g(s)f(E, d, A, L) \tag{4}$$

3. Dynamically, the growth of the spit and platform structures are in general inversely related. Thus as one grows faster, the other grows slower. This growth occurs in alternating cycles.

4. With continued growth, the spit-platform beach slope increases in steepness until it equals the headland beach slope. This change advances progressively along the spit-platform beach from the headland beach, thus extending the headland beach.

CHAPTER SIX

CONCLUSIONS

Introduction

The spit-platform concept presented here considers spits both morphologically and dynamically. It also attempts to indicate the role of spits in the development of shorelines. As a result of this research, several suggestions for further study are offered.

The Spit-Platform Concept

The spit-platform concept presents a view of the sequential development of spits. It is based on experimental results, in part readily observable, and in part statistically substantiated. The spit-platform concept may be stated as follows:

1. Morphologically, a spit is a partially emergent ridge which forms on the top of the larger submarine embankment, the platform. Both are formed principally due to beach drifting of the eroded material from the headland beach.

The platform is the structure, an embankment in character, adjacent to, and connected with the headland. The platform is elevated above the shelf, but below the mean low water level.

The spit is the structure, a ridge in character, located on the upper surface of the platform and partially emergent above the mean high water level.

The general spit-platform structure is a large scale primary sedimentary structure composed of a platform and a spit. The structure is connected at one end with the land, terminates in open water, and is elongate in character and straight to curved in plan.

2. Dynamically, the growth of the spit and platform structures are in general inversely related. Thus as one grows faster, the other grows slower. This growth occurs in alternating cycles.

The slope of the headland beach is a function of grain size of beach material, wave steepness, mass transport, wave energy, wave height, and wave length as follows:

$$y = g(s)\, f(S, q, E, A, L) \qquad (3)$$

The slope of the spit-platform beach is a function of grain size of beach material, wave energy, the depth ratio, wave height and wave length as follows:

$$y = g(s) f(E, d, A, L) \qquad (4)$$

The initial prerequisite for the development of a spit is the prior existence of a platform, except for the rare case when the depth of water on the shelf is so shallow so as to not require the formation of a platform.

3. With continued growth, the spit-platform beach slope increases in steepness until it equals the headland beach slope. This change advances progressively along the spit-platform structure from the headland beach, thus extending the headland beach. The development of the spit-platform structure is a constructive phase in the extension of coastal beaches.

Recommendations for Further Study

As a result of the research undertaken, several specific projects are suggested for future study. From these studies and others to come in the future, a suggestion is offered as to the reclassification of shorelines, based on the results of this general type of research.

It would be of interest to study the sedimentary characteristics of the spit and platform structures. This could be accomplished by using the same equipment and procedures as outlined in this research. Colored sands of different sizes could be employed and then successive cross-sections of the structures could be taken. This would provide information as to the exact nature of the bedding, sorting and distribution of the sediments in the spit-platform structure.

Dynamically it would be interesting to study the exact relationship between the different wave parameters as they effect the slope of the beach. Certainly the final equation expressing the relationship is more complex than just grain size and wave energy.

With additional study of the dynamics of shore processes, it is suggested that the concept of a shore profile of equilibrium be reconsidered as consisting of two parts: (1) individual beach profiles of equilibrium, and (2) coastal profiles of equilibrium. The Beach profile is principally a short term, dynamic equilibrium dependent on a local set of wave conditions, whereas the coastal profile is principally a long term, morphologic equilibrium based on the efficient transport of sediments.

The classification of shorelines, which at present is based on geomorphology should also be considered dynamically with respect to wave conditions and the erosion, transportation, and deposition of sediments. It is suggested that the classification of shorelines should be based on both geomorphology and dynamics of shore processes.

REFERENCES CITED

Bagnold, R.A., 1946, Motion of waves in shallow water, interaction between waves and sand bottoms: Proc. Roy. Soc. London, Ser. A., v. 187, p. 1-18.

Bascom, W., 1964, Waves and beaches: New York, Anchor Books, Doubleday & Co., Inc., 265 p.

Boldyrev, V.L., and Nevesskiy, Y.N., 1964, Sand transport by the Tyemruk current (Kerch Strait): International Geology Rev., v. 6, no. 2, p. 228-237.

Bridgman, P.W., 1922, Dimensional Analysis, New Haven, Yale University Press, 113 p.

Bruun, P., 1954, Use of small-scale experiments with equilibrium profiles in studying actual problems and developing plans for coastal protection: Amer. Geophys. Union Trans., v. 35, p. 445-452.

_____, 1962, Sea level rise as a cause of shore erosion: Jour. Waterways and Harbors Div., Amer. Soc. C.E. Proc., v. 88, p. 117-130.

Bruun and Kamel, A., 1964, Prototype experiments of littoral drift in wave tank: Proc. 9th Conference on Coastal Engineering, no. 2, 6.

Canadian Hydrographic Service Chart, 3449, Race Rocks to Turn Point.

Carr, A.P., 1965, Shingle spit and river mouth, short term dynamics: Inst. of British Geographers, Trans. and Papers, no. 36, p. 117-129.

Einstein, H.A., 1948, Movement of beach sands by water waves: Amer. Geophys. Union Trans., v. 29, p. 645-655.

Evans, O.F., 1939, Mass transportation of sediments on subqueous terraces: Jour. of Geol., v. 47, p. 325-334.

Gulliver, F., 1899, Shoreline topography: Am. Acad. Arts and Sciences Proc., v. 34, p. 151-258.

Harrison, W., Pole, N.A., and Tuck, D.R., 1965, Predictor equations for beach processes and responses: Jour. of Geophys. Research, vol. 70, no. 24, p. 6103-6109.

Hubbert, M.K., 1937, Theory of scale models as applied to the study of geologic structures: Geol. Soc. Amer. Bull., v. 48, p. 1459-1519.

Inman, D.L., and Bowen, A.J., 1963, Flume experiments on sand transport by waves and currents: Proc. 8th Conference on Coastal Engineering, Chap. 11, p. 137-150.

Johnson, D.W., 1919, Shore processes and shoreline development: New York, John Wiley and Sons, 583 p.

Johnson, H.W., 1949, Scale effects in hydraulic models involving wave motion: Amer. Geophys. Union Trans., v. 30, p. 517-525.

Kemp, P.H., 1961, The relationship between wave action and beach profile characteristics: Proc. 7th Conference of Coastal Engineering, v. 2, p. 262-277.

King, C.A.M., 1959, Beaches and coasts: London, Edward Arnold Ltd., 403 p.

Krumbein, W.C., 1944, Shore currents and sand movements on a model beach: Beach Erosion Board, Tech. Memo., no. 7, 44 p.

Nartov, L.G. and Levehenko, S.P., 1962, Wave generators used in wave basins and hydraulic flumes, in: Oceanographic methods and instruments; Trudy, Morskoi Gidrofizicheskii Institut, (Academiia Nauk SSSR.), v. 26, (Translated from Russian for Amer. Geophys. Union, 1963) p. 1-2-109.

Rector, R.L., 1954, Laboratory study of equilibrium profiles of beaches: Beach Erosion Board, Tech. Memo., no. 41, 38 p.

Relief Form Atlas, 1956, Institut Geographique National, Paris, France, 179 p.

Reynolds, K.C., 1933, Investigation of wave-action on sea walls by use of models: Amer. Geophys. Union Trans., v. 14, p. 512-516.

Samarin, V.G., 1962, New naval hydrodynamics laboratory in England, in: Oceanographic methods and instruments; Trudy, Morskoi Gidrofizicheskii Institut, (Academiia Nauk SSSR.) v. 26, (Translated from Russian for Amer. Geophys. Union, 1963) p. 110-119.

Saville, T., 1950, Model study of sand transport along and in definitely long straight beach: Amer. Geophys. Union Trans., v. 31, p. 555-565.

Schwartz, M., 1965, Laboratory study of sea-level rise as a cause of shore erosion: Jour. of Geol., v. 73, no. 3, p. 528-534.

Shepard, F.P., 1963, Submarine geology, 2nd Ed.: New York, Harper and Row, 557 p.

Silvester, R., 1960, Stabilization of sedimentary coastlines: Nature, v. 188, p. 467-469.

Sivakov, N.R., 1963, Wave action and the development of the outer foreshore margin: International Geological Review, v. 5, no. 11, p. 1440-1445.

Strahler, S.N., 1963, Physical geography, 4th Ed.: New York, John Wiley and Sons, 442 p.

Thornbury, W.D., 1954, Principles of geomorphology: New York, John Wiley and Sons, Chap. 17, p. 417-458.

Twenhofel, W.H., 1932, Treatise on Sedimentation, 2nd Ed., v. 1 & 2: New York, Dover Publications, 926p.

United States Coast and Geodetic Survey Chart 265, Nantucket Island.

United States Coast and Geodetic Survey Chart 269, Block Island.

Wiegel, R.L., 1964, Oceanographical Engineering: London, Prentice-Hall Inc., 532 p.

Wulf, 1963, Bars, spits, and ripple marks in a Michigan Lake: A.A.P.G. Bull., v. 47, pt. 1, p. 691-695.

Yasso, W.E., 1964, Geometry and development of spit-bar shorelines at Horseshoe Cove, Sandy Hook, New Jersey: Technical Report, no. 5, Project NR 388-057, Office of Naval Research, Geography Branch.

APPENDIX ONE

Part A. Detailed Wave Parameters

The individual values of the wave parameters of wave height and wave steepness is presented.

Experimental Tests			Tank		Scaled Equivalents	
AMPLITUDE	WATER DEPTH	FREQUENCY	Wave Height (Ft.)	Wave Steepness	Wave Height (Ft.)	Wave Steepness
H	H	H	0.0701	0.0366	15.82	0.0366
		M	666	208	13.32	208
		L	384	027	7.68	027
H	M	H	0.0959	0.0444	19.18	0.0444
		M	656	205	13.12	205
		L	250	017	5.00	017
H	L	H	0.0895	0.0414	17.90	0.0414
		M	708	221	14.16	221
		L	218	015	4.36	015
M	H	H	0.0333	0.0154	6.66	0.0154
		M	396	124	7.92	124
		L	197	014	3.94	014
M	M	H	0.0427	0.0197	8.54	0.0197
		M	396	124	7.92	124
		L	213	015	4.26	015
M	L	H	0.0552	0.0255	11.04	0.0255
		M	542	169	10.84	169
		L	240	017	4.80	017
L	H	H	0.0306	0.0141	6.12	0.0141
		M	208	065	4.16	065
		L	136	009	2.72	009
L	M	H	0.0292	0.0135	5.84	0.0135
		M	167	052	3.34	052
		L	104	007	2.08	007
L	L	H	0.0246	0.0113	4.92	0.0113
		M	218	068	4.36	068
		L	083	005	1.66	005

APPENDIX ONE

Part B. Experimental Slope Data.

The individual slope data for each of the twenty-seven experimental tests is presented. For each five minute time interval, the circles (open or closed) represent an individual slope measurement (Positions 1,2,3,4,5,6) on the headland-spit-platform structure. The closed circles (black dots) indicate slumping at that position at that time. Experimental tests are indicated by High (H), Medium (M), and Low (L), and subscripts of (A) Wave height, (D) Water depth, and (F) Wave period.

APPENDIX ONE

Part C. Detailed Beach Slope Equations

HEADLAND BEACH

1. Tank

$$y = 15.7 - 139.5\ A/F^2 + 1622.3\ A^2/F - 2393.7\ A^2F^2 + 127.5\ A + 0.246\ F^2$$

(36.11%) (8.66%) (5.25%) (1.22%) (0.11%)

Total percentage of variance explained = 51.35%

Standard Deviation : 2.2°

2. Nature

$$y = 10.9 + 0.0069\ F^2 - 0.000455\ A^2F^2 - 138.4\ A/F^2 + 1.51\ A + 0.341\ A^2/F$$

(32.91%) (6.28%) (0.56%) (5.77%) (0.28%)

Total percentage of variance explained: 44.90%

Standard Deviation: 2.4°

SPIT-PLATFORM BEACH

1. Tank

$$y = 18.8 - 730.4\ A^2F^2 - 3.63\ F^2 - 29.4\ D/F^2 - 21.7\ A$$

(10.18%) (11.58%) (17.99%) (0.21%)

Total percentage of variance explained: 39.97%

Standard Deviation: 2.4°

2. Nature

$$y = 15.3 - 0.000179\ A^2F^2 - 19.5\ D/F^2 - 0.0106\ F^2 - 0.165\ A$$

(26.73%) (1.17%) (9.22%) (0.49%)

Total percentage of variance explained: 37.6%

Standard Deviation: 2.4°

PLATE 1 - Spits : Tank and Nature

A - Wave tank and wave generator.

B - Typical cross-section of spit-platform structure, taken from an actual experimental test.

C - Spit-platform developed in the tank during one experimental test.

D - Headland of Lloyd's Neck, Long Island, New York.

E - Longitudinal view from spit crest of Easton's Neck spit, Long Island, New York.

F - Lake spit in Prince Albert National Park, Saskatchewan, Canada.

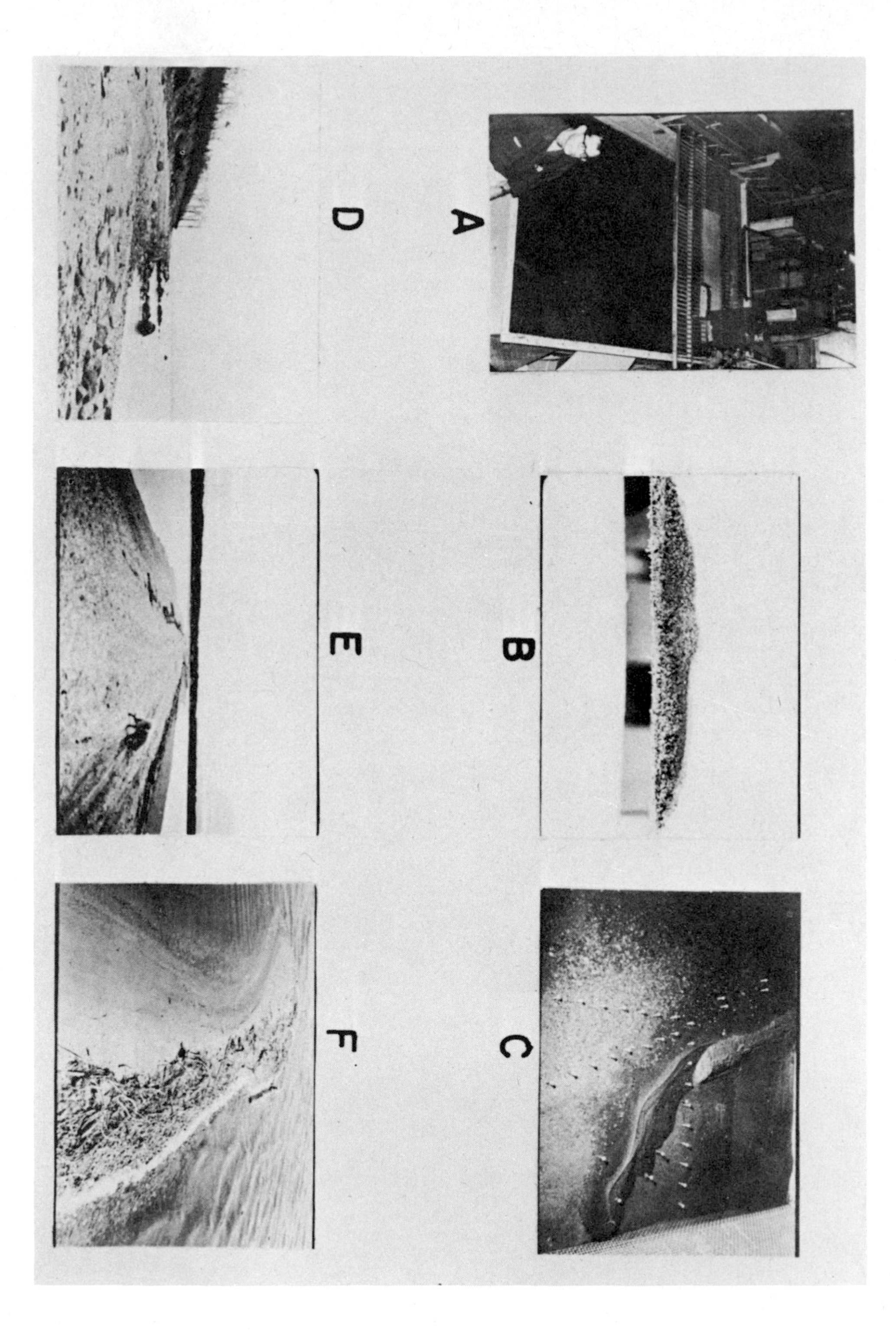
A
B
C
D
E
F

Editor's Comments on Paper 10

A "workbook" approach to spit and bar development is provided by Jean Tricart's article in *Photo Interpretation*. In this paper the author has provided stereoscopic photographs and a transparent overlay, accompanied by descriptive text, illustrating an interesting case of spit and bar development in Panama.

For those who wish, it is suggested that the photographs may be reproduced locally for positioning under a stereoscopic viewer. The page 33 overlay, once duplicated on clear acetate, may then be employed as was originally intended.

Abundant river deposits, a large tidal range, and strong Pacific swell, all combine to develop pronounced spits and bars along the Panamanian Pacific Coast. A tropical climate contributes to the development of mangrove swamps, the only discussion of these present in this collection of spit and bar papers.

Jean Tricart was born in 1920 in Montmorency, France. He studued at the Lycee Rollin and the Facultee des Lettres de Paris, and holds the degrees of Agrege d'histoire et geographie and Docteur des lettres. He is a professor at the University of Strasbourg; among his many organizational positions he was president of the Commission on Applied Geomorphology of the International Geographic Union. Dr. Tricart was also founder and director of the Revue de Geomorphologique Dynamique. His scientific interests are allocation of natural resources, and the development and utilization of natural resources.

FORMES LITTORALES D'ACCUMULATION TROPICALES

(Côte Pacifique de Panama)

TROPICAL LITTORAL ACCUMULATION SHAPES

(Pacific Coast of Panama)

FORMAS LITORALES TROPICALES DE ACUMULACIÓN

(Costa Pacífica del Panamá)

Revue "PHOTO-INTERPRETATION" © 2-1967 (2e tr.) Editions TECHNIP

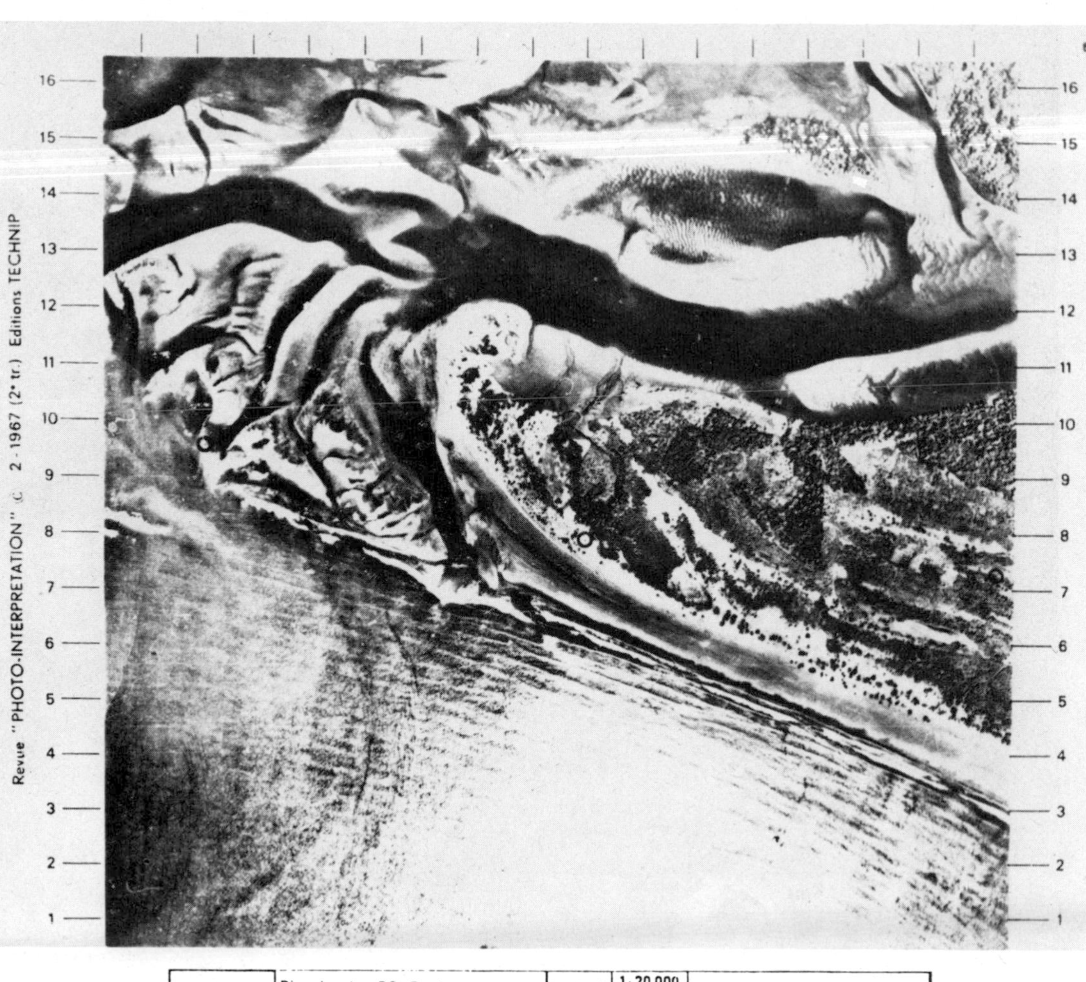

PANAMA	Direccion de Cartografia	DC - Pan 1 11 - 3 V - 5	n° 539	31 - 12 - 62 12 h 40	1 : 20.000 réduction f = 152,8	PANAMA 1 : 500.000

	1	2	3	4	5	6	7	8	9	H
0	ANTE PRIMAIRE PRIMAIRE	SECONDAIRE	TERTIAIRE	QUATERNAIRE PALEOLITHIQUE	NEOLITHIQUE PROTOHISTOIRE — 500	500	1 000	1 500	TEMPS MODERNES	0
0		COLONISATION	HABITAT RURAL		FORTIFICATIONS GUERRES	HABITAT URBAIN		LOISIRS LIEUX PUBLICS	RELIGIONS NECROPOLES	0
0	EXPLOITATION SOUS SOL	PRODUCTION D'ENERGIE	INDUSTRIE TRANSFORMATION	INDUSTRIE CONSOMMATION	DISTRIBUTION		COMMUNICATIONS AERIENNES	COMMUNICATIONS MARITIMES	COMMUNICATIONS TERRESTRES	0
0	COUVERT VEGETAL	EXPLOITATION DIRECTE DE LA VEGETATION	ORIENTATION ALIGNEMENTS	AMENAGEMENT IRRIGATION DRAINAGE CONSERVATION	GENIE RURAL REMEMBREMENT	STRUCTURES AGRAIRES	EXPLOITATION AGRICOLE		MONDE ANIMAL ELEVAGE CHASSE PECHE	0
7	LIGNES	POLYGONES CARRES	CERCLES RAYONS	METHODES EMULSIONS	REPARTITION CARTOGRAPHIE		ETUDE DYNAMIQUE	DEGRADATION		7
6	RESEAU HYDROGRAPHIQUE	VERSANTS	MICRORELIEF		DOCUMENTS ANCIENS	FORMATIONS SUPERFICIELLES	SOLS BRUTS NON OU PEU EVOLUES	SOLS EVOLUES		6
1	RELIEF NUL	RELIEF FAIBLE	RELIEF FORT				VENTS	EAUX	NEIGES ET GLACES	1
7	ARCTIQUE SUB ARCTIQUE	TEMPERE CONTINENTAL	TEMPERE OCEANIQUE	MEDITERRANEEN	PREDESERTIQUE	DESERTIQUE ARIDE	INTERTROPICAL	EQUATORIAL TRES HUMIDE	MICROCLIMATS	7
P 5	CRISTALLIN	EFFUSIF FILONIEN	METAMORPHIQUE	SEDIMENTAIRE STRATIGRAPHIQUE	MERS EAUX SALEES	STRUCTURES QUASI MONOCLINALES	STRUCTURES PLISSEES	STRUCTURES FAILLEES	STRUCTURES POLYGENIQUES DISCORDANCES	5

67-2 6

D 65 mm — Cliché D.C. PANAMA — G 65 mm

67-2
5

Revue "PHOTO-INTERPRETATION" © 2-1967 (2e tr.) Editions TECHNIP

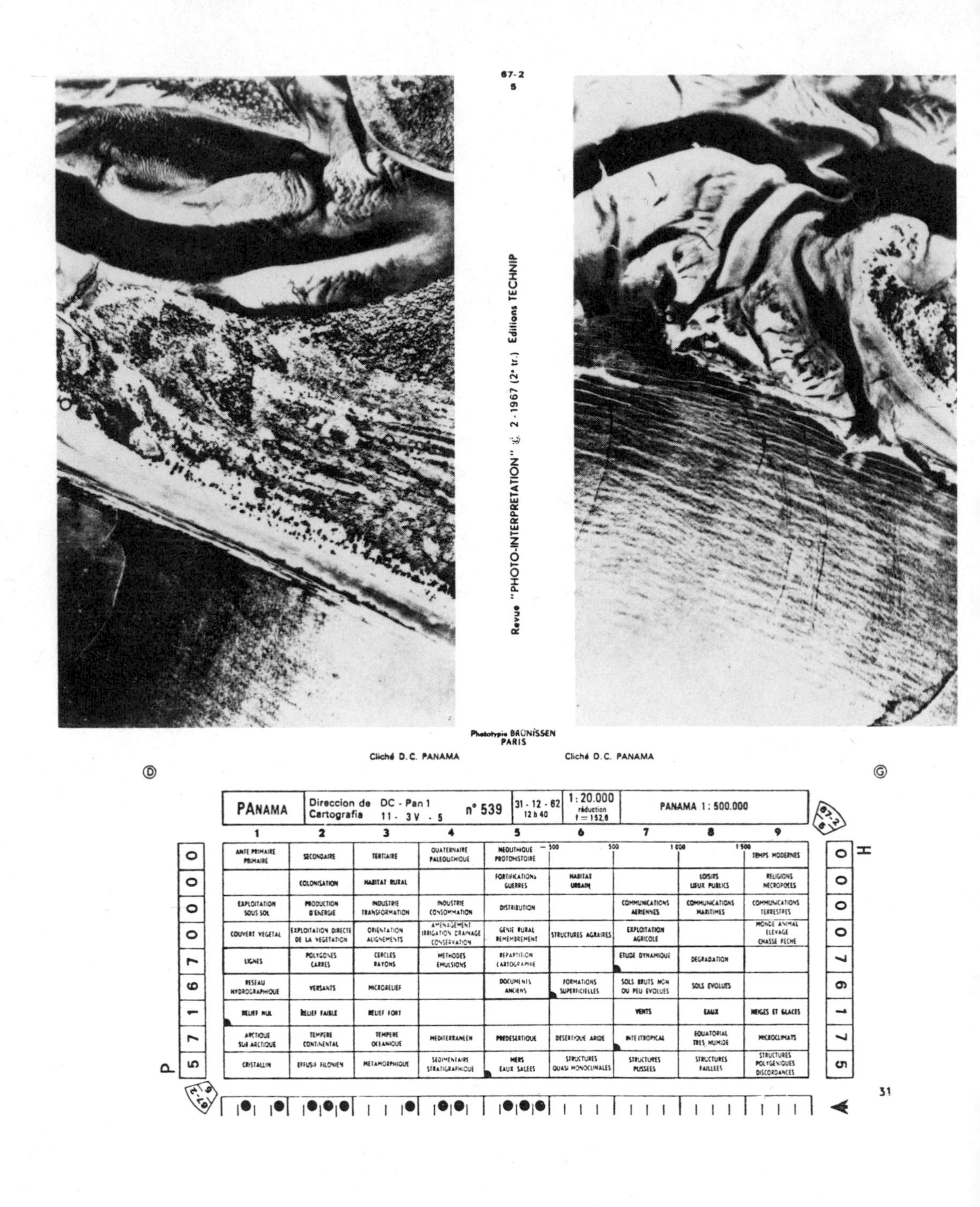

Phototypie BRUNISSEN
PARIS

Cliché D.C. PANAMA   Cliché D.C. PANAMA

Ⓓ   Ⓖ

PANAMA	Direccion de Cartografia	DC - Pan 1 11 - 3 V - 5	n° 539	31 - 12 - 62 12 h 40	1 : 20.000 réduction f = 152,8	PANAMA 1 : 500.000

P	1	2	3	4	5	6	7	8	9	H
0	ANTE PRIMAIRE PRIMAIRE	SECONDAIRE	TERTIAIRE	QUATERNAIRE PALEOLITHIQUE	NEOLITHIQUE PROTOHISTOIRE — 500	500	1 000	1 500	TEMPS MODERNES	0
0		COLONISATION	HABITAT RURAL		FORTIFICATIONS GUERRES	HABITAT URBAIN		LOISIRS LIEUX PUBLICS	RELIGIONS NECROPOLES	0
0	EXPLOITATION SOUS SOL	PRODUCTION D'ENERGIE	INDUSTRIE TRANSFORMATION	INDUSTRIE CONSOMMATION	DISTRIBUTION		COMMUNICATIONS AERIENNES	COMMUNICATIONS MARITIMES	COMMUNICATIONS TERRESTRES	0
0	COUVERT VEGETAL	EXPLOITATION DIRECTE DE LA VEGETATION	ORIENTATION ALIGNEMENTS	AMENAGEMENT IRRIGATION DRAINAGE CONSERVATION	GENIE RURAL REMEMBREMENT	STRUCTURES AGRAIRES	EXPLOITATION AGRICOLE		MONDE ANIMAL ELEVAGE CHASSE PECHE	0
7	LIGNES	POLYGONES CARRES	CERCLES RAYONS	METHODES EMULSIONS	REPARTITION CARTOGRAPHIE		ETUDE DYNAMIQUE	DEGRADATION		7
6	RESEAU HYDROGRAPHIQUE	VERSANTS	MICRORELIEF		DOCUMENTS ANCIENS	FORMATIONS SUPERFICIELLES	SOLS BRUTS NON OU PEU EVOLUES	SOLS EVOLUES		6
1	RELIEF NUL	RELIEF FAIBLE	RELIEF FORT				VENTS	EAUX	NEIGES ET GLACES	1
7	ARCTIQUE SUB ARCTIQUE	TEMPERE CONTINENTAL	TEMPERE OCEANIQUE	MEDITERRANEEN	PREDESERTIQUE	DESERTIQUE ARIDE	INTERTROPICAL	EQUATORIAL TRES HUMIDE	MICROCLIMATS	7
5	CRISTALLIN	EFFUSIF FILONIEN	METAMORPHIQUE	SEDIMENTAIRE STRATIGRAPHIQUE	MERS EAUX SALEES	STRUCTURES QUASI MONOCLINALES	STRUCTURES PLISSEES	STRUCTURES FAILLEES	STRUCTURES POLYGENIQUES DISCORDANCES	5

67-2 6   31   A

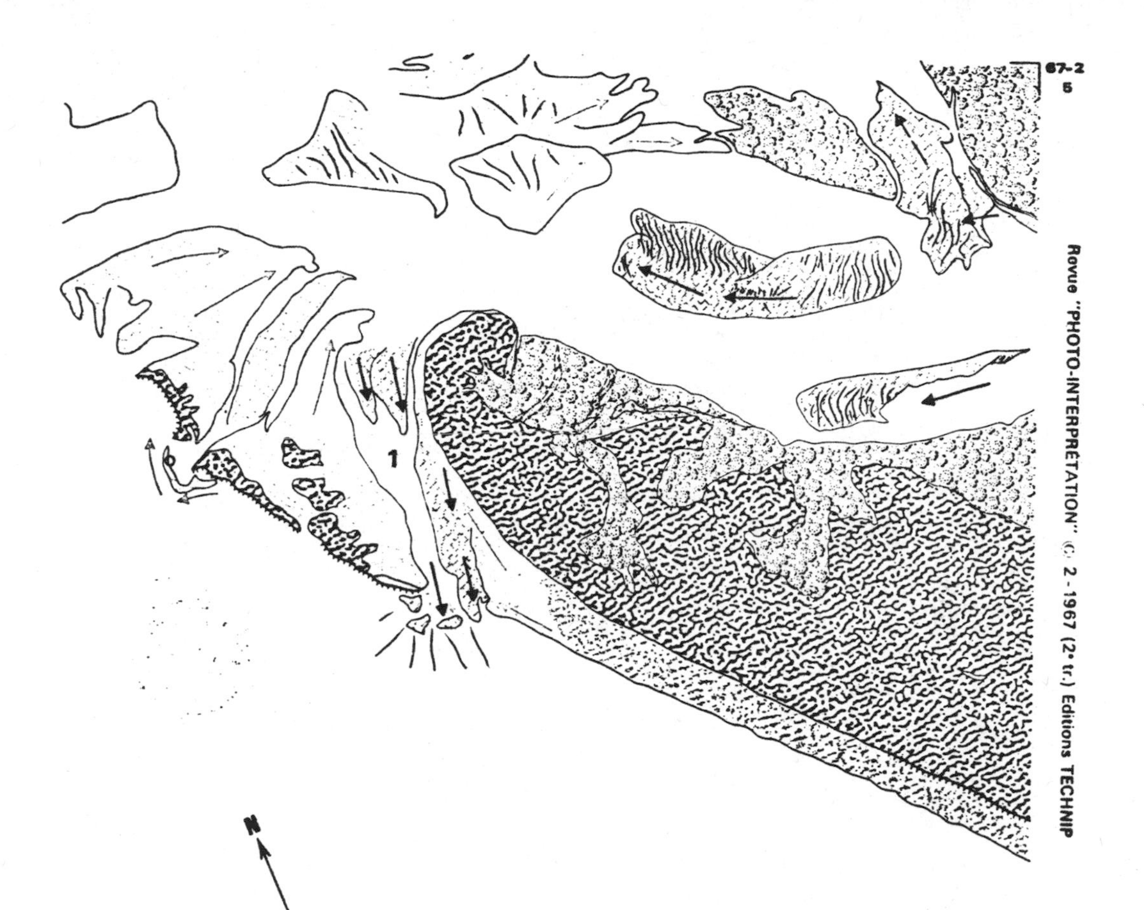

Cordon littoral ancien (dunkerquien)

Cordon dunkerquien démantelé

Bancs et courants de jusant

Bancs et courants de flot

Mangrove

Rides

33

MISSION DIRECCION DE CARTOGRAFIA DC - PAN-1 11-3V-5 Cliché n° 539 du 31/12/1962 (12 h 40)
Échelle 1 : 20000 focale 152,6 mm Carte 1 : 500 000 PANAMA

J. TRICART
Centre de Géographie Appliquée
Université de Strasbourg

FORMES LITTORALES TROPICALES
(Extrémité W de l'I. Sevilla, Chiriqui, Panama)

Le triplet montre une flèche littorale à l'extrémité d'un cordon sableux qui ferme partiellement un golfe. Des mangroves l'ont colonisé en partie. La dynamique est intense, par suite de la combinaison de :

— courants de marée violents qui résultent à la fois de la hauteur du marnage (4 m environ) et du volume considérable d'eau qui pénètre dans le golfe lors du flot,
— houle forte, typique du Pacifique, avec une direction à peu près parallèle à la côte lors de la prise de vue. Les brisants sont violents au contact d'une plage peu inclinée.

Les unités géomorphiques sont les suivantes :

1. Un cordon littoral stabilisé, qui n'est plus fonctionnel, qui s'est construit alors que le niveau marin était un peu supérieur à l'actuel (dunkerquien), à la fin de la transgression flandrienne. Il est colonisé par la végétation et on y observe même quelques petits champs. Les rides successives apparaissent en plus clair, car elles sont plus sèches et plus sableuses, les éléments fins du sable ayant été lessivés. Les creux intermédiaires sont plus foncés, car plus humides et, aussi, un peu plus riches en particules fines, qui s'y sont concentrées. L'un d'entre eux est occupé par une mare allongée. Ce cordon s'étendait autrefois davantage vers l'W mais a été coupé par rupture et un chenal s'est installé dans la coupure. A l'W de celle-ci, l'extrémité du cordon est soumise à une violente érosion et se réduit à des bancs de sable démantelés submergés à haute mer, mais qui fournissent des matériaux qui sont repris par les courants de marée.

2. Un cordon récent et actuel, en forme de poulier, qui entre dans le golfe sur le bord E du chenal interrompant l'ancien cordon dunkerquien. On peut nettement observer l'érosion du flanc de ce dernier sur la haute plage à l'E et le mouvement du sable poussé dans le golfe par le flot Le sable se déposant en eau plus calme, la flèche prend une forme recourbée.

3. Une série de bancs vifs, arasés par les courants de marée formés par du sable que ces derniers déplacent. La tendance prédominante est à l'entrée de sable dans le golfe lors du flot, et à la sortie, mais en moindre quantité, lors du jusant.

a) à marée montante, deux phénomènes se produisent :

— une migration du sable le long de la plage du cordon dunkerquien, permettant l'édification de la flèche qui pénètre dans la lagune,

— une intense érosion des bancs de l'extrémité W démantelée de ce cordon, qui alimente une série de bancs pénétrant dans la lagune.

b) à marée descendante, il se produit :

— à l'E, dans la partie aval de l'estuaire des mangroves, l'accumulation d'une série de bancs de sable avec de grandes rides, qui forment un véritable petit delta sous-aquatique,
— à l'W, il se creuse des chenaux de jusant dans les bancs de sable qui se déposent à marée montante. C'est notamment le cas du chenal qui coupe l'ancien cordon dunkerquien et à l'extrémité extérieure duquel se construit un petit delta sous-aquatique.

4. La mangrove envahit les surfaces suffisamment hautes pour ne pas rester inondées trop longtemps à chaque marée. On les rencontre sur le bord du cordon dunkerquien, dans les dépressions, et sur les parties les plus hautes des accumulations plus récentes qui bordent l'estuaire derrière le cordon, ainsi que sur un banc de sable stabilisé du delta construit à son débouché.

5. Une accumulation de vase s'édifie à l'abri du poulier, dans un site protégé des courants de marée. Elle est encore trop basse pour être colonisée par la mangrove, mais celle-ci s'avance déjà le long d'une petite ride.

Revue " PHOTO INTERPRÉTATION " N° 2 - 1967 fascicule 5

TROPICAL LITTORAL SHAPES (W EDGE OF SEVILLA ISLAND, CHIRIQUI, PANAMA)

The triplet shows a littoral spit running off the end of a sandy cordon which partially closes off a gulf. The mangrove swamp covers a good part of the gulf. The dynamics is intense as the result of the combination of:

— violent tidal currents which stem both from the height of the tidal flow (about 4 m) and the large volume of water running into the gulf with the flood tide;

— heavy swell typical of the Pacific, having a direction more or less parallel with the coast when the photo was taken. The surf is violent when it hits a slightly inclined beach.

The geomorphic units are as follows:

1. A stabilized littoral cordon which is no longer functional, built when the sealevel was slightly higher than at present (Dunkerquian) at the end of the Flandian transgression. It is covered by vegetation and there are even a few small fields. The successive wrinkles show up in a lighter color, because they are drier and sandier since the fine parts of the sand have been leached. The hollows in between are darker, because damper, and also slightly richer in fine particles which have concentrated there. One of them is occupied by an elongated pond. This cordon used to extend further westward but was broken off and a channel was dug into the cut. To the west of this cut, the end of the cordon undergoes violent erosion and is reduced to dismantled sand bars which are submerged at high but which furnish material that is taken off by the tidal currents.

2. A recent cordon now having the shape of a spit entering into the gulf on the E edge of the channel and cutting across the old Dunkerquian cordon. The erosion of the flank of the latter cordon can be clearly seen on the upper beach to the E along with the extensive sand movement in the gulf as the result of the tidal flow. The sand is deposited in calmer water, thus giving the spit a recurved shape.

3. A series of small bars worn down by the tidal currents but formed by the sand carried in these currents. The predominating trend is for sand to enter into the gulf with the flood tide and for less sand to run out with the ebb.

a) two phenomena occur with the rising tide:

— sand migrates along the beach of the Dunkerquian cordon and builds up the spit sticking out into the lagoon,

— intense erosion occurs on the bars at the dismantled W end of the cordon, and this supplies a series of bars penetrating into the lagoon.

b) the ebb tide brings:

— to the E in the downstream part of the mangrove swamp estuary, a series of sand bars accumulate with large wrinkles, forming a small sub-aquatic delta,

— to the W ebb channels are hollowed out in the sand bars deposited by the rising tide. This is especially the case of the channel cutting across the old Dunkerquian cordon at the outer end of which a small sub-aquatic delta is being built.

4. The mangrove swamp is invading sufficiently high areas so as not to remain inundated too long at each tide. This swamp is also found on the edge of the Dunkerquian cordon, in hollows and on the highest parts of the most recent accumulations on the edge of the estuary behind the cordon, as well as on the stabilized sand bar on the delta built up at its mouth.

5. An accumulation of mud is being built up in the shelter of the spit where it is protected from the tidal currents. It is still too low for the mangrove swamp to become implanted, but this swamp is already advancing along one small wrinkle.

FORMAS LITORALES TROPICALES (EXTREMO O DE LA I. SEVILLA, CHIRIQUI, PANAMA)

Sobre el trío de fotografías se observa una flecha litoral al extremo de un cordón arenoso que cierra parcialmente un golfo. Los manglares la han ocupado parcialmente. La dinámica es intensa, a causa de los efectos conjugados de:

— las corrientes de marea, violentas, que resultan a la vez de la altura de las margas (unos 4 m) y del volumen considerable de agua que penetra en el golfo durante el flujo;

— oleaje fuerte, típico del Pacífico, de dirección aproximadamente paralela a la costa en el momento de la fotografía. Los rompientes son violentos al contacto de una playa poco inclinada.

Las unidades geomórficas son las siguientes:

1. Un cordón litoral estabilizado, que dejó de ser funcional, que se formó en una época en que el nivel marino era algo superior al actual (Dunkerquiense), a fines de la transgresión flandriense. Está recubierto de vegetación pudiéndose observar algunos campos pequeños. Los pliegues sucesivos son más claros, ya que son más secos y arenosos pues los elementos finos de la arena han sido arrastrados por el agua. Los huecos intermediarios son más obscuros, ya que más húmedos y, también, algo más ricos en partículas finas, que allí se han concentrado. Una charca alargada ocupa uno de ellos. Antaño dicho cordón se extendía mucho más hacia el O, pero se ha visto cortado por ruptura instalándose un canal en el corte. Al O de este, el extremo del cordón se encuentra sometido a una erosión violenta viéndose reducido a una serie de bancos de arena desmantelados sumergidos en alta mar, pero que constituyen una fuente de materiales que las corrientes de marea transportan.

2. Un cordón reciente y actual, en forma de banco, que entra en el golfo por la parte E del canal interrumpiendo el antiguo cordón dunkerquiense. Se distingue perfectamente la erosión del flanco de este último sobre la playa alta al E, y el movimiento de la arena empujada hacia el golfo por el flujo. Como que la arena se deposita en las aguas más tranquilas, la flecha toma una forma encorvada.

3. Una serie de bancos vivos, arrasados por las corrientes de marea, formandos por la arena que estas últimas arrastran. La tendencia predominante es la entrada de arena en el golfo durante el flujo, y la salida de la misma, pero en menor cantidad, al reflujo.

a) al flujo, tienen lugar dos fenómenos:
— una migración de la arena lo largo de la playa del cordón dunkerquiense, que permite la formación de la flecha que penetra en la laguna,
— una erosión intensa de los bancos de la extremidad O desmantelada de este cordón, que alimenta una serie de bancos que penetran en la laguna.

b) al reflujo, se produce:
— al E, en la parte aguas abajo del estuario de los manglares, la acumulación de una serie de bancos de arena con grandes pliegues, que constituyen un verdadero pequeño delta subacuático.
— al O, se abren canales de reflujo en los bancos de arena que se depositan durante el flujo. Este es precisamente el caso del canal que corta el antiguo cordón dunkerquiense y en la extremidad exterior del cual se construye un pequeño delta subacuático.

4. El manglar invade las superficies suficientemente elevadas para que no queden sumergidas demasiado tiempo a cada marea. Se encuentra sobre el borde del cordón dunkerquiense, en las depresiones, y sobre las partes más altas de las acumulaciones más recientes que orlan el estuario detrás del cordón, así como sobre el banco de arena estabilizado del delta construido a su desembocadura.

5. Al abrigo del banco se constituye una acumulación de cieno, en un sitio al abrigo de las corrientes de marea. Es aún demasiado baja para que el manglar la colonice, pero este ya avanza siguiendo un plieguecito.

TROPICAL LITTORAL ACCUMULATION SHAPES

(Pacific Coast of Panama) (1)

The Republic of Panama has undertaken a survey of its land distribution and natural resources for which 1/16 000 aerial photographic coverage has been made. Unfortunately, these photos only cover the Pacific watershed in western Darien.

This scale is a particularly good one for studying the geomorphological phenomena which can only be captured in full detail by photos on a scale greater than 1/20 000. The comments on the 6 triplets in this special issue will clearly illustrate this fact.

The Pacific littoral of Panama runs along the edge of a volcanic region made up of Mesozoic and Cenozoic series of sandstones, schists and moderately rough lava (basalts and andesites). All of this material has been strongly transformed into clays. There is also some Quaternary volcanic evidence, especially near the border of Costa Rica (Bauru volcano) where it is of the explosive type. Considerable cinderfalls occur at various intervals and set off hydrovolcanic catastrophes which, on the Pacific coast have formed large expanses plunging under the ocean.

The climate is of the tropical type with an extended dry season lasting four or five months and with total annual rainfalls varying from 1 500 mm (Los Santos Province) to more than 4 000 mm (the David and Chiriqui regions) depending on the relief orientation. Rainfall is very intense in the entire region. Such conditions are favorable for an active morphogenesis. There is considerable run-off, even under natural conditions. The rivers are subject to powerful flooding and carry off much silt. Degradation of the fragile vegetal covering by grazing and crops has led to intense anthropic erosion which serves to bring out these natural features. In the coastal regions, accumulation phenomena resulting from this erosion are very highly developed.

In addition, the Pacific littoral of Panama is influenced by strong tides which rise and drop as much as 5 m at the end of the Gulf of Panama and which average at least 3 or 4 m.

The littoral accumulation shapes are mainly influenced by the following three factors :

— abundant river deposits resulting from the accentuated dryness at various times because of Quaternary climate variations, from hydrovolcanic catastrophes and from intense anthropic erosion ;

— big tides which cause strong currents, plus the strong Pacific swell which causes considerable littoral drift ;

— a tropical climate with a long dry season which is sometimes increased to allow for the development of mangrove swamps and which causes a high degree of evaporation during several months.

These factors are typical of tropical regions with dry seasons, although they are slightly modified by the big tides which are exceptional for this type of region.

The accumulation shapes occurring there are characterized by the following features which are typical of tropical littorals (2) :

a) the great abundance of fine material, such as clays and sands, which have been freed from rocks by powerful alteration and can move about between the rivers as the result of the vegetal covering. The sands form into littoral bars, and the silt settles in damp depressions ;

---

(1) The photos were chosen with the help of A. R. HIRSCH, and the interpretation overlays were chosen with the help of C. DEMENU, both of whom are technicians at the *Center of Applied Geography*. The work as a whole was carried out as part of the bilateral technical cooperation between the *University of Panama* and the *Center of Applied Geography* under the auspices of the *Ministry of Foreign Affairs*. Warmest thanks are extended to Professor RAQUEL DE LEON, director of the *Institute of Georgraphy of the University of Panama*, and L. GUARDIA, of the *National Cartography of Panama*, without whom this survey could never have been made.

(2) TRICART (J.). — 1965. — Le Modelé des Régions Chaudes, Savanes et Forêts. Treatise on Geomorphology by J. TRICART and A. CAILLEUX, *Sedes*, Paris, V, 322 p., 64 fig.

b) the extensive development of littoral bars, as the result of the abundance of sand and the strength of the coastal drift (and, in this case, from the violence of the tidal currents). Most of this sand comes from the continental shelf and has been pushed back by the Flandrian transgression, which has enabled Dunkerquian cordons to form. These cordons show frequent recent erosions. which go to feed the littoral bars. Sand is still arriving in abundance, carried down by the streams, and goes to build levees, lagoon deltas and estuary banks ;

c) the large size of the lagoons formed by depressions in the littoral plain which were transformed into gulfs during the Flandrian transgression and which were barred off by cordons in the final stage of this transgression in the Dunkerquian period. The Dunkerquian cordons play an important part in the tropical littoral accumulations as will be described hereunder ;

d) the growth of mangrove swamps in the lagoons is governed by the climate since mangroves need high temperature, briny water and lots of sunlight. These swamps are closely associated with a certain degree of saltiness of the water and a certain length of submersion by the tide, as can be seen by examining the photos presented here ;

e) the rapid filling in of the lagoons as the result of the growth of the mangrove swamps attenuates the wave action and slows down the currents. Thus helping along the depositing and then the settling of material (mainly fine sand and mud). Veritable levees are formed along the channels underneath the mangrove swamps ;

f) the development of littoral sibkhas similar to those in the delta of the Senegal River [3] in the highest parts of the lagoons. These sebkhas appear where the lagoons are only occasionally submerged during major floods and equinoctial tides and even during seismic tidal waves (tsunamis). After submersion, part of the water evaporates and the brine becomes concentrated. As a result, veritable salt marshes (sebkhas) are formed naturally which are typical of tropical climates with a long dry season ;

g) when the streams flow into the lagoons, their large overflow and the big solid load they carry lead them to form deltas. The only deltas on this coast are lagoonal, because on the outer shore the swell is too strong to allow them to form. The photos shown here illustrate how these lagoon deltas are formed and the complexity they can have ;

h) the lagoon estuaries are a final feature of this type of littoral. They are especially highly developed here as the result of the size of the tides. The lagoons make up large reservoirs which fill up with the flooding tide, in particular in the dry season when the rate of flow of the streams is at a low level, and which empty out with the ebb tide. The ebb tide flows are especially powerful (in volume and duration) in time of flood. The larger the area of the lagoons the more the currents are active and capable of holding open a breach in the littoral cordons which have an outer end in the shape of a recurved spit. This tends to form a dynamic complex in which current reversals under tidal influences create channels and bars operating with the tidal flood and ebb. The triplets shown here give characteristic examples of this.

J. TRICART,
Centre de Géographie Appliquée,
Université de Strasbourg.

(3) TRICART (J.) — 1961. Notice Explicative de la Carte Géomorphologique du Delta du Sénégal. *Mémoires du Bureau de Recherches Géologiques et Minières*, Paris, 8, 137 p., 9 pl., 3 color maps in appendix (1/100 000).
TRICART (J.) and CAILLEUX (A.) — 1961. Le Modelé des Régions Sèches, *C.D.U.*, Paris, 2 vol. 126 and 179 p.

Editor's Comments on Paper 11

Shifting offshore bars greatly affect the use of coastal areas by man. The limitations imposed by these bars and harbor shoaling require investigation and understanding if they are to be combatted. L. Bajorunas and D. B. Duane took as their study area a newly completed harbor on the southeastern shore of Lake Superior. Bar movement and harbor shoaling were delineated with respect to littoral drift, waves, currents, ice cover, and man-made structures. Not only have they compiled an empirical report on changes at Little Lake harbor, but they have, in their conclusion, outlined the means of forecasting such changes at other projected coastal installations.

Leonas Bajorunas was born on October 23, 1910 at Buiviskiai, Lithuania. He pursued studies at the Universities of Berlin and Vienna, and received a degree of Graduate Engineer in the field of hydraulics from the latter institution in 1936. The University of Kaunas, Lithuania, awarded him a Doctor of Engineering degree in 1943. Dr. Bajorunas came to the United States in 1949, and joined the U.S. Army Corps of Engineers as a hydraulic engineer in 1950. After transferring to the U.S. Lake Survey District in Detroit, Dr. Bajorunas continued working as a hydraulic engineer on various Great Lakes problems. In 1956 he was appointed Chief of Lake Survey's Special Studies Section. In 1962, as public interest in water resources research increased, Bajorunas was designated Special Advisor to the District Engineer. Two years later, he was appointed Chief of Lake Survey's then newly established Research Division. With the expansion of its research activities, the Research Division was redesignated as the Great Lakes Research Center in 1966, and Dr. Bajorunas was named as director.

Dr. Duane is Chief of the Geology Branch, Engineering Development Division, U.S. Army Corps of Engineers' Coastal Engineering Research Center, Washington, D.C. Dr. Duane received his Bachelor's degree from Dartmouth College (1952–1956). He continued his education at the University of Kansas (1957–1962) where he conducted research on ancient and recent sediments, receiving the Ph.D. degree in 1963.

After several years experience in the mid-continent as a research and exploration geologist with Mobil Oil Company and its predecessor companies, Dr. Duane joined the Great Lakes Research Center of the U.S. Lake Survey, Detroit, Michigan. At the Lake Survey he rose to Chief of the Shore Processes Branch before transferring to the Coastal Engineering Research Center as Geology Branch Chief. His responsibilities at CERC concern planning and carrying out research into geologic processes in, and engineering geology of, the coastal zone and inner continental shelf.

JOURNAL OF GEOPHYSICAL RESEARCH VOL. 72, NO. 24 DECEMBER 15, 1967

11

Shifting Offshore Bars and Harbor Shoaling

L. BAJORUNAS

Great Lakes Research Center, Corps of Engineers, Detroit, Michigan 48226

D. B. DUANE

Coastal Engineering Research Center, Corps of Engineers, Washington, D. C.

Movement of offshore bars in Lake Superior produces alternating accretion and erosion of the nearshore area and causes pulsation in littoral drift. Extensive formation of ice foots contribute to significant erosion in front of them. Harbor structures extending into the lake shift offshore bars, modify coastal currents, and cause shoaling in the lee. The pattern of harbor shoaling was established, and its relationship with the shifting of offshore bars and existing currents was analyzed. The bar in shallow water was feeding material to the shoal and the currents determined the pattern of deposition.

INTRODUCTION

Much of the southeastern shoreline of Lake Superior is covered by a sand bottom that is subject to movement during severe spring and fall storms.

During the summer of 1964 a survey of currents and bottom conditions was undertaken to learn the effect on the coastal environment of a newly completed harbor. That harbor built in 1963–1964 was located at Little Lake on the southern shore of Lake Superior 33 km west of Whitefish Point and 48 km east of Grand Marais harbor, the nearest man-made structure.

Little Lake harbor (Figure 1) consists of a small natural lake connected to Lake Superior by a 3.7-meter-deep channel dredged through a sand bar that was a natural separation between the two lakes. The channel and harbor entrance are protected from prevailing west and northwest waves by a dogleg breakwater. A short breakwater on the easterly side of the channel creates an entrance facing northeast, the direction of infrequent storms. Both breakwaters are rock mound rubble with end cells constructed of steel sheet piles. At the harbor site maximum waves are produced by the prevailing northwest winds, which blow along a 300-km fetch with the easterly long-shore wave-energy component exceeding the westerly component by a ratio of nine-to-one. The result is a strong predominance of longshore currents and littoral movement from west to east. The Lake Superior level is controlled and shows about 10-cm variation during the months June–November and a 30-cm difference between low in March and high in September.

OFFSHORE BARS

Sounding surveys in the vicinity of the harbor were made in 1940, 1958, 1964, 1965, and 1966. Depending on period and purpose, the surveying methods varied. Detailed shallow-water surveys, using a precision level and rod, extended from the beach to 1.5-meter water depth. Lead-line soundings and acoustic methods were employed to extend surveys to deeper water. Detailed shallow-water and fathometer surveys during 1964 and 1966 were made on fourteen to sixteen ranges extending from 1350 meters west of the harbor to 1125 meters east.

The offshore bottom is characterized by a multi-bar morphology with three main bars: one in shallow water, a second bar at an intermediate depth, and a third in deeper water (Figure 2). In addition there are some minor bottom undulations between the bars. Aerial photographs show that the multi-bar morphology, as surveyed in the Little Lake harbor area, extends over the entire eastern part of the southern shore of Lake Superior. Characteristics of the bars, as discovered by the surveys, are described below.

Shallow-Water Bar

Detailed surveys of the shallow-water bar were made along fourteen ranges perpendicular to the shoreline during the period July–October

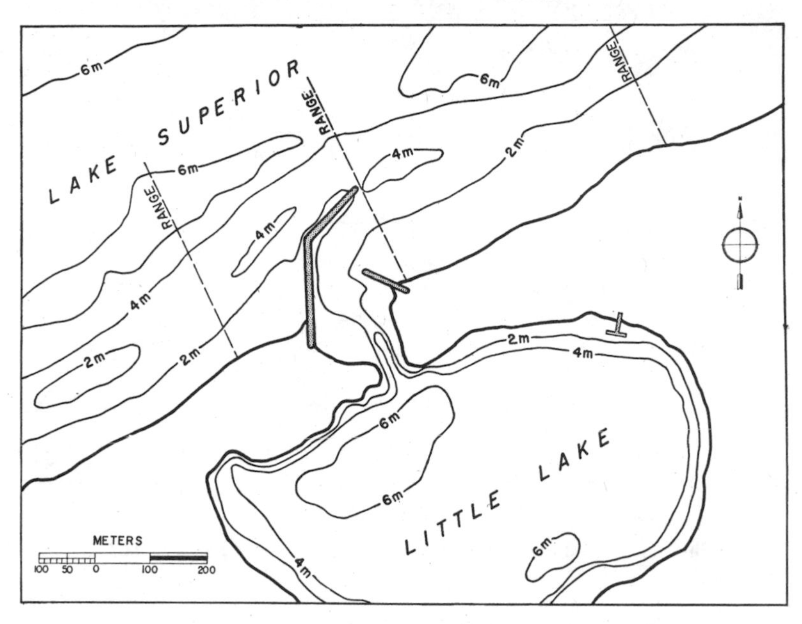

Fig. 1. Layout of Little Lake harbor.

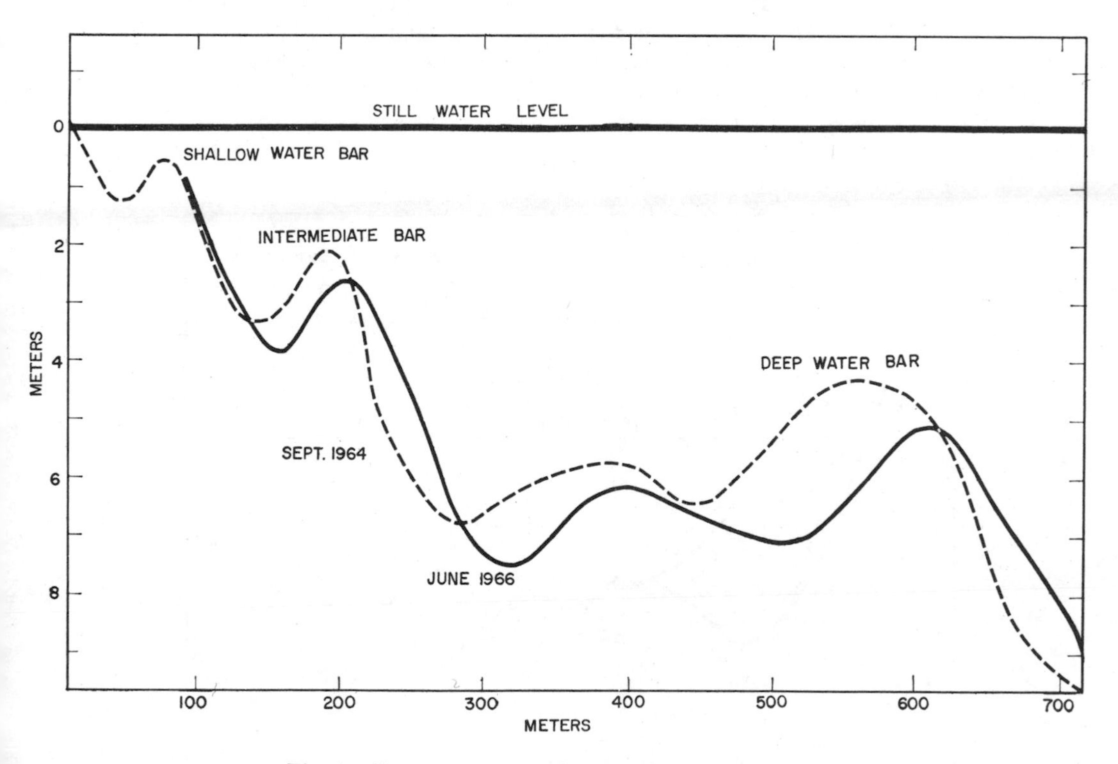

Fig. 2. Types and nomenclature of bars at study site.

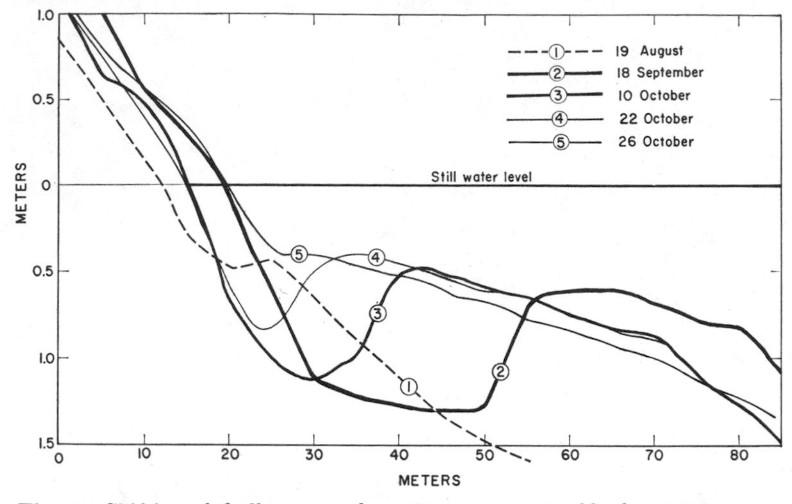

Fig. 3. Shifting of shallow-water bar, 300 meters west of harbor, 1964 survey.

1964. The survey methods, although allowing extremely high accuracy, limited soundings to maximum water depth of 1.5 meters. Bottom changes over three ranges are shown in Figures 3, 4, and 5. Location of ranges is indicated on Figure 1, and they depict conditions 300 meters updrift (on the west side) of harbor, at the harbor, and 525 meters downdrift of harbor.

Repetitive surveys indicated that the bar is highly mobile; movement is almost unidirectional, advancing from deeper water toward the shore. Near the shore at a depth of approximately 0.5 meter the bar loses its identity and merges with the shoreface completing the advance. After the bar and the shoreface coalesce, an erosion phase occurs within a band approximately 40 meters wide. For the most part, the advancing sand wave in the shallow-water zone is well-sorted sand, with median size in the range of 0.35 to 0.25 mm. The beach face and plunge point are sand mixed with, or armored by, granules and pebbles with a mean diameter

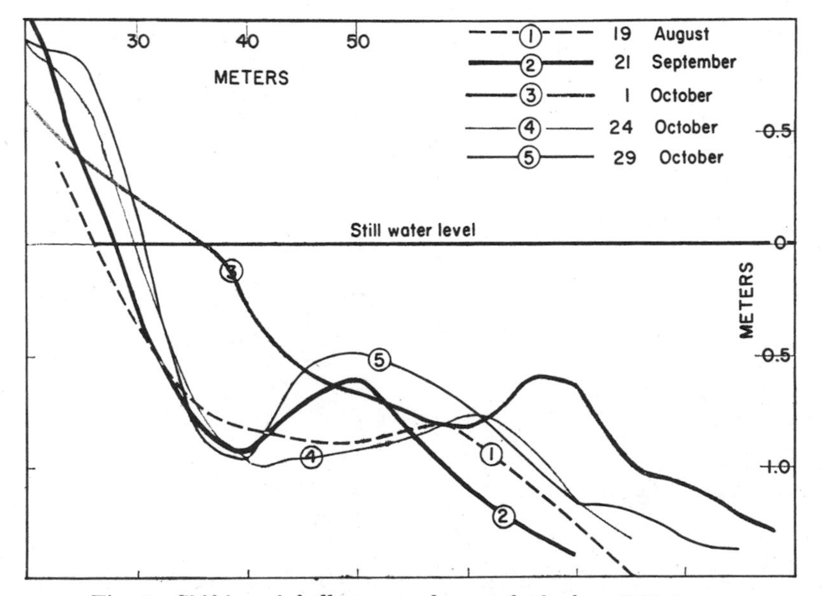

Fig. 4. Shifting of shallow-water bar, at the harbor, 1964 survey.

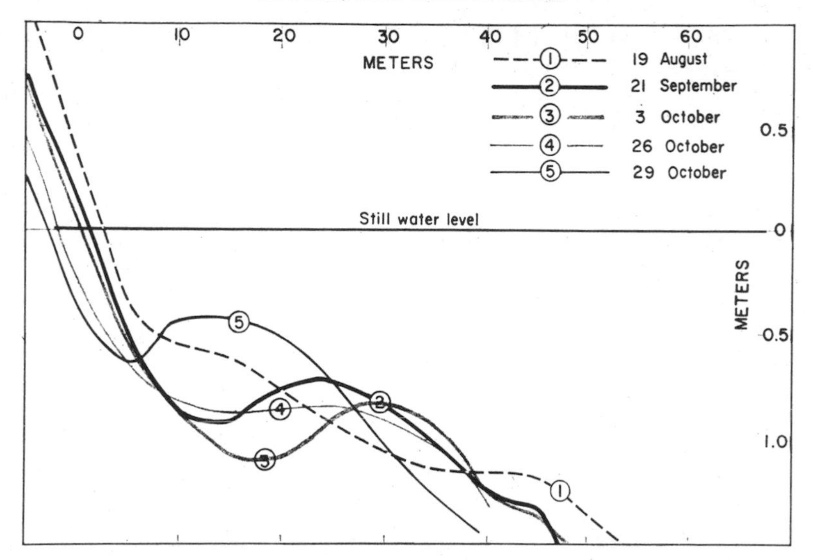

Fig. 5. Shifting of shallow-water bar, 525 meters east of harbor, 1964 survey.

of approximately 8.0 mm. Some areas are covered by stones with a diameter larger than 25 mm.

The sand wave approaches the shore at a small angle (compare also Figure 7); therefore, the two phases of sand wave movement, accretion and erosion, can be observed along the shoreline at the same time. The October accretion is shown in Figures 3 and 5 and erosion in Figure 4. Observations do not indicate a general change of underwater profile with the season.

The shallow-water bar disappears along the north edge of the west breakwater, appearing again in the vicinity of the harbor entrance from which point it continues in the downdrift direction through the reach where shore erosion takes place. There is a distinct difference in general nearshore slope on opposite sides of the harbor. On the updrift side under accretion conditions that slope is 0.008, and on the downdrift side, where extensive erosion takes place, a much steeper slope of 0.027 exists.

Intermediate Bar

At a depth of about 2 meters and at a distance of 120–260 meters from shore, surveys indicate another bar. Surveys of this bar were less frequent. They were of lower accuracy than the surveys of the shallow-water bar, because the techniques here employed acoustic equipment.

Available profiles of the range 825 meters downdrift of the harbor, under conditions considered to be unaffected by construction, indicate movements of this intermediate bar (Figure 6). The July 7, 1964, survey depicts a profile that is rather low with a small bar near the shoreline. Erosion of the small bar and formation of a new bar at greater depth is shown by the September 9 and 16, 1964, soundings. The June 6, 1966, survey would indicate movement of the bar toward the shore. Because the time interval between this and the previous surveys includes two winter periods, however,

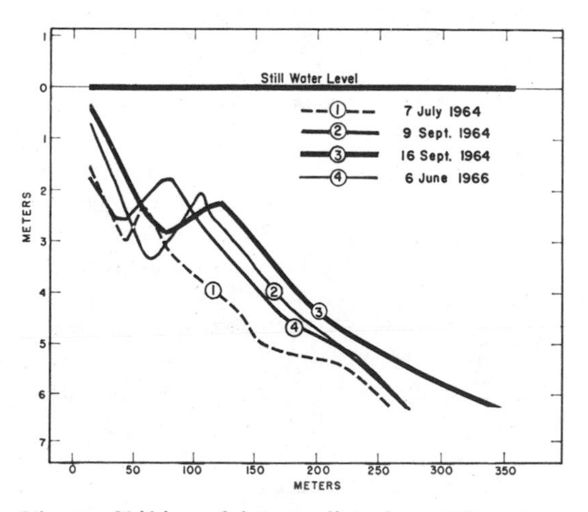

Fig. 6. Shifting of intermediate bar, 825 meters east of harbor.

the position of the bar in 1966 cannot be used for conclusions on the pattern of bar movement.

Deep-Water Bar

A third bar, located about 600 meters offshore with 5 meters of water over the crest, is referred to as the deep-water bar; relief of the bar is approximately 3 meters. During the period of the 1964 surveys storm waves reached 2.4 meters in height; yet, only local erosion and deposition in the vicinity of the bar is indicated and with only a limited movement extending over the rather broad 400-meter-wide base of the bar. As depicted in Figure 2, the 1966 survey indicates some movement of the bar toward deeper water. Sediment on the crest of the deep-water bar has a median diameter of 0.61 mm with finer material on the lower sections of front and back slopes having a median diameter of 0.25 mm.

Littoral Drift

The volume of littoral drift in the easterly direction was forecast as 130,000 m^3 per year based on relationship between wave energy and littoral drift [*Bajorunas,* 1961]. The data on volumetric sediment changes during 1963–1964 surveys confirmed the forecast value.

The sources of littoral drift are the sandy shore and nearshore as well as the bluffs fronting the lake. The process of transferring material from shore into littoral drift is complex; however, some insight can be obtained by analyzing the movement of the shallow-water bar. Two phases are recognized in the bar movement, the build-up and the erosion. During the build-up phase, the bar moves toward the shore, the shore material being transferred to the bar. Once the bar merges with the beach face, impinging waves are able to expend more of their energy directly on the beach, thus producing erosion, as shown by the October 1 and 24 surveys depicted in Figure 4. The fine-to-medium sized sand, comprising the sand wave now merged with beach, is maintained in suspension in the turbulent swash zone and is then moved by the longshore currents. At this erosive stage, the littoral drift rate should reach its largest value. As the movement of the sand wave is cyclic, the maximum drift occurrence would be cyclic. Littoral transport, then, is subject to pulsations. The minimum drift occurs during the time when the shallow-water bar has a dominant shoreward movement and is, in effect, storing material; maximum transport is taking place during the erosion phase. Under the wave conditions prevailing during the 1964 survey the cycle had a period of 2 to 3 months. It seems that more frequent storms tend to reduce the period; however, there is at present no evidence to indicate whether larger storms shorten the period or whether they cause the sand wave to reach farther into the lake at the onset of the advance phase.

The pulse-type movement of sediment in the surf-zone was reported by *Taney* [1963] in connection with the results of shore-process model tests utilizing radioactive sand tracers. Furthermore, the radioactive label clearly indicated that in the surf zone the pulse-type sediment movement is in the along-shore direction.

Although the shallow-water bar is directly involved in the littoral drift process, the role of the bars in deeper water is not clear. Some shuffling back and forth of the deeper-water bars and some minor bottom undulations point to a possible mechanism for the transfer of sediment between deeper-water and shallow-water bars. Any contribution to littoral drift from the intermediate and deep-water bars is, however, unknown.

Shifting of Bars

Under conditions not affected by man-made structures the shifting of bars is limited to rather definite zones. *Davis and McGeary* [1965], investigating bars in Lake Michigan, found the bars stable over one season of detailed surveys and by examining aerial photos stated the bars had been stable for a 10-year period. On the other hand, present investigations in Lake Superior indicated that the bars are not stable but shift in well defined pattern. Figure 7 depicts natural, prior to any construction, lake bottom changes in the vicinity of Little Lake, based on October 1940 and October 1958 surveys. The linear movement of bars at a small angle to the shoreline is clearly shown with an unstable zone through what is now the harbor entrance and its protective structures. After harbor construction additional shift of bars in the vicinity of structures was observed. The interaction of several factors is involved in these bar movements. The predominant factors

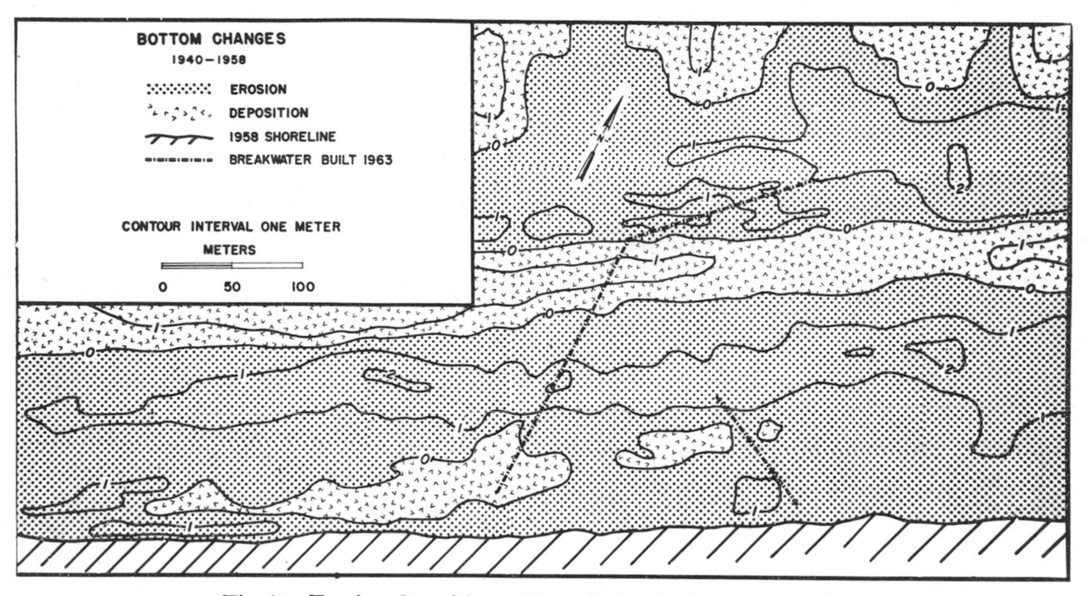

Fig. 7. Erosion-deposition patterns before harbor construction.

here are waves, currents, ice, and man's activity.

Waves and currents. There are no direct measurements of the effects of waves and currents; however, the character of sediment across the bars provides some clues as to the effect of wave action. The distribution of sediment size across the bar in deep water is opposite the distribution across the bar in the shallow water and is interpreted to indicate different mechanisms producing sediment movement in the two zones. The sand wave in shallow water consists in its entirety of medium to fine-grained sand moving on a base of much coarser material. This leads to the interpretation that the waves and currents are able to move the entire shallow-water bar over the coarse, rather stable bottom material. In deep water, however, the orbital movement of waves momentarily suspends the fine-grained sand by a winnowing action; currents sweep the material from the crest depositing it on the lower sections of the bar with the more coarse material remaining on the crest.

Ice cover. Ice cover is another important factor in the shifting of the bars. During the ice season, which in Lake Superior occurs from December through April, much of the shoreline is affected by formation of an ice foot. An ice foot is a mass of grounded ice and frozen slush forming a narrow ridge parallel to the shore [*Zumberge and Wilson*, 1954; *Marshall*, 1966]. This grounded mass of ice may accumulate to a height of 6 meters (Figure 8) above the normal lake-ice surface. In the course of the ice season, multiple ice foots may form at increasing distances from shore. These ice foots closely follow trends of the sand bars in the shallow-water zone and the intermediate zone. The ice foot, in effect, acts as a seawall; some of the forces of impinging waves are deflected downward by the ice foot and scour the lake bottom, placing bottom material in suspension. Suspended material is carried along shore or offshore or is thrown up on the ice by succeeding waves and thus incorporated in the ice foot. The effect of scour is to oversteepen

Fig. 8. Ice foot at Little Lake harbor, March 1966.

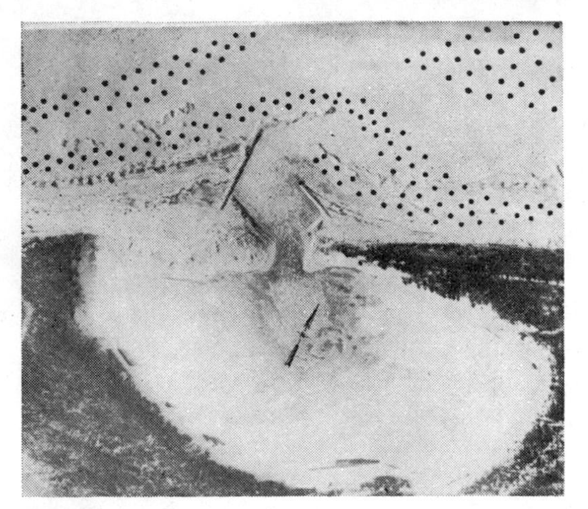

Fig. 9. Ice foot and erosion (stippled areas) March 1966.

the lakeward slope of the offshore bar or to deepen the associated trough, depending on the location of the ice foot. With the removal or melting of the ice, these bottom features become unstable and are subject to modification by waves. Bottom erosion shown in Figure 9 can be largely attributed to the location of the ice foot. Erosion off the west breakwater is assumed to be the result of interaction of wave forces deflected by the ice foot with reversing currents through the harbor entrance.

Breakwaters. By construction of breakwaters and dredging operations at the harbor site, the continuity of the shallow-water and intermediate bars was destroyed, as were parts of the bars themselves. After construction, physical and dynamical processes regenerated the intermediate bar in deeper water, yet in continuity with extension of the bar on both sides of the breakwaters. The shallow-water bar, however, is not discernible along the northern edge of the west breakwater, but it appears again just past the entrance to the harbor on the easterly side. Both these bars are shifted lakeward by the breakwater.

Figure 10 depicts this shift of the bars at the base range which passes through the end cell of the west breakwater. The shift of both the shallow-water and intermediate bars was uniform and of the order of 60 meters. The movement of the deep-water bar cannot be observed over as long a time span as the movement of the shallow and intermediate-depth bars, because the preconstruction surveys did not extend to deep water. In comparison to the

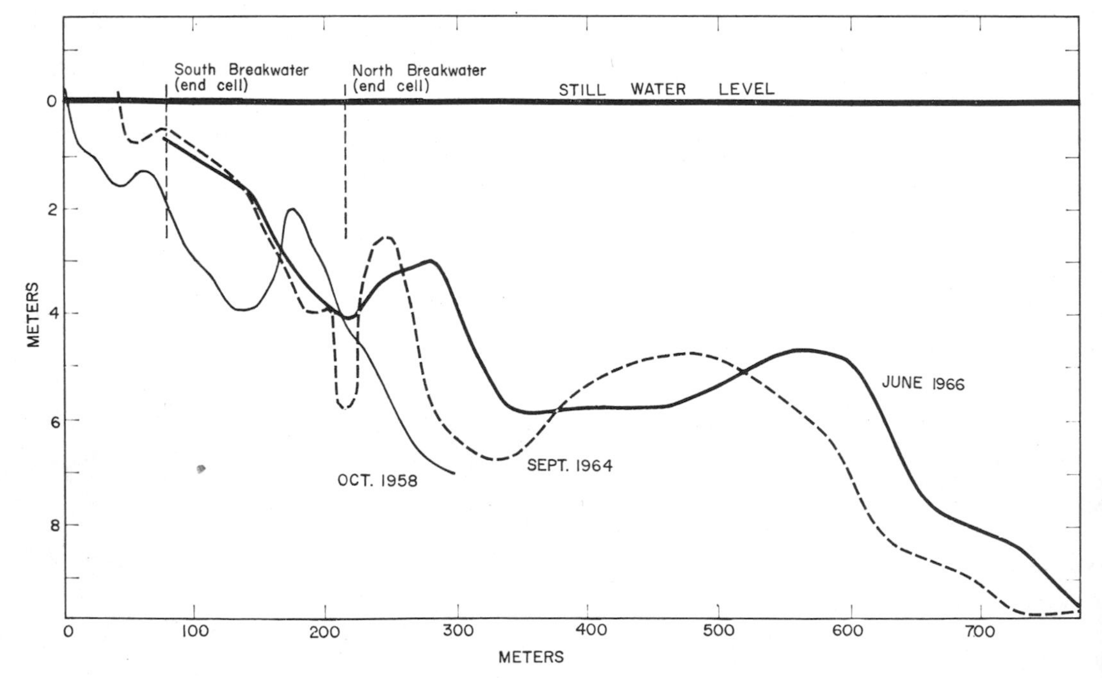

Fig. 10. Shifting of bars at breakwater.

1964 survey, however, the 1966 survey does indicate some lakeward movement of the deep-water bar. At this time it cannot be ascertained if that movement is temporary or is in response to the harbor breakwater.

Harbor Shoaling

Currents. In addition to changes in the offshore bottom, surveys conducted at the Little Lake harbor area determined the pattern of shoaling at the harbor entrance. It was recognized that currents are the main force in the shoaling process. Currents at Little Lake and adjacent coastal areas were measured during 1964 by tracing drogue and dye movement. *Saylor* [1966] recognized several types of currents: coastal (wind driven and residual), longshore, eddy, and channel currents. During the surveys, winds from the west-southwest through north-northwest were observed to induce strong easterly currents, which are deflected away from shore and around the harbor entrance by the dogleg breakwater. A large clockwise eddy exists in the lee of the breakwater-protected harbor entrance. Almost continuous water level oscillations in Lake Superior create water level differences between Lake Superior and Little Lake and induce currents in the entrance channel to the harbor, which can reach high velocities before decreasing to zero and changing direction. The water level oscillations are forced by wind and barometric pressure variations and continue as free oscillation (seiches) with periods determined by geometry of the lake. The first five modes of seiches with periods 7.8, 4.0, 3.3, 2.7, and 2.2 hours can readily be identified. Amplitudes of 35 cm were observed at the frequency of one per week.

The reversing channel currents affect the circulation pattern of the eddy current. The composite effect of outflow and inflow coupled with the eastward longshore current is shown in Figures 11 and 12. The eddy at the harbor entrance and the reversing current in the navigation channel interact in such a way that flow of the reversing current is not distributed uniformly across the harbor entrance. Under outflow conditions, current is compressed to a narrow stream following the west breakwater, and under inflow conditions the flow is confined to the southern parts of the breakwater.

Maximum current speeds measured were the same as the reversing currents. Measured velocities in the harbor entrance reached 60 cm/sec; however, currents greater than 200 cm/sec may be postulated once per week, based on statistics of water level oscillations [*Saylor*, 1966]. Under the wind conditions encountered during the survey, measured coastal currents in Lake Superior near the harbor were of the order of 20–25 cm/sec at the surface and 5–20 cm/sec at the 5-meter depth. The eddy currents were

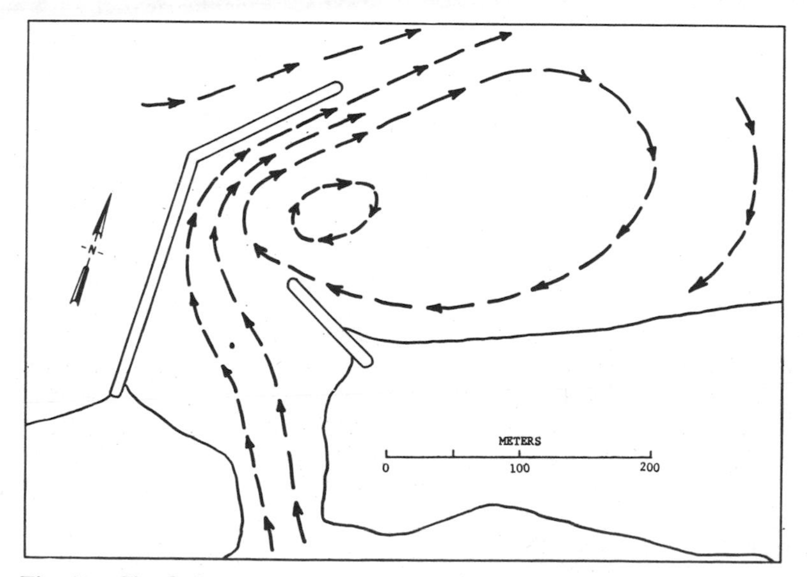

Fig. 11. Circulation model during westerly storm and outflow [*Saylor*, 1966].

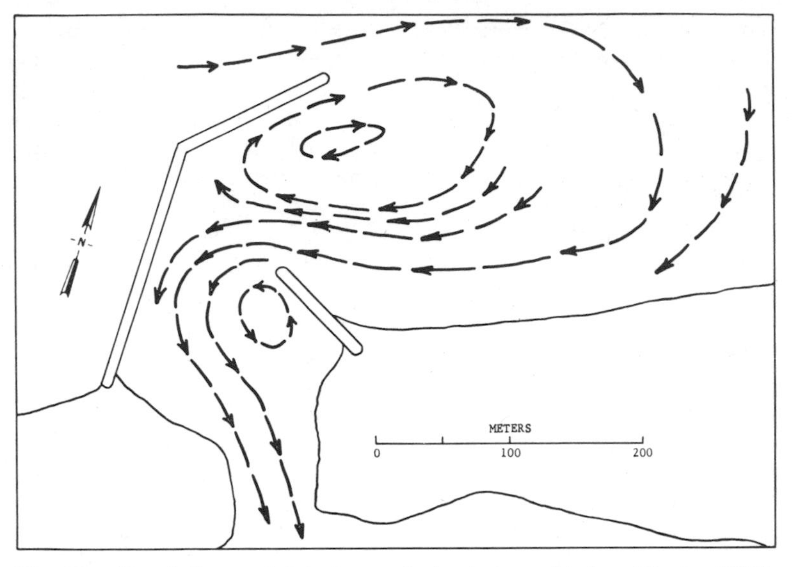

Fig. 12. Circulation model during westerly storm and inflow [*Saylor,* 1966].

found to be extremely variable, ranging from 5 to 20 cm/sec. For the most part, the measured velocities exceed the settling velocities of quartz spheres having a diameter equivalent to the median diameter of sand in the shifting bars.

Shoaling pattern. Figure 13 illustrates the change in bottom elevations between two surveys slightly less than one year apart. The shoal growth, depicted by increase in bottom elevations, is in a northerly and westerly direction from the east breakwater across the harbor into the dredged channel. There is a tendency for the currents to maintain that dredged channel with a greatly reduced width by creating some erosion along the west breakwater.

Generally, shoaling is caused by two factors: availability of sediments from any source and the means to transport that sediment into the harbor. Comparison of current patterns (Figures 11 and 12) with the shoaling pattern (Figure 13), indicates that the current circulation under inflow conditions produces most shoaling. Because the shallow-water bar disappears along the west breakwater, the movement of sediment past the harbor is interpreted as the result of transport in suspension by the longshore current outside the harbor. Under inflow conditions the current is partly deflected into the harbor, permitting sediment from outside the harbor to be moved into the harbor, where, as the current wanes and decreases to zero, it is deposited. Observations on the shoal indicate that after deposition some redistribution of material takes place. Sand was seen to be moved on the shoal in phase with the current. Movement was by traction (sometimes the formation of a current vortex placed material in suspension), and the moving sediment created a bed form of asymmetrical sand waves (transverse dunes) moving in the direction of current flow. The formation of the shallow-water bar on the downdrift side of the harbor is another factor that affects the shoal. As shown previously, the bar transfers sand to littoral drift, and, as a result, the shallow-water bar at the harbor entrance tends to prevent the shoal from growing up to the water surface. The outflow pattern, Figure 11, with the concentration of flow lines along the inner edge of the west breakwater, carries out some of the deposited material along with limited additional sediment eroded from the entrance to the harbor. The erosion process is increased during winter months because of the higher current velocities that result from channel area restriction by ice cover or ice foot development.

The intermediate bar, displaced lakeward following construction of the breakwaters, gradually trends back to its natural offshore position approximately 300 meters east of the harbor,

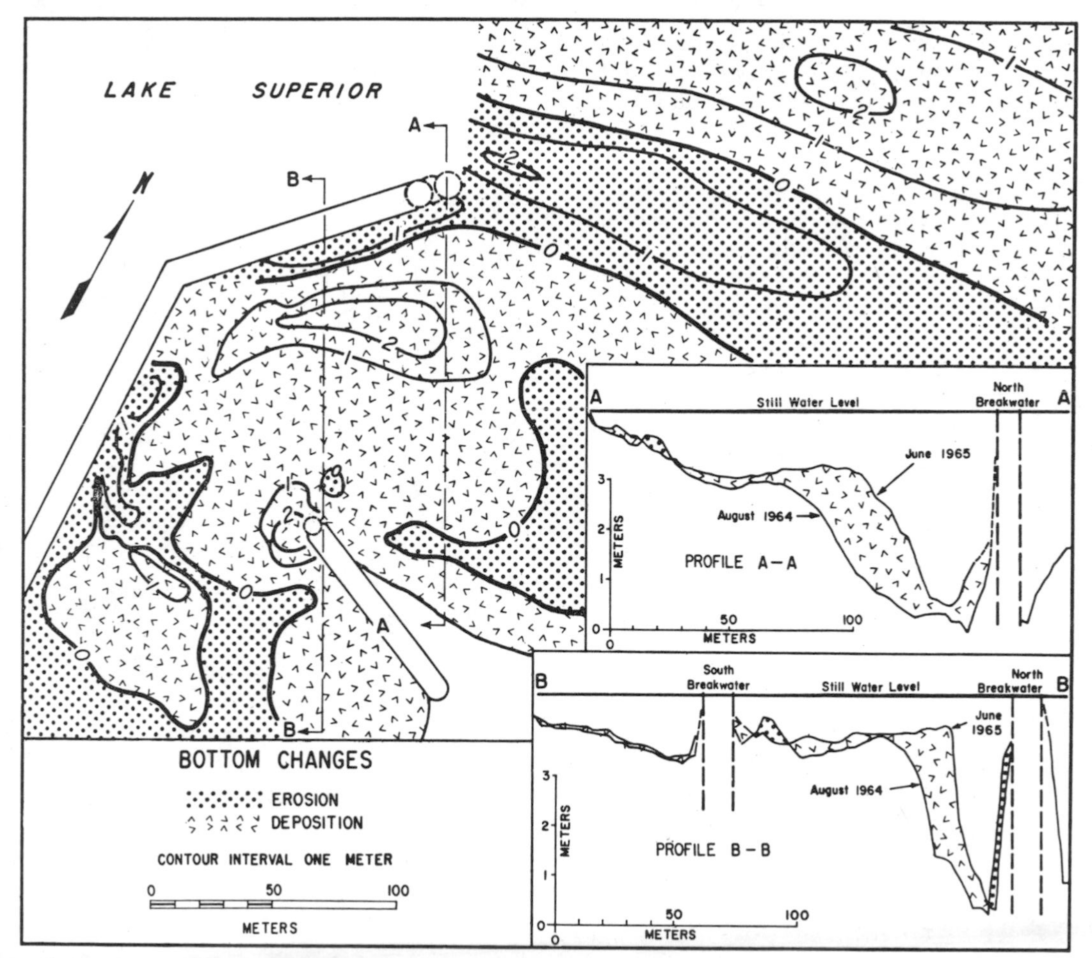

Fig. 13. Pattern of shoaling at harbor entrance.

an area marking the eastward limit of the eddy current. At the harbor entrance the trend of the intermediate bar is interrupted by the erosion observed in that vicinity. It is judged, therefore, that sediments from the shallow-water and intermediate bars, when placed in suspension, may be carried past the harbor or into the harbor, depending on the direction of the reversing current, either outflowing or inflowing, respectively.

Contrary to the above, the interpretation is that the deep-water bar does not play a direct-role in shoaling of the harbor. Trajectories of coastal currents in deeper water indicate sediment of the deep-water bar, when placed in suspension by wave action, would not be transported directly toward the harbor.

Conclusions

The existence of offshore bars and their shifting patterns indicate mobile bottom material and potential for shoaling. The shifting pattern of the bars can be determined by periodic surveys before any construction. The rate of littoral transport estimated from the wave energy and other factors would point out the magnitude of problem. The shallow-water and intermediate bars, when shifted by the construction of breakwaters, maintained the same distance between the bars as before construction. This observation would provide useful means of forecasting the location of erosion and deposition areas associated with the construction of offshore structures, if the maintenance of distance between bars is caused by the character of waves in the vicinity of a harbor and not by the configuration of a particular harbor site. Data from other locations under different wave conditions are needed.

A close relationship exists between currents and shoaling pattern. Three types of currents are determining the shoaling pattern: longshore, channel-reversing, and eddy currents. Before construction, an estimate of the resulting current structure can be made from field surveys of longshore currents and of water-level fluctuations of the main water body at the proposed harbor site.

The ice foot induces erosion in front of it. An understanding of the conditions for formation and knowledge of the location of the ice foot might permit the designing of structures in a matter to benefit from the erosion-inducing capacity of the ice foot.

Acknowledgments. Data collections and analysis were made in connection with the general research program of the Great Lakes Research Center, U. S. Lake Survey, Corps of Engineers. Grateful acknowledgment is made to the Detroit District of the Corps of Engineers for providing much of the survey data. Permission of the Chief of Engineers to publish results of this study is appreciated.

References

Bajorunas, L., Littoral transport in the Great Lakes, *Proc. 7th Coastal Eng. Conf., Council on Wave Res., Univ. Calif., Berkeley,* 326–341, 1961.

Davis, R. A., Jr., and D. F. R. McGeary, Stability in nearshore bottom topography and sediment distribution, southeastern Lake Michigan, *Great Lakes Res. Div., Univ. Michigan, Publ. 13,* 222–231, 1965.

Marshall, E. W., Air photo interpretation of Great Lakes ice features, *Special Rept. 25, Great Lakes Res. Div., Univ. Michigan,* 92, 1966.

Saylor, J. H., Currents at Little Lake Harbor, Lake Superior, *Res. Rept.* 1-1, *U. S. Lake Surv.,* Detroit, Mich., p. 19, 1966.

Taney, N. E., Laboratory applications of radioisotopic tracers to follow beach sediments, *Proc. 8th Coastal Eng. Conf., Council on Wave Res., Univ. Calif., Berkeley,* 279–303, 1963.

Zumberge, J. H., and James T. Wilson, Effect of ice on shore development, *Proc. 4th Conf. Coastal Eng. Council on Wave Res., Eng. Foundation, Univ. Calif., Berkeley,* 201–205, 1954.

(Received April 27, 1967;
revised July 17, 1967.)

Editor's Comments on Paper 12

One type of spit or bar is that which connects an island to the mainland or another island. This depositional link is called a tombolo. The variation in kinds of tombolos is considerable. O. C. Farquhar has undertaken the task of outlining the stages of tombolo growth and describing the possible variations. Thus we have formation of islands, erosion, sediment transport, and widening of the connection to form tombolos; then the potential of single, forked, complex, double, parallel, or multiple tombolo forms. Also included in this paper are two case studies and reclamation, engineering, and classification considerations.

Oswald C. Farquhar was born in Great Britain in 1921. He received his B.A. from Oxford in 1940 and after a stint in the RAF, his M.A. in 1947; he then went on to a Ph.D. in geology in 1951 at Aberdeen, Scotland. Dr. Farquhar has taught at Aberdeen (1948 to 1953), Kansas (1954 to 1957), the University of Massachusetts (1957 to 1961), and Wellington, New Zealand (1963 to 1964). He is presently Professor of Geology at the University of Massachusetts in Amherst.

Oceanogr. Mar. Biol. Ann. Rev., 1967, **5**, 119–139
Harold Barnes, Ed.
Publ. George Allen and Unwin Ltd., London

12

STAGES IN ISLAND LINKING

O. C. FARQUHAR
University of Massachusetts

There are three things that can happen to islands. They may remain about the same size for a long time; they can be eroded gradually until they cease to exist; or they may become joined to each other or to an adjacent coast. Where they are joined to other areas of land, the link is usually provided by sand or shingle bars. These are called tombolos, which are notable features along many coastlines. Several varieties of tombolos have been recognized in all parts of the globe, linking islands along the edges of lakes as well as oceans. Parts of a tombolo may be single, forked, or complex. Where two or three join neighbouring land areas together, tombolos may be called double, parallel, or multiple. Tombolos between islands can be of considerable benefit. Not only do they improve communication, but the water on one or both sides is usually sheltered, providing quiet anchorage. Where more than one tombolo exists, the water enclosed may form a lagoon. Some lagoons of this type have been drained and used for building and agriculture, while others have been changed from salt water to fresh water storage. Also, sand may be blown into the lagoonal area, parts of which may then emerge above sea level. Tombolos are clearly of importance in studies of inshore hydrography and of considerable interest in littoral marine biological work.

This paper deals first with the origin and occurrence of tombolos and includes a brief review of selected examples. The rest of the paper is concerned mainly with shoreline topography and land reclamation in two areas with double tombolos, Quebec and New Zealand. The tombolos seem to result from normal wave action, and in both areas long stretches of sand running side by side in pairs tie groups of rocky islands together.

ORIGIN AND OCCURRENCE OF TOMBOLOS

The term tombolo is a descriptive one, referring to the geometry of the beach itself rather than to its origin. Tombolos are features of an irregular shoreline but have the general effect of straightening the coast. They can develop wherever the water behind an island is not too deep, provided sufficient sediment of the right particle size is available and provided the wave direction and wind strength favour beach development. Tides and currents are important as transporting agents, especially for the medium and coarse sands of which such beaches usually are composed. The processes that determine the formation of tombolos are similar to those needed for the growth of spits, bars, and similar features along a typical irregular coast.

Shoreline submergence is not, as sometimes stated in the older literature, a factor in the formation of tombolos, which can occur just as readily along a suitably indented coast that is either static or emerging. Raised tombolos are found, for instance, around the central west shores of Baffin Island, just south of Eqe Bay and north of Longstaff Bluff (C. A. M. King, pers. comm.). The coast here has emerged since the time when offshore rock islands became tied to the land by ridges of coarse shingle, and the features concerned are now some 80 metres above modern sea level. Tombolos and other beaches may be terraced by a falling level or destroyed by a rising level, but their initial development depends upon numerous other factors.

In glaciated areas drumlins provide the ideal material for the growth of tombolos. The drumlins are eroded by waves, which leave large boulders on the shore, while the finer material is moved by longshore drift and currents to build up such depositional features as tombolos on the lee shore. Good examples of boulder lags and the related sand accumulations formed in this way can be seen just offshore in many harbours and inland waters along the Canadian and New England seaboards. Examples in Massachusetts Bay are discussed in a separate article (Farquhar, in press).

Where the climate is warm enough, sand sedimentation may be assisted by the growth of mangroves and other types of vegetation which act as a baffle or trap, especially within tidal inlets. Sometimes the bars increase after some freak event like a shipwreck. Between 1816 and 1959 a total of 343 wrecks took place around St Pierre and Miquelon, the French islands off the south coast of Newfoundland. Of these nearly 100 were on the Isthme de Langlade, a tombolo linking the two main islands of Miquelon (Fig. 1A). Most of this number occurred along the western shore of the tombolo, as shown on a map published in Paris (Institut Géographique National, 1960). The shipwrecks were caused by the tombolo and also helped it to grow temporarily. The forked tombolo at the north end of the shoal is submerged at some high tides. The same is true of the breached bars connecting Holy Island to the Northumberland coast in northern England (Fig. 1B).

The southwest tip of Puerto Rico, called Cabo Rojo, consists of forked tombolos tying two separate limestone knobs to the mainland (Kaye, 1959). These knobs are also tied to each other by a narrow beach, a complex arrangement that results in the enclosure of a small triangular lagoon (Fig. 1C).

Tombolos are not confined to coasts with a large tidal range. They occur around the shores of such virtually tideless waters as the Great Lakes of North America (Fig.1D). Good examples also exist around the shores of the Mediterranean, where the maximum tidal range is no more than about three feet. Tombolos are equally prominent on coasts with tidal ranges varying from normal as at Chesil Beach in the English Channel (Fig. 2A) to extreme as in the Bay of Fundy. In these two areas some of the beach material is cobble-sized, mainly because of storm waves.

Much of the sediment forming tombolos is either derived from longshore

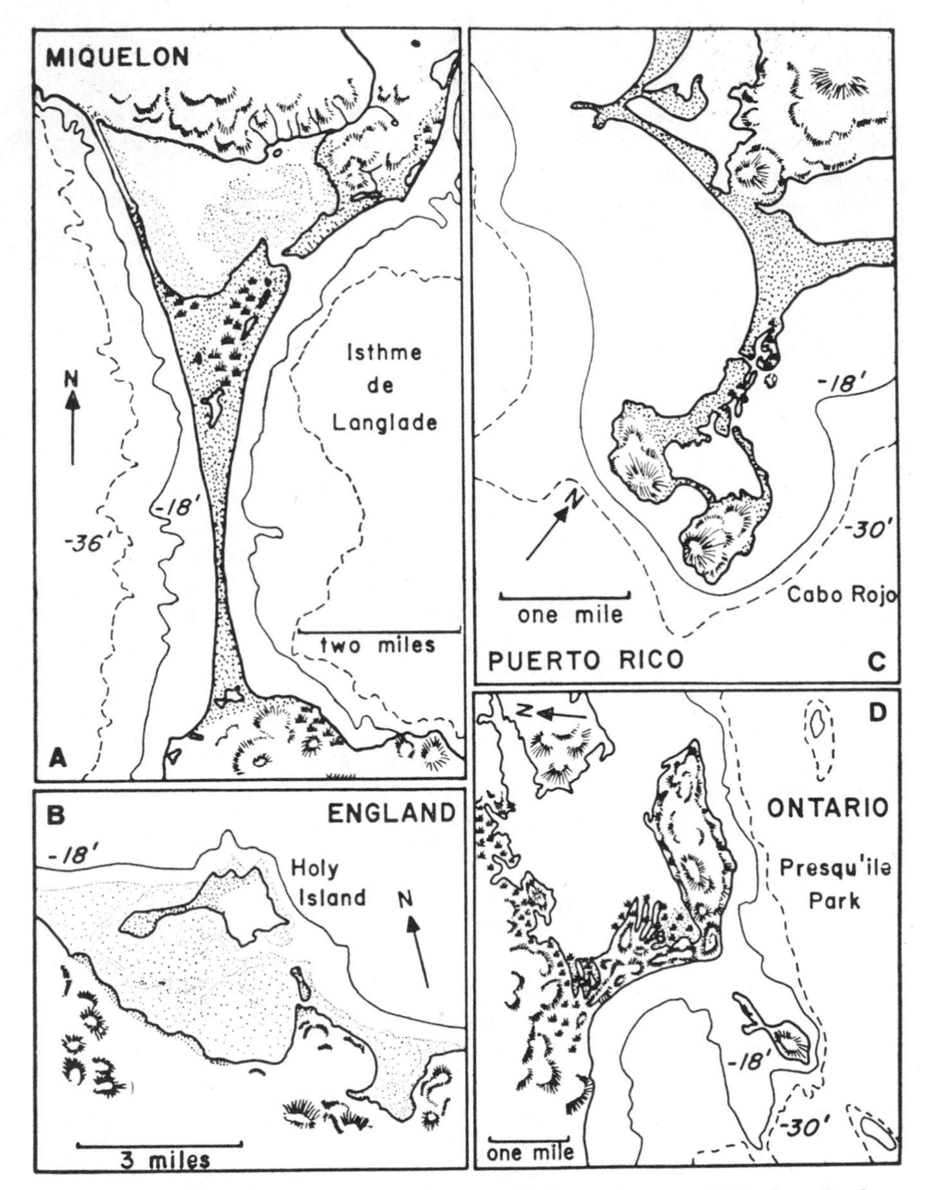

Fig. 1.—Links between islands. A. Miquelon off the south coast of Newfoundland, comprising two islands joined by a long tombolo; this is forked at the north end to enclose a shallow lagoon. B. Holy Island on the northeast coast of England; sand spits and sand bars between the island and the coast are prevented from forming complete tombolos by tidal streams. C. Cabo Rojo, the southwest tip of Puerto Rico, consisting of limestone knobs; these are tied to the mainland and to each other by a series of tombolos. D. Presqu'île, an island linked to the north shore of Lake Ontario by a sand beach or tombolo facing west; the sheltered area behind the beach is occupied by swamp.

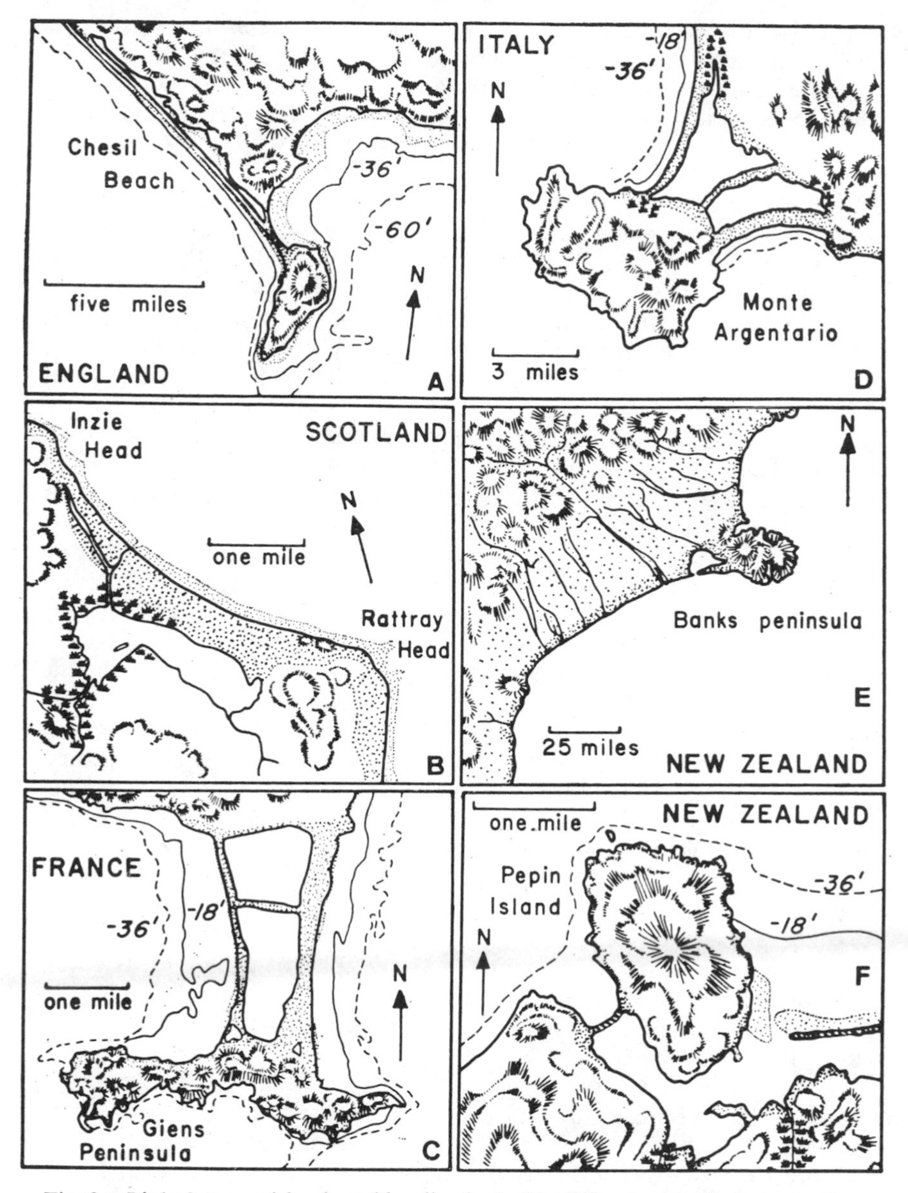

Fig. 2.—Links between islands and headlands. A. Chesil Beach, a tombolo connecting Portland to the mainland; the beach is composed mainly of cobble-sized material. B. Inzie Head, joined to Rattray Head by a four-mile sand beach, behind which lies a loch or pond; this is a baymouth bar, formed by the same type of wave action as tombolos. C. Giens peninsula, one of the Hyères Islands, tied to the Mediterranean coast of France by two parallel tombolos; between the tombolos is a lagoon, the north end of which is used as a salt evaporating basin. D. Monte Argentario, linked to the coast of Tuscany by a pair of tombolos; a third, partly man-made link, and also a canal, cross the enclosed lagoon, called the Stagno di Orbetello. E. Banks Peninsula, a Pleistocene volcano joined to the coast by later glacial outwash gravels; the nearest rock headlands on either side are about 30 and 90 miles up and down the coast. F. Pepin Island, near Nelson, attached on its west side by a single tombolo, which lies behind the general line of the shore; east of the island two spits partly block the entrance to a lagoon.

drift or else, having been washed out to sea as the result of land erosion, is subsequently carried back to the shore by waves and coastal currents. In other instances material transported directly from the land may be emplaced along the coast in such a way as to produce features closely resembling tombolos. Thus valley train material extending seaward off Alaska has prograded almost to the headlands. The resulting features may become offshore or baymouth bars, and, if an island is joined to other land, tombolos. Wave and current action here has only a secondary influence on the shape of material deposited to form such bars. Again, along the Aberdeenshire coast of Scotland a four-mile sand beach with a loch or pond behind it joins two rocky headlands together (Fig. 2B). This is a baymouth bar: except that the headlands are not islands but are parts of the mainland, the beach between them is similar to a tombolo and shaped by the same type of wave action. This example of a baymouth bar is used for comparison with tombolos partly because of personal acquaintance with the area but chiefly because, as pointed out by Walton (1956), the evolution of the coastline in the vicinity of Rattray Head illustrates the dynamic effect of wind and waves on a low sandy coast particularly well.

Although single links are more common, some coasts are noted for pairs of tombolos behind nearshore islands. Double tombolos may be seen around the French coast, both in the Bay of Biscay and on the Mediterranean (Fig. 2C). On the Adriatic coast of Italy nearshore islands protect the adjacent mainland from marine encroachment, the double tombolos on each side stretching for several miles up and down the coast. At Orbetello a third, partly man-made tombolo crosses the centre of the lagoon, which is about five miles long and three miles wide (Fig. 2D). The lagoon here lies behind a prominent rock island, seven miles long and over six hundred feet high. The two outer tombolos run for nearly ten miles. Between the tombolos that join another former island to the Italian coast lie the Pontine Marshes, a huge area reclaimed from the sea and overlooked by the World War II beachhead of Anzio. The recent history of this area in terms of sea levels and the sequence of coastal deposits is given by Zeuner (1959). The work of Sion (1934) provides a classic reference to the geography of Italy and its coast. The Geographical Handbook Series (issued during the years of World War II) includes volumes on Italy and France, with special attention to the coastal areas.

The Banks peninsula near Christchurch, New Zealand, is flanked by a shoreline which is remarkably similar in shape to the Italian examples mentioned above, although entirely different in origin. Fluviatile materials, which are mainly glacial outwash gravels, extend from the mainland to a large volcanic island (Fig. 2E), while the shore on either side consists of concave beaches that terminate at rock headlands many miles away. These beaches are due to erosion, probably begun at a lower level of the sea, unlike tombolos which are depositional features.

From studies of laboratory models, Sauvage de Saint Marc and Vincent (1955) have distinguished tombolos according to the different types of current required for their formation. Their main conclusions are that, in the case of simple tombolos, two phenomena are dominant. These are, first, local swell caused by diffraction as the waves reach the shore and, secondly, the relative calm that exists in the shelter of an island. Simple tombolos may be symmetrical, or not, depending upon whether the waves and currents approaching from one direction are stronger than those approaching from the other. Submarine topography is clearly an important influence here.

Both in the Magdalen Islands, Quebec, and on North Cape, New Zealand, the tombolos consist of sediment supplied by longshore currents moving tangentially past islands of rock. They have been formed by accumulations of sand as tails in the shelter of islands. Sand may also be deposited on the opposite shores. The resultant beaches, or tombolos if they extend far enough to link the two shores, tend to become orientated at right angles to the direction and approach of the locally predominant waves.

Double tombolos are formed by waves approaching the lee side of the tied islands at different angles of incidence on each flank. In either single or multiple examples, longshore drift of material continues only until the amount of erosion virtually equals the amount of deposition. Sand movement slows down as the supply decreases. Except where they curve to join a headland or islands, the longest beaches become almost straight. Their development has been discussed most recently by Davies (1959) on the evolution of shoreline curves, by Russell (1959) on long, straight beaches, and by Steers (1964) on the orientation of sand beaches. The various conclusions apply equally to tombolos, which are simply a type of beach. Davies in particular has shown that the plans of some Australian beaches may be interpreted in terms of the constructive effect of the refracted swell and are the result of the wave plan. Wave action along the coast has been the subject of a theoretical study by Le Méhauté (1961) and, more recently, has been discussed in a full treatment of the fluid mechanics of oceanographical engineering by Wiegel (1964) and several papers on estuary and coastline hydrodynamics (Ippen, 1966).

Double tombolos and the land they join to the coast usually form a triangular section of the shoreline. Almost the same shape is characteristic of cuspate spits reaching out from an angular, commonly convex, shore toward each other. Like tombolos, such spits are formed by waves and currents, but without an island at the point where they would meet if sufficiently extended. Spits, therefore, may be somewhat different in origin. Even so, tombolos and spits can develop in the same environment where the coast is irregular, and may, therefore, be found adjacent to each other, as is the case near Nelson, New Zealand (Fig. 2F). A tombolo there connects Pepin Island with the shore on the west, thus forming a lagoon around the south side of the island. The entrance to this lagoon is between double spits that project between the island's east coast and the shore.

The rest of this paper will be concerned mainly with two areas, the Magdalen Islands in the Gulf of St Lawrence and North Cape, the northernmost peninsula of New Zealand. Each consists of several former islands, now linked by long stretches of sand. In both areas the tombolos are double for several miles and run more or less parallel, with lagoon or swamp between. Comparisons are based upon certain aspects of geological history, such as the origin of the islands and long term changes in sea level, and upon the processes of tombolo formation, including sediment supply, wave action, and wind transport. The Magdalen Islands, which resemble an earlier stage in development than North Cape, will be described first.

MAGDALEN ISLANDS

The position of the Magdalen Islands in the Gulf of St Lawrence is shown in Figure 3. The islands consist of almost a dozen rock areas linked by dune-covered tombolos (Fig. 4). The whole group is quite narrow, averaging only four miles across, and extends for about forty-five miles from southwest to northeast. The tombolos are mainly in parallel pairs or twins enclosing lagoons and swamps which comprise more than one-third of the total area of 140 square miles. In length the tombolos occupy two-thirds of the island group.

The latest geological report on the Magdalen Islands is that of Sanschagrin (1964) incorporating the results of a survey for the Quebec Exploration

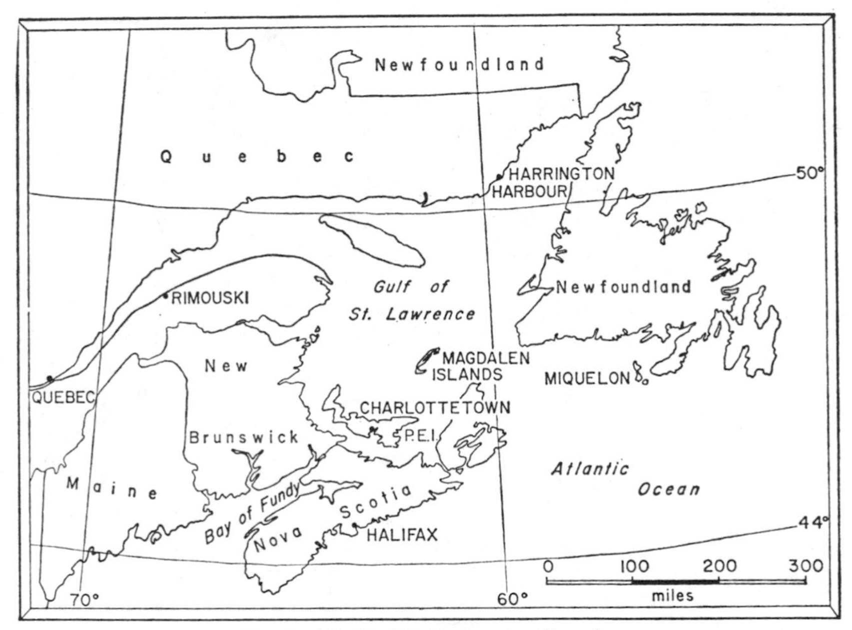

Fig. 3.—Gulf of St Lawrence, Canada.

Service. Among earlier accounts are those by Hamelin (1959) on the geographical aspects, Falaise (1954) chiefly on the shores and currents, Alcock (1941) on the mineral resources, Goldthwait (1915) on the Pleistocene history and Clarke (1910) on the physiography. More general works include those by Smith (1932) and by Linton (1959).

The tombolos linking the principal islands of the archipelago are formed of grey sand made up mostly of quartz. About 75% of the quartz grains are clear, the remainder being coated with a thin film of haematite. Much of the sand has been derived from the red sandstone of a Mississippian formation eroded along coastal cliffs and stream banks. Some has been carried offshore before being washed back to the beach. From the beach, large quantities have been swept by the wind into dunes which rest behind the shoreline, commonly upon marsh deposits.

Each lagoon has been closed off, or nearly closed off, from the sea by pairs of tombolos and tends to become filled with sediment carried in by the tides and by the wind. Where there are channels through the tombolos, tidal deltas have formed on both the sea and the lagoon sides. The lagoons themselves are shallow, generally not more than eight or ten feet in depth and their length affords a considerable fetch for the winds which travel along them. Further, being well separated from the nearest land, the entire island group is approached on all sides by winds of moderately long fetch. The winter winds are westerlies and northwesterlies with only 1–4% calms, while the summer winds are predominantly southwesterlies with some quite strong southeasterlies and with 5–7% calms. Individual sand dunes are often about 30 feet high, but blown sand locally forms hills at least 100 feet above the general surface. Sand also rises over the highest hills along the coast, and pockets are found several hundred feet above sea level. The dunes are not quite continuous, but, where present, the main belt on the west coast varies in width from a few feet to more than one mile. The dunes provide a temporary resting place for material that may be subsequently blown into the lagoons.

RECENT GEOLOGICAL HISTORY

Ganong (1964) includes some interesting fifteenth century charts on the St Lawrence but gives little information about the Magdalen Islands themselves. Falaise has, however, presented diagrams suggesting that the island group was at one time considerably smaller because the principal members were not connected. Some of the sediment later used by waves in building the tombolos was brought into the area by tidal streams or currents. According to the latest edition of the St Lawrence Pilot (1963), tidal currents seldom amount to one knot except close inshore or around the points; since currents of one knot can move only silt, this means that movement of larger material is confined for the most part to areas near the shore. Mean tides have a range of almost 13 feet. The stream of flood tide from the Atlantic, entering the Gulf through the straits on both sides of Newfoundland, flows past the Magdalen Islands in a southwesterly direction. On returning, the ebb waters sweep along the island coasts in the reverse direction, toward the northeast. In the largest lagoon the currents are mainly southwest and northeast, channels across the tombolos allowing tidal streams to pass through. The flow directions depend to some extent on the wind.

In the Gulf of St Lawrence, sea-level data come from four permanent tide gauges maintained by the Canadian Hydrographic Service. The locations are shown in Figure 3. Tabulated elevations in 1963 from Father Point (Rimouski) and Harrington Harbour show that mean sea level has not appreciably changed there since 1910 and 1940, respectively. At Charlottetown and Halifax, however, mean sea level has risen approximately 0·5 feet and 0·4 feet since 1920 (N. G. Gray, pers. comm.). Since the Magdalen Islands are in this southern part of the Gulf, nearer to Charlottetown than the other points of observation, some rise in sea level probably has also occurred there. Water depth is one factor in the development of tombolos as well as other types of beach. Further work is, however, needed to show the precise relationship between water depth and beach development in terms of local coastal history.

Tombolos can form regardless of whether submergence or emergence is taking place. As the figures for part of this century indicate, it appears that the Magdalen Islands are submerging. Other evidence of this has been presented by Falaise who, in an examination of the bathymetric maps of the sea around them, discerned several shallow preglacial valleys which he considered extend at least 10–15 miles from the present shoreline. Sanschagrin (1964) comments that, if these valleys exist, they provide evidence of considerable lowering of sea level during the Pleistocene period. Submarine contours around the Magdalen Islands show that the sea floor is now at rather shallow depth, reaching 20 fathoms an average distance of almost 12 miles from the coast.

At lower levels of the sea there may have been bars beyond the existing coastline which have since migrated shoreward. Sandbars run parallel to

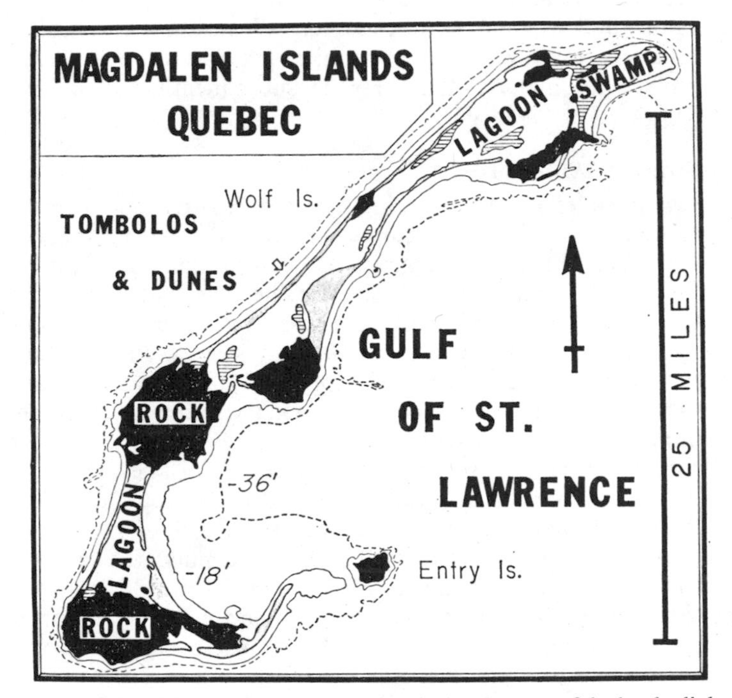

Fig. 4.—Magdalen Islands, Quebec, showing isolated areas of bedrock, linked by dune-covered tombolos, with lagoons between.

many of the present beaches and under certain wave conditions act as a source of supply. These bars are usually from 300–600 feet offshore and at depths of about ten feet. Between the bars and the shore the water is about 20 feet deep. For the crews of vessels washed onto the bars this deep inshore water makes landing attempts extremely treacherous, especially with heavy surf. At any rate, offshore bars are part of the equilibrium conditions under which the tombolos or beaches are maintained.

Although, relative to sea level, the islands are now being submerged, the probability that postglacial emergence has also occurred is indicated by the different heights of old terraces above the present shoreline. Aerial photographs show that extensive strand plains, marked by beach ridges fringing the present lagoons, have been partly destroyed by marine erosion. Thus, the coast is a compound one with a history of both submergence and emergence. In fact, a postglacial uplift of about 15 feet has been suggested, although the actual glaciation of the Magdalen Islands has not been generally accepted. Erosion of the nearby Cabot Strait has been attributed to glacial action but the immediate area of the Magdalen Islands is shown on a glacial geology map in the 1957 Atlas of Canada as "unglaciated." The islands are near the margin of the glacial north, but both they, as noted above, and the mainland shores of the Gulf of St Lawrence are marked by emerged beaches, indicating probable late-glacial rebound.

Recent level changes in the southern part of the Gulf are partly due to gradual melting of the ice caps. The water released now causes a general rise of some 4¾ in. (120 mm) per century. However, world sea level has been essentially stable for the past 6000 years, and this rise is only a minor feature of climatic change. As Fairbridge (1961) comments, such a rise is a normal aspect of the short-period, low-amplitude, climatic oscillations of the Holocene epoch. But the rate is of the same magnitude as the tabulations at the Charlottetown and Halifax tide gauges, quoted above, though not those at Father Point and Harrington. The assumption is that the changes in level are effected by some other cause, such as isostasy.

Apparent differences between sea level and parts of the lithosphere, as recently pointed out by Vella (1962), can result from at least six processes, namely, eustatic sea-level change, vertical tectonic movements, deposition, compaction, decompaction, and erosion. Also, as determined by Bloom (1966), "the postglacial eustatic rise of sea level introduced . . . a wedge-shaped load [of water] onto all continental margins. Moderate isostatic adjustment in response to the water load along the present shoreline of northeastern United States would account for the observed differences in submergence histories at several localities."

Assuming that erosion of the Magdalen Islands will continue, Falaise (1954) has made some predictions about the future situation. Based on an analysis of the currents around the shores and in the lagoons, one of several possibilities is that an island on the midwest coast (Wolf Island), standing less than 100 feet high, will be eroded away and its place taken at sea level by sand. This sand could form a bar joining the two tombolos that at present link this island with others to the north and south. Another prediction is that the prominent sandy hook in the southeast will link up with the island off its eastern tip (Entry Island). But for dredging operations to maintain a navigation channel, this could have already happened.

NORTH CAPE

Some of the most magnificent beaches in the world occupy the long stretches between rock headlands on North Cape, the peninsula which forms the northern tip of New Zealand (Figs 5 and 6). As in the Magdalen Islands, there are about a dozen former islands linked by pairs of dune-covered tombolos. These tombolos run almost parallel along the main peninsula for about 65 miles. There are similar shorter pairs in the southeast, northeast, and northwest corners of the area.

Compared with the Gulf of St Lawrence platform on which the Magdalen Islands rest, the sea surrounding North Cape is quite deep. Nearby submarine contours are shown on Figure 5; these are approximate only, being sketched in from the few soundings available. The bathymetry of the New Zealand region is described by Brodie (1965). In contrast to the Magdalen Islands, much of the area between the North Cape tombolos is filled above sea level by Recent sediment. Part of the surface is still occupied by small lagoons or swamp. In other respects the shape, size, and overall environment of the rock 'islands' and tombolos in the two areas are remarkably similar.

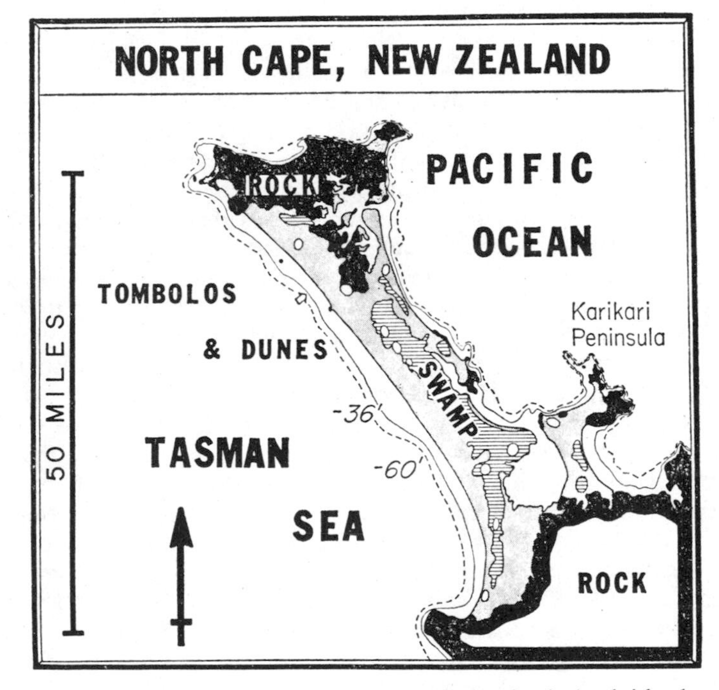

Fig. 5.—North Cape, northernmost New Zealand, showing bedrock islands and the mainland, linked by dune-covered tombolos, with swamp between.

Among earlier reports on New Zealand geology are those of Smith (1881) and Maxwell (1897) on alterations in the coastline of the North Island, and that of Cockayne (1911) on the dune areas. As pointed out more recently by Kear and Hay (1961), North Cape is tied to the mainland by a long and narrow tombolo-like isthmus. The isthmus consists of parallel sand spits joining lines of rocks, one series facing the Tasman Sea and another the Pacific Ocean. The climatic evidence to be obtained from a study of sea-level fluctuations, partly in reference to northern New Zealand, has been discussed by Schofield in several papers (e.g. 1963). Shifts in coastal currents and the formation of coastal dunes are among the types of evidence considered, but ecological changes also result. Thus, in studies of the species distribution of land snails, Powell (1947) has shown that a large rock area on the Pacific side, forming the end of the present Karikari Peninsula, was at one time an island. Dune-covered tombolos now provide links to the mainland. Without such links, large areas in the extreme north, composed mainly of volcanic rocks, would be cut off from the rest of New Zealand.

Along the west coast there is a steady movement of sand which drifts up from the south. Much of this is volcanic material brought down by the Waikato River, which carries huge volumes of sediment derived from the North Island's central volcanic region along a mainly steep 200-mile course and across a large delta to empty into the Tasman Sea. As a consequence, grains of pumice are found among the sand that reaches the long northern tombolos, although the main constituent is quartz. The general effect of sediment deposition from the sea, resulting in formation of the tombolos, has been to straighten the western coastline. Gaps between rock islands are filled in by the tombolos, and baymouth bars are built. Marine erosion of the islands is slowed down by the abundant supply of sand that locally has formed protective beaches.

A few small lagoons remain in the swamp area between the tombolos. None of these is more than a mile or two across at the present time, although they were more extensive in the past. Several areas which formerly carried shallow lakes have now dried up, and others are occupied by water only on a seasonal basis. On the east coast three gaps in the line of tombolos each mark the mouth of an estuary.

Much of the peninsula has already become productive farm and forest land, and other sections are being treated in different ways. For instance, lupins are used to stabilize drifting dunes as a first step in land development, the seeds being scattered from the air. Also, certain areas have been oiled first and then grassed over.

RECENT GEOLOGICAL HISTORY

Bars of sand may have existed below the present sea level at former times when the sea was lower than now. Also, there are today, as Shepard (1950) noted, long submerged bars offshore on the west side of North Cape. A few miles from the east coast local shallows are present, and these may have supplied sand for the tombolos. Charts of the sea floor around the peninsula indicate a number of submerged reefs, including Pandora Banks, which are about ten miles off the northwest coast and have often proved a hazard to shipping. As recently as May 1966 a collier was lost with all hands after breaking up on these rocks. Shallow deposits of sand undoubtedly lie around such areas and, under suitable conditions, some could be carried to the mainland shore.

The North Auckland area underwent partial submergence during the general postglacial rise of the oceans. Brothers (1954) concluded that dune formation during subsequent periods of shoreline progradation added great quantities of sand to the foreland around Auckland. There is other evidence, too, of some emergence during the Holocene, including the presence of shore platforms and two levels of marine terraces, the higher some 8 to 15 feet above the present sea level. The compound nature of the coast is due not only to eustatic changes of sea level but also to regional crustal movements. Although the whole North Auckland peninsula was, until recently, regarded

A

Fig. 6.—Tombolos and islands on the North Cape Peninsula, New Zealand.

A (*Above*) Double tombolos linking the northeastern extremity to the rest of the peninsula. North Cape itself is the point in the foreground.

B (*Over, upper*) Scott Point, with Cape Maria van Diemen beyond. These rock islands with sand connections occupy the northwest corner of the peninsula.

C (*Over, lower*) The northern third of Ninety Mile Beach, along tombolo facing the Tasman Sea. The rock in the foreground is the Bluff, tied by a Y-shaped tombolo to the main beach.

(Reproduced by permission of Whites Aviation Service, Auckland)

B

C

as relatively stable, a series of earth tremors occurring in 1963 has altered this view.

The various deposits and erosion levels observed indicate a long history of sedimentation between the parallel coasts of North Cape. Tidal currents flowing into the lagoons and strong winds off the ocean beaches were the main transporting agents. The dominant direction of winds of all speeds is from the south-southwest. Coastal sandhills have been built up to a few hundred feet above sea level. Their greatest extent follows the longest beach on the Tasman Sea coast, exposed to the prevailing wind. Parts of these dunes rest upon the tombolos, and parts upon older beach ridges above high water mark.

Locally the tombolos moved onshore or landward at intervals during the Holocene. There are indications that they extended in the opposite direction too, by seaward progradation, especially on the Pacific side. Particularly in the southeast many concentric dunes overlie the older tombolos behind the prograded shoreline. From the air these older tombolos are seen to form the well-marked accretionary ridges of a chenier plain. They are somewhat modified by blown sand, and small ponds occupy some of the angles between them. Many of the present hollows may be due to deflation by the wind, but infilling of former swamps has also taken place.

STAGES IN ISLAND LINKING

The Magdalen Islands and North Cape possess many features in common, both being island complexes with long sand connections between otherwise isolated rock areas. Adjacent beaches on opposite coasts stand back to back, concave toward the sea on either side of the peninsulas. Between them are sand ridges, lagoons, swamps, dunes and, along North Cape, sections that are being reclaimed to form productive farm and forest land.

While the rock islands were being eroded and were furnishing sand to the beaches, they also caused deposition of some material from currents impinging upon them. The pumice grains, transported to the western North Cape beaches from the south, provide evidence of an external, as well as a local, source. Sedimentation between the tombolos of North Cape has been assisted by the growth of warm climate vegetation, such as mangroves, which help to trap the sand. The main stages in development of a coast with tombolos comprise: (1) formation of islands; (2) erosion; (3) movement of material between the islands; and (4) widening of the link.

FORMATION OF ISLANDS

The islands of the Magdalen group may be remnants from gentle upwarps or, more likely, diapiric blocks uplifted along axes of salt or gypsum movement. Evaporite sediments of Mississippian age have caused similar uplift in nearby Nova Scotia and are major structural elements elsewhere, notably in Axel Heiberg Island in Arctic Canada. Further displacement of such areas that have once been disturbed could recur due to the same cause.

The bedrock 'islands' of North Cape are mainly Mesozoic volcanics, some silicic but most basaltic. Early Tertiary sediments are locally associated with the volcanic rocks, which formed seamounts on the floor of a Cretaceous geosyncline (Quennell, Hay and Farquhar, 1965; Farquhar, 1966). Geophysical evidence provisionally suggests that the lava massifs rise from 1000 to 3000 feet below sea level.

The islands in the two areas thus had different origins, although their shore profiles were not unlike: both the Magdalen Islands 'diapirs' and the North Cape 'seamounts' must have had rather steep sides initially.

EROSION

The presence of islands is a factor in altering the rate of currents moving near the shore. Such currents may move faster through the narrow straits around headlands or be locally slowed down by a single island obstruction. In the Magdalen Islands and North Cape the latter effect caused deposition of sediment which was built out into spits extending far enough to form complete links or tombolos from one island to the next.

As already noted, the islands themselves may have furnished much of the sediment composing the tombolos, some coming from erosion of the shoreline and some transported from the hinterland by streams. Nearby land areas, underwater reefs, and offshore glacial deposits are other sources of supply. The material may have originated as river sand and then belonged to an offshore, beach, or dune environment. Modern techniques of tracing nearshore sediment transport, for example using fluorescent grains (Ingle, 1966), would be needed for a complete analysis of beach sand movement here. Sand movement by wind is another subject that has received recent consideration (e.g. see Belly 1964).

So far as the various environments along the shore are concerned, beach sands and dune sands have been distinguished by several authors, including Friedman (1961). As pointed out, however, by Schlee, Uchupi and Trumbull (1964), the results of tests carried out by different methods are not consistent. Using the statistical parameters of grain size they found no significant difference between the beach and eolian sands of Cape Cod, Massachusetts. One explanation, put forward by Hayes (1965) in regard to the southern Texas coast, concerns the mixing of sizes from more than one river source and the consequent obscuring of any variations due to the dune or beach environments. For surface material the best observation of environment is at the sample site itself: sand blowing across a dune is clearly being conditioned in an eolian environment. The nature of the older materials below the surface of a beach or dune probably is best determined on some basis other than size. Virtual removal of shell particles by abrasion is one of the results of wind action. Studies of rounding are useful also, and examination of surface texture with the electron microscope is now especially important.

MOVEMENT OF MATERIAL BETWEEN THE ISLANDS

The tombolos are composed of sand which drifted along the shore and was deposited by waves on the sheltered sides of the islands. Although waves normally approach the shore somewhat obliquely, due to refraction the wave fronts become more parallel to the beach. Each part of the tombolo assumes an orientation almost at right angles to the direction from which the dominant waves approach and is maintained by continuous exposure to such waves. Sand from offshore is carried by these formative waves to the beach from the breaker zone: there is thus a close relationship between varying beach configuration and wave conditions. Further, the offshore bars on the west coast of North Cape and the probable submarine valleys around the Magdalen Islands are features which affect the shape of the shoreline. In these two cases locally shallow and locally deep water, respectively, change the energy density of the swell and the height of the resultant waves. In general, coarser-grained material is found on beaches exposed to larger waves, the particle size and the steepness of the beach being somewhat interdependent.

WIDENING OF THE LINK

Winds from the sea pick up material from the tombolos, dropping much of their load inland behind the beaches. Both in the Magdalen Islands and on North Cape the effect of this aerial transport is to fill in the lagoons between the double tombolos. The tombolos usually become wider, and those on opposite sides tend to join in one expanse above sea level, as the lagoons become shallower. This process of widening and infilling depends upon the continued availability of sand from cliff erosion, offshore deposits, and longshore drift.

In the Magdalen Islands the lagoons are connected to the sea by channels which cut across the tombolos. As tidal waters move in and out, deltas form on both sides. On North Cape, former lagoons between the tombolos are now mostly built up above sea level. Mangroves, which continue to flourish in existing estuaries there, appreciably aided the filling-in process by serving as sand traps. At an intermediate stage parts of the lagoons may have been converted into salt marsh. Reclamation of salt marsh can result in fertile soil and productive land. Varying mixtures of silt, sand, vegetation, and salt are subject to different degrees of compaction, a fact which, to some extent, complicates the filling in of the lagoons.

SIGNIFICANCE OF TOMBOLOS

The significance of tombolos will be discussed next in terms of reclamation, engineering, and classification.

RECLAMATION ASPECTS

The degree of island-linking noted in the Magdalen group appears to represent an early mature stage in coastal evolution. The effect of the tombolos, formed primarily by wave action as accretions to the land, is to smooth out a generally irregular shoreline.

North Cape has probably reached a later stage in development. At one time, like the Magdalen Islands, the New Zealand peninsula consisted of long tombolos separated by almost continuous lagoonal areas. The zone between the main tombolos, once below sea level, has now been filled up to a large extent by sediment. Some of this was transported into former lagoons by tidal currents and some was first brought to the beaches by waves and subsequently distributed over the surface by wind action, strongest from the southwest quarter.

The Magdalen Islands may be merely erosional remnants of large upfaulted blocks, now constituting a shallow marine platform of which the greater part has been worn away. Large quantities of sand are being moved by the sea, as evidenced by the deltas built up on either side of entrances to the lagoons. Differences in sediment distribution even in the short interval between 1950 and 1956 are evident from two sets of aerial photographs taken in those years. Other evidence of sand movement is provided by the dredging needed to keep a ship passage open into the bay in the southeast. Whether in the long run the tombolos are prograding or retrograding is not apparent, but it seems unlikely that they can remain stable for any lengthy period. It is more likely that they build up under certain wave conditions and are eroded at other times.

Tombolos in both areas have a general curvature, concave toward the sea, with the radius increasing with distance from the headland. Where the radius is long, this is because the tombolos are long and, therefore, nearly straight. In a recent account of headland-bay beaches Yasso (1965) found that their plan geometry, which results from wave movements, closely fits a log-spiral. If the curvature becomes such that tombolos from opposite shores cut back into each other, this may destroy the island links completely, with or without a rise of sea level. Before this stage is reached there may well be a pause, during which the land area of the Magdalen Islands increases by the partial infilling of the lagoons.

Natural reclamation of land from the lagoon and swamp areas between the tombolos would be of great benefit in islands such as the Magdalen group. The population there is about five to an acre, particularly dense for a rural community. The 1957 Atlas of Canada records over 10,000 people of whom 80% were counted as farm population, so that land is by far the most important of the Islands' assets.

After detailed studies, measures might be taken in the Magdalen Islands to assure the protection and even the growth of the shoreline. It is important in any case that undue land wastage be avoided. Dredged material could be placed directly in the lagoons or in a position where it should eventually be swept into the lagoons, rather than being dumped out to sea. If material dredged from river mouths and other ship channels is dumped incorrectly, some may drift back. On the other hand, dredged material can be dumped at selected points offshore so that currents and waves will carry some of it back to part of the shoreline where sand nourishment is desired.

Along North Cape, except for the removal from one of the east coast beaches of a pure white quartz sand for glassmaking, there is virtually no human interference with natural conditions. The changes which occur are due to currents and waves acting on the coast. Winds are also instrumental in shaping the peninsula, as evidenced by the presence of extensive dunes.

Substantial reclamation has been possible on North Cape, and huge tracts of land have been brought under cultivation by the government and to a smaller extent by private owners. After the addition of suitable chemicals to first condition the ground, some areas have been turned into forest plantations, while others, with regular aerial topdressing of the soil, support large numbers of beef cattle and crossbred sheep. Yet the human population in the peninsula is extremely small, totalling about 1000, only one tenth of the population in the Magdalen Islands. It is, of course, much easier for people on a peninsula like North Cape to come and go than for those whose whole life has been spent in an island territory like the Magdalen Islands. The patterns of farming are also quite different. In the Magdalen Islands, production is from smallholdings, but along North Cape it is mainly from government 'blocks' many hundreds of acres in extent.

ENGINEERING AND OTHER CONSIDERATIONS

Although this paper is concerned mainly with the geographical nature of particular sections of the ocean margin, the subject of island linking by tombolos is also pertinent to some other aspects of marine science treated in the volumes of this Review.

(i) Tombolos may be able to furnish unique data on inshore wave mechanisms. Unlike most other beaches their existence depends upon contact with more than one separate body of moving water.

(ii) Both sides of a tombolo-like isthmus may be affected at different times by winds of extremely long fetch across water, as in the Magdalen Islands and on North Cape. By contrast, the majority of beaches and bars are directly marginal to the land or lie just offshore.

(iii) If the material derived from the erosion of a series of offshore islands is of sufficient particle size to form tombolos between the islands, destruction of the shore itself may be greatly reduced. Boulder trains between drumlins along the New England coast are effective means of such protection.

(iv) In studies of coastal history tombolos are apt to be dismissed as unimportant. They can, however, provide a partial record of wave action and wind conditions in the immediate past. Full descriptions of tombolos require sampling of the sediment composing them, not only along their length but also across surface profiles and at various points in depth below the surface. Where tombolos form behind islands there may be areas of scour, the pattern of littoral drift being correspondingly altered. Beach ridges and terraces associated with tombolos indicate earlier levels of the lake or sea around which they occur.

(v) From the engineering viewpoint tombolos may help to enclose areas just offshore that can be reclaimed as land. A proposal is now under consideration to fill in part of the area behind the long Nantasket tombolo in Massachusetts, the evolution of which was described by Johnson years ago (1925). Tombolos are also the natural equivalent of tidal dams or barrages, and they may have some value in feasibility studies for man-made structures of this type. Consultants have recently been engaged to report on a proposal for an embankment across the river Dee in Liverpool Bay, which would cut off between 12 and 20 square miles from the sea. Again, where tombolos, or such related features as spits, bars, and headland-bay beaches, project from the shore, they can serve some of the same purposes as costly barrages or groins. The Pool at Biddeford, Maine, lies between twin tombolos which provide both coastal protection and enclosure for an area of quiet water. Further, a natural tombolo can be reinforced by an artificial superstructure to form a permanent embankment, particularly in harbour improvement works. A sand dyke reinforced by a timber pile jetty has been constructed between a natural beach and an island at Chatham, Massachusetts. In summary, the functions of coastal barrages, whether they include parts of natural beaches or are entirely engineered structures, comprise: (*a*) land reclamation; (*b*) freshwater storage (the huge, new Plover Cove project in Hong Kong is an example); (*c*) flood control and coastal protection; (*d*) salt recovery; (*e*) oyster cultivation; (*f*) tidal power generation; (*g*) better harbour and river navigation; (*h*) provision of new amenities like water skiing and fishing; and (*i*) improved communication by means of the barrage or tombolo itself, to be discussed next.*

(vi) There are many examples from all around the world in which natural tombolos provide the best means of connection between an island and the shore. Locally they may be the most important element in the coastal geography. On the Adriatic coast of Yugoslavia a tombolo not only ties the rock island of Primosten, with its church and village, to the mainland but also provides a sheltered harbour. Mount Maunganui in New Zealand benefits in the same way, and the natural three-mile bar or tombolo linking Presqu'ile de Quiberon to the Biscay coast of France is stable enough to carry a railway.

(vii) The margins of tombolos are subject to the same type of survey and interpretation as those of any other tideland or beach. Because they are links between islands, tombolos may have special legal significance, especially if they become liable to destruction and possible replacement by an artificial causeway. The entire subject of shore and sea boundaries in the United States has recently received comprehensive treatment by Shalowitz (1962–64).

(viii) Part of each modern tombolo is within the intertidal zone and, like other beaches, this part may be inhabited by distinctive animals and plants. One tombolo-like peninsula may face two oceans or two parts of the same ocean, each side with a different fauna and flora: North Cape confronts both the Pacific Ocean and the Tasman Sea. Once a tombolo is established between two land areas, it can provide a type of land bridge, as indicated by the work of Powell in New Zealand, already discussed. Where tombolos, particularly double tombolos, have closed off parts of the former intertidal zone, their presence may radically alter the ecology. Trapping of sediment together with the growth of grasses and other plants may result: a recent account of such events, as they pertain to a tropical delta, is given by Allen (1965). A system of tombolos with an opening to the sea may form a lagoon. Without such an opening the water enclosed may increase in salinity and a salt marsh develop. Whether a particular lagoon or salt marsh originates by the growth of tombolos or through some other cause, the new environment is bound to change the animal and plant distribution on the sea boundary. Only one item in the comprehensive literature on this subject, a standard work on marine ecology, will be mentioned here (Hedgpeth, 1957).

SHORELINE CLASSIFICATION

The tombolo is recognized as a distinct littoral form, but cannot provide the sole basis for classification of a particular coast. Different varieties include single, forked, complex, and double or multiple tombolos, to which may now be added the parallel series exemplified by the Magdalen Islands and North Cape, twin lines of bedrock islands being joined by tombolos in each case. Tombolos are depositional in character, as are the bars and barriers that Shepard (1952) included in a revised nomenclature of such coastal features.

* Scientific aspects of some of these subjects are presented in publications issued over the past ten years by the International Institute for Land Reclamation and Improvement of Wageningen in the Netherlands.

A classification that takes into account certain of the principal coastal processes has been proposed by Valentin (1952). This depends upon the stage of evolution already reached and particularly upon whether land is being gained or lost. The main criteria are the relative movements of land and sea in the immediate past, as well as the balance between erosion and construction at any point along the coast. Tombolos are features of construction, new land being obtained by local outbuilding, and parts of the coast with growing tombolos are clearly advancing. Elsewhere, of course, parts of the same coast, especially the cliffs which supply material to the tombolos, are retreating because erosion and submergence occur together. The growth of tombolos and related features at the expense of rock cliffs undergoing erosion is emphasized by Valentin's classification. An example on the Irish coast, recently described by Guilcher and King (1961), shows how tombolos and spits have been built above sea level by constructive waves. Cotton (1954) has found that Valentin's classification applies well to certain sections of the New Zealand coast south of Auckland. Because it takes into account the conditions needed for coastal accretion, the same classification is also useful for the tombolos built out as links between the former islands of North Cape. While Valentin's main parameters of description are only two in number, erosion or deposition and submergence or emergence, a recent paper by Bloom (1965) stresses the additional need for a genetic and historical classification; as he indicates, the factor of time provides a third dimension in coastal geomorphology.

SUMMARY AND CONCLUSIONS

Four aspects of the linking of islands by double tombolos are considered: (1) formation of islands; (2) erosion; (3) movement of material between the islands; and (4) widening of the link. The main examples described are the Magdalen Islands and North Cape. These are two groups of isolated rock areas, islands or former islands, now linked by pairs of almost parallel tombolos. The space between is occupied to varying extents by lagoon, swamp, and sediment. The Magdalen Islands include about 12 rock islets linked by twin, partly dune-covered tombolos which are nearly parallel for 30 miles. The lagoons between the islets comprise about one-third of the total 140 square miles and are locally converted to swamp. The surrounding sea is rather shallow. North Cape also consists of about a dozen former islands linked by twin, dune-covered tombolos that nearly parallel each other for 60 miles along the peninsula. Swamp and sand lie between. The surrounding sea is quite deep. Although these two island complexes possess many features in common, the parallel coasts of the Magdalen group are separated from each other along most of their length, mainly by the lagoons. North Cape represents a later stage of shore development, the area between the twin tombolos there being filled above sea level by sediment, except for three large estuaries on the east.

Comparison may be useful for two reasons. First, the area between the parallel coasts of the Magdalen Islands could also become filled above sea level by sediment, depending upon climatic and other changes. Such natural reclamation would greatly benefit an island territory with a particularly

E*

dense rural population. Secondly, shoreline formation in the present-day Magdalen Islands may resemble an earlier stage in the geological history of North Cape, providing some insight into coastal evolution there. Sedimentary materials between the shores of the New Zealand peninsula now form large tracts of reclaimed land producing timber, meat, and wool.

Factors involved in the growth of tombolos and of other beaches are coastal and offshore topography, the type and quantity of sediment available, and wind and current conditions. Tombolos are not confined to coasts with a large tidal range or to coasts with any particular history of level change. They are varieties of beach that can be formed along irregular shorelines of both seas and lakes, joining islands together or to the mainland wherever sufficient sediment exists.

Reference is made to Valentin's classification, depending upon the balance between erosion and deposition for each coastal feature, and to Bloom's, in which the factor of time is also considered. It is concluded that tombolos are features of construction. Like other bars and barriers, they are commonly important elements in shore protection and even in land reclamation, and they may also serve as engineering models for comparison with artificial barrages.

ACKNOWLEDGEMENTS

Visits to the Magdalen Islands were in 1962 initially on vacation and to North Cape in 1944–45 during war service and in 1963–64 during tenure of a D.S.I.R. (New Zealand) Fellowship. The author wishes to thank C. A. Kaye, I. S. Evans, and A. L. Bloom for helpful comments on earlier drafts of this paper. Main sources for the figures are British Admiralty charts 111, 2615, 2608, 158 (and 1719), 2529, and 2525 for Figs 1B, 2A, 2C, 2D, 2F, and 5; Canadian Hydrographic Service charts 4626, 2061, and 4451 for Figs 1A, 1D, and 4; and U.S. Coast and Geodetic Survey chart 901 for Fig. 1C. The New Zealand aerial photographs are reproduced by permission of Whites Aviation Service, Auckland.

REFERENCES

Alcock, F. J., 1941. *Trans. Can. Inst. Min. Metall.*, **44**, 623–649.
Allen, J. R. L., 1965. *Bull. Am. Ass. Petrol. Geol.*, **49**, 547–600.
Belly, P. Y., 1964. *U.S. Army Coastal Eng. Res. Center*, Tech. Mem., 1, 38 pp.
Bloom, A. L., 1965. *Z. Geomorph.*, **9**, 422–436.
Bloom, A. L., 1966. *Geol. Soc. Am.*, abs. Phila. meeting, 1 p. only.
Brodie, J. W., 1965. *Bull. N.Z. Dep. scient. ind. Res.*, **16**, 56 pp.
Brothers, R. N., 1954. *N.Z. Geogr.*, **10** (1), 47–59.
Clarke, J. M., 1910. *Bull. N.Y. St. Mus.*, **149**, 134–156.
Cockayne, L., 1911. *N.Z. Dep. Lands*, C-13, 76 pp., folio.
Cotton, C. A., 1954. *Geogrl J.*, **120**, 353–361.
Davies, J. L., 1959. *Geogrl Stud.*, **5**, 14 pp.
Fairbridge, R. W., 1961. *Ann. N.Y. Acad. Sci.*, **95**, 542–579.
Falaise, N., 1954. *Revue can. Géogr.*, **4**, 63–80.
Farquhar, O. C., 1965. *Geol. Soc. Am.*, Spec. Pap. 82, abs., p. 367 only.
Farquhar, O. C., 1966. *New Scient.*, **29**, (487), 698–699.

Farquhar, O. C., in press.
Friedman, G. M., 1961. *J. sedim. Petrol.*, **31,** 514–529.
Ganong, W. F., 1964. *Spec. Publs R. Soc. Can.*, No. 7, 511 pp.
Geographical Handbook Series, Naval Intelligence Division, BR 503, France, **1,** 1942, 279 pp., and BR 517, Italy, **1,** 1944, 595 pp., H.M.S.O., London.
Goldthwait, J. W., 1915. *Mus. Bull. Can. geol. Surv.*, No. 14, 11 pp.
Gray, N. G., 1964. *Dom. Hydrogr.*, *Canada*, letter of October 22, File 1384–2–2.
Guilcher, A. and King, C. A. M., 1961. *Proc. R. Ir. Acad.*, 61B, **17,** 283–338.
Hamelin, L-E., 1959. *Sables et mer aux Îles de la Madeleine*. Quebec Ministere de l'Ind. et du Comm., Quebec, 66 pp.
Hayes, M. O., 1965. *Geol. Soc. Am.*, abs. Kansas City meeting, 1 p. only.
Hedgpeth, J. W., 1957. Editor, *Treatise on Marine Ecology and Paleoecology*, Vol. 1, Geol. Soc. Am. Mem., 67, 1296 pp.
Ingle, J. C., 1966. *The Movement of Beach Sand*, Elsevier, Amsterdam, 221 pp.
Institut Géographique National, 1960. *Carte des naufrages survenus sur les côtes des Îles St. Pierre et Miquelon de* 1816 *à nos jours.* 1: 50,000, Paris.
Ippen, A. T., 1966. Editor, *Estuary and Coastline Hydrodynamics*, McGraw-Hill, New York.
Johnson, D. W., 1925. *The New England-Acadian Shoreline*, Wiley, New York, 608 pp.
Kaye, C. A., 1959. *Prof. Pap. U.S. geol. Surv.*, 317–B, 49–140.
Kear, D. and Hay, R. F., 1961. *N.Z. Geological Survey Map.* 1: 250,000, Sheet 1, North Cape, 1st ed.
Le Méhauté, B., 1961. *J. geophys. Res.*, **66,** 495–499.
Linton, L. R., 1959. *Can. geogr. J.*, **58,** 10–15.
Maxwell, C. F., 1897. *Trans. N.Z. Inst.*, **29,** 564–567.
Powell, A. W. B., 1947. *Rec. Auckland Inst. Mus.*, **3,** 173–188, and later papers.
Quennell, A. M., Hay, R. F. and Farquhar, O. C., 1965. *Geol. Soc. Am.*, Spec. Pap. 82, abs., p. 158 only.
Russell, R. J., 1959. *Eclog. geol. Helv.*, **51** (3), 591–598.
Sanschagrin, R., 1964. *Quebec Geol. Expl. Serv.*, Rep. 106, 58 pp.
Sauvage de Saint Marc, G. and Vincent, G., 1955. *Proc. Fifth Conf. Coastal Eng., Berkeley, Calif.*, 296–328.
Schlee, J., Uchupi, E. and Trumbull, J. V. A., 1964. *Prof. Pap. U.S. geol. Surv.*, 501–D, D118–122.
Schofield, J. C., 1963. *Proc. N.Z. ecol. Soc.*, **10,** (no pagination on reprint).
Shalowitz, A. L., 1962–64. *Shore and Sea Boundaries*, U.S. Dept. Comm., Washington, D.C., Vol. I, 420 pp., Vol. II, 749 pp.
Shepard, F. P., 1950. *Tech. Memo. Beach Eros. Bd U.S.*, No. 15, 32 pp.
Shepard, F. P., 1952. *Bull. Am. Ass. Petrol. Geol.*, **36,** 1902–1912.
Sion, J., 1934. *Géographie Univers.*, **7,** 235–394.
Smith, E., 1932. *Can. geogr. J.*, **4,** 331–348.
Smith, S. P., 1881. *Trans. N.Z. Inst.*, **13,** 398–410.
Steers, J. A., 1964. *The Coastline of England and Wales*, Cambridge University Press, Cambridge, 2nd ed., 750 pp.
St. Lawrence Pilot, 1963. The Canadian Hydrographic Service, Ottawa, 543 pp., 2nd ed.
Valentin, H., 1952. *Die Küsten der Erde*, Petermanns Geographischen Mitteilungen Ergänzungsheft 246, Justus Perthes, Gotha, 118 pp.
Vella, P., 1962. *Trans. R. Soc. N.Z.*, Geol., **1** (6), 101–109.
Walton, K., 1956. *Scott. geogr. Mag.*, **72,** 85–96.
Wiegel, R. L., 1964. *Oceanographical Engineering*, Prentice-Hall, New York, 532 pp.
Yasso, W. E., 1965. *J. Geol.*, **73,** 702–714.
Zeuner, F. E., 1959. *The Pleistocene Period*, Hutchinson, London, 447 pp.

Editor's Comments on Paper 13

This review article takes up the difficult question of just what defines a bar and, once defined, whether or not these structures build up through sea level. The reader will find support for, or disagreement with, Price in the views put forth by the other authors included here. As stated in the preface to this volume, there has not been complete accord on many basic concepts among the workers in this field. Therefore it seems appropriate to pause at this point and consider where the study of bars (and spits) had come to in 1968. Price's review is recommended both for its content and bibliography.

W. Armstrong Price is now a consulting geologist and geological oceanographer, and lives in Corpus Christi, Texas. He was born in Richmond, Virginia, in 1899, received the A.B. degree from Davidson College in 1909, and the Ph.D. degree from Johns Hopkins University in 1913. Price worked with the Maryland, West Virginia, and U.S. Geological Surveys between 1910 and 1918; and has been an oil and gas exploration consultant since 1919. He has also taught at West Virginia University and Texas A & M. Dr. Price is an active member and officer of many geologic organizations; and has published extensively, half of his papers being on coastal problems. The interested reader is referred to a planned volume in this series on cuspate coastal forms for other works by W. A. Price.

13

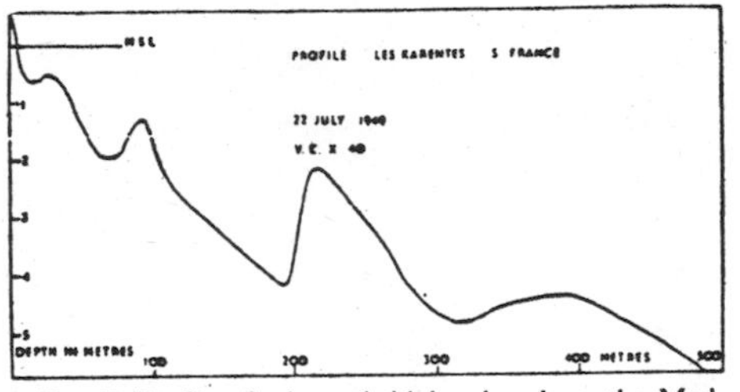

FIG. 2. Profile of a barred tideless beach on the Mediterranean (after Shepard).

BARS

Submerged ridges of detrital sediments which are larger and less regularly spaced than ripple marks are termed *bars*. Formed typically in shallow epicontinental or shelf waters by waves and currents, they are found singly or together, and internally laminated. Active quartzitic sand and calcareous sand bars are typically unconsolidated, shallow-based and highly mutable with changes in environmental factors (Fig. 1). The laminae show the directions of the formative currents, and cross-beds aid in determining directions of bar migration. Some bar-like accumulations of carbonate sand (Rusnak, Bowman and Ostland, 1964) show superficial encrustation dated from less than 200–1000 years ago.

Bars may be classified as *longitudinal* or *transverse* to a dominant current or to an associated shore line. The underwater extensions of sand spits are bar-like. Long, narrow, highly stable, bar-like ridges lying in parallel series in shelf waters beyond the surf zone have been called "tidal current ridges" by Off (1963) and occur where tidal ranges are 10 feet or more. Similar large, broad, submerged accumulations of sand with surfaces singly or numerously ridged, and related to bars, are given names such as subaqueous dunes, barchans, sand waves and sand bores (cf. Vol. I, *Subaqueous Sand Dunes*). So-called giant ripple-marks of carbonate sands on Bahamian platform shoals are so regular in their spacing and parallelism as to suggest that they are formed by somewhat the same process as normal ripple marks (Newell and Rigby, 1957).

Bars longitudinal to confined currents include stream channel bars and some bars of river mouths, funnel estuaries, straits and tidal inlets. River

FIG. 1. Circulation of water in surf as steep waves transform dashed profile to solid profile by moving sand from berm to bar (Bascom, 1964). (By permission of Doubleday & Co., N.Y.)

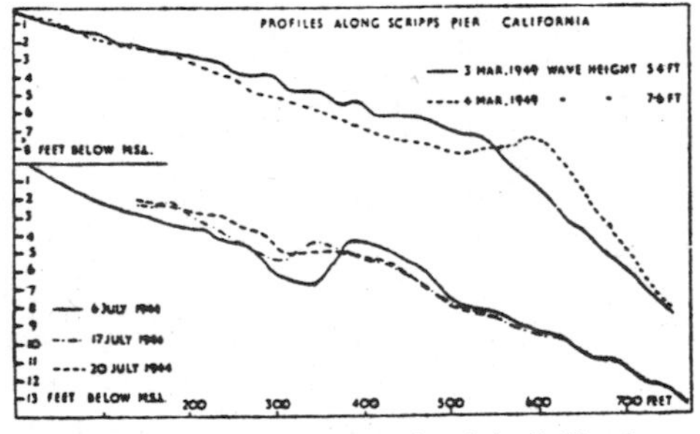

FIG. 3. Profiles of the Scripps beach in California to show bars and their removal (after Shepard).

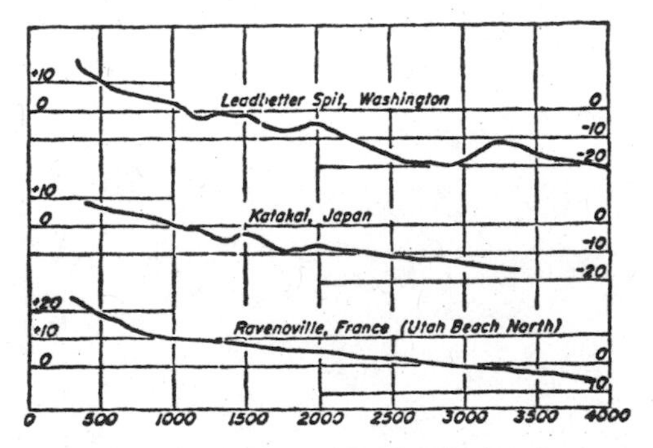

FIG. 4. Profiles of flat beaches. Note the three bars on the beach at Leadbetter Spit (slope exaggerated 1:5; Bascom, 1964). (By permission of Doubleday & Co., N.Y.)

55

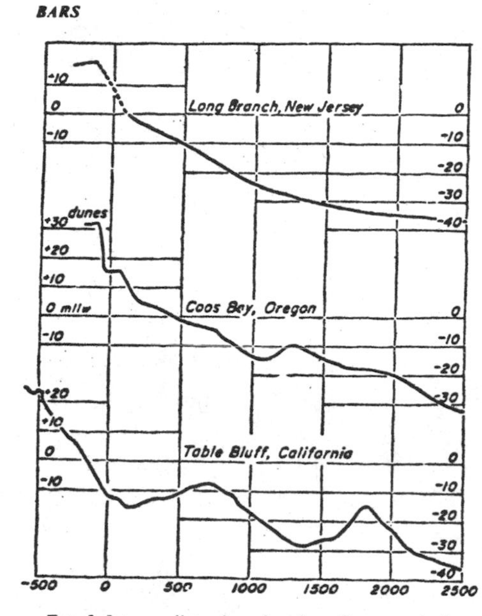

FIG. 5. Intermediate slope beaches. Contrast the bar-less beach profile at Long Branch with the two huge bars at Table Bluff (slope exaggerated 1:5; Bascom, 1964). (By permission of Doubleday & Co., N.Y.)

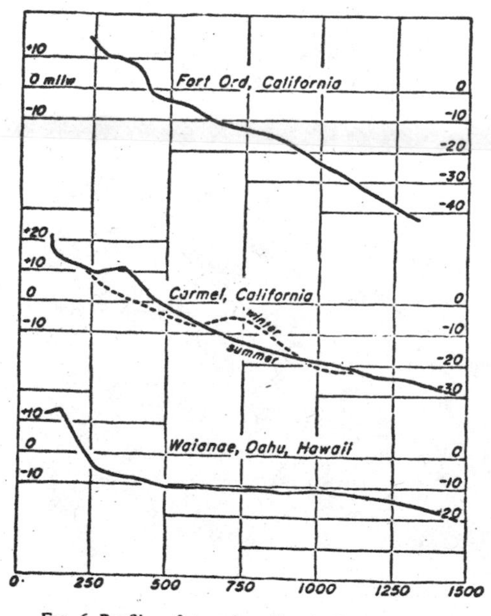

FIG. 6. Profiles of steep beaches. At Waianae only the beach face is sand. The flat area is a coral reef over which great breakers form and where the surfing is excellent (for an expert) (exaggerated 1:2.5; Bascom, 1964). (By permission of Doubleday & Co., N.Y.)

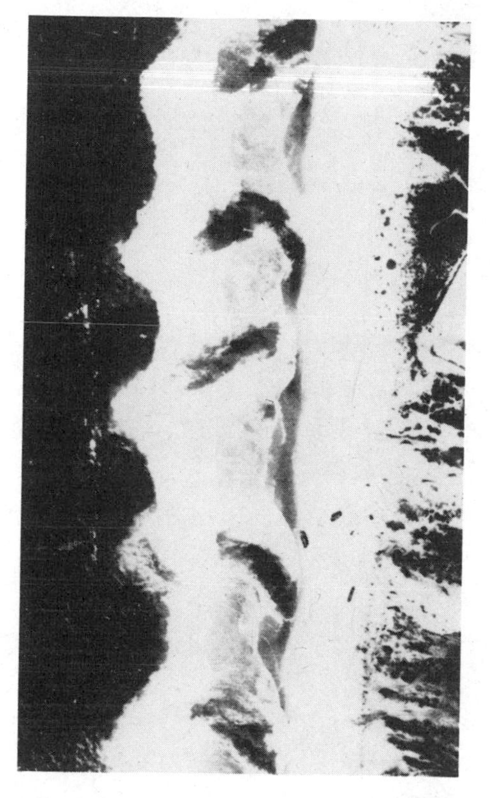

FIG. 7. Rip channels cutting a "terrace-like" bar, Fort Ord, California. Bar and ocean (dark) at left, beach (white) and vegetated land (dark) at right, with trough between. (Bascom, 1964). (By permission of Doubleday & Co., N.Y.)

channel bars may become channel islands at lowered stream stages. Off (1963) considers his "tidal current ridges" to be longitudinal to the dominant currents. Smaller ridges of similar nature on coasts with smaller tidal ranges—e.g., off the Rio Grande delta, Texas, and the coast of Georgia—are diagonal to the nearby shore lines. The huge Bahamian "sand bores" and giant subaqueous barchans of calcareous sands seem to be longitudinal to the dominant current. They may have small bars and aqueous dunes on their surfaces.

Bars transverse to confined currents include crescentic bars of some river mouths and tidal inlet channels. There are sand waves and aqueous dunes which are transverse to the current.

Longshore Bars

Longshore bars occurring singly or in parallel series of 2 or 3 off sandy beaches are formed in the surf zone along the seaward flanks of plunge troughs excavated by the swirling motion of breakers (Figs.

FIG. 8. Oblique air photograph of the Bahama Banks, showing patterns of huge calcareous sand bars at shallow depth beneath the water. (Courtesy F. Goro and *Science*.)

2–7). The excavated sands rise in the turbulent water, some accumulating as a series of bars along the seaward flanks of troughs, while the rest is carried shoreward in the rush of the solitary wave (*wave of translation*) formed by the water from the tops of the breakers (see *Beach*). Currents from both flanks later build up the longshore bar, both trough and bars being affected also by longshore currents, and the troughs are scoured by water escaping laterally behind the bars from the beach zone and longshore drift-currents bringing sand.

Longshore bars—straight or looped—are not built up above stillwater (half-wave) level because the breakers plunge on and over them, keeping their summits planed down. The building of these bars up to emergence is described in the article *Barriers*. In some experiments these bars are topped at low tide level, in others at half-wave height.

The term bar is only appropriate to features below swash limits. The criterion of submergence at normal high tide advocated by Price (1951) and Shepard (1952) eliminates from the category *bar* such proposed emergent forms as the swash "bar," bay "bar," marsh "bar," "flying bar" and "offshore bar" (the term formerly used for "*barrier*" q.v.). The terms *swash ridge, bay barrier, marsh beach ridge* and *flying spit* may be employed in such cases, while a true *offshore bar* should be kept for any bar lying well offshore.

The above definition does not embrace shallow, bar-like sedimentary structures of large size, the emergent parts of which are not regarded here as bars. Such structures include "distributary-mouth bars," "bar fingers" of the Mississippi River mouth (Fisk, 1961, p. 30, Fig. 2) and "point bars" of the inner side of river meanders which are composed of a crescentic beach plain, a true current bar and a "point" beach. "Barrier bar" is also used by some for a buried barrier or the submerged parts of an existing barrier.

W. Armstrong Price

References

True Bars

Bascom, W., 1964, "Waves and Beaches; The Dynamics of the Ocean Surface," pp. 191–204, Pls. 18, 19, Figs. 60–62, Garden City, N.Y., Anchor Books, Doubleday & Co., 267pp.

Evans, O. F., 1942, "Origin of spits, bars and related structures," *J. Geol.*, **50,** 846–865. (The "low and ball" of Evans and certain British writers are the longshore trough and bar of present usage.)

Guilcher, A., 1958, "Coastal and Submarine Morphology," London, Methuen; New York, John Wiley & Sons, 274pp. (translated by B. W. Sparks and R. H. W. Kneese).

Keulegan, G. H., 1948, "An experimental study of submarine sand bars," *Beach Erosion Board, Chief of Engineers, U.S. Army, Tech. Rept. 3*, 40pp.

McKee, E. D., and Sterrett, T. S., 1961, "Laboratory Experiments on Form and Structure of Longshore Bars and Beaches," in (Peterson, J. A., and Osmond, J. C., editors) "Geometry of Sandstone Bodies," pp. 13–28, American Association of Petroleum Geologists.

Price, W. A., 1951, "Barrier island, not offshore bar'," *Science*, **113,** 487–488.

Shepard, F. P., 1950, "Longshore bars and longshore troughs," *Tech. Mem. 15, Beach Erosion Board, Chief of Engineers, U.S. Army*, 31pp.

Shepard, F. P., 1952, "Revised nomenclature for depositional coastal features," *Bull. Am. Assoc. Petrol. Geologists*, **36,** 1902–1912.

Bar-like Ridges and Larger Bar-like Sand Bodies

McManus, D. A., and Creager, J. S., 1963, "Physical and sedimentary environments on a large spitlike shoal," *J. Geol.*, **71,** 498–512.

Newell, N. D., and Rigby, J. K., 1957, "Geological studies on the Great Bahama Bank," in (LeBlanc, R. J., and Breeding, J. G., editors) "Regional aspects of carbonate sand bodies," *Soc. Econ. Paleontologists Mineralogists, Spec. Publ.*, **5,** 15–79 (see pp. 56, 57 and Pls. 10, 11, 13).

Off, T., 1963, "Rhythmic linear sand bodies caused by tidal currents," *Bull. Am. Assoc. Petrol. Geologists*, **47,** 324–341.

Rich, J. L., 1948, "Submarine sedimentary features on Bahama Banks and their bearing on the distribution patterns of lenticular oil sands," *Bull. Am. Assoc. Petrol. Geologists*, **32,** 767–779.

Rusnak, G. A., Bowman, A. L., and Ostlund, H. G., 1964, "Miami natural radiocarbon measurements III," *Radiocarbon*, **6,** 208–214.

Stratigraphic "Bars"

Fisk, H. N., 1961, "Bar-finger sands of Mississippi Delta," in (Peterson, J. A., and Osmond, J. C., editors) "Geometry of Sandstone Bodies," pp. 29–52, American Association of Petroleum Geologists.

Frazier, D. E., and Osanik, A., 1961, "Point-bar deposits, Old River locksite, Louisiana," *Trans. Gulf Coast Assoc. Geol. Soc.*, **11,** 121–137.

Cross-references: *Barriers*; *Beach*; *Cuspate Foreland or Spit*; *Sand Dunes*; *Swash, Swash Mark*; *Tidal Inlet.* Vol. I: *Subaqueous Sand Dunes.*

Editor's Comments on Paper 14

George deVries Klein was born in The Hague, Netherlands, in 1933. He received his primary education in Australia and has been a resident of the United States since 1947. Klein received the B.A. degree in Geology from Wesleyan University, Connecticut, in 1954, the M.A. degree in Geology from the University of Kansas in 1957, and the Ph.D. degree in Geology from Yale University in 1960. Since then, he has been employed as a research sedimentologist by Sinclair Research, Inc. (1960 to 1961) and as a faculty member by the University of Pittsburgh (1961 to 1963) and the University of Pennsylvania (1963 to 1969). In the fall of 1969, he was a Visiting Fellow of Wolfson College, Oxford University, and during the spring of 1970, he was a Visiting Associate Professor of Geology at the University of California at Berkeley. Since February 1970, Klein has been Associate Professor of Geology at the University of Illinois in Urbana where he also serves as Placement Officer for the Geology Department, and is an elected member of the University Senate. Dr. Klein is the recipient of two awards from the Society of Economic Paleontologists and Mineralogists: (1) the Outstanding Paper Award, 1970, *Journal of Sedimentary Petrology* for his paper "Depositional and dispersal dynamics of intertidal sand bars" which is presented here, and (2) Honorable Mention, Outstanding Paper, 1971 Convention for his paper "An environmental model for some sedimentary quartzites." His research interests are in clastic sedimentology, emphasizing tidalites, tidal sediment transport, sedimentary structures, and sandstone petrology.

This award-winning paper by one of today's foremost investigators of intertidal sand bars thoroughly covers the subject. Under the headings of depositional processes, sedimentology, paleocurrent analysis, origin of bedforms and cross stratification, and recognizing fossil intertidal sand bar environments Klein gives us a comprehensive insight into the complexities of this bar form. With regard to Klein's choice of an appropriate field site the reader should note that as is pointed out on the second page of his paper, the Minas Basin tidal range reaches 18 meters. What better place could there be to study intertidal sand bars?

JOURNAL OF SEDIMENTARY PETROLOGY, VOL. 40, NO. 4, P. 1095–1127
FIGS. 1–34, DECEMBER, 1970

DEPOSITIONAL AND DISPERSAL DYNAMICS OF INTERTIDAL SAND BARS[1]

GEORGE DEVRIES KLEIN
Department of Geology, University of Illinois, Urbana, Illinois, 61801

14

ABSTRACT

The intertidal sand bars of the Minas Basin, Nova Scotia, are asymmetrical in cross-section and linear in plan. They are formed and reworked by tidal currents which are characterized by an asymmetric time-velocity profile. A causal connection between time-velocity asymmetry of tidal currents and bar topography is demonstrated. Average bottom *ebb* current velocities (90 cm/sec) were found to exceed average bottom flood current velocities (65 cm/sec) over gently-sloping (2 to 3 degree) bar surfaces, whereas over steep-sloping (8 degree) bar surfaces, average bottom *flood* current velocities (90 cm/sec) were found to exceed average bottom ebb current velocities (65 cm/sec).

Both bar topography and zones of flood- or ebb-dominated bottom tidal currents control the distribution of sediment texture and sedimentary structures. On steeper bar slopes, the sediment is fine-grained sand and is characterized by airholes, planar lamination, and surface dunes and ripples. On the gentler bar slopes, the texture of the sediment is medium-grained and coarse-grained sand which has been fashioned into current ripples, simple dunes, complex dunes, simple sand waves and complex sand waves. The internal organization of some dunes and sand waves is extremely complex, inasmuch as thickness of internal sets of cross-stratification is considerably *less* than dune and sand wave height. A total of 14 sedimentary bar facies are defined from combinations of texture, internal cross-stratification and surface bedforms.

Both topography and zones of ebb- and flood-dominated bottom tidal currents control the orientation of directional current structures. Over steep faces of bars, the slip faces of sand waves are oriented in the same direction as flow directions of bottom flood tidal currents, although they do show evidence of reworking by ebb currents. Over gently sloping bar surfaces, slip faces of both dunes and sandwaves are oriented parallel to the direction of flow of bottom ebb current systems. Because dunes and sand waves migrate only during a single phase (ebb *or* flood) of a tidal cycle, their orientation is generally unimodal. Current ripples are formed in depths of water less than 0.6 meters, during late-stage sheet-like runoff that is controlled by local slope changes on the bar. Individual tracer grains are dispersed in a radial-elliptical pattern form a point source.

The size of directional properties is hierarchically sensitive to dispersal history. Tracer grains (modeling mineral indicator grains) are dispersed in a radial-elliptical pattern from a point source. Trimodal, highly variant, current ripple slip face orientation develops during late stage sheet runoff at low water depths; this flow is controlled by local bar slope. Unimodal, low variant dune slip faces, sand wave slip faces and associated maximum dip direction of cross-stratification are aligned parallel to tidal current flow and basinal topographic trend. The sand bars are aligned parallel to basinal topographic trend and in turn segregate zones of flood- and ebb-dominated bottom current systems.

The tidal sand bars of the Minas Basin are equilibrium forms. Sand is dispersed alternately around the bar through zones of ebb- and flood-dominated bottom tidal currents. As a consequence, the sedimentary facies distribution on the bars has remained essentially unchanged since 1938. Comparison of low-tide air-photos taken in 1938, 1947 and 1963 in one of the study areas shows a slight erosional regime. This erosional regime controls the extreme reworking of sediments, and the formation of an *erosional* sand wave bedform by wave processes.

Physical criteria characteristic of tide-dominated sand bodies include: sharp erosional contacts between sets of cross-stratification; rounded upper set boundaries of cross-stratification; unimodal and bimodal distributions of orientation of cross-stratification; bimodal distributions of set thickness and dip angles of cross-stratification; orientation of dunes, sand waves and cross-stratification in the dominant direction of flood or ebb tidal flow, basinal topographic trend and sand body axis; trimodal orientation of current ripples; oblique or 90° superposition of smaller current ripples on larger current ripples; double-crested current ripples; superposition of current ripples at 90° or 180° on slip faces and crests of dunes and sand waves; complex organization of internal cross-stratification in sand waves; etch marks on slip faces of dunes and sand waves; and alignment of the long axis of sand bodies parallel to tidal current flow, basinal topographic trend and basin axis.

INTRODUCTION

During the past 20 years, studies of ancient sedimentary rocks have included the mapping of directional properties to delineate trends of sediment dispersal. Complementary mineralogical, textural and isopach data supplemented such directional data to delineate models of basin geometry, paleocurrent flow, depositional paleoslope and probable location of source areas from which the sediments were derived (Potter and Pettijohn, 1963). A causal connection between orientation of directional properties and sandbody trend was also established (Potter, 1962; Potter and others, 1958; Potter and Mast,

[1] Manuscript received October 27, 1969; revised May 20, 1970.

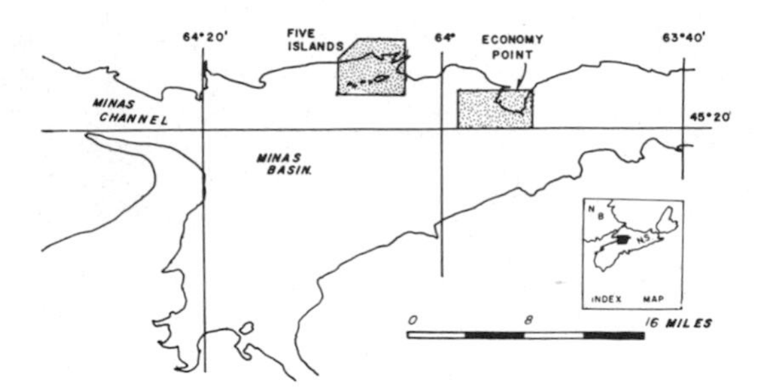

FIG. 1.—Map of Minas Basin, Nova Scotia, showing locations of areas of investigation.

1963). In fluvial sedimentary rocks and trubidites, directional current structures have been used to delineate paleoslope inasmuch as such sediments are deposited by slope-controlled processes. In modern marine environments and fossil equivalents, environmental distinctions in paleocurrent modal patterns emerge that are independent of slope (Tanner, 1959; Klein, 1967; Selley, 1968) and in fact across-slope and upslope flow systems are common (Hayes and others, 1969). Many of the earlier paleocurrent models need reappraisal and testing, therefore, particularly by comparison to dispersal models from Recent sedimentary environments.

Additional problems posed by paleocurrent analyses include: (1) What are the relationships between flow direction of depositional current systems and the orientation of directional current structures in modern sedimentary environments? (2) To what extent do directional current structures in Holocene sediments indicate all possible flow directions to which sediment is subjected? (3) How does the direction of dispersal of individual sedimentary particles compare with the larger-scale directional current structures that are used to map paleocurrent systems? (4) Does the detailed organization of flow over a bedform control orientation more than major flow systems, as suggested by Allen (1966)?

This paper aims to examine some of these problems in the *intertidal environment*. The intertidal environment was selected because logistically, it permits rapid comparison of sedimentary features to flow directions and other parameters of depositional currents. At low tide, a variety of sedimentary properties can be mapped by conventional means and compared to hydro-dynamic data obtained at buoys moored at critical geological locations.

The area selected for this study was the Minas Basin of the Bay of Fundy, Nova Scotia. Its sediments were discussed in reconnaissance by Klein (1963). Regional geological studies by Belt (1964, 1965), Klein (1962) and Weeks (1948) in Carboniferous and Triassic rocks, and by Swift and Borns (1967) in Pleistocene sediments established a depositional framework which permits comparison of directional properties of intertidal sediments to patterns of dispersal from rocks of different sources.

Two areas were selected for study on the north shore of the Minas Basin (fig. 1). The sand bars at Five Islands and Economy Point were selected because they are exposed for longer intervals at low tide than sand bars occuring on the south side of the Minas Basin (Swift and McMullen, 1968). Field observations were confined to the summer months of 1965, 1966, 1967 and 1968.

DEPOSITIONAL PROCESSES

Depositional processes in the intertidal zone at Five Islands and Economy Point include tidal currents, wind-generated waves, wave-generated currents, low-tide open-channel flow, and ice. Each agent transports and redistributes sediment and, as summarized in Table I, controls the gross textural characteristics.

Tidal Currents

Tidal currents dominate the summer sediment transport regime. The tidal range in the Minas Basin is the highest recorded in the world, reaching 18 meters at Burntcoat Head, Hants County (Canadian Hydrographic Service, 1968; Klein, 1963; Swift, 1966). Swift (1966, p. 118) summarized the origin of the high tidal range, indicating that the dimensions of the Bay of Fundy produce a natural period of oscillation that is identical to the semidiurnal component of the tides. Consequently, resonant amplification of these periods of oscillation gives rise to a high volume inflow of water which produces the high vertical range.

Within the two study areas, tidal currents reach maximum surface velocities of 280 cm/sec. Bottom current velocities reach maxima of 110 cm/sec. Bottom flood current velocities average 65 cm/sec, whereas bottom ebb current velocities average 90 cm/sec. Tidal current velocity and directional data reported in this paper were obtained by a calibrated Pritchard-Burt Current Meter (Pritchard and Burt, 1951).

The tidal currents are thoroughly mixed from the bottom to the surface as indicated by the vertical distribution of water temperature (fig. 2) and vertical profiles of current velocity and current directions (fig. 3). The flood and ebb tidal currents are dynamically similar to open-channel flow or sheet flood (cf. Leopold and others, 1964, p. 153–156).

TABLE 1.—*Process-response relations between sediment texture and sedimentary processes, Intertidal Zone, Minas Basin, Bay of Fundy, Nova Scotia (after Laub, 1968: Klein, 1968)*

Sediment Type	Occurrence	Source	Erosional Processes	Depositional Processes
Marginal Gravel	Along shore at high-water line	Seacliffs of bedrock and glacial sediments	Wave action	Longshore drift
Sandy Gravel	Bars oriented parallel and normal to shore	Seacliffs of bedrock and glacial sediments	Low-tide, open-channel flow, wave action, ebb tidal currents	Low-tide, open-channel flow, wave action, tidal currents
Sand	Linear sand bars in low tidal flats	Seacliffs of bedrock and glacial sediments	Tidal currents	Tidal currents
Mud	High tidal flats, estuaries	Seacliffs of bedrock and glacial sediments	Wave action	Suspension sedimentation during slack periods at high tide

The flow directions of bottom tidal currents at Five Islands and Economy Point are summarized in Figures 4 and 5. Flood bottom currents flow east to east-northeast in both areas, whereas ebb currents flow west to west-southwest.

Locally, counter eddy flow systems develop in response to flow separation of tidal currents around coastal irregularities. At Five Islands (Fig. 4), a counter eddy forms during the flood phase of the tidal cycle on the northeast side of Pinnacle Island, whereas two counter eddies form on the northwest side of Long and Pinnacle Island during the ebb phase of the tidal cycle. Both Long and Pinnacle Islands obstruct the flow of tidal currents and thus produce a flow separation and associated counter eddies on the down-current end. Similar counter eddies occur on the southwest side of Economy Point during the ebb phase of the tidal cycle (fig. 5) where Economy Point juts into the main system of ebb flow and induces a small counter eddy. All these eddies decrease in size during the lowering of water level associated with ebb tide.

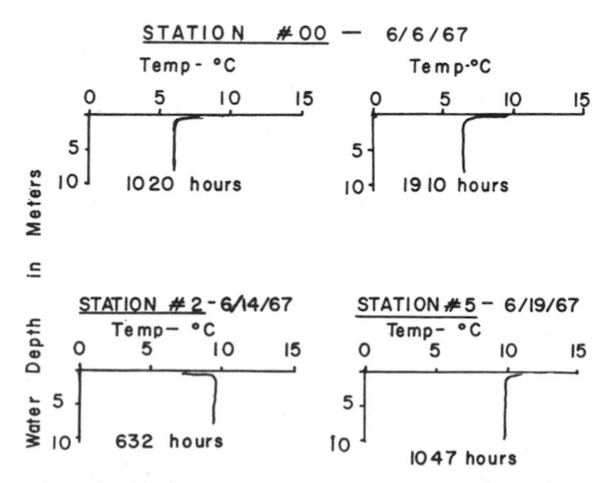

FIG. 2.—Bathythermograms at Stations #00, #7–2 and #7–5, Economy Point. All show that water is thoroughly mixed from bottom to surface, and that surface temperatures are in equilibrium with air temperatures.

Wind-generated Waves

Wind-generated waves are the second-most important agent of sedimentation in the intertidal zone during the spring and summer months. The prevailing summer winds are directed predominantly from southwest to northeast (Annual Metereological Survey, 1966; fig. 6), parallel to the axis of both the Bay of Fundy and Minas Basin. Because of this parallel alignment, 15 to 30 knot winds generate waves ranging from 60 cm to 3 m in height at Five Islands and Economy Point.

Wave action directed against seacliffs undermine and erode bedrock and drift, redistribute it and leave a lag gravel at the high water line. Unconsolidated sand and clay are removed by

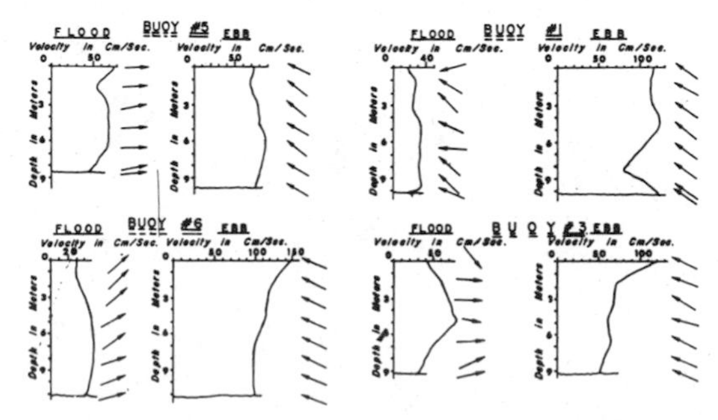

FIG. 3.—Vertical profiles of current velocity and flow direction at Stations #3–1, #3–3, #3–5 and #3–6, Pinnacle Flats, Five Islands, during flood and ebb phase. Uniform flow direction and velocity profile indicate tidal currents are unidirectional flow systems.

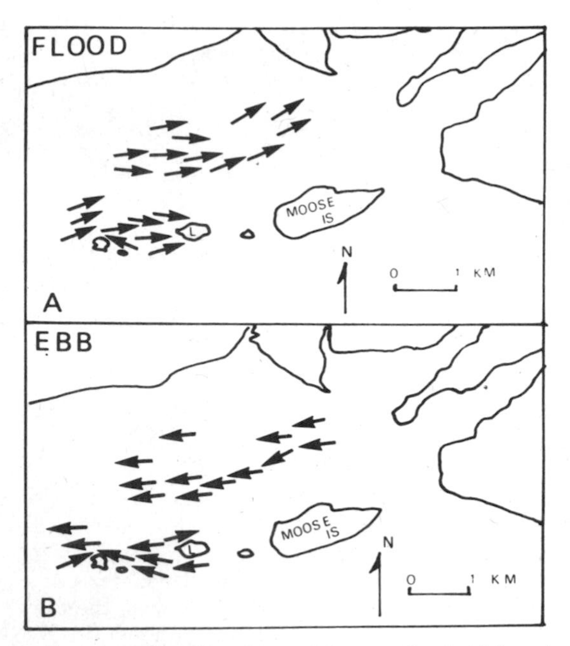

FIG. 4.—Flow directions of bottom flood (A) and bottom ebb (B) currents Five Islands (Abbreviation: L—Long Island).

tidal action after the initial undermining (Klein, 1968; Laub, 1968; R. S. Smith, 1969).

Wave-generated Currents

Currents generated by wave motion are confined to the upper surface zone (maximum of three meters below surface) when wind velocities exceeded 15 knots. Current velocity measurements during slack tide showed that wave-generated currents reach a maximum velocity of 20 cm/sec, averaging 10 cm/sec.

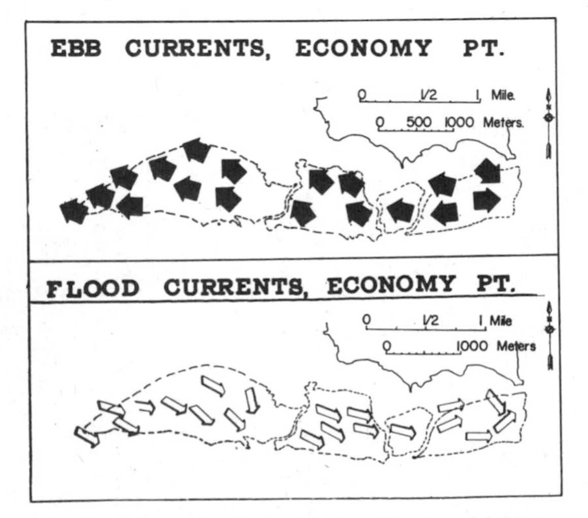

FIG. 5.—Flow direction of bottom flood and bottom ebb currents, Economy Point.

Low-tide Open-Channel Flow

In the late stages of ebb-current runoff, and after emergence, open-channel flow transports sediment in the intertidal zone. This open-channel flow occurs either in large, highly ordered, semi-permanent channels on the sand bars and mudflats, or in troughts of progressively emergent dunes and sand waves (Klein, 1963; Imbrie and Buchanan, 1965). Semi-permanent, meandering and straight channels are developed on mudflats on the northern side of the Five Islands area. On the sand bars, small semi-permanent channel systems occur in areas of lowest elevation or between sand bars.

As tidal currents ebb and progressively expose the bars, predominantly westerly ebb current flow ceases. The tidal currents are then deflected around the exposed bars and also drain off the bar as a sheet flood in the direction of local bar slope (Klein, 1963). As water level continues to fall, slope-controlled flow is confined to the semi-permanent channel systems. Sheet runoff becomes subdivided into a series of parallel, downslope-flowing channels within the troughs of dunes and sand waves as these bedforms emerge. As the bar becomes completely drained, this system of channel flow maintains itself by dewatering of the bar sediment.

Ice

Although field observations were confined to the late spring and summer months, ice transport appears to be of some significance. During extremely cold winters, the Minas Basin is known to be completely frozen. Local polygonal patches of ice-rafted gravel occur on all the sand bars. The gravel is redistributed by polygo-

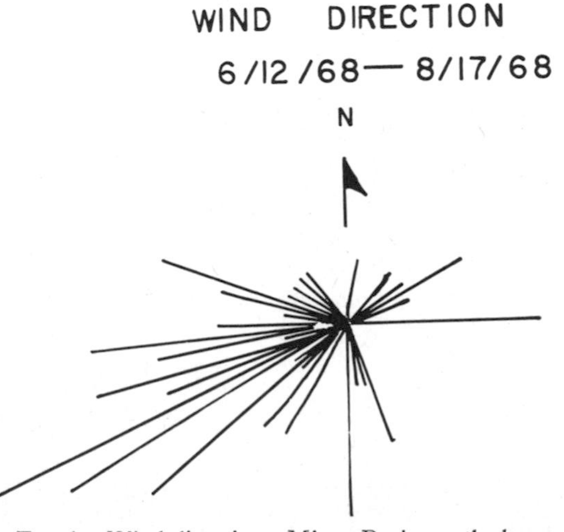

FIG. 6.—Wind directions, Minas Basin north shore; data collected at Parrsboro, Nova Scotia, by author, June 12-August 17, 1968.

nal ice floes during the spring ice break-up (cf. Bancroft, 1902; Dionne, 1968). Between 1965 and 1968, isolated boulder and pieces of tidal marsh have appeared and then were removed by ice-rafting.

SEDIMENTOLOGY OF THE SAND BARS

Geology (figs. 7 and 8)

The intertidal-zone geology at Five Islands and Economy Point (figs. 7 and 8) show a coarsening-seaward textural distribution similar to that reported from intertidal sediments elsewhere (Van Straaten and Kuenen, 1957, 1958). Clayey and silty muds occur on both the high and part of the low tidal flat environment. The seaward low-tide zone is dominated by sand bars comparable to Van Straaten's (1952, 1954) low tidal flat environment.

The muddy tidal flats are drained by meandering tidal creeks. These migrate laterally resulting in alternating cutting and back-filling (cf. Van Straaten, 1952). The tidal channels are floored with a mixture of coarse shelly debris and gravel derived from outcrops on shore (Klein, 1963).

Within the estuaries of larger streams entering the intertidal zone, mud sedimentation has proceeded at a sufficiently rapid rate so that mud flats are built high enough to be submerged only during spring tides. Tidal marshes have developed on these flats.

Because the Minas Basin shoreline is flanked by seacliffs of Carboniferous and Triassic bedrock and Pleistocene glacial drift, wave action on shore has formed a marginal gravel at approximately high water. These marginal gravels are produced by a combination of wave action, longshore drift and tidal current redistribution (Laub, 1968; Klein, 1968; R. S. Smith, 1969). During slack water at high tide, mud suspension settlement occurs within this marginal gravel zone, leaving both a mixture of gravel with a muddy matrix, and a thin veneer of mud on the gravel surface.

Sand bars and sandy deposits occur in the lowest areas of the low tidal flat environment.

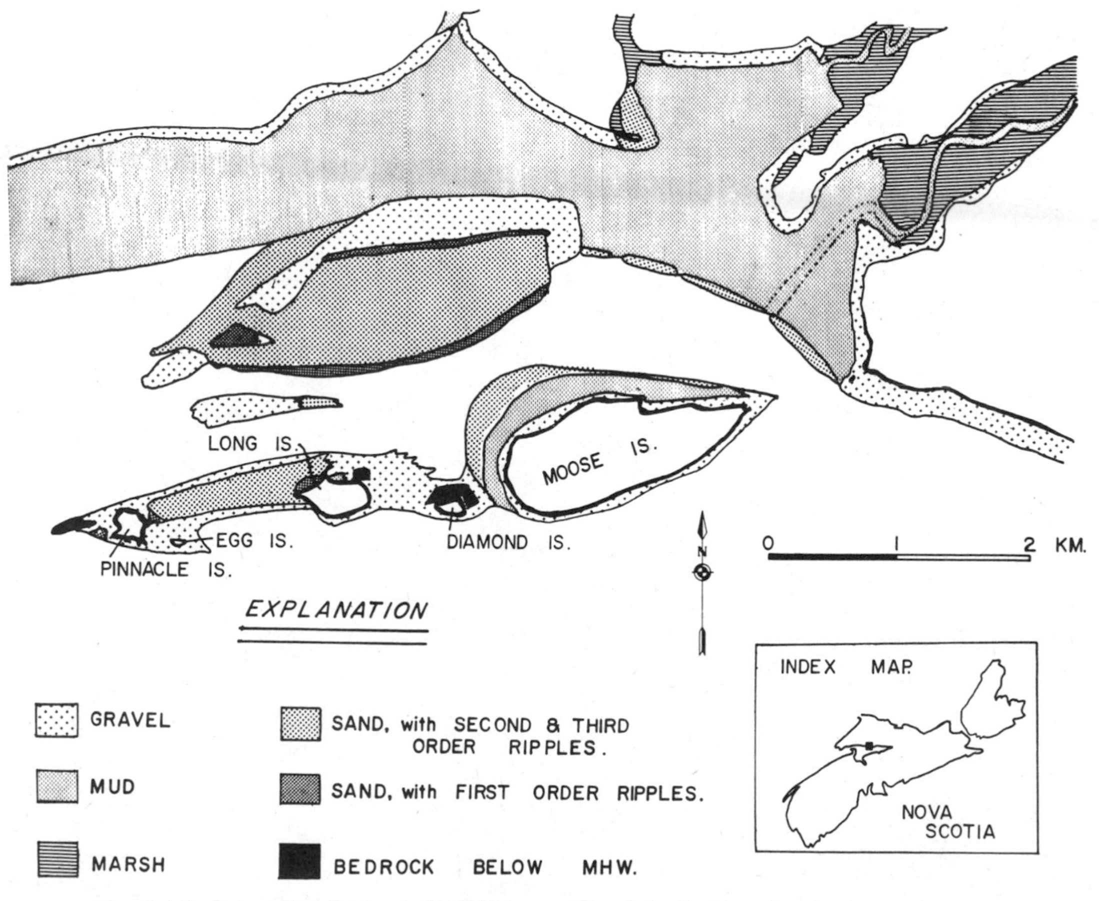

FIG. 7.—Sediment distribution in intertidal zone, Five Islands, Nova Scotia. BB (Big Bar). PF (Pinnacle Flats).

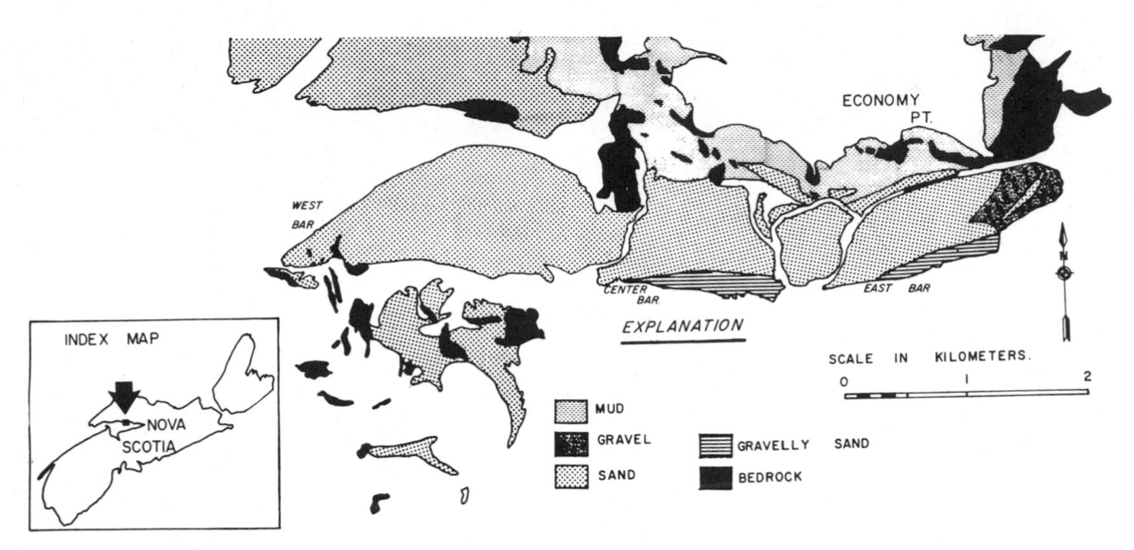

FIG. 8.—Sediment distribution in intertidal zone, Economy Point, Nova Scotia.

At Five Islands, two major sand bars occur. The bar in the center of the area is referred to hereafter as "Big Bar." The second occurs as a linear sand belt connecting Long and Pinnacle Island and is referred to hereafter as "Pinnacle Flats" (Fig. 7). Three large sand bodies occur immediately south and west of Economy Point are are referred to from west to east as "West Bar," "Center Bar" and "East Bar" (Fig. 8).

Topography (fig. 9)

The topography of Big Bar, Pinnacle Flats and West Bar is linear in plan and asymmetric in cross-section. Center Bar is trapezoidal in plan and asymmetric in section and East Bar is linear and nearly symmetrical (fig. 9). The bar crests are displaced toward the south side of all the bars except West Bar where the crest is displaced toward the north side. The steep slopes resulting from the bar asymmetry average 8° in slope angle, whereas the gentler slopes average 2°.

Continuous 13-hour observations of bottom current velocities, flow direction and water depths from high tide to succeeding high tide (fig. 10; Table 2; Table 9[2]) shows that the time-velocity profile for bottom tidal currents is asymmetric. Over the steep faces at West Bar, Big Bar and Pinnacle Flats, flood bottom current velocities are dominant, whereas over the gently-sloping bar surfaces, bottom ebb current velocities are dominant.

The distribution of the time-velocity asymme-

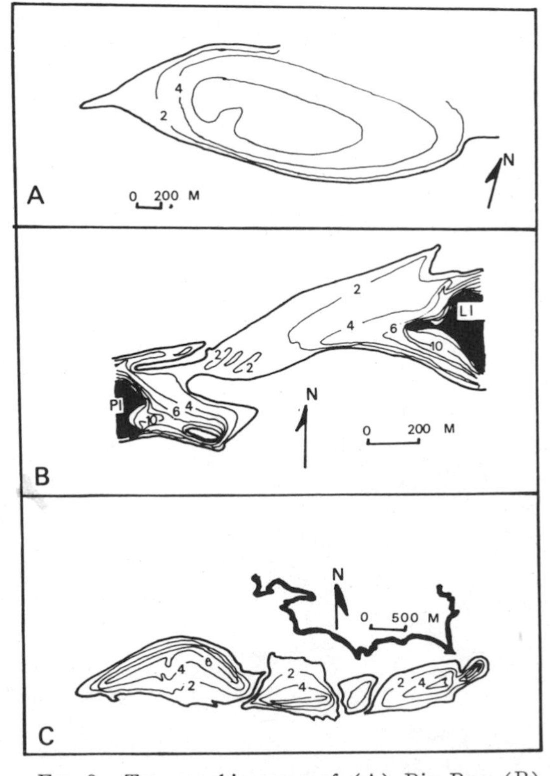

FIG. 9.—Topographic maps of (A) Big Bar, (B) Pinnacle Flats, and (C) Economy Point. Contour interval is 2 meters. Datum is mean low water on (A) August 1, 1965, (B) July 17, 1965, and (C) August 19, 1966.

[2] For Table 9 order Document Number NAPS 01062, from ASIS—National Auxiliary Publications Service, c/o CCM—Information Corporation, 909 Third Avenue, New York, New York 10022; remitting $2.00 for each microfiche, or $5.00 for each photocopy. Advance payment is required.

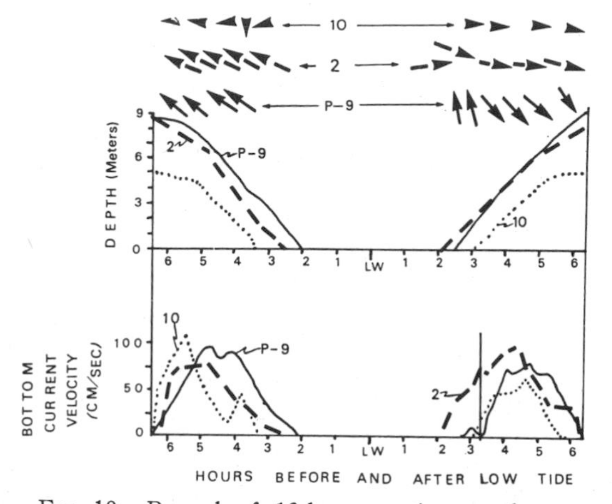

FIG. 10.—Record of 13-hour continuous observations of bottom current velocity, water depth and flow directions at Station #2 (Steep face), West Bar, Station #P-9 (gentle face), West Bar, and Station #10 (gentle face), East Bar. Station localities shown in Fig. 12. Vertical line shows time of submergence of West Bar.

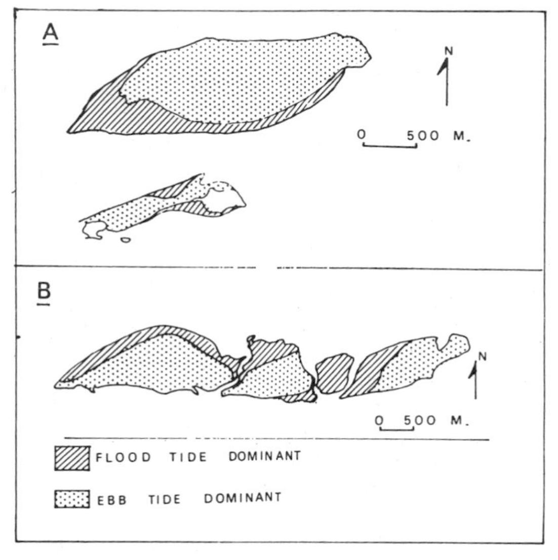

FIG. 11.—Distribution of flood- and ebb-dominated bottom current velocity zones at (A) Five Islands, and (B) Economy Point.

try of bottom tidal currents was mapped over the bars (fig. 11) both from bottom current velocity data (spot checks) and from 13-hour continuous observations (Table 2, 9). The distribution of zones of flood- and ebb-dominated bottom tidal currents coincides with bar topography (cf. figs. 9 and 11). At Pinnacle Flats, a local area of flood-dominated bottom current velocities occurs northwest of Long Island where ebb current systems converge after streamlining around Long Island in a westerly direction. The convergence of these ebb currents generates a shock wave which reduces the effective bottom current velocities during ebb tide. This convergence zone is flood-dominated, therefore, even though the maximum velocities of the flood bottom currents in it (station No. 11) are less than over the steep face (Station No. 7) at Pinnacle Flats (Table 2).

The steep-sloping, flood-tide dominated sediments of Big Bar, Pinnacle Flats, West Bar and Center Bar act to reduce the effective flood-tide, bottom current scour on the gently-sloping portions of the bars (see fig. 10, Station No. p.-9). In this way, the bar crests and the flood-dominated steep slopes shield the gently-sloping, ebb-dominated part of the bar. Such shielding areas are referred to as *Flood Shields* as defined by Hayes and others (1969, p. 296) and Daboll (1969, p. 341). The shielding areas reduce the length of time during which Flood tide bottom scour can rework the gently-sloping portions of the sand bars.

TABLE 2.—*Summary of maximum bottom current velocities (and associated water depths) and difference in maximum velocity of flood and ebb bottom currents, at stations where 13-hour continuous observations of velocity and water depth were obtained. Bar facies at station also indicated (Abbreviations as per Table 3). Location abbreviations: BB-Big Bar; PF-Pinnacle Flats; EP-Economy Point.*

Station No.	Facies	Flood Tide		Ebb Tide		Velocity Difference, Max. Flood and Max. Ebb (Cm/sec)
		Max. Vel. (cm/sec)	Assoc. Depth (m)	Max. Vel. (cm/sec)	Assoc. Depth (m)	
2 (EP)	MSSD	100	5.25	77	6.0	23
P-9 (EP)	CSSD	75	6.0	95	6.5	20
10 (EP)	CSCSW (CSSD)	45	3.75	95	5.0	50
16 (BB)	MSSD	60	2.0	75	5.0	15
21 (BB)	MFS	65	4.0	53	4.75	12
11 (PF)	FSSSW	53	6.0	47	5.0	6
7 (PF)	MSCR	75	5.0	.30	8.3	45

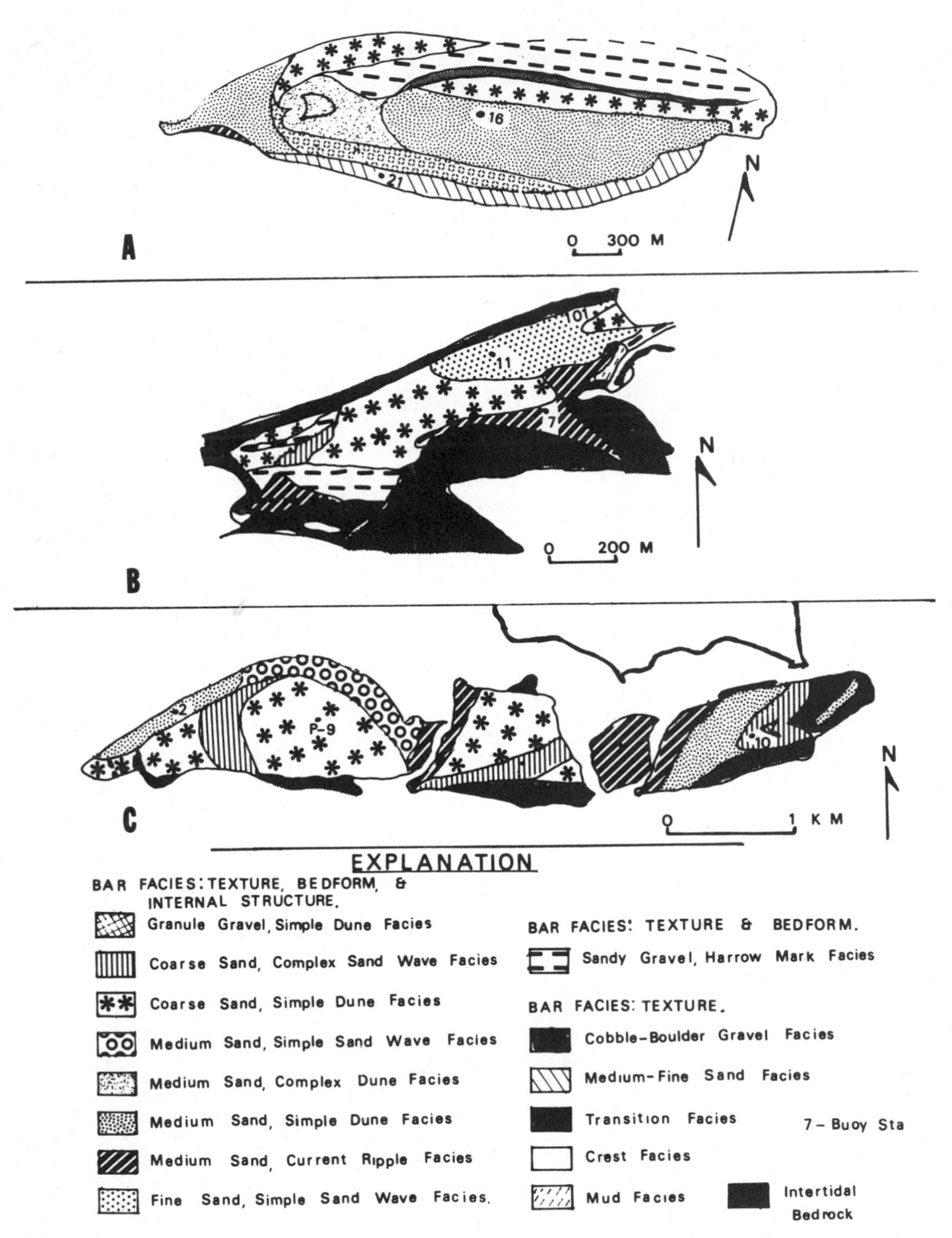

FIG. 12.—Distribution of sedimentary bar facies at (A) Big Bar, (B) Pinnacle Flats, and (C) Economy Point. Numbers refer to stations described in figures 10 and 34.

Sedimentary Facies (fig. 12)

On all the sand bars at Five Islands and Economy Point, sand bar facies can be discriminated according to sediment texture, a combination of texture and external bedform, or a combination of texture, bedform morphology and internal organization of cross-strata into simple or complex sets. The distribution of these facies (fig. 12) shows good agreement with topography (fig. 9) and the distribution of flood- and ebb-dominant bottom current velocity zones (fig. 11).

The facies distribution can be summarized in the following way. Over steep faces of sand bars and in channels separating sand bars, the texture is dominantly fine- to medium-sand. Associated bedforms are dominantly current ripples and dunes. Only one sand wave facies occurs on a steep slope. Over gently-sloping bar surfaces, however, the sediment texture is dominantly medium or coarse-sand, with accessory pebble gravel. Associated bedforms are simple and complex dunes and sand waves. Gently-sloping areas are flanked by gravel deposits at the low water line.

Texture.—Textural distinctions for facies identification are based on megascopic and sieve-size analysis. The diagnostic textural name is either the median diameter, the modal class or a significant secondary mode. The sands of all facies are dominantly medium sand, but important field distinctions of bedform type and internal structure coincide with the presence of a significant secondary mode, rather than a primary modal class. Hence the use of secondary modal classes to distinguish sedimentary facies.

Bedform morphology.—Surface bedforms were found to be of two dominant types: harrow marks (Karcz, 1967) or rippled bedforms of varying length. The frequency distribution of bedform wave lengths (fig. 13a, c) is grouped into three major orders of increasing wave length at Pinnacle Flats and Economy Point, and two orders at Big Bar (fig. 13b). The boundary limits between first order (Current ripples), second order (dunes) and third order (sand waves) bedforms are shown in Figure 13. Some overlap exists between orders when comparing the three areas from which measurements were obtained.

The external morphology of the ripples is highly variable. Linguoid, lunate and straight-crested current ripples (first order) are common. Sinuous dunes and sand waves also occur. The sinuosity is produced both by meandering

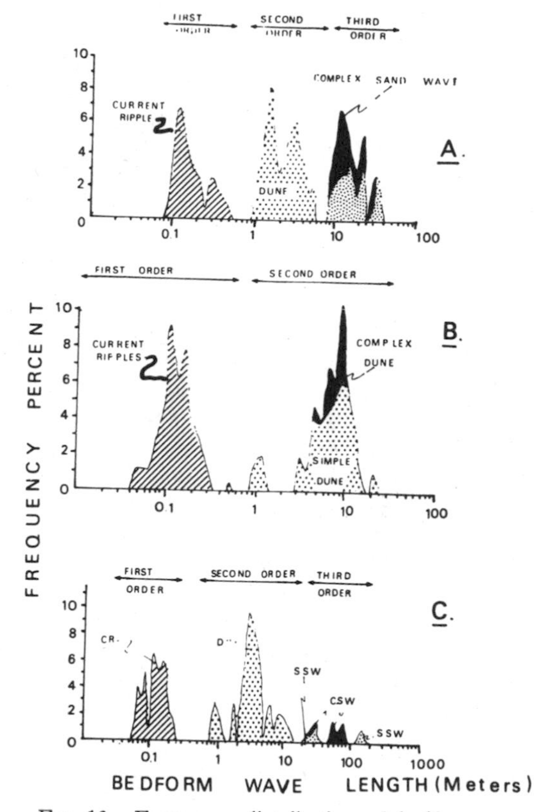

Fig. 13.—Frequency distribution of bedform wave lengths at (A) Pinnacle Flats, (B) Big Bar, and (C) Economy Point.

open-channel flow during emergence, and by differential rates of bedform migration.

The profile of most of the rippled bedforms, regardless of order, is asymmetric, with slip faces oriented in the down-current direction. Slope angles of slip faces average 26 degrees. However, in flood-dominated areas, slope angle of slip faces of flood-oriented sand waves average 18 degrees because the bedform is reworked by ebb currents. Nearly all dunes and sand waves are capped by sumperimposed lower order ripples and are therefore metaripples (Imbrie and Buchanan, 1965).

During emergence of some of the sand waves and dunes, the character of the flow was observed to show a progressive change from tranquil to rapid, particularly during the last 10 minutes prior to emergence. In dunes and sand waves of coarse-grained sand, sediment migrates as grain flow characteristic of the plane bed phase (Simons and others, 1965). Some of this sand is deposited on the slip face of the dune in a manner similar to delta building reported by Jopling (1965). Such outflow produces a planed ripple (cf. Boothroyd, 1969, p. 427) often superimposed on a sand wave. These planed ripples are merely a modification of existing bedforms by last minute flow when Froude numbers would be expected to increase. Because of the high turbulence, scour pits form on the downcurrent face of the bedform. The planing process is not a function of normal bedform migration, when Froude numbers are low (Table 9).

A third bedform, not mapped in facies distributions, is flute marks (fig. 14d). They occur on a semi-permanent channel, 7 cm deep, at the northeast margin of the medium sand, simple dune facies on Big Bar. The channel flow is transitional betweeen tranquil and rapid turbulent. Turbulence is high and rapid scouring and filling is common over the irregular channel floor. Turbulent scouring produces a deltoid, scalloped surface (fig. 14d). These "scallops" resemble flute marks because the narrow, upcurrent part of the structure is steeply-walled, whereas the downcurrent end of the structure slopes gently into the channel floor.

Internal organization.—Further subdivision of sedimentary facies is based on the internal organization of second and third order ripple bedforms. Two classes of internal organization are recognized: simple and complex. Simple dunes and sand waves are characterized by internal cross-stratification whose thickness is nearly the same as the vertical height of the dune or sand wave (fig. 15b). Complex dunes or complex sand waves are characterized by internal cross-stratification whose set thickness is *considerably less* than the height of the dune or sand wave in which it occurs. Such bedforms are characterized by a composite internal organization (Figs. 32b, 33c). Such complex internal organization of cross-stratification has been reported previously from sand waves by Reineck (1963) and Coleman (1969, p. 208–214).

Facies characteristics.—A total of 14 sedimentary facies are recognized on the sand bars at Five Islands and Economy Point. Diagnostic megascopic features of these facies are summarized in Table 3, whereas the average quantitative parameters of the sandy and gravelly facies are given in Table 4. The origin of some of the structures in these facies is discussed in a later section (p. 1115–1123).

Directional Properties (figs. 16, 17, 18)

The orientation of sedimentary bedforms and internal cross-stratification is summarized in Figures 16 and 17. Good agreement exists between orientation of sedimentary bedforms, sedimentary facies and bar topography. On steep faces of West Bar and the ebb-convergence zone at Pinnacle Flats, sand waves are oriented easterly, in agreement with flow directions of flood-dominated bottom tidal currents. On the gently-sloping sand bar surfaces at Big Bar, Pinnacle Flats, West Bar and Middle Bar, both dunes and sand wave slip faces are oriented to the west, in agreement with ebb-dominated tidal current flow. At East Bar, slip faces of both depositional sand waves (in the flood-dominated western part) and erosional sand waves (p. 1120–1122) are oriented to the east, whereas slip faces of dunes are oriented to the west, in agreement with ebb-dominated tidal currents. General agreement occurs between flood- or ebb-dominated flow and the orientation of internal cross-stratification (fig. 17). The agreement between orientation of bedforms, and flood- and ebb-dominated areas, compares favorably with similar observations by Van Straaten (1950, p. 78–79) from subtidal channels and adjoining subtidal flats in the Basin d'Arcachon of France.

Slip faces of current ripples are oriented parallel to the direction of ebb current flow, or at right angles or opposite to ebb current flow (figs. 14b, f, 16, 27b, c, f, g). Current ripples are oriented at right angles to dunes only in troughs (fig. 14f) and are formed by late-stage open-channel flow in the troughs as dune crests emerge. In the northeast part of Big Bar, late-stage sheet runoff flows to the east (in response to an easterly bar slope) and therefore, orientation of current ripples is also east (fig. 14b).

The directional data are summarized into rose diagrams (fig. 18) and the variance was determined by graphical methods outlined by Dennison (1962). These data show that current ripples are characterized by a high variance and are trimodal in their distribution, in response to their formation during slope-controlled, late stage sheet runoff. Low variant and unimodal or bimodal-bipolar (oppositely oriented) distributions characterize the orientation of dunes, sand waves and cross-stratification, and show good agreement with ebb, or both ebb and flood bottom current flow directions.

The vector mean and variance are tabulated in Figure 18. The vector mean shows agreement only with modal class in unimodal distributions. In bimodal distributions, the vector mean is of no value in delineating depositional current trends. These findings re-emphasize the importance of using modal patterns, rather than vector means, in determining depositional current trends in marine sediments (Tanner, 1955, 1959).

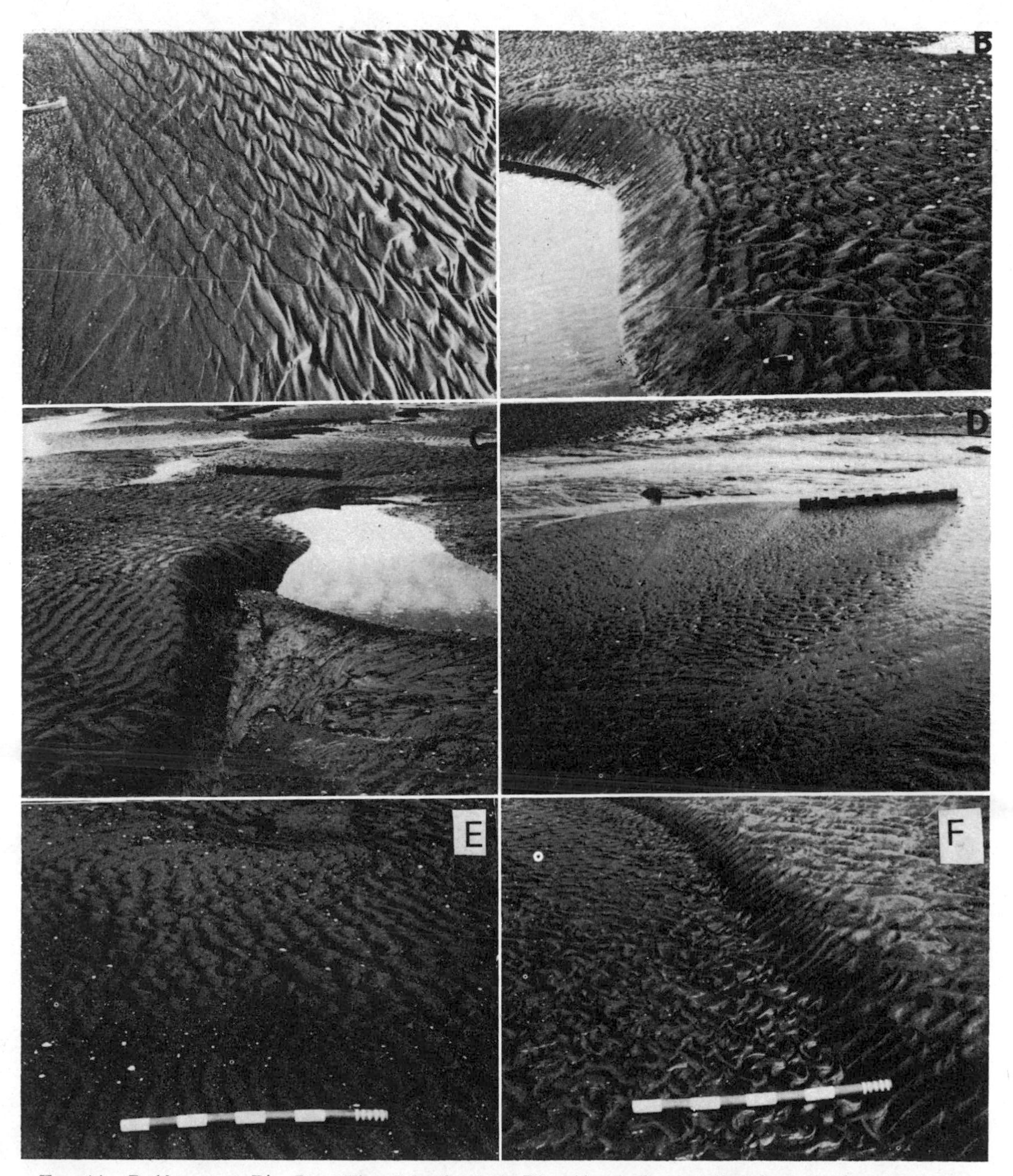

FIG. 14.—Bedforms at Big Bar, Five Islands. (A) Rhombic ripples, grading laterally into longitudinal ripples, Crest Facies, Big Bar. View to north. (B) Dune with superimposed oppositely oriented, double-crested current ripples, coarse sand simple dune facies, Big Bar. View to north. Ebb current flow to west (left). (C) Simple dune with superimposed current ripples, coarse sand simple dune facies, with scour pit. Scale in decimeters. View to south. Ebb current flow to west (right). (D) Flute-marked surface in tidal creek, at northern margin of medium sand, simple dune facies, Big Bar (Scale in cm and decimeters). View to north. (E) Lateral change from transitional facies with straight-crested current ripples (foreground) to straight and linguoid current ripples in coarse sand simple dune facies (background), Big Bar. Scale in cm and decimeters. View to south. Ebb current flow to west (right). (F) Simple dune with straight-crested current ripples superimposed at right angles on slipe face; and linguoid current ripples in trough, medium sand simple dune facies, Big Bar. Scale in cm and decimeters. View to north. Ebb current flow to west (left).

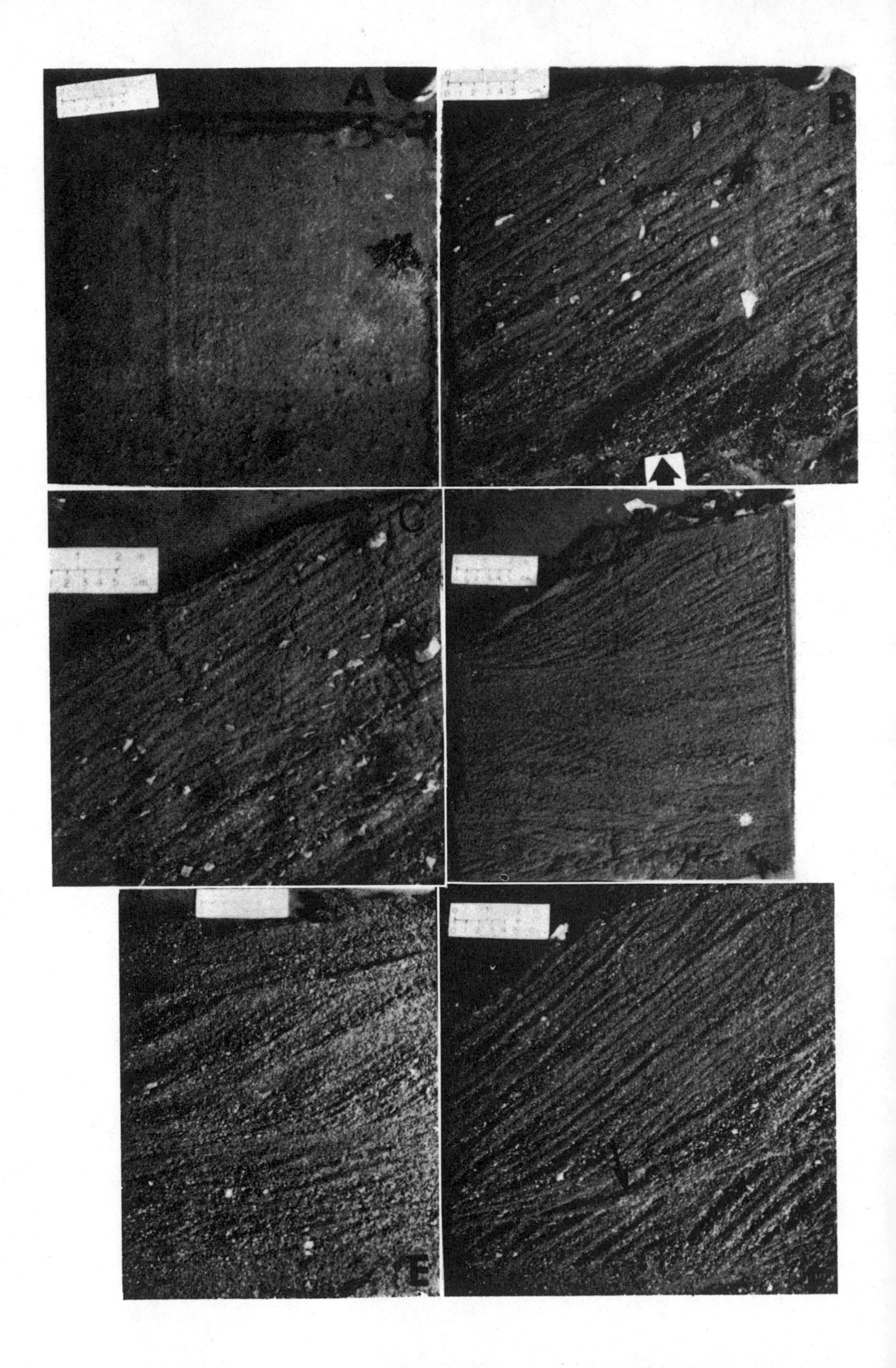
A
B
C
E

TABLE 3.—*Diagnostic megascopic features of sedimentary facies, Big Bar, Pinnacle Flats, West Bar, Middle Bar and East Bar*

Facies (and facies abbreviations for Tables 2, 4, 6, 7)	Diagnostic Features
(A) Facies classified according to texture, bedform and internal structure	
Granule Gravel, Simple Dune (GGSD)	Granule gravel, with secondary mode of very coarse sand. Planed dunes. Thin, planar cross-stratification with sharp contacts; trough-like toesets.
Coarse Sand, Complex Sand Wave (CSCSW)	Coarse sand, with secondary mode of medium sand. Sand wave crests splay into tier of dunes arranged in steps. Thin sets of planar cross-stratification with sharp contacts between sets. Trough-like toesets. Thickness of cross-stratification is less than sand wave height. Rounded upper sets. (Figs. 15e; 27d; 33a, b; 34b, c).
Coarse Sand, Simple Dune (CSSD)	Medium grain sand with secondary mode of coarse sand. Simple dunes. Trough and planar cross-stratification with sharp set contacts. Superimposed current ripples. Troughs lined with pebbles and shells of *Mya arenarea*, *Crepidula fornicata*, and plant stems. Rounded upper sets of cross-stratification. Metaripples. (Figs. 14b, c; 15b, c, f; 27c, g).
Medium Sand, Simple Sand Wave (MSSSW)	Medium sand with nearly equal proportion of fine sand. Simple sand waves. Planar cross-stratification, with sharp set boundaries. Rounded upper set boundaries. Superimposed current ripples oriented opposite to sand waves.
Medium Sand, Complex Dune (MSCD)	Fine sand with major secondary mode of medium sand. Complex dunes. Low-angle cross-stratification; consisting of micro-cross-laminated layers. Sharp set boundaries between cross-stratified sets. Superimposed current ripples.
Medium Sand, Simple Dune (MSSD)	Fine sand with major secondary mode of medium sand. Simple dunes. Planar cross-stratification with sharp contacts. Current ripples superimposed on crest and slip face at 90°, 180° or parallel to dune orientation. Metaripples. (Figs. 14f; 27a, b).
Medium Sand, Current Ripple (MSCR)	Medium sand with significant mode of fine sand. Straight, linguoid and lunate current ripples. Parallel laminae below surface sets of micro-cross-laminae in channel borders. Flute marks on surface of channel wall and floor.
Fine Sand, Simple Sand Wave (FSSSW)	Medium sand with significant secondary mode of fine sand. Simple sand waves with superimposed current ripples parallel, at right angles or opposite to sand wave orientation. Low angle, planar cross-stratification with sharp set boundaries. (Figs. 15d; 27f).
(B) Facies classified according to texture and bedform.	
Sandy Gravel, Harrow Mark (SGHM)	Pebble and cobble gravel with major secondary mode of medium sand. Harrow marks. Current crescents. Local areas of straight-crested, planed current ripples. Granule gravel mixed with sand in harrows.
(C) Facies classified according to texture only.	
Cobble-Bounder Gravel (CBG)	Pebble gravel with major secondary modes of cobbles and boulders.
Medium-Fine Sand (MFS)	Fine sand with secondary mode of medium sand. Airholes. Planar lamination (Fig. 15a).
Transition (T)	Fine sand with secondary mode of gravel. Straight-crested current ripples with granules and pebbles in troughs (cf. Hooke, 1968). Harrow marks. Current crescents.
Crest (C)	Fine Sand. Straight and linguoid ripples. Longitudinal ripples. Rhombic ripples. (Fig. 14a).
Mud	Silty clay. Parallel laminated. Burrowed by *Mya arenarea*, *Arenicola marina*, *Nereis virens*, *Heteromastus filiformis*.

FIG. 15.—Epoxy peels from Five Islands and Economy Point. (A) Medium-fine sand facies, Big Bar, showing parallel lamination and airholes (Peel #5-A). (B) Coarse sand simple dune facies, Big Bar, showing planar cross-stratification (Peel #5-DC). (C) Coarse sand simple dune facies, Center Bar, showing slight rounding and truncation of cross-stratification at bedform surface (Peel #7-37B). (D) Fine sand simple sand wave facies, Pinnacle Flats, showing trough cross-stratification in surface set and truncated cross-stratification in lower set (Peel #4-Bu-8A). (E) Coarse sand complex sand wave facies, Pinnacle Flats, showing thin sets of trough cross-stratification (Peel #3-1A). (F) Coarse sand simple dune facies, Pinnacle Flats, showing at arrow a rounded upper set boundary (reactivation surface) and planar cross-stratification with trough-like toesets (Peel #3-103A).

TABLE 4.—*Average Quantitative Parameters of sandy and gravelly Sedimentary Facies.* (Textural parameters after Folk and Ward, 1957)

Facies	Parameter													
	Modal Size (ϕ)	Mean Size (ϕ)	Percent Coarse Sand	Percent Medium Sand	Percent Fine Sand	Median Diameter (ϕ)	Deviation	Skewness	Sand-gravel ratio	Ripple Wave Length (M)	Ripple Height (M)	Dip Angle, Slip Face	Thickness of sets of cross-stratification (Cm)	Dip Angle, Cross-stratification
							Big Bar							
C	2.75	2.13	4	15	60	2.26	0.62	−0.22	0	—	—	—	—	—
MFS	2.75	2.35	0	7	81	2.35	0.31	−0.01	0	—	—	—	12.5	
MSSD	2.25	1.55	6	9	67	1.61	0.79	−0.09	6.	7.55	0.18	30°	12.0	23°
MSCD	2.25	1.55	12	23	50	1.67	0.86	−0.22	0	7.65	0.19	29°	5.2	16°
CSSD	1.50	0.96	14	28	29	1.20	1.19	−0.29	23.25	5.84	0.16	30°	12.0	27°
SGHM	−0.25	−0.13	16	21	9	−0.23	1.45	0.08	2.70	—	—	—	—	—
T	−1.75	0.52	8	15	27	2.21	2.09	−0.17	2.21	—	—	—	—	—
							Pinnacle Flats							
CBG	−2.24	0.47	7	17	12	−1.09	1.60	0.49	1.11	—	—	—	—	—
MSCR	1.75	1.23	21	33	22	1.29	0.69	−0.15	—	0.20	0.03	—	—	—
CSCSW	1.25	0.91	34	43	13	0.98	0.68	−0.16	0	13.2	0.77	20°	8.1	21°
CSSD	0.75	0.70	32	45	10	0.85	0.75	−0.11	4.0	2.44	0.17	31°	9.8	29°
FSSSW	1.75	1.30	16	53	26	1.33	0.64	−0.10	0	16.84	0.17	14°	14.3	23°
							Economy Point							
GGSD	−1.75	−0.89	22	8	1	−1.08	1.11	0.26	1.13	5.0	0.53	18°	12.9	32°
MSCR	2.25	1.60	4	35	57	1.69	0.58	−0.25	0	12.4	0.008	28°	—	—
MSSD	2.25	1.89	4	36	60	1.90	0.41	−0.35	0	5.54	0.27	28°	8.23	32°
MSSSW	1.75	1.24	4	29	30	1.16	0.58	0.01	0	47.36	0.69	18°	7.9	24°
CSSD	1.75	1.35	21	36	22	1.50	0.67	−0.27	16.1	3.58	0.21	27°	9.5	29°
CSCSW	1.75	0.81	24	38	13	1.00	0.77	−0.24	10.2	50.7	1.17	30°	7.9	27°
SGHM	−1.75	−0.27	18	17	11	0.63	1.36	0.10	3.2	—	—	—	—	—

Sand Dispersal Systems

A series of grain dispersal experiments were completed at Big Bar in cooperation with C. L. Sandusky (1968), using sand stained with an acrylic lacquer (See Yasso, 1966, for a discussion of such techniques). A total of eight stations were occupied.

The dispersal pattern of grains from a point source was found to be radially-elliptical (fig. 19). The direction of *maximum distance* of grain transport from a point source shows excellent agreement with the flow direction of flood-dominant bottom tidal currents on the steeper bar slopes (fig. 11a) and with flow direction of ebb-dominant bottom tidal currents on the gentler bar slopes (Fig. 20a).

These data suggest a generalized transport model (fig. 20b) which is elliptical in a counter-clock-wise manner around Big Bar. This elliptical transport pattern is a response by sediment to alternate transport through flood- and ebb-dominated portions of Big Bar. A similar pattern of sand transport was documented by Philipponneau (1955; cited in Verger, 1968, p. 184–188) from sand bars in the Baie de Mont Saint Michel in France, and appears also to

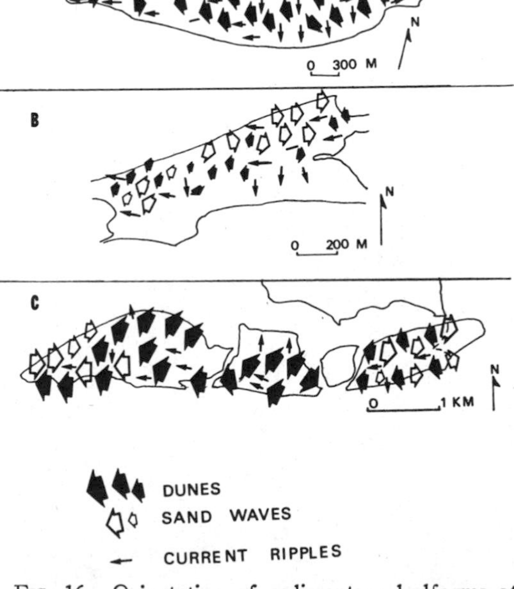

FIG. 16.—Orientation of sedimentary bedforms at (A) Big Bar, (B) Pinnacle Flats, and (C) Economy Point.

characterize other sand bars as seen by comparing aerial photographs and tidal current flow maps shown by Verger (1968, p. 67–68, 104–105, 205–207) from the Basin d'Arcachon, Les Pertuis d'Oleron and the Baie de Somme along the French coast (See also Van Straaten, 1950, p. 78–79). Newton and Werner (1968) documented a similar dispersal system of sand transport on intertidal sand bars near Cuxhaven, West Germany, and Reineck (1963) proposed a similar dispersal system of sand transport for a subtidal sand bar in the deeper channels of the Jade. A comparison of Newton and Werner's (1968) observations with airphotos of intertidal sand bodies elsewhere on the North Sea Coast of West Germany (Gierloff-Emden, 1961, p. 35–36, 43–45, 49, 53, 55, 58–59) suggests a history of sand transport identical to that at Big Bar.

In sub-tidal, tide-dominated regimes, a similar system of sand transport also is known. Houbolt (1968) inferred such a dispersal system from dune orientation on the Well Bank of the North Sea (p. 257), whereas James and Stanley (1968) documented a similar transport pattern by heavy mineral studies from the Scotian Shelf near Sable Island. J. D. Smith (1969) reported a similar circulation pattern of tidal flow over

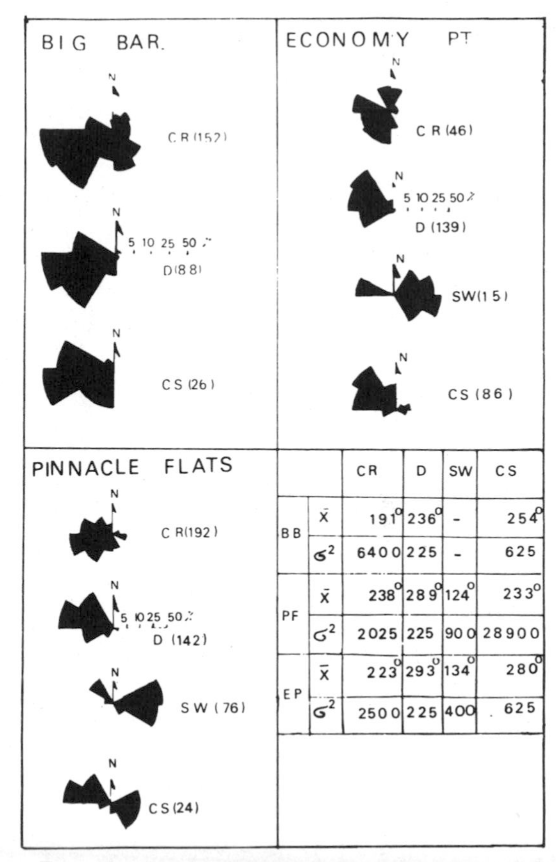

		CR	D	SW	CS
BB	$\bar{X}$	191°	236°	-	254°
	σ^2	6400	225	-	625
PF	$\bar{X}$	238°	289°	124°	233°
	σ^2	2025	225	900	28900
EP	$\bar{X}$	223°	293°	134°	280°
	σ^2	2500	225	400	625

FIG. 18.—Summary rose diagram of orientation data for current ripple (CR), sand wave (SW), dune (D), and cross-stratification (CS) at Pinnacle Flats, Big Bar and Economy Point. Table summarizes vector mean ($\bar{X}$) and variance (σ^2) for each property at each study area (BB—Big Bar, PF Pinnacle Flats, EP—Economy Point).

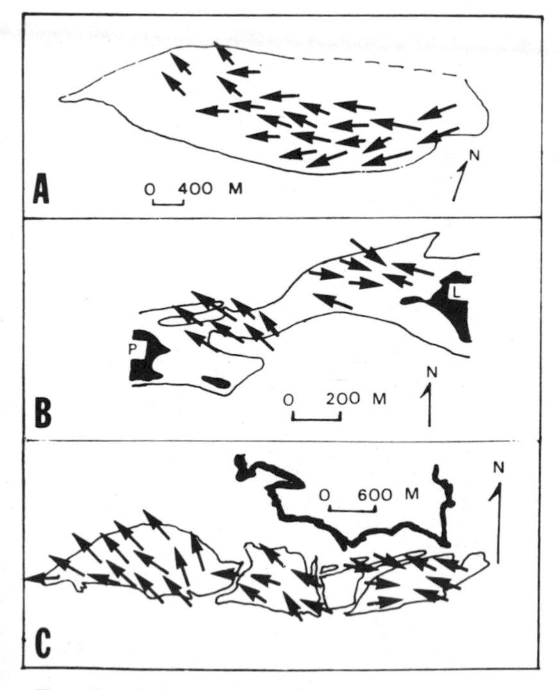

FIG. 17.—Orientation of maximum dip direction of cross-stratification at (A) Big Bar, (B) Pinnacle Flats, and (C) Economy Point. (Abbreviations: P—Pinnacle Flats; L—Long Island).

tide-dominated sand bars off the coast of Massachusetts, as did Verger (1968, p. 165) from subtidal sand bars off the Loire River Delta, France. The conclusion is therefore inescapable that this elliptical pattern of sand transport through both flood- and ebb-dominated portions of tidal sand bars is a characteristic transport pattern for intertidal and subtidal sand bodies.

Examination of low-tide airphotos taken in 1947 (Canadian Air Photo Library Photo No. A11790-66) and in 1963 (Canadian Air Photo Library No. A18061-21) at Big Bar shows that both the sedimentary facies and morphology there has remained unchanged since 1947. The elliptical sand transport system must be an equilibrium system, and consequently, Big Bar is an equilibrium intertidal sand bar. The equilibrium exists between tidal current sand transport, tidal current time-velocity asymmetry, bar topography, distribution of sedimentary facies, and ori-

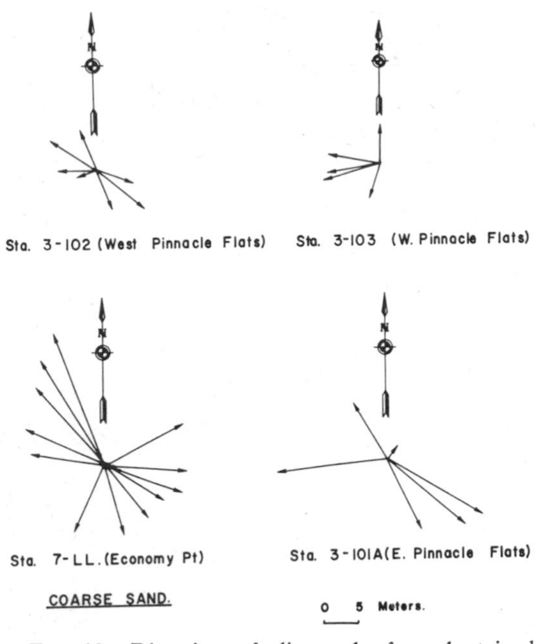

FIG. 19.—Direction of dispersal of sand stained with an acrylic lacquer. Distances from point source measured after two tidal days (four tidal cycles).

entation of directional current structures.

At Pinnacle Flats, the same 1963 low-tide air photo shows that the topography and distribution of sedimentary facies observed between 1965 and 1968 has existed as far back as 1963. Only three stations were occupied for grain dispersal experiments, all indicating major dispersal parallel to ebb-dominated bottom tidal flow. At one station (Station No. 3-103) orange sand, originally placed in 1966, was observed there again in 1967, 1968 and 1970 when this station was re-occupied. Pinnacle Flats must be an equilibrium sand bar also.

Airphoto analysis at Economy Point indicates that the facies distribution mapped by the author in 1966, 1967 and 1968 (Fig. 12) is identical to that shown on an airphoto taken in 1963 (fig. 21). A somewhat similar facies distribution was present as far back as 1947 (fig. 22) and 1938 (fig. 23). The sand bars at Economy Point were definitely subjected to an erosional regime between 1938 and 1963 (figs. 21, 22, 23), however. In the axial portion of East Bar, a prominant "V"-shaped depression occurs in 1963 (fig. 21) but was absent in 1938 (fig. 23) and small in 1947 (fig. 22). An estimated 5,000 cubic meters of sand was eroded from this part of East Bar since 1938.

Center Bar and West Bar were one continuous sinusoidal bar in 1938 and 1947 (figs. 22, 23). Between 1947 and 1963, the two bars became separated by the development of a major channel south of a bedrock ledge just northwest of what is now Center Bar. An estimated 160,000 cubic meters of sand was removed between 1947 and 1963 by this channel.

Figures 21, 22 and 23 show that the present crestal elevation of Center Bar is higher than in 1938 and 1947. In 1938, a small creek ("s" on fig. 23) drained Center Bar, flowing east. This creek reversed its flow direction by 1947 (fig. 22) and drained into a channel south of the bedrock ledge on the northwest side of Center Bar. The creek contributed an increased discharge into the larger channel and probably caused it to breach the connection between what is now Center and West Bar. The increase in crestal elevation of Center Bar between 1938 and 1963 was caused by sand supply from East Bar and erosion of the channel separating West and Center Bar. The channel between West and Center Bar disperses sand south of the bars where flood-dominated currents (flowing on the south side of Center Bar) redisperse some of this eroded sand back onto Center Bar by northeasterly transport. An estimated 100,000 cubic meters of sand was restored onto Center Bar by flood tidal current transport. The remaining 65,000 cubic meters of sand represents the total volume of sand loss from the bars at Economy Point between 1938 and 1968.

This airphoto analysis shows that although each of the sand bars at Economy Point probably are characterized by an elliptical dispersal model similar to Big Bar, a slight erosional regime exists on some of the sand bars. The ero-

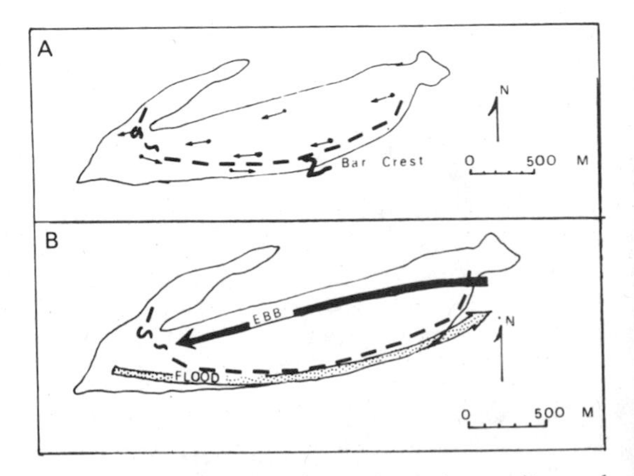

FIG. 20.—(A). Directions of maximum dispersal of sand from a point source after one tidal day (data in part from Sandusky, 1968), Big Bar. (B). Generalized circulation model of sand transport transport through zones of flood- and ebb-dominated tidal currents. Dashed line shows position of bar crest at Big Bar.

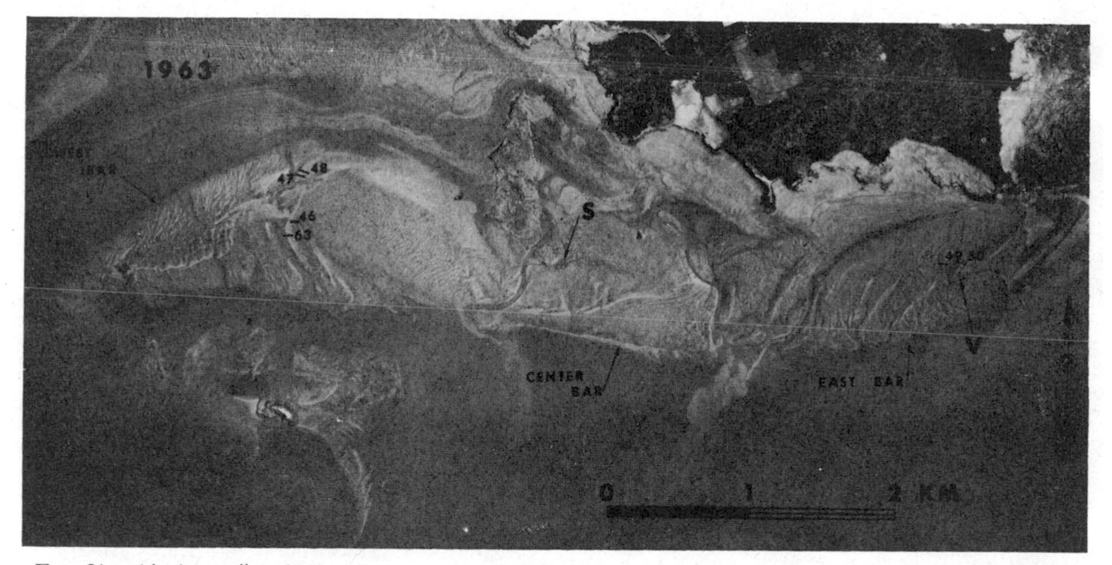

FIG. 21.—Airphoto #A-18061-16 taken at low tide in 1963, intertidal sand bars at Economy Point (Abbreviations: E-esker, V—'v'-shaped depression; S—stream). Numbers refer to trench numbers. (Reprinted by permission of the National Airphoto Library, Surveys and Mapping Branch, Canadian Department of Energy, Mines and Resources).

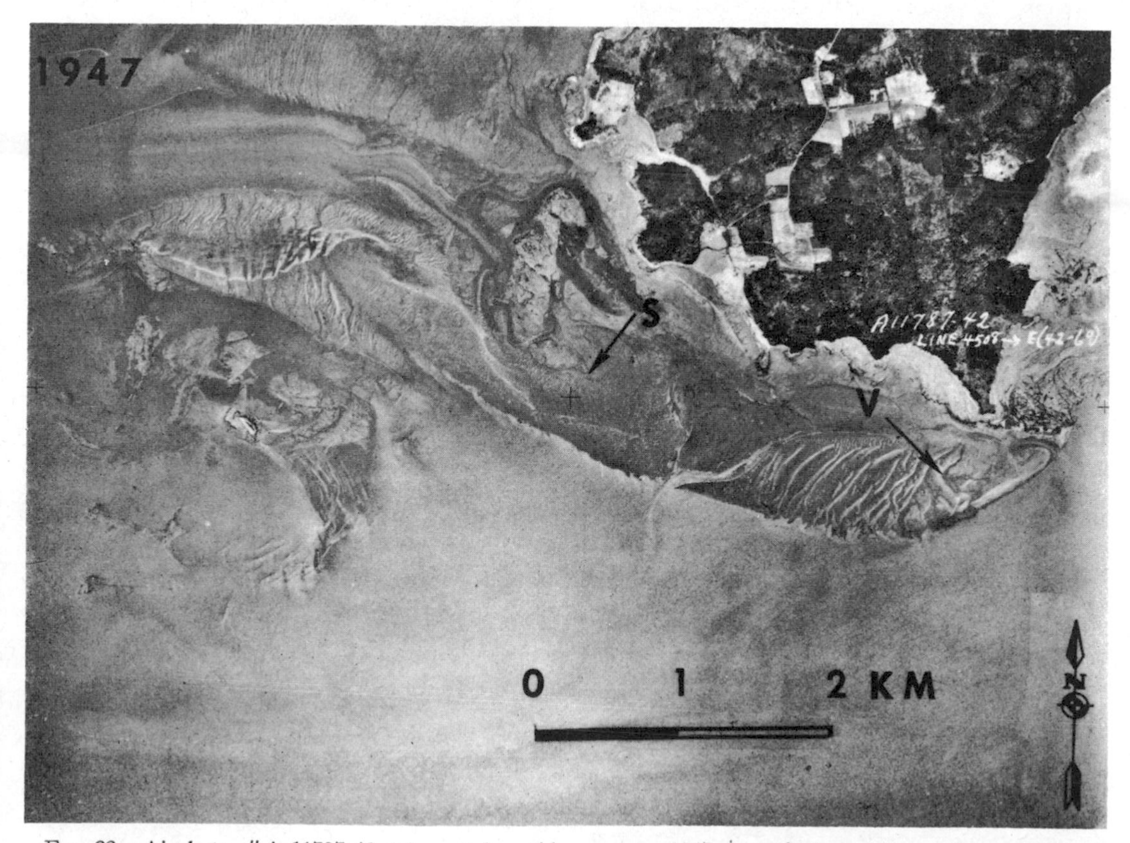

FIG. 22.—Airphoto #A-11787-42 taken at low tide, summer 1947, intertidal sand bars, Economy Point. Abbreviations as in Fig. 21. (Reprinted by permission of the National Airphoto Library, Surveys and Mapping Branch, Canadian Dept. of Energy, Mines and Resources).

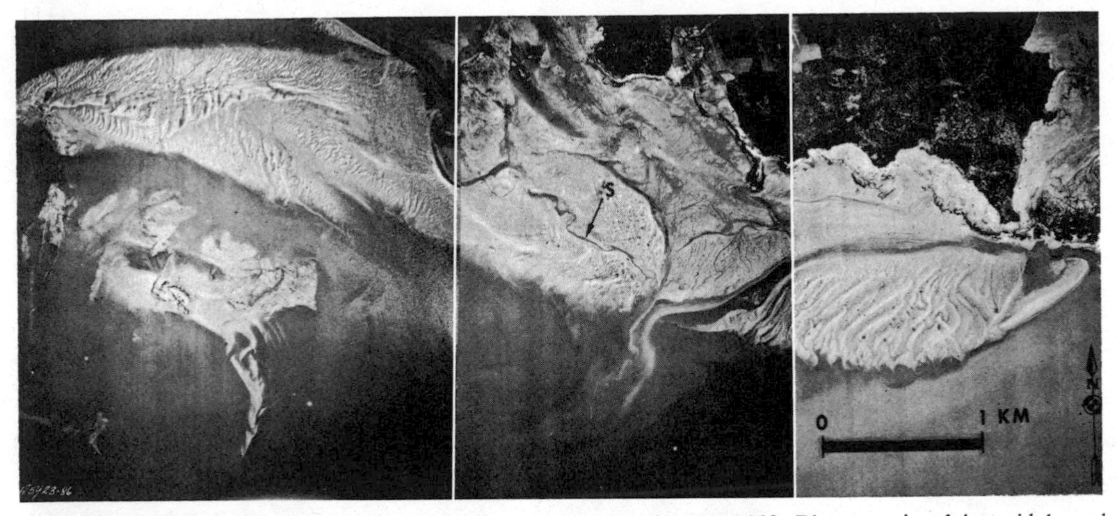

FIG. 23.—Airphotos #A-5923-86, 88, 90 taken at low tide, summer, 1938. Photomosaic of intertidal sand bars at Economy Point. Abbreviations as in Figure 21. (Reprinted by permission of the National Airphoto Library, Surveys and Mapping Branch, Canadian Dept. of Energy, Mines and Resources).

sional regime accounts for extreme reworking of the bars (Dinkelman, 1968) and appears to be removing sand from ebb-dominated segments of East and Center Bars. Tidal current transport in flood-dominated zones is returning some of this eroded sand onto the bars, such as Center Bar. The 65,000 cubic meters of sand eroded by this regime appears to have been dispersed in a westerly direction.

Provenance of the Sediments

The sediments on the intertidal sand bars are derived from two major sources, Triassic sedimentary rocks occuring both as intertidal ledges, and sea cliffs (Klein, 1962), and Pleistocene fluvio-glacial sediments of the Five Islands Formation (Swift and Borns, 1967) which occur both on shore in sea cliffs and in the interidal zone. Swift and McMullen (1968) demonstrated that most of the sea bed of the Minas Basin is underlain by nearly equal proportions of Triassic sedimentary rocks and Pleistocene sediments which are available for reworking and redistribution by tidal currents.

Pleistocene sources.—The Pleistocene fluvio-glacial deposits of the Five Islands Formation are exposed on shore. They consist mostly of gravelly sands (fig. 24). Their texture is bimodal, with the two modal classes falling at +2 Phi and −2 Phi. They are medium reddish brown and consist of subarkosic and quartzose sand mixed with pebbles and cobbles of granite, rhyolite, metaquartzite and recognizable Pennsylvanian sedimentary rocks known to occur in the Cobequid Highlands to the north. The sand fraction is similar in composition to the Triassic (Klein, 1962) and in this case is subarkose and orthoquartzite (Folk, 1954, classification). The sand fraction is stained with hematite, derived from the Triassic.

Much of the Pleistocene sandy gravel is recognized in reworked form by megascopic comparison between intertidal sediments and the Five Islands Formation. The coarse sand, simple dune facies, the coarse sand, complex sand wave facies, the sandy-cobble gravel facies, and the sandy gravel, harrow mark facies are similar in color and composition to the Five Islands Formation.

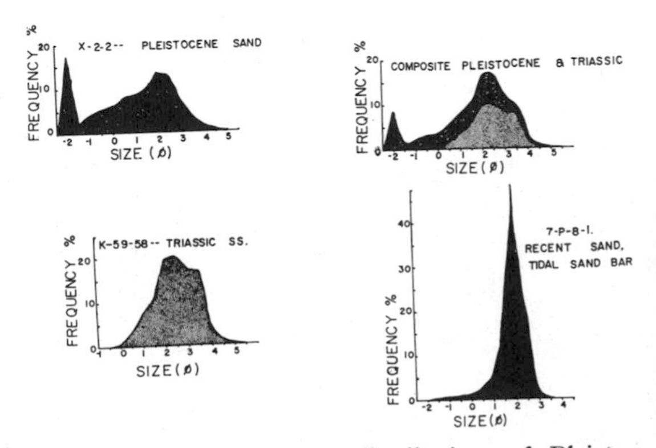

FIG. 24.—Size-frequency distributions of Pleistocene fluvio-glacial sand, Five Islands Formation (upper left), Triassic Wolfville Formation (lower left). composite of equal propertions of Triassic and Pleistocene sand (upper right), and Holocene sand from intertidal sand bar at Station #7–P–8, West Bar, Economy Point. Size analyses taken from representative samples.

The eastern third of East Bar at Economy Point is an esker of Pleistocene age cropping out in the intertidal zone ("E" in fig. 21) and is being reworked by tidal currents. The esker consists of a pebbly-cobbly sandy gravel; most of the sand is medium- and fine-grained. The surficial part of the esker is covered by a sand-free gravel lag-concentrate which grades laterally into a fine- and medium-sand. This sand rests as a surface veneer on the pebbly-cobbly sandy gravel of the esker (fig. 25a, c).

The esker is being reworked by tidal currents and modified by ice-rafting. In August, 1966, a thin strip of sand occurred on the crest of the esker and was observed to grade laterally into a gravel lag (fig. 25a). By June, 1967, the surficial sand was removed by tidal currents and the esker was completely armored with gravel (fig. 25b). In June, 1968, conditions were similar to 1966 (Fig. 25c). During the winter of 1967-68, the Minas Basin was frozen and the 1967 gravel lag deposit was removed by ice-rafting in the spring of 1968. The pebbly-cobbly-sandy gravel of the interior of the esker was exposed again and then reworked by tidal currents. By June, 1968 (Fig. 25c), sand was winnowed into a linear strip on the crest of the esker.

These observations reconfirm that the tidal currents of the Minas Basin transport, redistribute and localize sand on sand bars and leave gravel behind as a lag concentrate.

Triassic sources.—Triassic sandstone ledges are common in the intertidal zone of the two study areas. At Economy Point, they occur as a substrate for the sand bars, whereas at Five Islands, Triassic bedrock underlies the south side of Pinnacle Flats and crops out on all the islands (Klein, 1960, 1962).

The Triassic bedrock is both scoured, striated, and polished. No organic growth, such as algae or barnacles, occur on the polished ledges. Some of the polished bedrock ledges are similar to bedrock surfaces scoured by sand flow in submarine canyons (Dill, 1964).

The surfaces of Triassic bedrock are also straited by tidal current scour and sand blasting (fig. 25d, e, f). The striae are oriented parallel to present ebb and flood current flow directions and the orientation of slip faces of current ripples and dunes (fig. 25f). The striae also cut Holocene clam shells which bored into Triassic sandstone (fig. 25e).

Some workers (Swift and Lyall, 1968) proposed a Pleistocene glacial origin for the east-west oriented striae. Hickox (1962) demonstrated from petrologic study of till boulders that Wisconsin glacial transport was directed from north to south across the Minas Basin, however. The striae orientation do not agree with known directions of glacial transport. Agreement between striae orientation, tidal current flow directions, and orientation of bedforms indicates an orign by tidal current scour. Cutting of Holocene clams by striae confirms their Holocene age.

Provenance summary.—Recognition of both Triassic and Pleistocene sediments as sources for the intertidal sand bars places known limits on the particle sizes available for tidal current transport. Figure 24 shows the textural distributions for representative samples of Pleistocene sand collected from the Five Islands Formation and of Triassic sandstone collected from Economy Point. The modal class of Triassic sandstone is + 2 Phi. The Pleistocene sample is bimodal; the primary mode being − 2 Phi and the secondary mode being + 2 Phi.

Swift and McMullen (1968) showed that the Minas Basin subtidal and intertidal zone is underlain by nearly equal proportions of Triassic and Pleistocene sediment. A composite frequently distribution can be constructed representing the particle sizes available for transport (fig. 24); the composite size distribution is bimodal with a modal class of + 2 Phi. Comparison of the composite size frequency distribution and percentages of coarse, medium and fine sand (Table 4) on the sand bars indicates that the sand bars are enriched in sand. A representative frequency distribution for the medium sand simple dune facies (fig. 24) confirms the enrichment. These data demonstrate sand enrichment on the bars by tidal current transport and redistribution from both Triassic and Pleistocene sources. Gravel is left as a lag concentrate (fig. 25b) for redistribution by wave action, longshore drift and ice-rafting (Laub, 1968; R. S. Smith, 1969).

INTERTIDAL MODEL FOR PALEOCURRENT ANALYSIS

The data on flow directions, orientation of directional properties, provenance and dispersal experiments are integrated into a model for paleocurrent analysis of ancient counterparts (Table 5). The model is hierarchically sensitive to flow directions. Grain dispersal data (modelling indicator grains) are sensitive to *all* flow directions occurring in the intertidal environment. Current ripple orientation data are sensitive to *slope-controlled* flow on sand bars. As an indicator of paleoflow of fossil tidal currents, current ripple orientation would prove unreliable. Form lines constructed parallel to crests of current

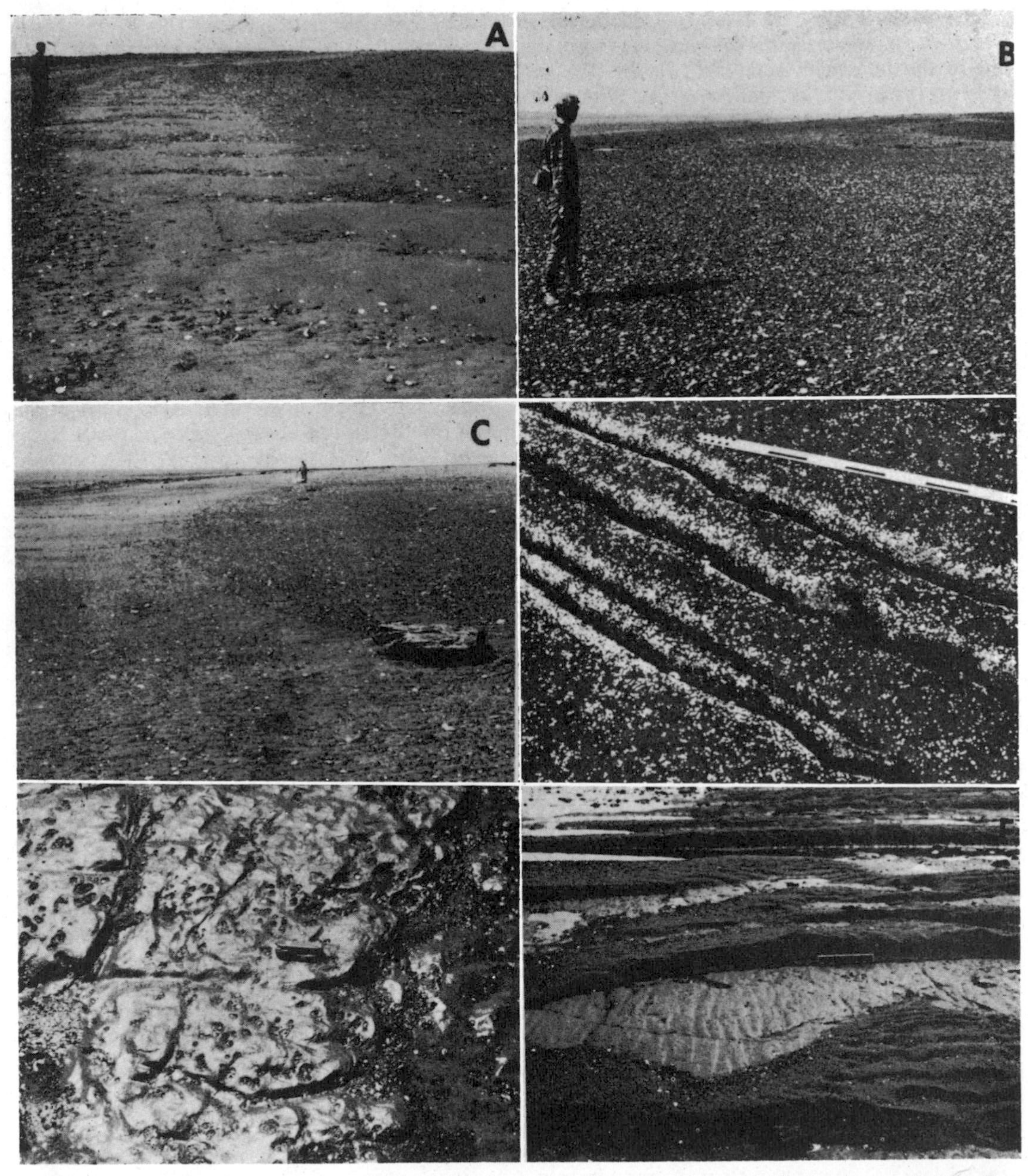

FIG. 25.—Sediment sources for intertidal sand bars. (A) Surface of reworked esker east of East Bar, August, 1966. Esker crest consists of mixed sand and gravel covered with strip of winnowed sand and gravel-lag pavement. View to southwest. (B) Surface of reworked esker east of East Bar, June, 1967. Esker crest consists entirely of gravel-lag pavement that inhibits further sand winnowing by tidal currents. View to southwest. (C) Surface of reworked esker east of East Bar, June, 1968. Esker crest has been eroded by ice-rafting, and subsequent tidal current reworking have winnowed a sand strip from sandy-gravel esker sediments. View to southwest. (D) Striated pavement, Triassic sandstone ledge, West Bar. Striation carved by tidal currents (Scale in cm and decimeter) View to northwest. (E) Striated pavement in bored ledge of Triassic sandstone, West Bar. Striae cut both bedrock and Holocene clams, proving Holocene age of striae. View to north. (F) Striated pavement of Triassic sandstone, western tip of West Bar, with thin veneer of sand fashioned into current ripples and dunes. Orientation of slip faces of ripples and dunes agrees with orientation of striae. Scale in inches. View to east.

TABLE 5.—*Intertidal Paleocurrent Model* (See Fig. 16, 17, 18)

Directional Property	Dispersal Pattern	Variance	Relation to Flow Systems	Relation to Basin Slope	Relation to Source	Ancient Analog
Grains	Radial-elliptical	Extreme	All flow directions	Indeterminant	Indeterminant	Mineral Grains
Current Ripples	Trimodal	High	Flow direction controlled by sand bar slope	Indeterminant	Indeterminant	Ripple Mark Micro-Cross-laminae
Dunes	Unimodal	Low	Parallel to tidal currents	Parallel to basinal topographic trend and sand bar alignment	Indeterminant	Dunes Cross-stratification
Sand Waves	Unimodal	Low	Parallel to tidal currents	Parallel to basinal topographic trend and sand bar alignment	Indeterminant	Sand waves. Cross-stratification.
Sand bars	—	—	Parallel to tidal currents but segregates currents into zones of ebb and flood domination	Parallel to basinal topographic trend and basin axis	Indeterminant	Sand Body

ripples could show the broad topographic trends of a fossil equivalent, however.

Dunes and sand waves (and associated cross-stratification) orientation data are sensitive to the *main flow direction* of tidal currents to which intertidal sand bodies are subjected. The long dimension of the sand bars are sensitive in part to tidal current flow directions, but also control the time-velocity asymmetry of tidal current flow over the bars. Slip faces of dunes and sand waves and maximum dip direction of cross-stratification are also aligned parallel to the long axes of the sand bars, basinal topographic trend and basin axis (fig. 26).

ORIGIN OF THE SEDIMENTARY BEDFORMS AND INTERNAL CROSS-STRATIFICATION

The previous descriptions of sedimentary facies included a brief discussion about bedform morphology. This section will review the origin of the bedforms and their internal cross-stratification (summarized in Table 6).

Tidal current bottom shear forms most of the bedforms on the sand bars. All orders of ripple bedforms form under flow conditions characteristic of the lower flow regime. The flow of the tidal currents is tranquil turbulent, with Froude numbers of bottom currents ranging from 0.01

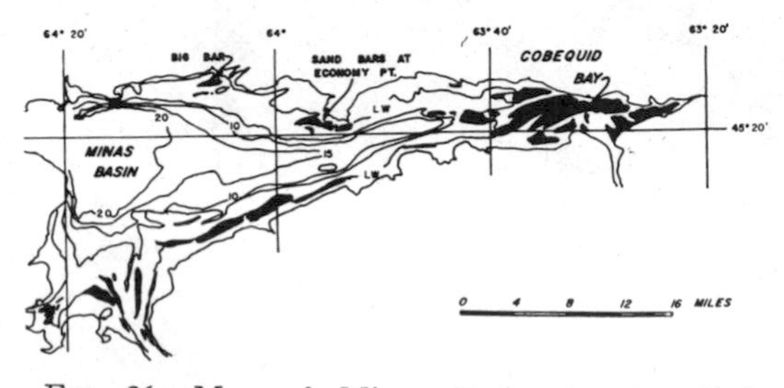

FIG. 26.—Map of Minas Basin showing basin bathymetry and locations and orientations of linear sand bars (black) generated by tidal currents in intertidal zone (contour interval is 5 meters).

to 0.577 (Table 9). Rapid turbulent flow is observed only when the water depths are less than 20 cm at which time some dunes are planed. Rapid flow also produces longitudinal and rhombic ripples on the crest of Big Bar.

Bedform Migration

Migration of dunes and sand waves was established by placing sand stained with acrylic lacquer on bedform slip faces. These bedforms were trenched after a tidal day, after which the stained sand was buried as a cross-stratified layer. Simple dunes and simple sand waves migrated an average horizontal distance of 24 cm per tidal day (two tidal cycles).

Bedform migration is controlled by the time-velocity asymmetry of bottom tidal currents. In ebb-dominated environments, the dunes and sand waves are oriented in a westerly (ebb) direction. Fathometer surveys at high water, slack tide over ebb-dominated zones of all the bars show that the slip faces of dunes and sand waves remain oriented to the west (Ebb) *after the completion of the flood tidal phase.* Such bedforms migrate *only* during the *ebb* phase of the tidal cycle and are stationary during the flood phase. Similarly, in flood-dominated environments, slip faces of dunes and sand waves are oriented east (Flood direction) after the ebb phase, indicating their migration *only* during the *flood* phase. The slip faces of these flood-oriented dunes and sand waves are eroded, rounded and covered with superimposed westerly-oriented current ripples (fig. 27f).

The minimal velocity and associated water depths in which dunes and sand waves migrate during a single phase of a tidal cycle can be determined from continuous 13-hour observations (Table 2) and are summarized in Table 7. This minimal velocity is taken as the maximum velocity of flood currents in ebb-dominated environments, and the maximum velocity of ebb cur-

D
G

TABLE 6.—*Origin of Sedimentary Bedforms, Intertidal Sand Bars, Minas Basin North Shore, Nova Scotia.* (Facies Abbreviations as per Table 3), Locality abbreviations—BB (Big Bar), PF (Pinnacle Flats), EP (Economy Point).

Sedimentary Bedform	Locality and Facies	Sand-Gravel Ratio	Process of Formation
Flute Marks	BB (in stream between CSSD and MSSD Facies)	—	Turbulent scour by open-channel flow
Harrow Marks	BB, SGHM BB, T EP, CBG	2.7 2.2 1.1	Tidal current winnowing of sand from sandy gravel.
Straight, Linguoid, Lunate Current Ripples	All areas, any facies with sand as dominant texture	—	Tidal current bottom shear in water depths less than 0.6 m; open-channel flow; sheet runoff; all under lower flow regime
Rhombic Ripples	BB, C	0	Sheet runoff under upper flow regime.
Simple Dunes	BB, MSSD BB, CSSD PF, CSSD EP, CSSD EP, MSSD	6.0 23.25 4.0 16.1	Tidal current bottom shear under lower flow regime. Water depths in excess of 3.75 m for CSSD and 2.0 m for MSSD.
Complex Dunes	BB, MSCD	0	Tidal Current bottom shear (lower flow regime).
Simple Sand Waves	PF, FSSSW EP, MSSSW	0 0	Tidal current bottom shear under lower flow regime in water depths greater than 5.0 m (for FSSSW).
Complex Sand Waves	PF, CSCSW EC, CSCSW (West Bar)	0 10.2	Vertical accretion of dunes by tidal current bottom shear in direction of upward slope.
	EC, CSCSW (East Bar)	10.2	Wave erosion by high waves.

rents in flood-dominated environments. This figure is justified because no dunes or sand waves migrate during the subordinate phase of the tidal cycle. The data in Table 7 suggest that the linear distance of dune and sand wave migration is minimal for the fine sand, simple sand wave facies and maximal for the coarse sand simple dune facies. Distances of bedform migration appears to increase with increasing grain size and increasing difference between maximum flood and maximum ebb velocity, but to decrease with increasing wave length. Data collected by McLain (1968) and Dinkelman (1968) and the writer confirm these relations.

Because dune and sand wave migration occur

FIG. 27.—Criteria of intertidal sand bar sedimentation. (A) Slip face of dune with horizontal etch marks, medium sand simple dune facies, Big Bar. Scale in cm and decimeters. View to north. Ebb current flow to west (left). (B) Superimposed small current ripples on straight-crested current ripples. Superimposed rippled trend at right angles to larger ripples. Medium sand simple dune facies, Big Bar. View to northwest. Ebb current flow to west (upper left corner). (C) Superimposed, oppositely oriented current ripples on dune. Coarse sand simple dune facies, Big Bar. View to north. Ebb flow to west (left). (D) Rounded erosional upper boundary (reactivation surface) of cross-stratification set, produced by destructional phase of tidal cycle. Coarse sand complex sand wave facies. Trench (No. 63 in Fig. 21) is oriented 300° from right to left. Carpenter scale is 1 meter long. View to north. Ebb current flow to west (left). (E) Rounded erosional crest of dune, coarse sand simple dune facies, Pinnacle Flats. Rounded erosional set boundary (reactivation surface) at arrow produced by destructional phase of tidal cycle. View to south. Ebb current flow to west (right). (F) Sand waves in fine sand simple sand wave facies, Pinnacle Flats showing eastward-oriented slip face and westward-oriented superimposed current ripples. Curvature of crests of current ripples due to change in flow direction prior to emergence. View to north. Flood current flow to east (right). (G) Superimposed small dune (east facing) on eroded westward-facing dune, coarse sand simple dune facies, Pinnacle Flats. Superimposed, eastward-facing dune formed by reversal of ebb current as crest of Pinnacle Flats became exposed. View to north. Ebb current flow to west (left).

TABLE 7.—*Minimal limiting velocities and associated water depths (based on 13-hour continuous data) in which dunes and sand waves migrate*
(Facies Abbreviations as per Table 3).

Facies	Bedform	Minimal Migration Velocity (Cm/sec)	Associated Minimal Depth (m)	Difference in maximum velocities between bottom flood and ebb tide (cm/sec)
MSSD (Ebb-dominated)	Dune	60	2.0	15–23
CSSD (Ebb-dominated)	Dune	45	3.75	20–50
FSSSW (Flood dominated)	Sand Wave	47	5.0	6

only during one phase of the tidal cycle, these bedforms are subjected to a *constructional event* (migration) and a *destructional event* (erosion, rounding, reworking). The destructional phase rounds the slip face and crests of the bedforms and produces a lower angle of dip on the resulting erosional slip face (fig. 28). These rounded bedform surfaces produce a sharp boundary between sets of cross-stratification (figs. 27d, e) and are identical to the reactivation surfaces described by Collinson (1970) and diastems described by Boersma (1969). Such reactivation surfaces are a criterion of reversing tidal scour over dunes and sand waves subjected to a combined constructional-destructional history. Such a combined constructional-destructional history also produces a unimodal orientation of bedforms and cross-strata (Figs. 16, 17, 18).

The constructional and destructional histories of intertidal dunes and sand waves controls some of the quantitative descriptive parameters of their internal cross-stratification. Frequency distributions of thicknesses of sets of cross-strata and of dip angles of cross-strata are shown in Figures 29 and 30. In areas of complex internal organization of bedforms in particular (which are characterized by extensive reworking), the frequency distributions for both parameters are bimodal. The two modes fall in the low (10° to 15°) and high angle (in excess of 25°) range, and in thin sets (3 to 8 cm) and thicker sets (in excess of 12 cm) of cross-strata. These distributions are at variance with published distributions of both set thickness and dip angle, where unimodal distributions are common (Loof and Hubert, 1964; Sedimentology Seminar, 1966; Cazeau, 1960; Pettijohn, 1957; Farkas, 1960; Schwarzacher, 1953). Only Pelletier (1958) reported a different type of distribution, a hyperbolic distribution, of set thicknesses of cross-strata, which were attributed to decreasing set thickness with increasing distance from source.

The bimodal data suggest two sample populations are present (fig. 31), a destructional population (low dip angles, thin sets of cross-strata) and a constructional population (dip angles approaching the angle of repose; thicker sets of cross-strata). Such bimodal distribution may identify sediments subjected to both constructional and destructional events during a complete tidal cycle.

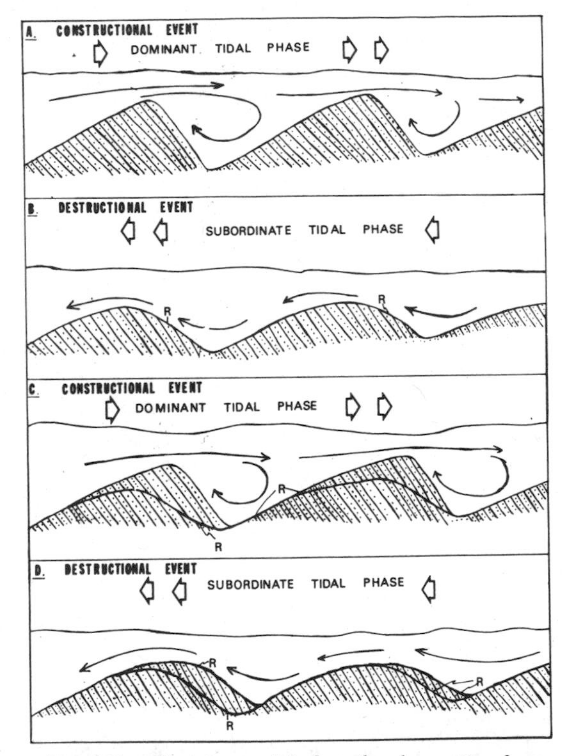

FIG. 28.—Genetic model for development of reactivation surfaces during alternation of dominant tidal phase (constructional event) with subordinate tidal phase (destructional event). During initial constructional event (A), dunes develop by dominant-phase tidal flow and produce internal avalanche cross-stratification. Reversed tidal flow during subordinate phase (B) destroys sharp asymmetry of dunes, producing a subdued, asymmetrical profile and a rounded reactivation surface (R). During the next constructional event (C), dunes build on reactivation surface (R) and develop a superimposed set of avalanche cross-stratification. Destruction of the dune profile occurs during reversed, subordinate-phase tidal flow (D) and produces a second reactivation surface (R).

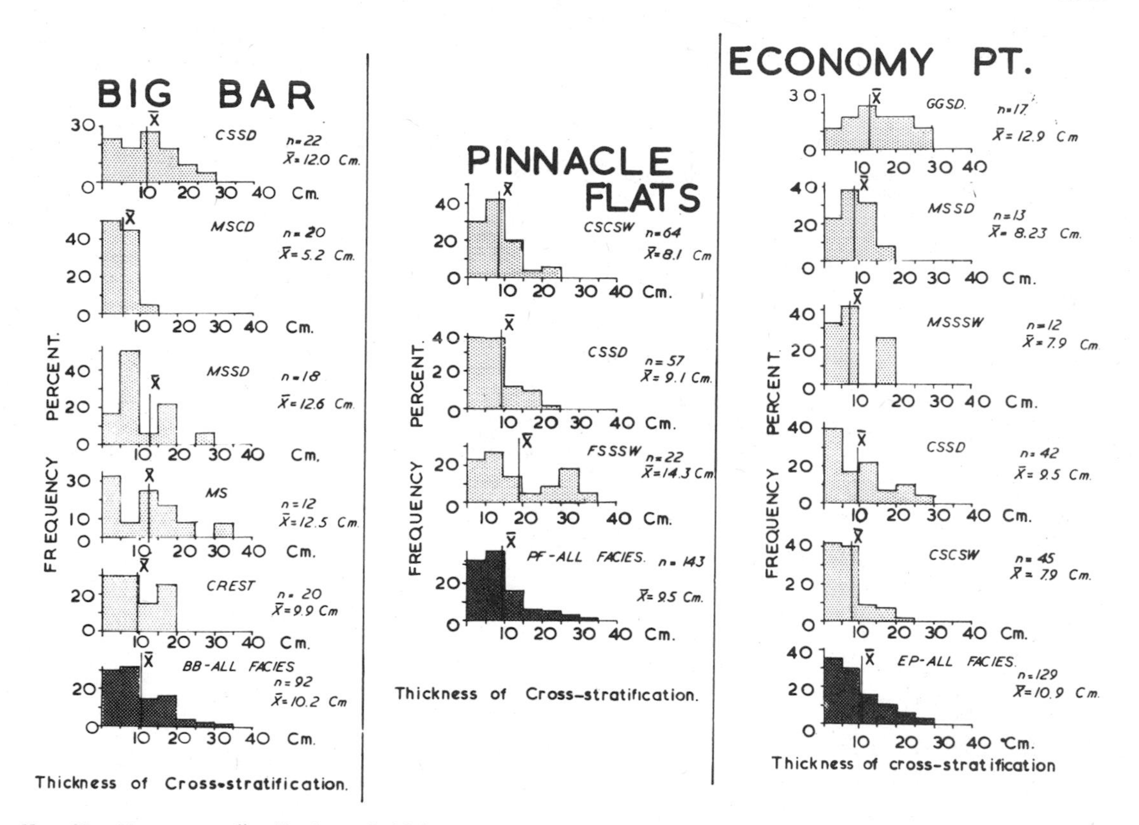

FIG. 29.—Frequency distribution of thicknesses of cross-stratification sets (in cm) for each study area and for each sedimentary facies (Facies code in Table 3).

Reworking of Sand Bars and the Origin of Herringbone Cross-stratification

The amount and depth of reworking of sand was determined in a preliminary experiment by Dinkelman (1968) at the northern end of the coarse sand, complex sand wave facies at West Bar. Trenching after a tidal day indicated that sand stained with an acrylic lacquer (red) was dispersed and reworked downward to a depth of 40 cm (fig. 32a). Only an accessory amount of herringbone cross-stratification was observed in this trench, and all of it was below the 40 cm depth of reworking. Most of the cross-stratification in the trench was oriented west.

The tide tables of the Canadian Hydrographic Survey (1968) indicate that the tidal range during Dinkelman's experiment (August 1, 1968) was 12.1 meters (range midway between neap and spring tide). The writer repeated the experiment, using sand stained with a green lacquer, during the tidal day which included the maximum spring tide of August 9, 1968 (range was 14.6 meters). The depth of reworking during this spring tide was 80 cm (fig. 32b). Depth of reworking in this environment therefore, may be a function of vertical tidal range. The reworked zone included some herringbone cross-stratification.

The paucity of herringbone cross-stratification in the trenches is noteworthy. In most trenches (fig. 27 d, e) good agreement occurs between the orientation of surface ripple bedform and the orientation of cross-stratification *below* the surface set. This relationship is characteristic regardless of trenching in flood- or ebb-dominated environments. Herringbone cross-stratification is rare. These data from the Minas Basin are at variance with observations from intertidal areas in Western Europe (Hulsemann, 1955; Reineck, 1963; Reineck and Wunderlich, 1968; Evans, 1965) where herringbone cross-stratification is common. In the Minas Basin, the paucity of herringbone cross-stratification is a response to the alternating constructional-destructional history of bedform migration and the intense reworking during a single tidal cycle. Herringbone cross-stratification was only observed in trenches during the reduction in tidal range during the lunar tidal cycle between spring and neap tide.

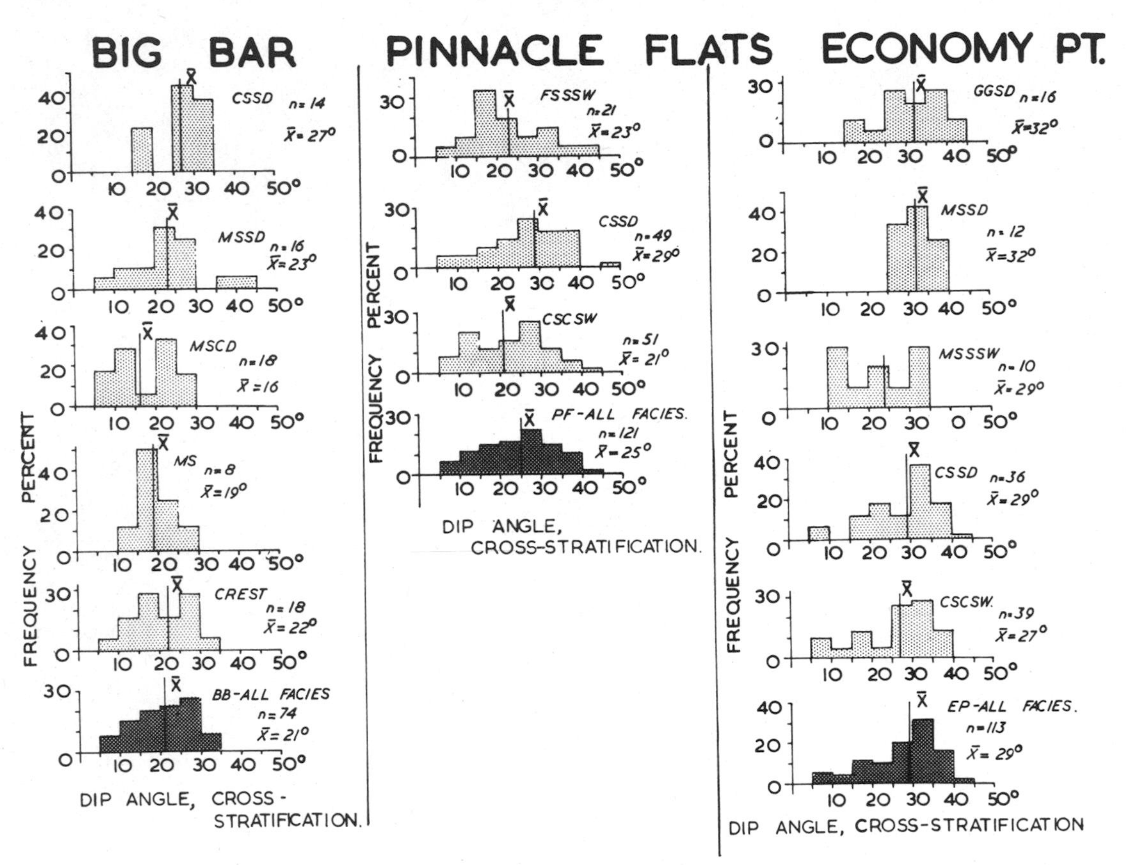

FIG. 30.—Frequency distribution of dip angle of cross-stratification for each study area and for each sedimentary facies (Facies code in Table 3).

Origin of Complex Internal Cross-stratification in Sand Waves

The complex organization of internal cross-stratification in complex sand waves is produced by tidal sediment transport in an upslope direction. The coarse sand complex sand wave facies at Pinnacle Flats and West Bar occur in areas where ebb currents flow to the northwest (figs. 4, 5) and where the upslope direction is also west (cf. figs 9 and 12). The effect of tidal current transport upslope is to reduce the spacing of the dunes in the coarse sand simple dune facies in much the same way as spacing of wave fronts is reduced where wave systems approach a beach. Eventually, the dunes are vertically accreted over each other and merge into a more complex sand wave form. The crests of such sand waves splay laterally into a series of simple dunes (fig. 33a) confirming such a process of sand wave formation. This mechanism is one way to produce sand waves characterized by larger wave heights and internally, complex thinner sets of cross-stratification (fig. 33b) Consequently, cross-stratification thickness in sand waves may not indicate the maximum height of the sand wave bedform in which the cross-stratification occurs.

Trenches (fig. 34c) in the coarse sand, complex sand wave facies at East Bar show that the orientation of internal cross-stratification is opposite to the orientation of the slip face of the surface bedform. The bedform is complex, with an easterly oriented slip face. The bedform crest splays into westerly oriented simple dunes (fig. 34b) which form by bottom shear of ebb-dominated tidal currents (fig. 10; Station No. 10). The sand wave crests, moreover, are oriented subparallel to topographic contours of East Bar and splay into swash bars at the low water line (fig. 34a).

The East Bar at Economy Point occurs 27 km east of the Minas Channel, from which the prevailing winds generate high waves during windy periods. In late August, 1968, unusually strong winds (45 knots) formed waves whose average height was 2 meters. These waves

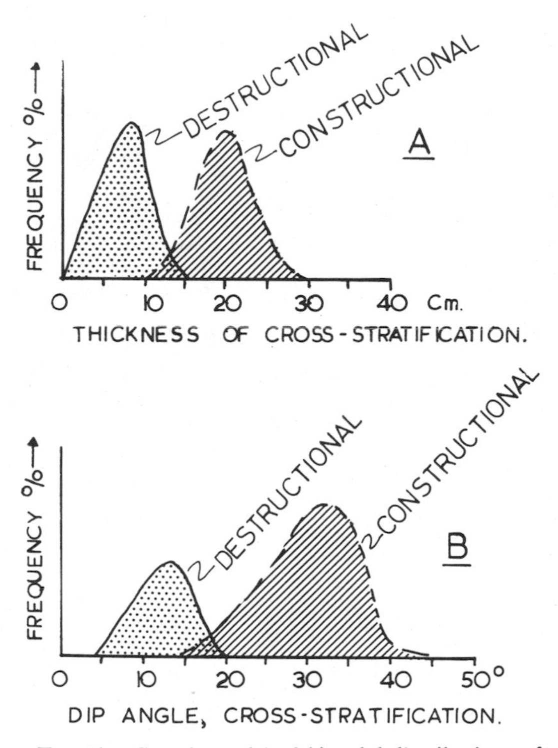

FIG. 31.—Genetic model of bimodal distributions of set thicknesses and dip angles of cross-stratification in tidal environment where bedform migration is subjected to a constructional and destructional cycle.

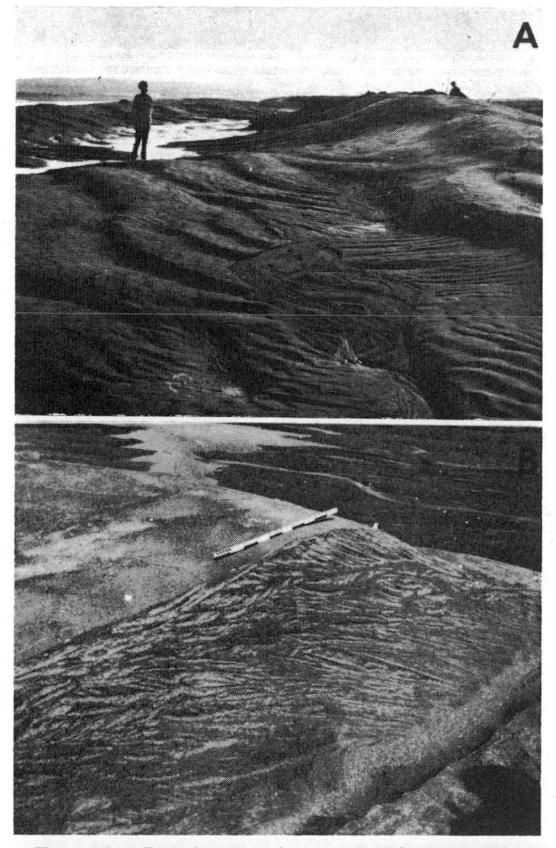

FIG. 33.—Complex sand waves, Pinnacle Flats, Coarse sand complex sand wave facies. (A) Sand wave crest grades laterally into dune crests, indicating accretionary origin of sand waves. View to north. (B) Trench in same sand wave showing that sand wave height is greater than set thicknesses of internal cross-stratification. Internal organization is classified as complex. Scale in cm and decimeters. View to south.

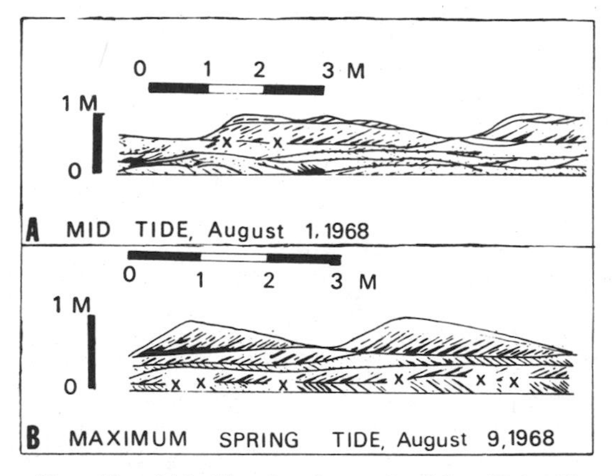

FIG. 32.—(A) Sketch of trench (No. 47 in Fig. 21) in coarse sand complex sand wave facies, West Bar. Trench orientation from left to right is 130°. Depth of reworking of red acrylic lacquer-stained sand during two tidal cycles is marked by "x"s. Middle tidal range, August 1, 1968. (B) Sketch of trench (No. 48 in Fig. 21) in same facies as (A). Trench orientation from left to right is 140°. "x"s mark depth of reworking after two tidal cycles during the period of maximum spring tide, August 9, 1968.

moved in an easterly direction and generated surface wave currents flowing east. These east-flowing wave currents erode the dunes into a sand wave bedform (with slip faces oriented east). Scouring by surface currents occurs only when water depth is less than three meters. The eastward-oriented sand waves formed by this process are thus an *erosional* bedform. Additional evidence confirming their origin by wave processes includes the sub-parallelism of sand wave crests to bar topography built by wave refraction, and the splaying of sand waves into wave-generated swash bars at the level of low tide (fig. 34a). The eastward-oriented slip face possibly is maintained by a combination of wave currents and flood tidal currents. The opposite orientation of the slip faces of the sand waves and their internal cross-stratification at East Bar is produced by a combination of tidal

FIG. 34.—Erosional sand waves, East Bar, Economy Point. (A) Oblique airphoto of East Bar, July, 1966, showing ebb-oriented dunes (d), wave-scoured sand waves (w) which are oriented subparallel to bar topography and which grade laterally into swash bars (b). "X" marks location of Station #10 (See fig. 10 for 13-hour continuous record). Line marks position of trench in C. View to north. Ebb current flow to west (left). (B) Eastward-oriented erosional sand wave that grades into crests of simple dunes; coarse sand complex sand wave facies. View to north. (C) Trench (No. 49–50 in Fig. 21; Line in Fig. 34a) through erosional sand wave, coarse sand complex sand wave facies, East Bar. Slip face dips east (right), whereas internal cross-stratification dips west (left). Scale in cm and decimeters. Trench orientation is 280° from right to left. View to north. Ebb current flow to west (left).

currents (cross-stratification) and wave erosion (slip face of sand wave).

Lateral Change in Bedform Size

Both size and type of bedform are found to change laterally over short distances in the intertidal zone. However, comparison of bottom current velocities (at identical tidal stage) between observation stations shows that lateral changes in bedform size occur over *shorter* distances than lateral changes in bottom current velocities (Table 9).

Table 6 compares the sedimentary bedform type to the ratio of sand and gravel of the sedimentary facies in which it occurs. The harrow marks occur only in sediments where the sand-gravel ration ranges from 1.1 to 2.7. Current ripples, dunes and sand waves are more common where the sand-gravel ratio exceeds 3.2. Straight-crested ripples are more common where the sand-gravel ratio exceeds 3.3, such as in the transition facies at Big Bar. Linguoid and lunate current ripples occur only if the sediment is characterized by sand-gravel ratios in excess of 3.3.

These data, although preliminary, indicate that differing ratios of textural classes control the occurrence and type of sedimentary bedform in the intertidal zone.

Modification of Bedforms by Water Level Fluctuations

Because the intertidal zone is subjected to a combination of changing bottom shear and water level fluctuations, dunes, sand waves and their internal cross-stratification are subjected to minor modifications that produce features not common to other environments. Water level fluctuation (particularly during the ebb phase) produces horizontal etch marks on the surface of intertidal sand bars and slip faces of dunes and sand waves (fig. 27a). These etch marks may extend for linear distances of up to 10 meters.

The etch marks are formed in the following way. As the tide ebbs and progressively exposes different parts of the sand bar, small or large pools form locally. Low-velocity open-channel flow also occurs in troughs of larger bedforms. A slight increase in surface wind velocity forms a small wave pulse on the pool surface. This wave pulse laps against the slip face of the bedform and erodes part of the bedform at water level to an average depth of 3 mm. Combination of continued water level reduction and repeated microwave pulses produces additional erosion. Continuation of the process produces a series of parallel etch marks.

Reduction of water depth controls the size and orientation of current ripples. Current ripples migrate in water depths less than 0.6 meters, whereas dunes and sand waves migrate in greater water depths. With the decline of water level during the ebb phase, the size of current ripples that are able to migrate decreases. Commonly, smaller ripples will be superimposed on slip faces, crests and lee surfaces of dunes and sand waves in response to water level reduction (fig. 27c, f). Because of later changes in flow direction controlled by bottom slope, the slip faces of superposed ripples are oriented parallel, 90 degrees or 180 degrees to the slip face of the larger dunes and sand waves. Such relations are especially well demonstrated by current ripples (wave length averaging 2 cm) superimposed on larger small current ripples (wave lengths averaging 5 cm).

The reduction of water depth also produces paired, or double-crested ripples (Fig. 14b), in which a secondary parallel ripple occurs in the trough of a current ripple. The secondary ripple occurs on the upcurrent side of the trough of an initially-formed, larger current ripple. The spacing between primary and secondary current ripples is identical. The secondary ripple forms after the slightly-larger current ripple forms. Tidal current bottom shear forms the initial current ripple until a critical water depth is reached which prevents further migration. The shear stresses are maintained, however, and are applied within the troughs of the initial ripple, where water depths are slightly greater. The trough depth is sufficiently large so that these shear stresses form the secondary current ripple. The secondary ripple has a smaller wave height. A similar observation of formation of paired ripples was reported by McMullen (1964).

Such paired ripples can form in a second way. On days when wave action is strong, wave currents flow across the intertidal zone during ebb tide and generate paired ripples in response to water level reduction.

PHYSICAL CRITERIA BY WHICH TO RECOGNIZE FOSSIL INTERTIDAL SAND BAR ENVIRONMENTS

The preceeding discussion concentrated on the effect of tidal currents and water level fluctuations on the origin of intertidal bedforms and internal sedimentary structures. Inasmuch as these features are indicators of the intertidal sand bar environment, they may serve as criteria for fossil equivalents. The combination of features which, in the author's view, are criteria of tide-dominated sand bodies are summarized in Table 8.

CONCLUSIONS

This study has demonstrated the following concerning intertidal sand bar deposition on the Minas Basin north shore:

(1). Tidal current transport is the dominant process of sand deposition. The tidal currents form intertidal sand bars and rework them by both tidal current bottom shear and water level fluctuations. Wave action reworks some of the sand and forms large-scale erosional bedforms.
(2). Tidal currents flow eastward during flood phase and westward during ebb phase. Locally, islands or coastal indentations create counter eddies.
(3). Bottom tidal currents are segregated into flood- and ebb-dominated zones over the intertidal sand bars. These zones are defined by the asymmetry of the time-velocity curve of bottom tidal currents during a complete tidal cycle. Flood currents are dominant over the steep facies of the sand bars, whereas ebb currents are dominant over gently sloping bar surfaces. The steep faces act as a flood shield.
(4). Each of the sand bars is subdivided into distinct sedimentary facies defined according to particle size, external bedform and internal organization of cross-stratification. The facies distribution is controlled by bar topography and associated zones of ebb- and flood-dominated bottom tidal currents.
(5). The intertidal sand bars are equilibrium forms and are subjected to a dispersal system of alternating flood and ebb transport. This elliptical pattern of dispersal involves alternate trans-

TABLE 8—*Sedimentary Criteria of Intertidal Sand Bars*

Criterion	Origin
(1) Sharp erosional set boundaries of cross-stratification. (Figs. 15b–f; 27e; 32b; 33c).	Constructional and destructional phases of tidal currents. Stream erosion in troughs of dunes and sand waves at low tide.
(2) Rounded upper set boundaries of cross-stratification (Reactivation Surfaces). (Figs. 15f; 27d, e).	Destructional phase of tidal currents reworking slip faces of dunes and sand waves.
(3) Bimodal frequency distribution of dip angles of cross-stratification (Fig. 30).	Constructional and destructional histories during bedform migration.
(4) Bimodal frequency distribution of thicknesses of cross-stratification sets. (Fig. 29).	Constructional and destructional histories during bedform migration.
(5) Unimodal orientation of dune and sand wave slip faces, and cross-stratification in direction of dominant flood or ebb tidal flow direction, basinal topographic trend and sand body axis. (Fig. 16, 17).	Constructional phase of tidal current flow in ebb- or flood-dominated environment. Migration only during single phase of tidal cycle.
(6) Trimodal distribution of current ripple orientation. (Fig. 18).	Late stage sheet runoff down bar slope.
(7) Small current ripples superimposed at 90° or obliquely on crests of larger current ripples. (Fig. 27b).	Changes in flow direction by late stage sheet runoff while water depth decreases.
(8) Double-crested ripples (Fig. 14b).	Tidal current or wave current bottom shear combined with water level reduction.
(9) Current ripples superimposed at 90° or 180° on crests of dunes and sand waves. (Fig. 14b, f).	Changes in flow direction by late stage sheet runoff prior to bar emergence.
(10) Current ripples superimposed at 90° or 180° on slip faces of dunes and sand waves (Fig. 14f; 27c, f).	Destructional phase of tidal cycle or late stage sheet runoff and open-channel flow across or in troughs of dunes and sand waves.
(11) Complex organization of sets of cross-stratification in sand waves. (Fig. 27d, 32b).	Upslope migration of dunes
(12) Etch marks on slip faces of dunes and sand waves. (Fig. 27a).	Water level reduction associated with erosion by small wave pulses.
(13) Alignment of bar axis with basinal topographic trend and basin axis. (Fig. 26).	Alternate sand transport through flood- and ebb-dominated zones of tidal current transport.

port of sand over steep and gently sloping bar surfaces. Comparison of air photos (low tide) taken in 1963 with ground studies in 1965-1968 confirms the equilibrium distribution of sedimentary facies.

(6). The size of directional properties is heirarchially sensitive to flow directions of tidal currents. Grain dispersal studies show that indicator grains are dispersed in a radial-elliptical fashion from a point source. The longest axes of this radial-elliptical dispersal pattern parallel flood and ebb directions. Current ripples are oriented in response to late stage sheet runoff flow down the slope of sand bars. Such flow produces trimodal distributions of ripple orientation, characterized by a high variance. Slip faces of dunes and sand waves and maximum dip directions of internal cross-stratification are aligned parallel to the main direction of tidal current ebb or flood flow. The frequency distribution of these directional properties is unimodal or bimodal, the two modes being 180° apart.

The long axes of sand bars are oriented parallel to dominant ebb or flood currents, slip faces of dunes and sand waves and maximum dip direction of cross-stratification, basinal topographic trend and basin axis.

(7) The utility of directional properties of the intertidal sand bars for paleocurrent analysis in ancient equivalents reflects this hierarchial sensitivity to flow directions. Small-scale properties such as individual mineral grains are dispersed in all directions of flow in the intertidal environment. Consequently, sand particles become thoroughly mixed into a relatively constant mineral population. Current ripples are characterized by trimodal distributions which are unreliable as indicators of paleocurrent flow, but may

aid in delineating form lines that would outline the external dimensions of the bar.

Dunes, sand waves and internal cross-stratification would delineate paleotidal flow, the long axes of sand bodies, basinal topographic trend and basinal axes. Their across-slope orientation can be used to delineate broad aspects of basin geometry.

The intertidal sand bars, although aligned parallel to tidal current flow, divert and segregate tidal currents into ebb- and flood-dominated zones.

No relationship could be established between the orientation of directional current structures and the location of sediment sources in this particular case.

(8). Dunes and sand waves were observed to migrate only during a single (ebb *or* flood) phase of a tidal cycle, depending on their occurrence in an ebb- or flood-dominated environment. These bedforms are characterized by both a constructional and a destructional history, which produces bimodal distributions of cross-stratification dip angles and set thickness, and rounded upper sets of cross-stratification (reactivation surfaces). It also produces a unimodal distribution of the orientation of dunes, sand waves and internal cross-stratification.

(9). Complex sand waves represent a composite structure produced either by merging of dunes into a sand wave by upslope dune migration, or by resculpturing of bedforms by wave erosion.

(10). Physical criteria which in combination may indicate the intertidal sand bar environment include: sharp erosional contacts between sets of cross-stratification; rounded upper set boundaries of cross-stratification; unimodal and opposite bimodal orientation of cross-stratification; bimodal frequency distributions of set thickness and dip angles of cross-stratification; orientation of dunes, sand waves and cross-stratification in the dominant direction of flood or ebb tidal flow, basinal topographic trend and sand body axis; trimodal orientations of current ripple distributions; oblique or 90° superposition of smaller current ripples on larger current ripples; double-crested ripples; superimposed current ripples at 90° and 180° on dunes, and sand wave, crests and their slip faces; complex organization of internal cross-stratification in sand was; etch marks on slip faces of dunes and sand waves; and alignment of the long axis of the sand bodies parallel to tidal current flow, basinal topographic trend and basin axis.

ACKNOWLEDGEMENTS

Support for this research came from three grants from the National Science Foundation (GA-407; GA-1583; GA-21141) for three summers (1967–68; 1970) of field work and three years of academic-year support. Field work in the summers of 1965 and 1966 was supported through the Visiting Scientist Program of Hudson Laboratories, Columbia University, and was sponsored by the Office of Naval Research through its contract Nonr-266 (84). Special appreciation is extended to Dr. J. E. Sanders for arranging support through Hudson Laboratories as part of their program on accoustical geological mapping.

Many individuals aided in this project. Outstanding boat services and assistance was rendered by Adrien LeBlanc and Wyman Walton of Five Islands, Nova Scotia. Capable field assistance was given by K. R. Aalto, P. H. Close, P. R. Dahlgren, J. R. Hartung, H. H. Hatfield, N. J. O'Shaugnessy, J. H. Way, Jr., and Michael Winters. Mary G. Laub served as Research Fellow and completed all textural analyses. A. R. Smith and Richard McAvoy are thanked for their photographic assistance.

The writer is indebted to members of the NSF-sponsored Advance Science Seminar on intertidal sedimentation in 1968 (NSF Grant GZ-1032) for their stimulus and field discussions. Special thanks are extended to G. V. Middleton for his critical discussions and suggestions and to C. L. Sandusky, M. G. Dinkelmann, J. L. Barr and J. R. McLean for contributing additional data and discussion during the project.

The manuscript benefitted considerably from the editorial suggestions of L. L. Sloss and Keene Swett, who are thanked for their suggestions.

Publication costs were contributed from NSF Grant GA-1583.

REFERENCES CITED

Allen, J. R. L., 1966, On bedforms and paleocurrents: Sedimentology, v. 6, p. 153–190.

Annual Meterological Summary, 1967, Dept. of Transport: Meterological Branch, Halifax, Nova Scotia, 39 p.

Bancroft, J. A., 1902, Ice-borne sediments in Minas Basin, Nova Scotia: Nova Scotian Inst. Sci., Proc., v. 11, p. 158–162.

Belt, E. S., 1964, Revision of Nova Scotia Middle Carboniferous units: Am. Jour. Sci., v. 262, p. 653–673.

——, 1965, Stratigraphy and paleogeography of Mabou Group and related Middle Carboniferous facies, Nova Scotia, Canada: Geol. Soc. America Bull., v. 76, p. 777–802.

Boersma, J. R., 1969, Internal structure of some tidal megaripples on a shoal in the Westerschelde Estuary, the Netherlands: Geol. en Mijnb, v. 48, p. 409–414.

Boothroyd, J. C., 1969, Hydraulic conditions controlling the formation of estuarine bedforms, p. 417–427: *in* Hayes, M.O., editor, 1969, Coastal environments of northeastern Massachusetts and New Hampshire: Eastern Section, Soc. Econ. Paleontologists and Mineralogists Guidebook, 462 p.
Canadian Hydrographic Service, 1968, Canadian Tide and Current Tables, v. 1 (Atlantic Coast and Bay of Fundy): Ottawa, Dept. of Energy, Mimes and Resources, 70 p.
Cazeau, C. J., 1960, Cross-bedding direction in Upper Triassic sandstone of West Texas: Jour. Sedimentary Petrology, v. 30, p. 459–465.
Collinson, J. D., 1970, Bedforms of the Tana River, Norway: Geog. Annaler, v. 52, p. 31–56.
Daboll, J. M., 1969, Holocene sediments of the Parker River estuary. p, 337–355: *in* Hayes, M. O., editor, 1969, Coastal environments of northeast Massachusetts and New Hampshire: Eastern Section, Soc. Econ. Paleontologists and Mineralogists Guidebook, 462 p.
Dennison, J. M., 1962, Graphical aids for determining reliability of sample means and an adequate sample: Jour. Sedimentary Petrology, v. 32, p. 743–750.
Dill, R. F., 1964, Contemporary erosion in the heads of submarine canyons (Abs): *in* Abstracts for 1963; Geol. Soc. America Spec. Paper 76, 341 p.
Dinkelmann, M. G., 1968, Migration of bedforms: Unpub. Report, Advance Science Seminar on Intertidal Zone Sedimentation, 9 p.
Dionne, J. C., 1968, Morphologie et sedimentologie glacielles, littoral sud du Saint Laurent: Annales de Geomorphologie, v. 7, p. 56–84.
Evans, Graham, 1965, Intertidal flat sediments and their environments of deposition in the Wash: Geol. Soc. London Quar. Jour., v. 121, p. 209–245.
Farkas, S. E., 1960, Cross-lamination analysis in the Upper Cambrian Franconia Formation of Wisconsin: Jour. Sedimentary Petrology, v. 30, p. 447–458.
Folk, R. L., 1954, The distinction between grain size and mineral composition in sedimentary rock nomenclature: Jour: Geology, v. 62, p. 344–359.
Folk, R. L., and Ward, W. C., 1957, Brazos River Bar: a study in the significance of grain size parameters: Jour. Sedimentary Petrology, v. 27, p. 3–26.
Gierloff-Emden, H. G., 1961, Luftbild und Kustengeographie am Beispiel der Deutsche Norseekuste: Landeskundliche Luftbildausweitung im Mitteleuropaischen Raum (Bad Godesburg), Heft 4.
Hayes, M. O., Anan, F. S., and Bozeman, R. N., 1969, Sediment dispersal trends in the littoral zone: a problem in paleogeographic reconstruction, p. 290–3p5: *in* Hayes, M, O., editor, 1969, Coastal environments of northeast Massachusetts and New Hampshire: Eastern Section, Soc. Econ. Paleontologists and Mineralogists Guidebook, 462 p.
Hickox, C. F., Jr., 1962, Late Pleistocene ice cap centered on Nova Scotia: Geol. Soc. America Bull., v. 73, p. 505–510.
Hooke, R. L., 1968, Laboratory study of the influence of granules on flow over a sand bed: Geol. Soc. America Bull., v. 79, p. 495–500.
Houbolt, J. J. C., 1968, Recent sediments in the southern Bight of the North Sea: Geol. en. Mijnb., v. 47, p. 245–273.
Hulsemann, Jobst, 1955, Grossrippeln und Schragschichtungs-Gefuge im Nordsee Watt und in der Molasse: Senck, Leth., v: 36, p. 359–388.
Imbrie, John, and Buchanan, Hugh, 1965, Sedimentary structures in modern carbonate sands of the Bahamas, p. 149–172: *in* Middleton, G. V., editor, 1965, Primary sedimentary structures and their hydrodynamic interpretation: Soc. Econ. Paleontologists and Mineralogists Spec. Pub. No. 12, 265 p.
James, N. P. and Stanley, D. J., 1968, Sable Island Bank off Nova Scotia: sediment dispersal and Recent history: Am. Assoc. Petroleum Geologists Bull., v. 52, p. 2208–2230.
Joplin, A. V., 1965, Hydraulic factors controlling the shape of laminae in laboratory deltas: Jour. Sedimentary Petorolgy, v. 35, p. 777–791.
Karcz, Iaakov, 1967, Harrow marks, current-aligned sedimentary structures: Jour. Geology, v. 75, p. 113–121.
Klein, G. deV., 1960, Stratigraphy, sedimentary petrology and structure of Triassic sedimentary rocks, Maritime Provinces, Canada: Unpub. Ph.D. dissertation, Yale Univ., 302 p.
———, 1962, Triassic sedimentation, Maritime Provinces, Canada: Geol. Soc. America Bull., v. 73, p. 1127–1146.
———, 1963, Bay of Fundy intertidal zone sediments: Jour. Sedimentary Petrology, v. 33, p. 844–854.
———, 1967, Paleocurrent analysis in relation to modern marine dispersal patterns: Am. Assoc. Petroleum Geologists Bull., v. 51, p. 366–382.
———, 1968, Intertidal zone sedimentation, Minas Basin north shore, Bay of Fundy, Nova Scotia, p. 91–107: *in* Margulies, A. E., and Steere, R. C., editors, 1968, National Symposium on Ocean Sciences and Engineering of the Atlantic Shelf: Marine Tech. Soc. Trans., 366 p.
Laub, Mary G., 1968, The origin and movement of gravel bars in the intertidal zone of Parrsboro Harbour, Nova Scotia; Unpub. M.S. thesis, Univ. of Pennsylvania, 41 p.
Leopold, L. B., Wolman, M. G., and Miller, J. P., 1964, Fluvial processes in geomorphology: San Francisco, Freeman and Co., 522 p.
Loof, K. M., and Hubert, J. M., 1964, Sampling variability in the Post-Myrick Station Channel Sandstone (Pennsylvanian), Missouri: Jour. Sedimentary Petrology, v. 34, p. 774–776.
McLain, J. R., 1968, Bedform migration, Pinnacle Flats, Nova Scotia: Unpub. Report, Advance Science Seminar on Intertidal Zone sedimentation, 16 p.
McMullen, R. M., 1964, Modern sedimentation in the Mawddach Estuary, Barmouth, North Wales: Unpub. Ph.D. dissertation, Univ. of Reading (UK), 399 p.
Newton, R. S., and Werner, Friedrich, 1969, Luftbildanalyse und Sedimentgefüge als Halfsmittel fur das Sandtransportproblem im Wattgebeit vor Cuxhaven: Hamburger Kustenforschung, Heft 8, 46 p.
Pelletier, B. R., 1958, Pocono Paleocurrents in Pennsylvania and Maryland: Geol. Soc. America Bull., v. 69, p. 1033–1064.

PETTIJOHN, F. J., 1957, Paleocurrents of Lake Superior Precambrian quartzites: Geol Soc. America Bull., v. 68, p. 469–480.
PHLIPPONNEAU, M., 1955, La Baie du Mont-Saint-Michel: Mem. Soc. Geol. et Min. de Bretagne, v. 11, 215 p. (Not Seen).
POTTER, P. E., 1962, Regional distribution patterns of Pennsylvanian sandstones in Illinois Basin: Am. Assoc. Petroleum Geologists Bull., v. 46, p. 1890–1911.
POTTER, P. E., AND MAST, R. F., 1963, Sedimentary structures, sand shape fabric and permeability: Jour. Geology, v. 71, p. 441–471.
POTTER, P. E., NOSOW, EDMUND, SMITH, N. M., SWANN, D. H., AND WALKER, F. H., 1958, Chester cross-bedding and sandstone trends in Illinois Basin: Am. Assoc. Petroleum Geologists Bull., v. 42, p. 1013–1046.
POTTER, P. E., AND PETTIJOHN, F. J., 1963, Paleocurrents and basin analysis: New York, Springer Verlag.
PRITCHARD, D. W., AND BURT, W. V., 1951, An inexpensive and rapid technique for obtaining current profiles in estuarine waters: Jour. Marine Research, v. 10, p. 180–189.
REINECK, H. E., 1963, Sedimentgefüge im Bereich der sudlichen Nordsee: Abh. Senck. Natursfors. Gesell., No. 50, p. 1–138.
REINECK, H. E., AND WUNDERLICH, FRIEDRICH, 1968, Zur unterscheiding von assymetrischen Oszillationsrippeln und Stromungsrippeln: Senck. Leth., v. 49, p. 321–345
SANDUSKY, C. L., 1968, Sand distribution from a point source: Unpub. Report, Advance Science Seminar on Intertidal Zone Sedimentation, 16 p.
SCHWARZACHER, W., 1953, Cross-bedding and grain size in the Lower Cretaceous Sands of East Anglia: Geol. Mag., v. 90, p. 322–330.
SEDIMENTOLOGY SEMINAR, 1966, Cross-bedding in the Salem Limestone of central Indiana: Sedimentology, v. 6, p. 95–114.
SELLEY, R. C., 1968, A classification of paleocurrent models: Jour. Geology, v. 76, p. 99–110.
SIMONS, D. B., RICHARDSON, E. V., AND NORDIN, C. F., JR., 1965, Sedimentary structures generated by flow in alluvial channels, p. 34–52: *in* Middleton, G. V., editor, 1965, Primary sedimentary structures and their hydrodynamic interpretation: Soc. Econ. Paleontologists and Mineralogists Spec. Pub. No. 12, 265 p.
SMITH, J. D., 1969, Geomorphology of a sand ridge: Jour. Geology, v. 77, p. 39–55.
SMITH, R. S., 1969, Sedimentology of the deltas of the Moose and Diligent Rivers, Minas Basin, Bay of Fundy, Nova Scotia: Unpub. MS thesis, Univ. of Pennsylvania, 112 p.
SWIFT, D. J. P., 1966, Bay of Fundy, p. 118–119: *in* Fairbridge, R. W., editor, Encyclopedea of Oceanography: New York, Van Nostrand.
SWIFT, D. J. P., and BORNS, H. W., JR., 1967, A raised fluviomarine outwash terrace, north shore of the Minas Basin, Nova Scotia: Jour. Geology, v. 75, p. 673–711
SWIFT, D. J. P., AND LYALL, ANIL, 1968, Origin of the Bay of Fundy, and interpretation from sub-bottom profiles: Marine Geology, v. 6, p. 331–343.
TANNER, W. F., 1955, Paleogeographic reconstructions from cross-bedding studies: Am. Assoc. Petroleum Geologists Bull., v. 79, p. 2471–2483.
———, 1959, The importance of modes in cross-bedding data: Jour. Sedimentary Petrology, v. 29, p. 221–226.
VAN STRAATEN, L. M. J. U., 1950, Giant ripples in tidal channels: Kon. Ned. Aadr. Genoots. Tijds., v. 67, p. 76–81.
———, 1952, Biogene textures and the formation of shell beds in the Dutch Wadden Sea: Woninkl. Ned. Akad. Wetenschap., Proc., Ser. B., v. 55, p. 500–516.
———, 1954, Composition and structure of Recent marine sediments in the Netherlands: Leidse Geol. Meded., v. 19, p. 1–110.
VAN STRAATEN, L. M. J. U., AND KUENEN, Ph.H., 1957, Accumulation of fine-grained sediments in the Dutch Wadden Sea: Geol. en. Mijnb., v. 19, p. 329–354.
———, AND ———, 1958, Tidal action as a cause of clay accumulation: Jour. Sedimentary Petrology, v. 28, p. 406–413.
VAN VEEN, JAN, 1950, Eb- en vloedschaar systemen in der Nederlandse getijwatern: Kon. Ned. Aadr. Gen. Tijds., 2 e reeks, v. 67, p. 43–65.
VERGER, FERNAND, 1968, Marais et wadden du littoral Français: Bordeaux, Biscaye Freres, 541 p.
WEEKS, L. J., 1948, Londonderry and Bass River map areas, Colchester and Hants Counties, Nova Scotia: Geol. Survey of Canada Mem. 245, 86 p.
YASSO, W. E., 1966, Formulation and use of fluorescent tracer coatings in sediment transport studies: Sedimentology, v. 6, p. 287–307.

Editor's Comments on Paper 15

These next two papers in our collection are authored by Cuchlaine A. M. King, noted British geographer and coastal geomorphologist. King received a Degree in Geography from Cambridge University in 1943; and was with the meteorological Branch of WRNS from 1943 to 1946. She returned to Cambridge in 1946, investigated the movement of beach sands and sand bar formation, and received her Ph.D. for this work in 1949. After a year as Demonstrator in Geography at Durham University, Dr. King was appointed to the Geography Department at Nottingham University in 1951. She was promoted to Reader in 1961, and has been Professor of Physical Geography there since 1969. In addition Dr. King has been Visiting Lecturer in Geography, Canterbury University, Christ Church, New Zealand, 1956; Visiting Professor, Univeristy of Colorado, Boulder, Institute of Arctic and Alpine Research, 1969; and NSF Senior Visiting Scientist, State University of New York, Binghamton, 1972. She is widely known too for her books: "Beaches and Coasts" (London, E. Arnold, 1959), "An Introduction to Oceanography" (New York, McGraw-Hill, 1962), and "Techniques in Geomorphology" (London, E. Arnold, 1966). A second edition of "Beaches and Coasts" will be available in 1972.

In this first paper C. A. M. King presents an eighteen year study of changes in the spit at Gibraltar Point, Lincolnshire, England. Documented observations of spit morphology and sediments from 1951 to 1969 reveal the speed with which major alterations of the coast have taken place. These changes point up the value of such studies as this by King and the one by A. P. Carr included earlier in the volume. After reading these two reports one wonders if it would not be possible to project spit development into the future by a consideration of all the variables involved in their growth, as King mentions having done in 1961. This naturally leads to our next paper wherein Dr. King attempts just that with the assistance of modern computer science.

CHANGES IN THE SPIT AT GIBRALTAR POINT, LINCOLNSHIRE, 1951 to 1969

CUCHLAINE A. M. KING

15

INTRODUCTION

In 1957 an account was published of the development of the spit at Gibraltar Point,[1] based on surveys covering the previous six years. The surveys have since been continued over a further 12 years, giving a continuous record that covers more than 17 years. It seems appropriate now to review the whole series of observations and to relate the changes recorded to the processes that bring them about.

Before the more recent observations are considered, the earlier results should be reviewed briefly. The spit prolongs the north–south coast where it turns sharply westwards into the embayment of the Wash. It is a small spit, and was only just over 1,800 feet long when first surveyed in 1951. C. Kidson[2] commented upon its small size, which can be accounted for by the general pattern of coastal change on this coast of accretion, which extends for about three miles from the southern end of Skegness to Gibraltar Point. Along this stretch of coast there are a number of different environments that depend on the dominance of different formative processes.

Offshore sand banks formed by the tidal streams play an important part in determining the details of the pattern of accretion on the foreshore, which in turn is closely related to the development of the spit. The offshore banks form between the interdigitating ebb and flood tidal stream channels along which sediment is carried in the direction of the residual of the tidal streams.[3] This sediment is eventually brought to the foreshore where it can be deposited by the waves, which are the dominant agents on the beach. The sediment is built up into beach ridges on those parts of the beach where there is a surfeit of material reaching the foreshore. These beach ridges move landwards up the beach under the influence of the waves as they diverge slightly from the coast in an offshore direction southwards, and move bodily southwards with the general direction of wave transport along the coast.[4] As they approach the upper beach they may become stabilised through the growth of further ridges to seaward which allows wind-blown sand to accumulate upon them round sand-loving vegetation. They then form foredunes, separated from the coast by muddy strips of salt marsh. Thus an eastward outgrowth of the coast is brought about in the area where most material is reaching the foreshore.

Records of past changes in the offshore banks indicate that their movement is associated with the changing pattern of accumulation along this stretch of coast. There is also evidence of the varying position of maximum accretion in the pattern of dune ridges that is revealed on vertical aerial photographs of the area. The spit is intimately associated with the pattern of accumulation, for it represents its southward extension, being built from the material that passes southwards from the growing ness of sandy foredunes. Therefore, as soon as a new major ness of accumulation starts to develop a new spit starts to form at its southern end and the present spit is only one of a series of such features that have developed during the last few centuries. Such a former spit was illustrated on Armstrong's map of Lincolnshire, drawn in 1779.

Occasional storms also leave their marks on the coast by breaking through old dune ridges and spreading sand over the marsh deposits behind the dunes. One such storm preceded the initiation of the present spit's formation in 1922, when the tip of the main eastern dune ridge was pierced and sand washed over the mature marsh that had formed in its shelter. Since that date further dune ridges have developed on the foreshore eastward of the main eastern dune ridge, and it is from the southern end of these ridges that the present spit extends. In its shelter the new marsh is developing.

Observed Changes in Spit Morphology

The spit is linked to the dune ridge by a broad low proximal section, across which sand is washed into the new marsh at very high tide. At its distal end the spit is higher and narrower. The distal end has been surveyed by plane table on 20 occasions between May 1951 and May 1969, mostly in the late spring, with the assistance of undergraduate students specialising in geomorphology at Nottingham. The surveys record the area of the spit above each one-foot contour from 10 feet O.D. (Liverpool) and the position of the contours relative to fixed posts. The relevant data are assembled in Table I. The areas above the various contours have been used to calculate the volume of the spit above the 10 feet contour. The table also records the length of the spit at the different levels, and the horizontal movement of the crest of the spit at the site of the fixed posts.

TABLE I

Measurement of spit growth

Date	Highest point, feet	Volume, 1,000 cubic feet	*Area in 1,000 square feet* Above 10 feet	Above 11 feet	Above 12 feet	Above 13 feet	Above 14 feet	Horizontal movement of crest, feet	*Length in feet from fixed pegs* To 10 feet	To 11 feet	To 12 feet
May 1951	12.8	105.9	128.5	44.3	6.7	0	0				
Aug. 1951	12.95			46.3	9.9	0	0				
April 1952	11.9	123.6	187.9	45.8	0	0	0				
Aug. 1952	11.7	120	207.2	43.8	0	0	0	0	240	120	
April 1953	12.5	168.6	179.3	78.5	8.3	0	0	10	460	300	130
April 1954	11.9			60.7	0	0	0	20	620	160	
June 1955	11.9	243.6	329.7	100.3	0	0	0	20	570	220	
April 1957	12	303.8	289.8	153.6	7.1	0	0	20	630	550	400
May 1958	11.4		298.6	153.1	0	0	0	0	610	430	
Nov. 1958	12	244.6	266.3	116.5	1.6	0	0	−20	570	390	40
May 1959	13.16	295.6	256.3	128.6	52.8	4.9	0	0	720	710	540
May 1960	12.44	289.9	238.1	144.9	36.1	0	0	50	620	440	280
June 1961	13.16	356.4	217.3	157.1	103.6	11.6	0	60	800	780	600
May 1962	13.07	348.7	250.0	155.4	87.2	13.0	0	110	760	480	350
April 1963	13	339.3	266.7	143.1	89.3	6.9	0	170	560	460	390
April 1964	12.6	457.4	389.0	203.0	107.0	0	0	150	990	590	320
May 1966	13.59	511.4	260.0	206.0	173.0	140.0	0	150	940	920	910
April 1967	14.23	483.0	226.5	176.0	136.5	58.8	15.6	180	830	810	790
April 1968	15.12	514.0	279.0	189.5	123.4	60.2	25.1	190	760	740	720
May 1969	14.26	436.5	229.1	163.7	130.0	74.5	0.3	210	820	790	716

During the years over which measurements have been made the distal end of the spit has moved a considerable distance westwards. This movement has been such that one fixed post which was originally on the landward side of the spit was buried by its landward movement and then later reappeared on its seaward side. New fixed posts have had to be put on the spit crest. The trends of the changes of the various measurements throughout the period are shown in Figure 1.

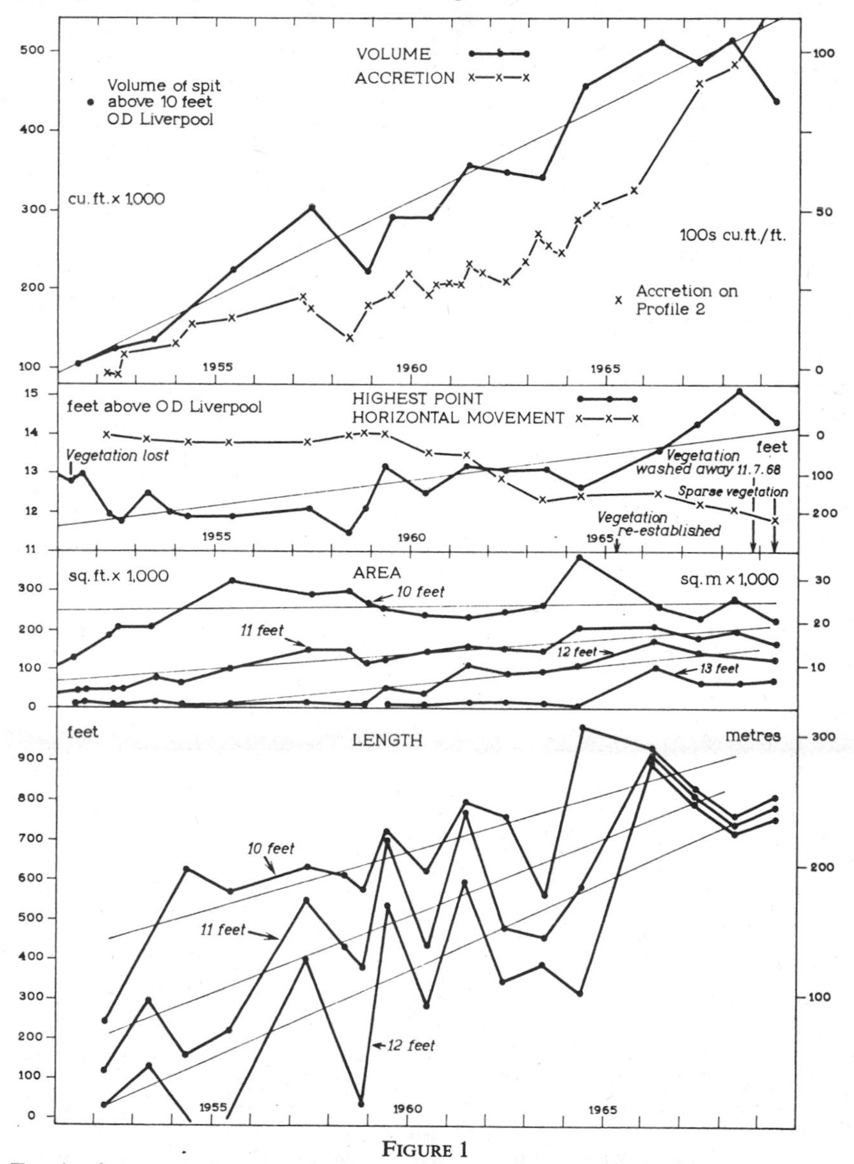

FIGURE 1

Trends of changes in the volume, area, length, height and horizontal movement of the spit at Gibraltar Point

Accretion on profile 2 is also shown. The trend lines have been calculated and their equations are given in Table 2. Measurements refer to the distal end beyond the line marking the top margin of Figure 2

21

The trend equations have been calculated for most of the changes shown in Figure 1. The equations are given in Table II. The trend equation is given in the form $Y = a + bX$. a is given by $\Sigma Y/n$, and b is given by $\Sigma XY/\Sigma X^2$, where X refers to the years numbered from the central year of the sequence, with negative values for the first half and positive values for the second half. The values for the a coefficient in the equation give the mean value for the variable in the middle of the period, which was mid-1960. The b coefficient gives the rate of change in the variable annually, as the trends have been calculated for the annual values. For the years 1956 and 1965, when observations were not made, interpolated values have been used. This method of trend analysis provides least square best fit trend lines, which have been entered on Figure 1.

TABLE II

Trends of spit growth and profile accretion

	Trend equation	
	1952–68	*1951–69*
Spit maximum height	Y = 12.8 + 0.154 X	Y = 12.85 + 0.134 X
Spit length		
at 10 feet ..	Y = 686 + 30 X	
at 11 feet ..	Y = 525 + 40 X	
at 12 feet ..	Y = 383 + 47.5 X	Y = 384 + 46.8 X
Spit area		
above 10 feet ..	Y = 269 + 2 26 X	Y = 259 + 1.56 X
above 11 feet ..	Y = 143 + 8.75 X	Y = 139 + 8.07 X
above 12 feet ..	Y = 62 + 11.6 X	Y = 62.3 + 9 9 X
Spit volume		
above 10 feet ..	Y = 315 + 24 9 X	Y = 310 + 24.4 X

Profile accretion 1953–69:

Profile 1	Y = −1 + 43.42 X
Profile 2	Y = 826 + 114.68 X
Profile 4	Y = 1504 + 190.71 X

All the trend coefficients given in Table II are positive, which indicates that the spit has been gaining in height, length, area and volume throughout the period of detailed surveys. The increase in height has not been uniform, but has occurred in steps. After being relatively high in level when surveys began, with vegetation growing on its crest, the spit was lowered, to reach a minimum elevation in 1958. Subsequently, it has increased in height, rapidly at first and then more slowly until a rapid upward growth started again in 1964. Small amounts of vegetation have been recorded on the spit each year since 1965; and all vegetation was washed from the spit during a very high tide on 11th July 1968.

The increase in height has been more rapid than the increase in area above 10 feet, which has been very slow, amounting to a mean of only 1.56 × 1,000 square feet. The increases in the areas above 11 and 12 feet have been faster and by 1966 there was a substantial area above 13 feet. There is a close correlation between the area above 11 feet and the volume of the spit, which has increased fairly steadily during the period of observation. The difference between the calculated trend value and the measured surveyed value has a mean of 22.8 × 1,000 cubic feet. This shows that material has been reaching the spit at a fairly constant rate, with rather more than the mean amount arriving during the periods 1955 to 1957, and 1964 to 1966, and slightly less than the mean amount in the intervening periods.

The actual form of the spit, however, as indicated by its length and shape, has varied considerably more. The trend lines for the length of the spit show that the higher contours have lengthened more rapidly than the lower,

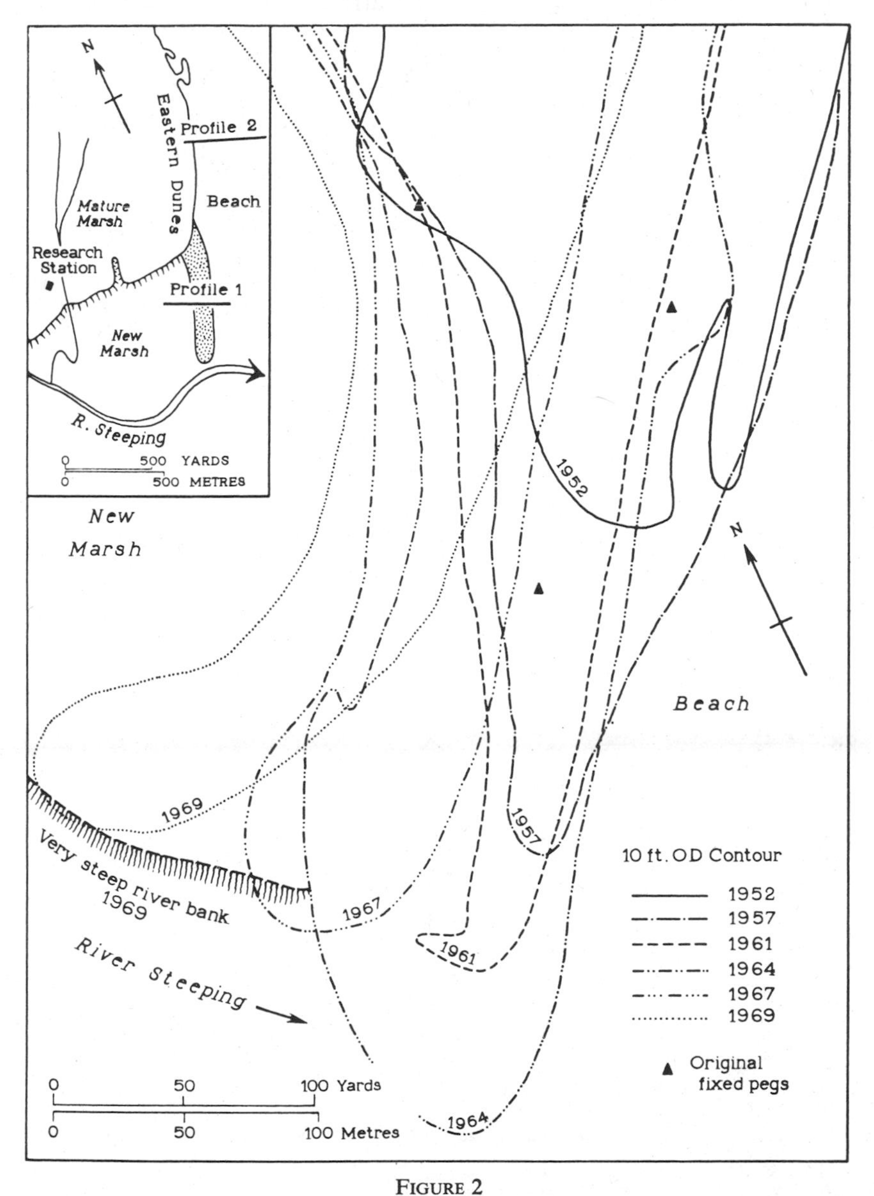

FIGURE 2

Selected 10-foot contours, surveyed in 1952, 1957, 1961, 1964, 1967 and 1969, to illustrate the changing form and position of the spit at Gibraltar Point

in the same way as the area. But there has been a much greater oscillation in the length of the distal portion and steepness of the distal point of the spit. The mean variation of the length at 11 feet from the trend value is 115.6 feet, and there have been rapid oscillations above and below the mean trend value. In general the three contours have varied together, and the trend lines approach each other towards the end of the period, reflecting an increase in the steepness of the distal end of the spit. During the three years 1965–68 the tip of the spit was very steep and was reduced steadily in length. This was caused by the swinging of a meander of the Steeping river across the marsh behind the spit until it actively eroded the distal tip of the spit. The river has now cut through the beach direct to the sea, washing the distal tip of the spit as it flows past. This reduction of length was more than compensated for by an increase in the height and area of the spit at the upper levels, so that the volume continued its upward trend until 1969, when the spit was lower and narrower once more.

The earlier extensions of the spit marked the arrival of ridges from the north and contractions with their passage further south. During these oscillations the spit varied in shape as indicated by the examples of its morphology, shown in Figure 2. At times the spit has had a recurved end, but at other times it has had a simpler form. The recurves probably form during periods of strong easterly winds, which at high tide can move material west from the tip of the spit.

THE NATURE OF THE SEDIMENT

The spit is formed predominantly of sand, but it also includes some shingle, while mud accumulates on the new marsh in its shelter and on the lower foreshore of the beach in front of it. Samples of sediment have been collected from the spit and along the neighbouring beach profiles. The details of the graphic moment measures[5] for these samples are given in Table III, and Figure 3 illustrates some points concerning the characteristics of the sediments.

TABLE III

Sediment samples, graphic moment measures (all values in Ø units)

	Mean	*Sorting*	*Skewness*	*Kurtosis*	*Median*
Profile 1					
Seaward side spit crest	1.89	0.755	−0.77	1.34	2.16
Swash slope	2.14	0.67	−0.74	2.24	2.27
Lower beach below swash slope ..	2.29	0.395	−0.05	1.52	2.30
Low water beach level	5.07	2.03	+1.04	1.51	4.12
Profile 2					
Dune crest with vegetation ..	1.55	0.80	−0.445	1.45	1.85
Beach ridge no vegetation ..	1.13	1.34	−0.55	0.815	1.70
Top of swash slope	1.98	0.45	−0.30	1.04	2.07
Seaward ridge (landward side) ..	1.81	1.23	−0.83	1.50	1.85
Seaward ridge (seaward side) ..	1.85	1.09	−0.59	2.74	2.13
Muddy runnel (landward side) ..	3.31	2.00	+0.77	1.52	2.40
Muddy runnel (seaward side) ..	4.36	2.285	+0.46	0.54	3.60

The sediments fall into three groups, on the basis of their characteristic moment measures. The first group comprises those sediments characteristic of the ridge crests, dune ridges and the spit crest. These are distinctly coarser than those of the other groups, less well sorted and with a stronger negative skewness, indicating a tail of coarse particles. These characteristics result from the fact that the sediment is brought by wave action, which can move the coarser particles to the ridge crests, where they accumulate amongst a considerable amount of finer wind-blown material. The lag deposit of coarser material gives the negative skewness and the large mean value.

The sample from the spit crest is less coarse than those from the other ridge crests because the shingle on the spit was not included in the sample. However, this shingle has been separately analysed for roundness, using stones with a mean length of 40 centimetres, which is characteristic of those on the spit crest. The mean roundness of the stones, using Cailleux's roundness index,[6] is 320, with a standard deviation of 125. The histogram of the distribution is shown in Figure 3. There are a number of very well rounded pebbles that are far-travelled erratics, derived from the till exposed on the foreshore to the north at Ingoldmells and Sutton-on-Sea and possibly also from further north, in Holderness. The pebbles of lower roundness, of which there are a fair number, are angular pieces of flint that have been shattered fairly recently by frost action. They also must have been largely derived from the drift, although some could have been brought down by the River Steeping or be material used in temporary sea defences in 1953. The pebbles are not as well rounded as many beach shingle stones, probably owing to the relatively weak wave action on this coast, which is protected from the most violent wave action by offshore sand banks, which dry out at low water.

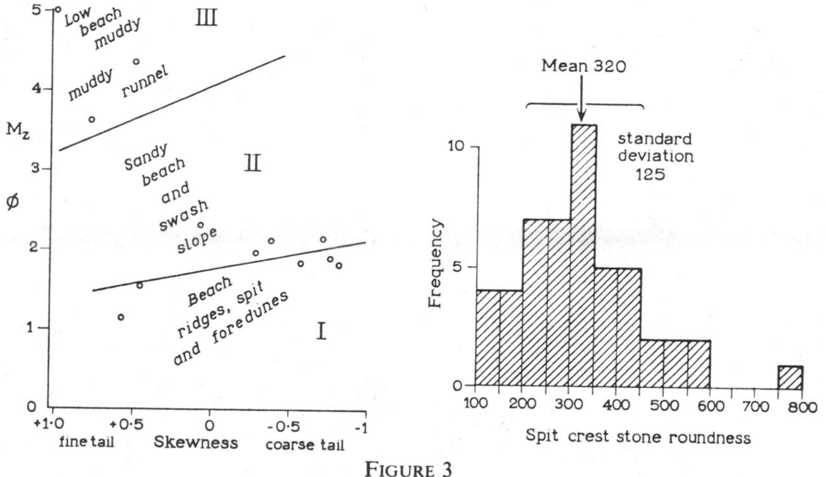

FIGURE 3

The graphic mean and skewness of the sediments at Gibraltar Point, showing the three main sediment groups

The histogram of the roundness of the pebbles on the spit crest is shown

The second group of sediments are those from the swash slope and sandy beach. They are intermediate in size, consisting of pure sand, with a mean size of about 2 ø.[7] They are better sorted and still have a negative skewness, but rather smaller than that of the first group. This signifies the sorting effect of sustained wave action which has removed the finer particles that cannot settle in the vigorous environment in which they accumulated.

The third group includes those sediments that have collected in the quietest conditions in the runnels, protected from waves by the beach ridges, and on the lower foreshore, which is protected from wave action at low water by the drying banks offshore. These sediments have a positive skewness and are much finer than those in the more vigorous environment. Their sorting is poorer than that of the swash slope samples, as revealed by the bimodal nature of the curves, which include both a silt and clay fraction and a sand fraction.

Accretion on the Foreshore

The intimate relationships between the build-up of the spit and the accretion on the foreshore to the north depends on the fact that material is supplied to the spit from the beach to the north by wave action. The trend in the accretion on the foreshore and its relationship to the form of the beach profile will, therefore, be discussed briefly. The trend of accretion of material on three of a series of long-surveyed profiles down the beach is shown in Figure 4. Profile 1 runs seaward from the crest of the spit; profile 2 lies about half a mile to the north; and profile 4 is at the southern end of Skegness in the zone where accretion is at a maximum at present. A new ness of foredunes and marsh slacks has developed in this area.

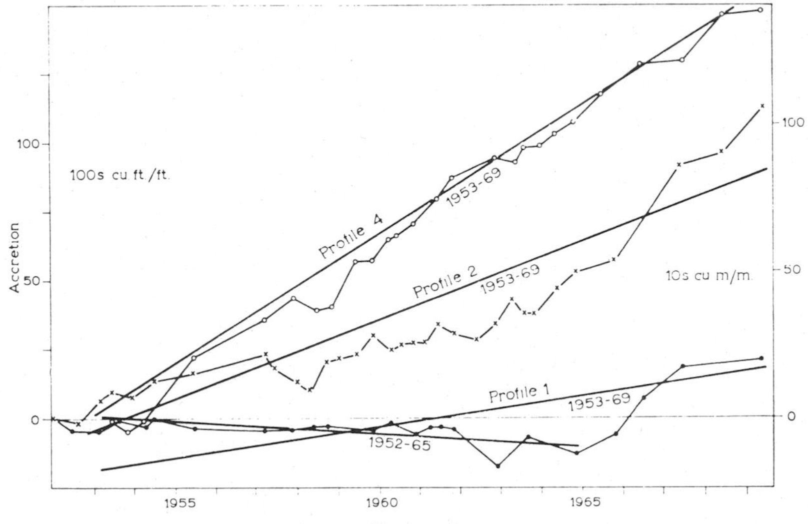

Figure 4

Trends of accretion on profiles 1, 2 and 4 for the period from 1953 to 1969. The trend lines have been inserted

The trend lines in Figure 4 show that profile 4 has gained a large amount of material, and profile 2 a smaller amount, but profile 1 has gained relatively little, an upward trend starting there only in 1966. The ridges are highest on profile 4 (Figure 5), and lowest on profile 1. This relationship is probably a consequence of the action of the waves, which are attempting to build up a gradient suitable to their size on an overall gradient that is flatter than the equilibrium swash slope gradient. The more material there is on the beach, the flatter the overall gradient, and hence the larger the size of the ridges built by wave action.

The relationship between the size of ridge and the amount of accretion is shown in Figure 5, on which the three profiles are differentiated by distinctive symbols. The regression lines for the three individual profiles are shown, together with that for the combined data, derived by analysis of co-variance. This technique allows the ridge height and amount of accretion to be correlated at the same time as variation due to the three different profiles is controlled. The co-variance test also allows the interaction between the three profiles to be assessed. If the interaction is not significant then the three profiles can be pooled to give a stronger correlation between the two variables.

Theoretically, as mentioned above, a correlation between the amount of accretion and the ridge size is to be expected. The mean ridge size differs between the three profiles; it is 4.4 feet on profile 4, 1.9 feet on profile 2 and 1.45 feet on profile 1. The mean accretion is 16.2 $\times$ 10^5 cubic feet/feet on profile 4, 9.7 on profile 2, and only 2.4 on profile 1. There is a fairly strong correlation between ridge height and accretion on profiles 4 and 2, the coefficient of correlation, *r*, having values of 0.747 and 0.754 for profiles 2 and 4 respectively. The correlation for profile 1 is not significant.

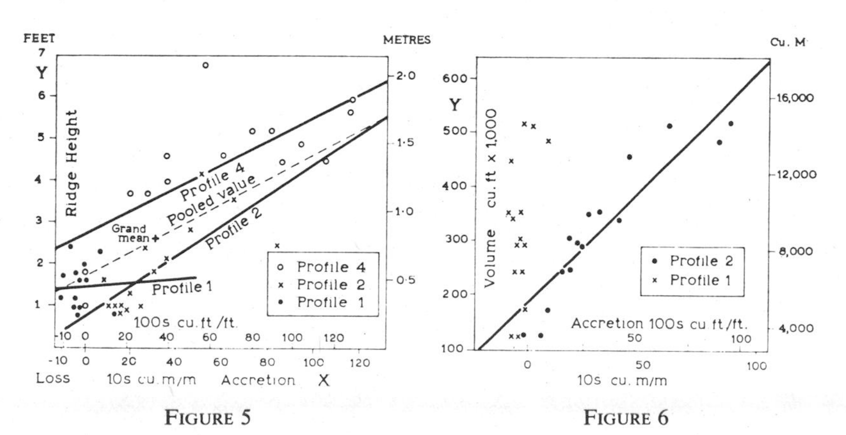

FIGURE 5

The relationship between ridge height and accretion on profiles 1, 2 and 4. The regression lines have been calculated by analysis of co-variance, which provides the pooled regression line

FIGURE 6

The relationship between volume of the spit above 10 feet O.D. Liverpool and accretion on profiles 1 and 2. The regression line for profile 2 and the spit volume is given

When the three sets of data are combined, controlling for the three profiles, the amount of variation in the data that is explained increases from about 50 per cent to about 80 per cent, taking the ridge height as the dependent variable that is explained in terms of accretion on the foreshore. The increase in percentage of variation explained is due to the control of the difference between the ridge heights on the different profiles. Both profiles react in the same way to accretion, but the ridges are higher on profile 4. This greater height is explained by the coarser grade of sand on profile 4 (mean 1.35 ϕ) as compared with that on profile 2 (mean 2.27 ϕ). The coarser sand requires a steeper equilibrium gradient and hence a higher ridge.

The development of the large ridges is the result of continued accretion of sand on the foreshore, and as the ridges become stabilised so the coast builds out eastwards. Some of the sand is, however, transmitted southwards along the coast, and this sand goes in part to build up the spit.

The Relation between the Accretion on the Spit and the Foreshore

The regular landward movement of the ridges on the foreshore indicates that movement southwards along this stretch of coast is also fairly regular. In order to establish the source of material building up the spit, the accretion on profile 1, in front of the spit, and profile 2, a little to the north, were correlated with the volume of the spit. The relationship between these variables is shown in Figure 6, which makes it clear that there is no significant correlation between the amount of sand on profile 1 and the changing volume of the spit. These ridges on profile 1 have never exceeded 2.5 feet in height and have never been large enough to provide protection to the spit, which, with only brief exceptions when minor ridges lay in front of it, has always been fronted by a steep swash slope.

However, Figure 6 shows that a much closer relationship exists between the volume of the spit and the accretion on profile 2. The correlation coefficient of 0.90 is a value that explains 80.8 per cent of the variability of the spit volume in terms of the amount of material reaching profile 2. The regression equation is $Y = 174.1 + 0.20\ X$, where Y is the spit volume and X is the accretion on profile 2. The reason for this relationship must be the southward transfer of material from profile 2 to the spit by wave action. It seems that most of the material moving south from profile 2 has been carried to the spit and not to the foreshore in front of it where profile 1 lies. This means that the material must move along the highest part of the foreshore and not along the ridges on profile 2, which run out on to the lower part of profile 1, where relatively little additional sand has been received until the last year or two of observations. The movement must, therefore, be accomplished by the waves at high tide. At such times the depth of the water offshore is at its greatest so that wave refraction will be less effective than at other times and the waves will approach the shore most obliquely and hence have their maximum capacity for longshore transport of material.

One note of caution should be added. There is a danger of auto-correlation in relating two time series of the type correlated in this instance, when the individual observations are not necessarily independent of those preceding and following them. The relationship can be demonstrated, but the connection may be the result of auto-correlation, rather than being a true causal relationship. In view of the close proximity of the two areas and the southerly movement of material, it seems likely that here the relationship is one of genuine cause and effect.

Conclusion

In an earlier paper[8] it was suggested that the spit in this vicinity would undergo a cyclic development related to the formation of the nesses of accumulation that are in turn related to the pattern of tidal offshore banks and their migration. The present spit has been shown to have been in the early stage of development throughout the period of detailed surveys, as it has been gaining material at a fairly steady rate. It has regained its former height, and vegetation has become re-established on its crest. However, it has moved over 200 feet bodily inland during the last 10 years and its distal end is now being eroded by the meander swing of the River Steeping, which made a permanent break through the foreshore to establish a new channel for the tidal flow up and down the river in the early summer of 1968. But, because the spit receives most of its material from the north, changes that are taking place at its southern end may not much affect its volume, although they are changing its morphology as indicated by the rapid swing inland of its distal tip between 1968 and 1969. The development of recurves could well be associated with changes on the foreshore near the distal end, where erosion would allow more effective waves to reach this point from the deeper water offshore.

The continued westward transgression of the spit over the New Marsh was also predicted in 1961.[9] A break-through of the River Steeping was also suggested as the possible cause of erosion at the distal end of the spit. Both of these possibilities have in fact occurred and the developing hook at the end of the spit, shown in the latest outline in Figure 2, may be fulfilling predictions made earlier.

Ultimately it is the major accretion that is proceeding to the north that is more likely to set a limit to the growth of the spit. When the ness has built out sufficiently far to divert material further offshore along the beach, then the spit is likely to be starved of sand and to become static as another spit develops seawards of it. There is some indication that this change is already beginning. In 1969 profile 2 had a large volume of accretion, and the beach ridge near the top of the profile is clearly becoming stabilised, for it now supports vegetation, and dune formation is beginning upon it. This development will prevent the passage of material southwards on the upper beach as this is raised above the effective level of the waves at high tide. The recent growth on the lower foreshore at profile 1, which for a long time was very static and muddy, suggests that much sand is now moving along shore at a lower level on the beach, rather than to the spit, the growth of which was reduced between 1968 and 1969.

The spit in fact is an ephemeral feature on this coast, which is building out eastwards rapidly as a result of beach ridge growth, following the accretion of sand determined by the pattern of tidal streams. The beach ridges in turn develop into dune ridges, separated by elongated marsh slacks, formed from the mud-filled beach runnels, At any one time a spit will be likely to prolong the outermost stabilised beach ridge which is being converted into foredunes. The speed with which these changes take place means that repeated surveys of the type carried out annually at Gibraltar Point provide data for a useful quantitative study of the particular coastal processes involved.

NOTES

[1] F. A. Barnes and C. A. M. King, The spit at Gibraltar Point, Lincolnshire, *East Midland Geographer* No. 8 (1957) 22–31.

[2] C. Kidson, Movement of beach material on the east coast of England, *East Midland Geographer* No. 16 (1961) 3–16.

[3] C. A. M. King, The character of the offshore zone and its relation to the foreshore at Gibraltar Point, Lincolnshire, *East Midland Geographer* No. 21 (1964) 230–243.

[4] C. A. M. King and F. A. Barnes, Changes in the configuration of the intertidal beach zone of part of the Lincolnshire coast since 1951, *Zeit. für Geomorph.* NF8 (1964) 105*–126*.

[5] Graphic moment measures, as devised by: R. L. Folk and W. C. Ward, Brazos River Bar: a study in the significance of grain size parameters, *Jour. Sed. Petrol.* 27 (1957) 3–26; are found from percentile readings on the cumulative frequency graph as follows:

Mean $$M_z = \frac{\emptyset_{16} + \emptyset_{50} + \emptyset_{84}}{3}$$

Median $$Md\emptyset = \emptyset_{50}$$

Sorting $$\sigma_1 = \left(\frac{\emptyset_{84} - \emptyset_{16}}{4}\right) + \left(\frac{\emptyset_{95} - \emptyset_5}{6.6}\right)$$

Skewness $$Sk_1 = \left(\frac{\emptyset_{16} + \emptyset_{84} - 2\emptyset_{50}}{2(\emptyset_{84} - \emptyset_{16})}\right) + \left(\frac{\emptyset_5 + \emptyset_{95} - 2\emptyset_{50}}{2(\emptyset_{95} - \emptyset_5)}\right)$$

Kurtosis $$K_g = \frac{\emptyset_{95} - \emptyset_5}{2.44(\emptyset_{75} - \emptyset_{25})}$$

[6] Cailleux's roundness index, R, is found by $R = \frac{2r}{a} \times 1000$, r = minimum radius of curvature in principle plane and *a* is length of long axis of stone.

[7] $\emptyset$ units provide a logarithmic size scale. $\emptyset = -\log_2$ millimetres. Coarse sediment has a negative value and the finer the sediment size the larger the $\emptyset$ value, for example:

$-1\ \emptyset$ = 2 millimetres
$0\ \emptyset$ = 1 millimetre
$+1\ \emptyset$ = 0.5 millimetre
$+4\ \emptyset$ = 0.0625 millimetre

[8] F. A. Barnes and C. A. M. King, *op. cit.* (1957).

[9] F. A. Barnes and C. A. M. King, Salt marsh development at Gibraltar Point, Lincolnshire, *East Midland Geographer* No. 15 (1964) 30.

Editor's Comments on Paper 16

Taking the variables involved in the Hurst Castle spit development, C. A. M. King and M. J. McCullagh have used a computer simulation model to first reproduce the field prototype then predict its subsequent growth. Identification of the variables, and their respective weighted importance, constituted a major task in the study. Built into the simulation model then were four different wave regimes, plus depth and refraction factors. The spit morphology reproduction obtained utilizing this technique was quite good; the projected spit development beyond that must be judged as most impressive. Application of the King–McCullagh method to other geomorphologic processes holds great promise for futher study.

Michael J. McCullagh was born in England in 1946 and attended Nottingham University where he received the B.Sc. degree in 1967, and the Ph.D. degree in 1970. Since that time he has been working as an invited Visiting Professor in the Department of Geography at The University of Kansas, and as a Research Associate in the Kansas Geological Survey. McCullagh's interests have been geomorphological and his inclinations computorial, therefore he is now engaged in developing techniques of quantitative analysis such as geologic simulation related to coastal and periglacial features. Computer mapping is another area of special interest, especially those aspects stressing the reliability of different algorithms and the reliability of the surfaces they create. Dr. McCullagh's interest in teaching has led to AAPG sponsorship of a series of short courses on the general subject of geomathematics held for university faculty and oil company representatives.

The following article by C. A. M. King and M. J. McCullagh is reprinted with permission of the *Journal of Geology*, as edited by Peter J. Wyllie and published by the University of Chicago Press.

Reprinted from THE JOURNAL OF GEOLOGY
Vol. 79, No. 1, January 1971

A SIMULATION MODEL OF A COMPLEX RECURVED SPIT[1]

16

CUCHLAINE A. M. KING AND M. J. MC CULLAGH[2]
Department of Geography, University, Nottingham, NG7 2RD, England

ABSTRACT

A simulation model of a particular coastal spit is presented to illustrate the value of such models in studying the processes that lead to spit formation. Hurst Castle spit in Hampshire was chosen for the model as it has a distinctive form due to the operation of several easily identified wave types. The processes forming the spit have been discussed by W. V. Lewis. The main ridge, which is composed of shingle, is prolonged by westerly waves, while storm waves build it up. The recurves are formed of material drifted round the end of the main ridge by waves from a southerly point. Waves from the northeast, coming down the Solent, build the lateral recurves. These become longer and more numerous toward the distal end of the spit. The computer program simulates the operation of waves from the west, storm waves, southeasterly waves, and waves from the northeast. In addition, a depth factor simulates the increasing depth of water offshore. This accounts for the increase in number of recurves toward the distal end. A refraction factor allows the curvature of the main ridge to be simulated. The proportion of random numbers allocated to these variables can be adjusted in the data input to elucidate the part they play in the formation of the spit. The close fit of the standard spit to the real spit suggests that the controlling variables are being correctly simulated.

INTRODUCTION

The formation of a complex recurved spit is the result of the operation of several distinct processes over a considerable span of time. A simulation model attempts to reproduce the stages of formation by simulating the different processes that are operating.

One of the main advantages of simulation models is that time can be speeded up. The formation of the feature can also be extrapolated into the future, once the past and present forms have been successfully reproduced. Another great advantage of simulation models is that the different process variables can be controlled and modified one by one. The effect of each process variable on the response variable, the form of the spit, can be ascertained in turn.

In a study of coastal features, simulation can be achieved in a number of different ways. The most common and efficient method of simulation, until computers became available, was by means of a hardware scale model. Model wave tanks large enough to overcome some of the scale problems attendant on the use of such models are, however, very expensive in space, time, and money. Increasingly, therefore, the newer type of computer simulation model is being developed to study the formation of suitable features.

This paper discusses an attempt to simulate a particular coastal spit. The method could, however, be applied to a wide variety of features of similar type. Suitable subroutines could be built into the computer program. Before writing the program, it is necessary to isolate the main process variables that are considered to be important in the formation of the spit being simulated. Then it is required to establish how these variables affect the growth and development of the spit, so that the correct response can be built into the program. The operation of the different process variables does not follow a predictable or uniform pattern in nature. This randomness in the order in which the processes operate is simulated in the program by use of random numbers to call the different process variable subroutines. It is possible, however, to control the proportion of random numbers that is allotted to each subroutine. This control allows the conditions of formation to be

[1] Manuscript received March 17, 1970; revised June 8, 1970.

[2] Present address: Department of Geography/Geological Survey, University of Kansas, Lawrence, Kansas 66044.

[JOURNAL OF GEOLOGY, 1971, Vol. 79, p. 22–37]

changed in different runs of the program. In this way the effect of different combinations of frequency of occurrence of the different variables can be simulated.

A whole run of the computer program including compilation only takes about $1\frac{1}{2}$ min of central processing unit (c.p.u.) time on an English Electric KDF9, once the program has been made operational. There is thus a great saving of time compared with the use of scale hardware models, in which one test could well take many hours to run. The recording of the results of the scale model also requires considerably more time, although the amount of detail that is possible is naturally greater.

One of the difficulties of the simulation model is, however, that the response of the feature to a given process variable is assumed to be known and to be always similar. This assumption may not be justified. The feature is certainly very drastically simplified in making the necessary assumptions that are built into the model.

The criterion of success must rest on the degree to which the model can correctly simulate the form of the prototype. When the proportion of random numbers allotted to the different processes is such that the final result is similar to the original feature, then it may be assumed that the processes that have been simulated are in fact those that are important in the formation of the feature. The model can then be used to predict future developments which are based on the assumption that similar processes continue to operate in the future. Thus the main advantages of simulation models of this type are their ability to control the variables and their speed of operation. Like all models, however, they are very simplified replicas of reality. Their success depends on the degree to which the important variables are selected and simulated correctly.

HURST CASTLE SPIT

The simulation model presented in this paper is based on Hurst Castle spit in Hampshire. This is a shingle spit just over 2 km long. It leaves the mainland coast, which runs in an east-southeast direction, and turns to the southeast. It then gradually bends round to run more eastward, before turning abruptly to the north-northwest in a major recurve. This recurve is the longest of a series that runs north-westward from the main line of the spit (fig. 4, inset).

Waves can approach the spit from the west along the English Channel. The largest waves reach the main part of the spit from the southwest, coming from between the Isle of Wight and the Isle of Purbeck. Waves can also reach the spit from the northeast down the Solent.

The formation of the spit has been studied by W. V. Lewis (1931, 1938). He has shown that the main line of the spit, where it leaves the coast, is aligned perpendicular to the direction from which the storm waves approach along the English Channel from the southwest. The main part of the spit faces 218°. This is also the direction of the greatest available fetch and from which the strongest winds and storm waves come. The lateral ridges, on the other hand, face the direction from which waves can approach from northeast along the Solent.

Lewis suggests that the spit builds out to the southeast in response to the movement of shingle by oblique waves from the west. They carry shingle that has been combed down by destructive waves along the beach. The main part of the ridge is built up by large waves approaching from the direction of maximum fetch. Constructive waves would build up to the high-tide level. Storm waves coming from the same direction would throw some of the shingle over the ridge crest, building up the height of the structure above the reach of normal waves. They would also comb down large amounts of shingle to and below low-water level. This shingle would be pushed up the beach by the action of constructive waves following the storm, thus consolidating the feature and perhaps driving it slowly to the northeast.

Lewis suggests that the prolongation of the spit is a discontinuous process. At the point at which recurves become numerous, the main line of the spit changes direction around a curve of large radius. Lewis considers that the rate of drift increases as the spit curves to the east, owing to the greater obliquity of wave approach.

It seems more likely, however, that the gentle curve in the spit is the result of wave refraction. The water depth offshore increases in this area. This increase would have the effect of making the wave fronts approaching from the southwest swing round to approach more from the south. At the same time the increasing depth of water would slow down the eastern growth of the spit. More and more shingle would be required to build up the spit above mean sea level, and later to high-tide level, as the water depth increases.

As the eastern growth of the spit is slowed down by the increase in water depth, so the recurves, built by the waves approaching from the northeast along the Solent, can increase in both number and length. The growth of these recurve ridges indicates that a considerable amount of shingle must pass around the sharp angle of the spit. This material must be derived from the west, and it seems likely that it travels along the spit under the influence of oblique westerly waves.

Lewis believes that the abundance of shingle at the sharp angle of the spit is due to the rapid movement along the far part of the main spit and its slow speed of progress along the recurves. He notes that waves 1.83 m high on the main spit are reduced to about 15 cm in height on the recurve. These low waves will approach the recurve at a much more oblique angle and thus be able to move some material northwestward along the recurves. The waves that cause movement of material along the recurves may be the refracted westerly waves observed by Lewis. Waves from the south and southwest will approach the recurves at a considerable angle. They will be short owing to the narrow fetch available in these directions. They will be little refracted as a result and will be able to cause considerable movement of material along the recurves. It is this movement that is simulated by the routine "seperp."

The sharp point of the spit, where the recurves join the main spit, is the result of the restrictions of ridge-building waves to two main directions of approach. The southwest storm waves build up the main spit, and the northeast waves build the recurves. The material that the Solent waves build onto the recurves is carried along them mainly by south and southeast waves, but these waves are not powerful enough, owing to the short fetch, to build a ridge. The south-east waves will become less effective as the spit lengthens because its tip passes more into the shelter provided by the Isle of Wight to the southeast.

The waves coming down the Solent will continue to build the recurves facing in the same direction. Each recurve must mark the position at which the end of the spit was stabilized for a period while material was moving around the sharp corner of the spit before storm waves from the northeast were sufficiently large to produce a permanent recurve ridge. As the main spit grew longer the periods of terminal stability increased in duration, allowing longer recurves to grow. The slower progress between periods of stability allowed recurves to form closer together.

Lewis makes some interesting comments on the effect of the tidal streams on the growth of Hurst Castle spit and the movement of shingle along it. He states that, at the ebb tide, there was a very strong tidal race off the sharp end of the spit. The surface velocity was 7.43–9.28 km/hour, within a few meters off the shore. This tidal stream, however, was not capable of carrying the shingle on the foreshore, probably because of a sharp decrease of velocity with depth.

When large waves were approaching the shore from the southwest, and the tidal streams were flowing the same way, the waves were surprisingly small. When the tidal stream changed direction, however, to flow against the waves, obliquity of the waves to the shore was greatly increased. Hence their capacity to transport shingle eastward along the spit was greatly enhanced. Thus the shingle moved more rapidly in a direction opposed to that of the more powerful ebb-tidal current that flowed westward along the foreshore. Lewis, therefore, concludes that little material is moved directly by the tidal currents and that the part they play in the growth of the spit is negligible.

The spit is still growing slowly to the southeast at its sharp point, but its speed of extension has slowed down rapidly as it has grown eastward. This is one of the features that is simulated in the model. The model includes the effect of four main types of waves. First, there are the waves from the west that cause material to drift along the main spit, prolonging it eastward. As the spit reaches deeper water a refraction factor is introduced. This produces the large radius curvature along the main spit where it turns slowly to run eastward. A depth factor is also introduced to slow down the eastward progress of the spit under the influence of westerly waves.

Second, there are the storm waves that build up the main ridge from the material drifted along the foreshore by westerly waves. These waves, as well as building up the main ridge, also comb material down from the foreshore as their action is partially destructive.

Third, there are waves from the south or southeast that drift material northwestward to form the recurve ridges. These waves become less effective as the spit is prolonged eastward. Their occurrence is, therefore, gradually reduced in the simulation model.

Fourth, there are the waves coming down the Solent from the northeast that build up the recurve ridges. In order to make the correct angle, the simulation model allows these ridges to build either diagonally or horizontally in a direction opposite to that of the westerly waves, but parallel to them. On the actual spit the recurves lie at an angle of about 35° to the proximal part of the main spit, although they lie at an angle of about 60° to the refracted distal part of the main spit. The details of the computer program subroutines that simulate these different waves, and the refraction and depth factors, are considered in the next section.

THE SIMULATION MODEL COMPUTER PROGRAM

A. INPUT

The spit is produced by placing numbers in a 50 × 60 matrix. A number of variables can be set in the program. The range of random numbers that call each of the four main wave subroutines can be specified. Thus subroutine "west" could be allocated from 0 to 60; subroutine "storm" from 61 to 70; subroutine "seperp," which produces the vertical recurves, from 71 to 80; and subroutine "noble," which produces the lateral recurves simulating the northeast Solent waves, from 81 to 100. The number of maps showing progressive stages of spit development is next specified, followed by the initial random number, from which others are generated as required. Two values, *A* depth and *B* depth, follow to determine the operation of the depth factor. This factor acts linearly between the beginning of the matrix and the *A* depth-factor column number and exponentially between the *A* and *B* depth-factor numbers. The depth becomes infinite at the *B* depth-factor column number, thus bringing the elongation of the spit to an end at this point. Finally the refraction factor is given. This value determines the likelihood of a jump occurring by which further spit growth moves up one row in the matrix, thus simulating the bending due to refraction. The higher the number specified, the less the chance of refraction taking place. The chance is exponential in its operation, thus making refraction increasingly likely as the spit becomes longer.

In the next section the different forms of spit produced by changes in all these controlling variables will be considered. Thus the effect of different proportions of waves coming from different directions and of different types can be examined. The effect of varying the depth factor can be analyzed, and also variation in the refraction factor can be examined. The effect of using different random numbers, but similar values for the other variables, can also be studied. By analysis of these different variables one by one, it is possible to come to some conclusion concerning the variables that are most important in controlling the growth of the spit. This is done by matching the simulated spit against the form of the real spit. The stages of growth of the spit under different conditions can also be simulated and future developments suggested by carrying on the program beyond the stage of best fit.

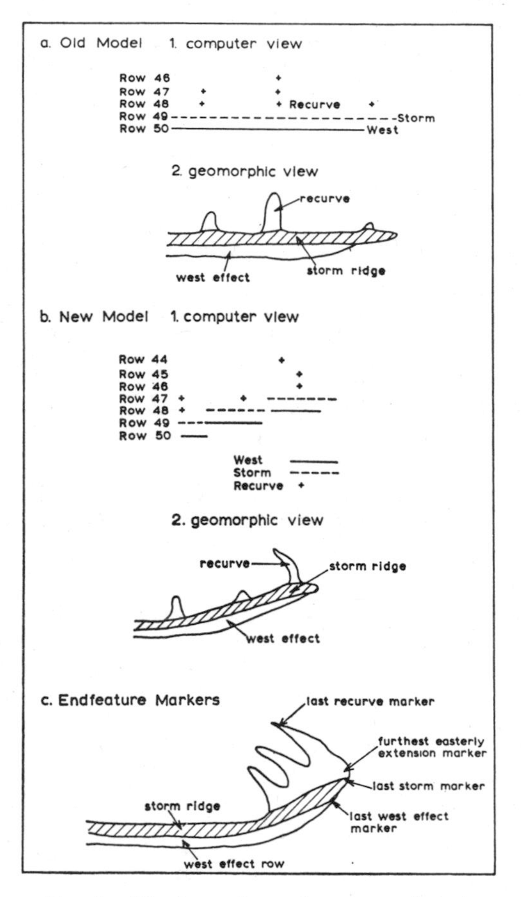

FIG. 1.—The figure shows the system of markers adopted in the program to provide end limits to the spit. At the same time the different features of the real spit are shown in relation to the simulated spit.

B. PROBLEMS OF MODEL DEVELOPMENT

Originally the spit model did not contain any provision for bending of the spit by refraction as it grew across the matrix. Hence all search procedures had only to look along a specified matrix row to be certain whether a particular spit feature had developed. It is possible to isolate three important rows which influence the development of the spit. Figure 1*a* shows these to be *West Row* (50), *Storm Row* (49), and *Recurve Row* (48). In the original model these rows were permanently rows 48, 49, and 50. When the model was extended to take into account the effect of refraction, however, the situation became much more complex. Figure 1*b* shows that any row of the matrix may now be a West, Storm, or Recurve row. This meant that the simple search procedure previously employed was ineffective and necessitated the introduction of an indexing system whereby the positions of potential growth (fig. 1*c*) on the spit are always known. The result has been a considerable increase in the speed of the program, and also a reduction of the number of cards in it.

Other difficulties have also occurred in the programming mainly concerned with the proper geomorphic representation of the processes at work, as modeled by the computer. Often the question of when a process is or is not allowed to act has not been specifically defined, and this has led to an incorrect simulation of reality. For instance, there is the question of whether it is possible for a vertical or a lateral recurve to be placed at the end of the spit following a storm that has removed some of the West Row's material. It was decided, from the geomorphic point of view, that material from the West Row would not alter significantly the growth of a recurve and, therefore, recurve action could take place in such circumstances. No easterly elongation of the spit, however, was to be allowed until the line of the West Row was full again up to and including the end of the storm ridge on the spit. This and similar difficulties led to increased complexity in the program.

C. PROGRAM OPERATION

A generalized flow diagram of the program is shown in figure 2. After the initial setting up of array space (about 4K is necessary), the initial setting of the markers, and reading in of data to the program, the main program begins the simulation at point *A* in the flow chart. At that stage, *I* starts increasing to indicate the number of times any routine has been called. This is followed by the calculation of a random number in the range 0–100 which is then checked against the input process routine proportions of that range. The appropriate routine is then called, and in its operation may use more random numbers to determine its particular course of action. At stage *B* in figure 2, after the operation of the process routine, a calculation is performed to see if a multiple of 50 routine calls (*I*) has been made. If this value is reached, the development of the spit up to that stage is printed out. If a map is printed, *J* is increased by 1 and tested (at stage *C* in fig. 2) to see if the total required number of maps has been printed. If that is the case, the program terminates; otherwise control returns to *A*.

In one run of the program producing 15 maps, at a standard 50 routine-call spacing, about 1,500 random numbers are likely to be used. These could have been read into the program at the beginning but there would have been a lack of variety in the sequences that could be employed. Instead it was thought better to incorporate a power residue-method random-number generator in the program. The resultant random numbers were acceptable, although the randomness attained was by no means perfect. This element of nonrandomness may be a contributory factor to the slightly differing shapes of the spit produced by altering the initial starting value of the generator, mentioned in the next section.

The best fit spit produced was that shown in figure 3. It represents the result of about 1,000 process routine calls and is a direct printout of the matrix mentioned at the beginning of this section. Every location in the matrix can be indexed by a row and column number. In this case, as only rows 26–50 are in effective use the remainder of the matrix is not given. The row numbers indicate the position of the land, and row 50 the point when the land stops and open sea begins.

Each process routine is represented by a different number in the matrix: 7's, 9's, and 3's indicate the action of the "noble" routine; 2's that of "storm"; the 1's along West

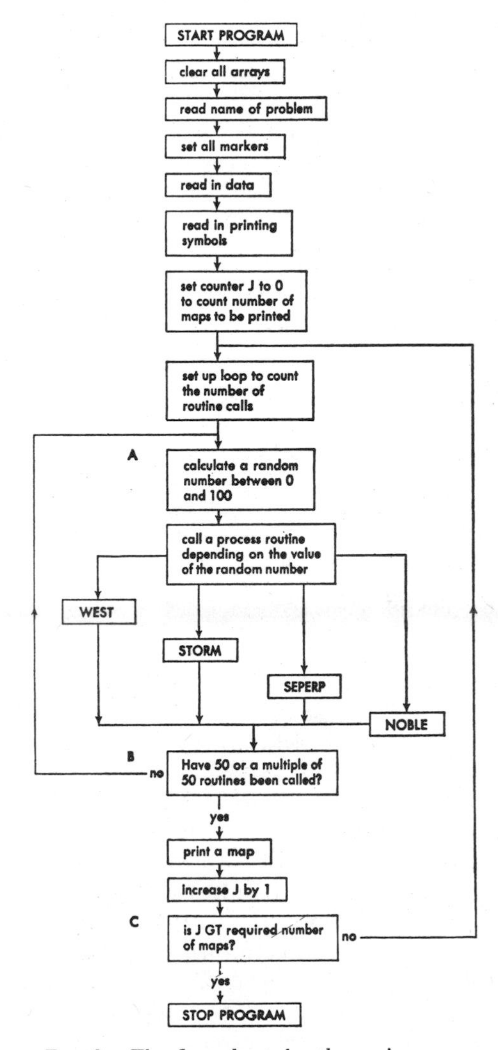

FIG. 2.—The flow chart for the main program indicates the relationship of the various process routines to the whole program and the control exerted by the program over those routines.

Row (see fig. 1) that of "west"; and those behind the West Row that of "seperp." From this information it is possible to build up a morphology of the spit so that in retrospect the factors causing particular parts to form can be noted. Thus in the present case it can be seen that the tip of the spit has largely been built up from the action of "noble," whereas earlier parts had only a small number of calls of "noble," but far more of "west" and successful "storms." In many places the action of "seperp" had been partially marked where the original "seperp" 1 has been changed to a 3 to indicate the takeoff point of a new "noble" recurve. A recurve formed by "noble" is made up of 7's and 9's. The reason for using both numbers is that the recurves do not grow back toward the matrix origin at 45°, but nearer 30°. This, in a rectangular computer matrix, means that it is impossible to lengthen the recurve on the principle of one row and one column less—every so often (40 percent of the time on average) the path of the recurve has to move one column back on the same row to bring the recurve back into line with the 30° requirement. In confused situations, such as at the tip of the spit, the introduction of the 9's to signify horizontal movements clarifies the track of any particular recurve.

```
ROW 26
ROW 27
ROW 28
ROW 29                                                                        9 7
ROW 30                                                                            9 7
ROW 31                                                                      7         7
ROW 32                                                                        9 7       7
ROW 33                                                                            7       7 7   7
ROW 34                                                                              7       7 9 7 3
ROW 35                                                                                7 7 7 7 3 7 3
ROW 36                                                                                  7 7 7 3 7 3
ROW 37                                                                                    7 7 3   3
ROW 38                                                                                      7 3 1
ROW 39                                                          9 7                           3
ROW 40                                                              7         7             2 1
ROW 41                                                                7         7 7       2 1
ROW 42                                            7                     7     9 7 3 7   2 1
ROW 43                                          9 7 3               9 7   3       3 3 3 1
ROW 44                                          9 7 3                   7 1 3   2 2 1 1
ROW 45                                            7 3                     3 3 3 1 1
ROW 46                          1                         7 3 7     1     2 2 2 2 2 1 1
ROW 47                        7 1                       2 2 3 3 3 2 2 2 2 1 1 1 1 1
ROW 48                      7 7 3   1 7   2 2 2 2 2 2 2 1 1 1 1 1 1 1 1 1
ROW 49          2 2 2 2 2 2 2 3 3 2 2 2 3 1 1 1 1 1 1 1
ROW 59          1 1 1 1 1 1 1 1 1 1 1 1 1
```

FIG. 3.—The simulated spit shown in this figure is the one nearest in resemblance to the real spit. The different numbers that make up the shape of the spit indicate the action of different process routines during the spit's formation.

For a detailed discussion of the operation of the program see McCullagh and King (1970), which gives full computational details and a listing of the FORTRAN program used.

THE RESULTS OF THE SIMULATION MODEL RUNS

A. SIMILARITY WITH PROTOTYPE

In order to test the validity of the assumptions made in the model, a series of runs was made with the same values for the variables but with different initial random numbers. This has the effect of altering the sequence in which the different subroutines are called, but not the frequency with which they are called. The depth and refraction factors also remained at the same values.

In this set of runs the subroutine "west" was allocated 65 random numbers, subroutine "storm" 15, subroutine "seperp" 8, and subroutine "noble" 12. This proportion makes allowance for the greater frequency of the prevalent westerly winds in the area. Storms operate more rarely than the other winds, but their effect is greater when they do operate. This was allowed for, since all the spit was affected by storm waves throughout the length of the main part of the feature, which is exposed to southwest storm waves. The other subroutines only added one element to the spit at each operation of the appropriate subroutine. The frequencies of winds in southwest England show a value of 21 percent for west winds, and 14 percent for southwest winds, of which 6 percent are force five or over. These strong southwest winds may be considered to generate the storm waves. Winds from the southeast have a frequency of 6.5 percent and those from the northeast of 10 percent. The values set for the standard spit correspond quite closely with these values, 15 being allocated matrix and exponentially between the 36 and the 55. Infinite depth did not allow the spit to grow beyond this point, which was in any case almost at the limit of the matrix. The scale of the computer spit is such that 10 columns of the matrix are approximately equal to 0.50 km. The total length of the spit, which is 2.75 km, would fit into the 55 columns of the matrix, which is where *B* depth prevents further elongation in the standard spit. The refraction factor was set at 2,300. This is an intermediate value be-

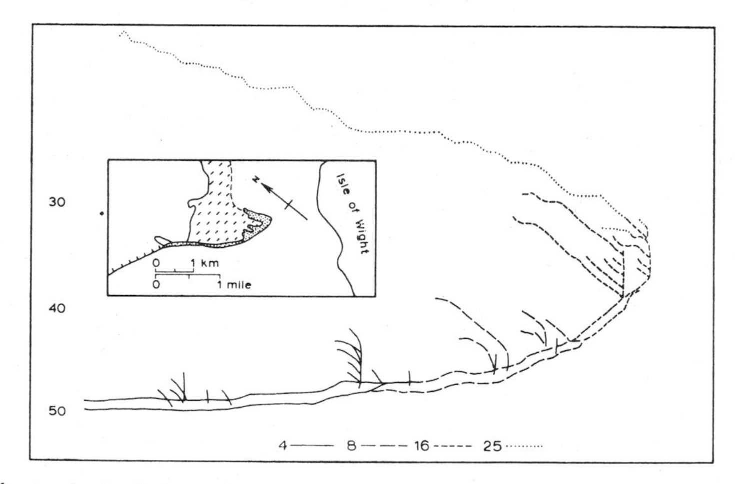

FIG. 4.—The standard spit shows the greatest similarity to the true spit. The ranges of random numbers allocated are: west 65, storm 15, southeast 8, northeast 12. Stage 16 is the best fit, and 25 suggests a possible future stage. The *A* depth value is column 35 on the matrix, which represents a point 12 m deep and 1.75 km from the spit root. The *B* depth is set at column 55 on the matrix, representing a point 20 m deep and 2.75 km from the spit root. The refraction factor is set at 2,300, which simulates a shoal area in front of the proximal two-thirds of the spit. This is in reality the Shingles shoal, while the deep water between it and the Isle of Wight allows the waves to bend to approach the distal part of the spit from the south. The inset shows the outline of the real spit.

to routine "storm," 8 to routine "seperp," and 12 to routine "noble." The frequency of west winds is set much higher in the standard spit input than the true frequency of west winds. The computer program, however, is so adjusted that only a proportion of the west routine values called can elongate the main spit, owing to the operation of the depth and refraction factors. A larger set of numbers must, therefore, be allocated to the west winds in the program.

The depth-factor values were set at column numbers 35 and 55 so that the depth increased linearly until the 35 column of the tween the maximum of 10,000, at which value refraction is very unlikely to occur, and a minimum low value at which refraction occurs at nearly every opportunity. Both the depth and refraction factors are incorporated with subroutine "west" because it is the westerly waves that are influenced by refraction and by the increasing depth offshore.

The spit produced with the variables set at the values given above was remarkably similar in outline to the real spit. This may be seen by comparing the inset of figure 4, which shows the outline of the real spit, and figure 4, which shows four stages of the development of the spit produced with the values of the variables set as given.

The curvature of the main line of the spit is very accurately reproduced, and the alignment of the recurves is also very similar to the natural ones. The increasing number and length of recurves is also well reproduced in the model spit. This close correspondence indicates that the values chosen for the variables were reasonable and closely similar to the real ones.

The runs made with different random numbers, but with the same values for the variables, also produced very similar shaped spits, although the curvature of the main spit varied slightly in the different runs. Figure 5 illustrates three spits produced with different random numbers. There are slight differences in the pattern of recurves, although they show the same tendency to increase in number and length toward the distal end of the spit. This is the most significant characteristic of the spit, together with the change in orientation along the main ridge. Thus different random numbers, although producing minor differences, do produce in general the correct major features of the spit outline. This leads to the conclusion that the order in which the subroutines are called does not influence the form of the spit, except to a relatively minor degree.

B. THE EFFECT OF SUBROUTINE CHANGES

The next step in testing the model was to vary the range of numbers allotted to the different subroutines. Three different propor-

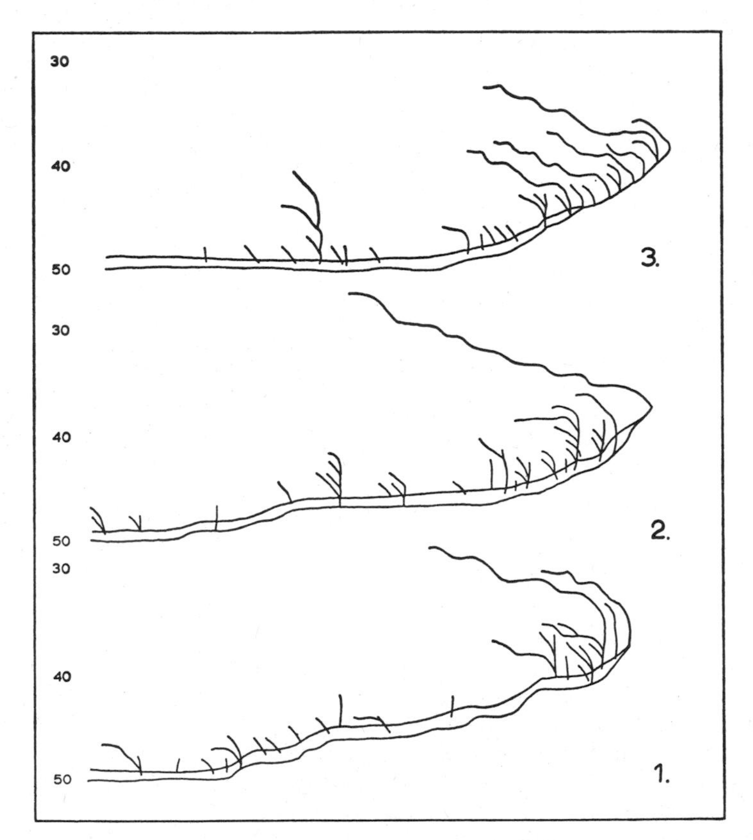

FIG. 5.—The same range of random numbers was allocated to each wave routine, and the depth and refraction values were also the same as those used for the standard spit, shown in fig. 4. The initial random number was different for each map. The results show that very similar spits are produced for the different random numbers.

tions are illustrated in figure 6 and given in table 1. The first change included more numbers in subroutine "west" with fewer storms, southeast, and northeast waves. The second had fewer numbers for the west waves, but more southeasterly waves, and the third had fewer west waves but more northeasterly waves. Several other combinations were also tested.

The effect of the increase in westerly waves is shown in figure 6*a* for three stages of development of the spit. The main difference between this spit and the standard one is the paucity of recurves. These do increase slightly toward the distal end of the spit, but are not nearly so numerous as in the real spit. The lack of recurves near the proximal end of the spit could be accounted for by the pushing of the spit to the northeast, over early short recurve ridges. They

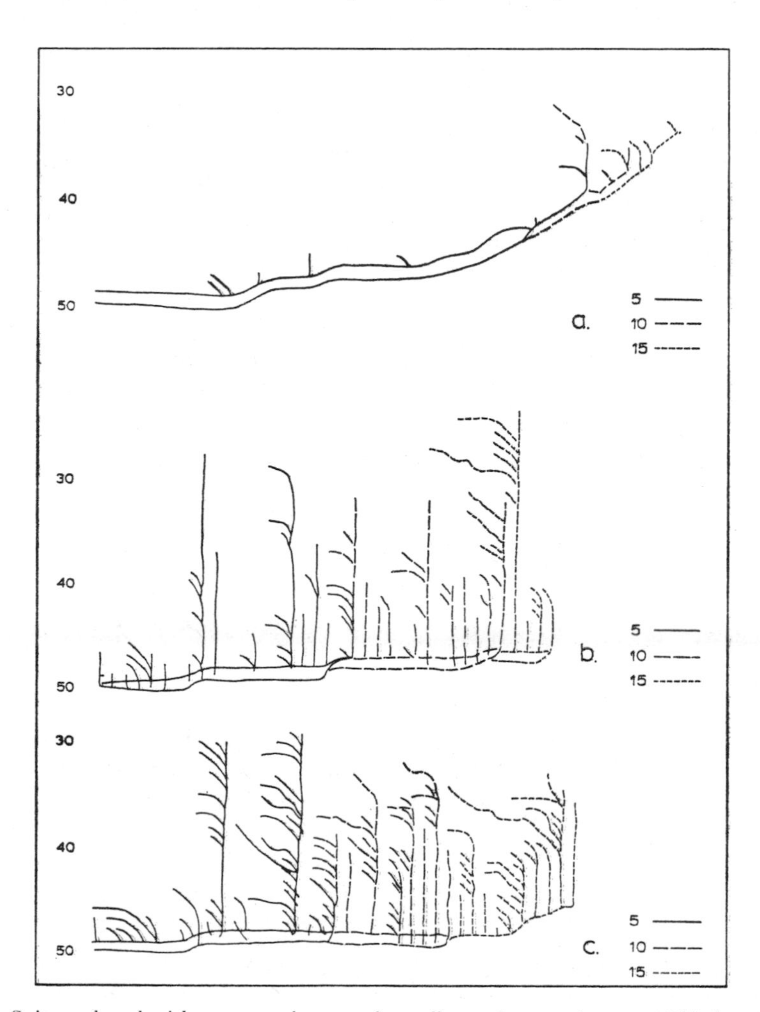

FIG. 6.—*a*, Spit produced with more random numbers allocated to routine west. This has the effect of elongating the spit and preventing a large number of recurves forming. *b*, Spit produced with fewer random numbers allocated to routine west and more to routine seperp. The shortness of the simulated spit is due to the low frequency of west waves, and the large number of vertical recurves is due to the excess of southeast waves, which in fact are of considerably lower frequency. *c*, Spit produced with fewer random numbers allocated to routine west and more to routine noble. A similar form to *b* is produced but the laterals on the recurves are now more in evidence, due to the greater occurrence frequency of the northeast waves which build these laterals.

would be short owing to more rapid growth of the main spit in the shallower water at this stage of growth.

As would be expected the second set of values, with fewer westerly waves, produces many more recurves, which form a very complicated pattern. Some of these are shown for different stages of development in figure 6*b*. The curvature of the spit is too slight and, compared with the real spit and the standard model spit, the recurves are far too numerous and form a much more complex pattern than the true spit outline. There is a predominance of vertical recurves owing to an excess of numbers allocated to "seperp." The conclusion may be reached that westerly waves, causing spit elongation, are considerably more important in the formation of the feature than the waves which cause the recurves to form.

TABLE 1

ALLOCATION OF RANDOM NUMBERS TO DIFFERENT SUBROUTINES

TEST	SUBROUTINE			
	West	Storm	Seperp	Noble
1..........	0–84	85–91	92–95	96–100
2..........	0–19	20–39	40–79	80–100
3..........	0–19	20–39	40–59	60–100
Standard...	0–64	65–79	80–87	88–100

The third test, in which the storms are fewer but the northeasterly waves more frequent, also shows far too many recurves, for the same reason. There is now, however, a preponderance of the lateral recurves. The small number of storms in the first test indicates that storm action need only be very infrequent to produce the major ridge. Changes in the frequency of waves from an easterly quarter, "seperp" and "noble", do affect the details of the pattern of the spit, but not its general form. The conclusion may be reached that the amount of westerly wave activity determines the form of the spit to a large extent, but it is not sensitive to minor changes in the frequency with which the other different wave types operate.

C. THE EFFECT OF DEPTH-FACTOR CHANGES

The major effect of the changes in depth-factor values is to speed up or reduce the eastern elongation of the main spit. The value determines the chance that a westerly wave random number has of elongating the spit. The greater the value of the depth factor, the greater the chance of spit elongation is and the shallower the simulated water depth. A sequence of four tests was illustrated, using different depth factors (fig. 7). The values used for the *A* depth and *B* depth settings were 10 and 20 for the maximum depth. This meant that the depth increase was exponential from the tenth column of the matrix. The next test used the values of 25 and 45, the third had 45 and 65, the fourth had 60 and 80, which means that in fact no depth factor operated in this test. The standard depth factor was 35 and 55, so that the first two had greater simulated depths and the last two had smaller simulated depth.

The *A* depth column, which represents the limit of linear depth increase, can be equated with a point 12 m deep. The *B* depth column is where the depth becomes virtually infinite, in that continued growth

FIG. 7.—The effect of variations in the depth factor is shown by the four spits. The uppermost spit was produced with very deep water, the next with fairly deep water, the next with fairly shallow water, and the lowest with very shallow water. The approximate depths would be as follows: *A* depth is set at about 12 m, and *B* depth at about 20 m at a point about 200 m normal to the shore. The positions of the *A* and *B* depth points differ in each run: they are as follows: (1) *A* depth 10 on matrix = 0.50 km; *B* depth 20 on matrix = 1.00 km from spit root. (2) *A* depth 25 on matrix = 1.25 km; *B* depth 45 on matrix = 2.24 km from spit root. (3) *A* depth 45 on matrix = 2.24 km; *B* depth 65 on matrix = 3.24 km from spit root. (4) *A* depth 60 on matrix = 3.05 km; *B* depth 80 on matrix = 4.00 km from spit root. The deeper the water the greater the volume of shingle required to enable the westerly waves to elongate the spit; hence, as the depth decreases, so the spit becomes more elongated. The standard spit, which fits best, has *A* and *B* depths intermediate between (2) and (3).

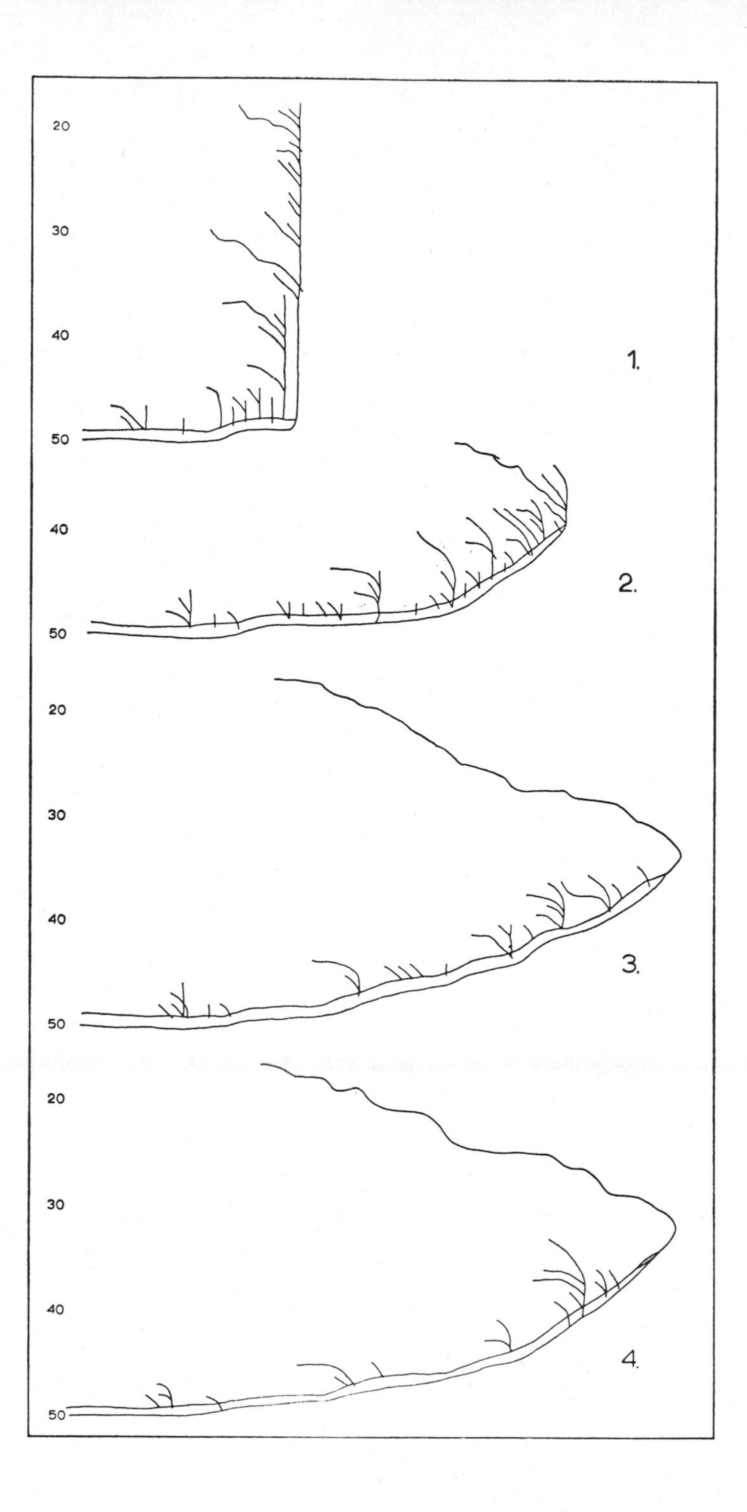
20
30
40
1.
50
40
2.
50
20
30
40
3.
50
20
30
40
4.
50

cannot occur beyond this point in the program. In reality this depth is about 20 m, a depth which occurs 154 m off the distal tip of the main spit. Beyond this, water deepens rapidly to over 57 m only 350 m further offshore. In the standard spit, *A* depth is at column 35, 1.75 km from the spit root, and *B* depth is at column 55, 2.75 km from the spit root. In the other spits, in figure 7, the values shown in table 2 are tested. According to depths recorded about 200 m offshore on the Admiralty Chart no. 2219, the increase in depth is still linear, at gradient of 1:260, for a distance of 2.10 km from the root of the spit. Thereafter the depth increases more rapidly. Thus the pattern of depth increase simulated agrees closely with reality. The depth values selected for the standard spit are realistic.

TABLE 2

SPIT VALUES

Spit	*A* Depth 12 m	Column	*B* Depth 20 m	Column
1........	0.50 km	10	1.00 km	20
2........	1.25 km	25	2.24 km	45
3........	2.24 km	45	3.24 km	65
4........	3.05 km	60	4.00 km	80

The first test, with the greatest depth, shows a very curtailed spit, with a very long vertical recurve and a large number of short laterals. In fact the program was stopped by the vertical recurve reaching the upper edge of the matrix. The second test spit is very similar to the best fit spit but slightly shorter owing to the greater depth preventing its elongation beyond the forty-fifth column. There are rather more recurves, as would be expected, but they are not very long. The third test shows the effect of a depth slightly less than the standard model. It resembles the standard model fairly closely and also fits the prototype well. The conclusion may, therefore, be reached that the standard depth factor best simulates the natural conditions, with a rapid increase of depth occurring nearly two-thirds of the way along the matrix. This makes the number of recurves increase at about the correct distance from the point where the spit leaves the mainland coast. The last two depth-factor tests, in which the depth increases very slowly or not at all, produce spits with a marked paucity of recurves. This is due to the rapid elongation of the spit, as the westerly waves have a much greater chance of effecting a lengthening of the spit under these conditions.

The curvature of the spit is still very similar to the natural spit, and the general outline fits well apart from the paucity of recurves. The results of these tests confirm that increasing depth, as the spit extends further offshore, is a very important factor in accounting for the patterns of the recurves. The depth is not so great, however, that elongation of the spit has been halted completely. It does, however, allow the recurves to increase in number and length as the depth increase becomes effective.

D. THE EFFECT OF CHANGES IN REFRACTION

The curvature of the main ridge of the spit is dependent in the model on the refraction factor which is part of the subroutine west. When a random number is selected that calls this subroutine, the spit can only be elongated, by the placing of a 1 at its distal end, if the random number is such that the depth factor allows the 1 to be placed. The program then chooses another random number, and if this number is greater than that set for the refraction factor, then the 1 is placed on the same row of the matrix. If it is less than the appropriate value, then the 1 is placed on the next row, above the row at which the spit currently terminates. Thus the spit can gradually become curved by a series of steps.

The refraction factor controls the frequency with which these steps are made. Five values for the refraction factor are illustrated in figure 8. These were 1,500, 2,000, 3,000, 5,000, and 10,000. With a low value the chance of refraction is high, with a large value it is low. The number used in the standard model spit was 2,300. With

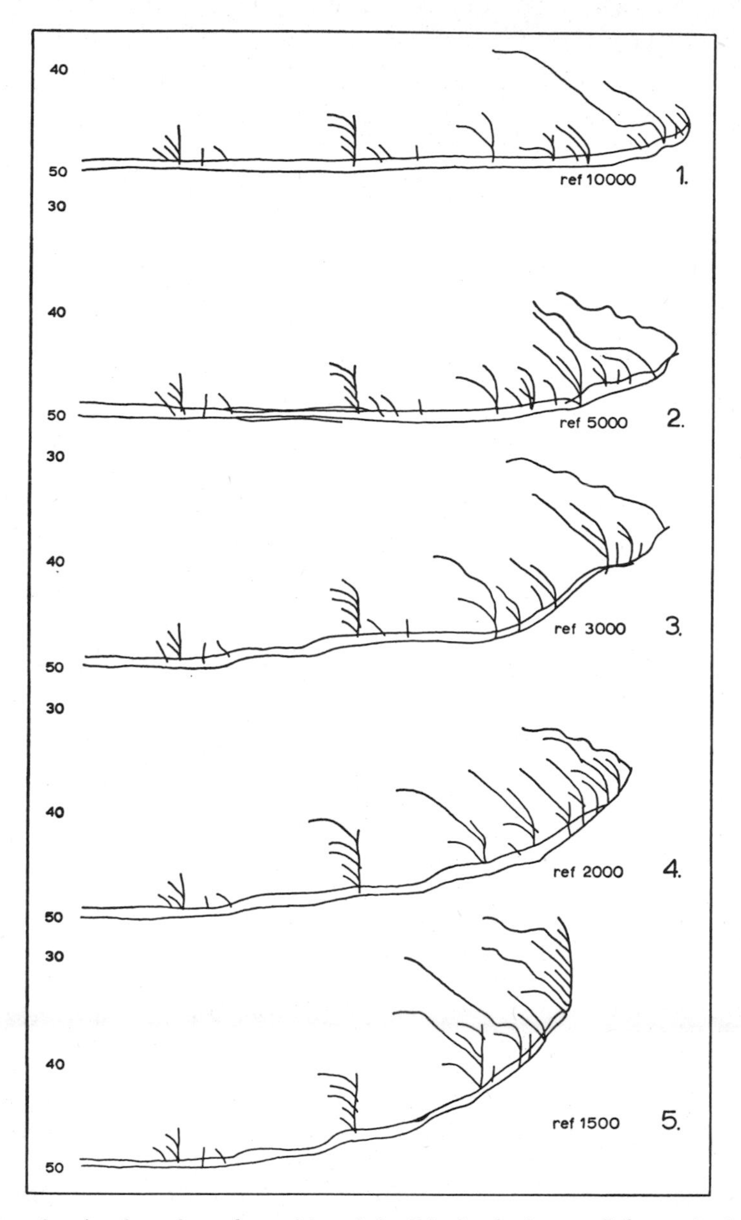

FIG. 8.—The refraction depends on the position of the Shingles shoal, now off the proximal two-thirds of the spit, and the deep water off the distal part of the spit, which allows waves to approach this part from further south. The upper spit shows what would happen if the shoal and deep water moved east so that no bending could occur. The lower spits show that if the shoal moved west and deep water occurred off the proximal part of the spit, bending would occur sooner.

the low values of the refraction factor the spit started to recurve very soon and continued almost diagonally across the matrix, recurving at almost every placing of a west 1 toward the distal end of the spit.

As already noted, the greater depth off the distal part of the spit allows waves to travel faster in this area, and those coming from the southwest swing round to approach the spit from the south. The proximal part of the spit, which swings out southeast from the west-northwest to east-southeast trending mainland, is influenced by the offshore shoal area, known as the Shingles. This area, which now lies off the proximal two thirds of the spit, causes the waves to be retarded. The refraction factor simulates variations in the relative position of this shoal and the deep water, lying between it and the Isle of Wight. When the refraction factor is low, the shoal and deep water are represented as being further west. When it is high, the shoal is represented as being continuous, preventing waves swinging round in deep water. Thus the effect of changing offshore relief can be examined. If the shoal were to retreat west of its present position, and the deeper water to move further west also, the spit would bend sooner. This situation is shown in the last example in figure 8. The first two spits indicate what would happen if the shoal moved east to the distal end of the spit and the deep water area were filled in. Bending of the spit would not then occur.

The series of spit outlines shown in figure 8 illustrates the gradual change of curvature produced by the different refraction factors. The curvature flattens off slightly when the value reaches 2,000, but it is still rather too sharp. The values between 2,000 and 3,000 all fit the natural curvature well, the 2,300 value being particularly close to the natural form of the main spit ridge. The curvature at 10,000 is much lower than that of the real spit near the distal end. It may be concluded, therefore, that the value selected for the standard model spit is a close approximation to the true control of the spit curvature. The effect of refraction may be considered to be an important element in accounting for the form of the major spit ridge.

CONCLUSIONS

The analysis by W. V. Lewis of the variables affecting the formation of Hurst Castle spit stressed the importance of waves coming from different directions in building the major ridge and the many recurves that join it at a sharp angle. The processes that are significant in the development of this spit are also important in the formation of many other spits. In shingle spits the material is usually brought to the spit by the waves of limited fetch, which are relatively short. They can, as a result, approach the spit at an angle sufficient to cause a significant amount of spit elongation by longshore beach drifting. In the case of Hurst Castle spit, these waves come from the west.

In shingle, the major features are formed by storm waves which can build the material, brought to the spit by the oblique waves, into permanent ridges above the reach of normal waves. At Hurst Castle spit these storm waves come from the southwest, approaching normal to the main line of the spit where it leaves the mainland near its root.

The effect of deepening water as the spit grows eastward is twofold. It slows down the rate at which the spit increases in length, because a greater volume of shingle is required to pass alongshore for each unit of elongation. It also allows the waves to become more strongly refracted. They swing round at their breakpoint to approach the coast with their crests aligned in a more east–west, rather than a northwest–southeast, direction. This process causes the bend of large radius of curvature along the main part of the spit.

The recurves that join the spit at a very acute angle near the major bend, and at a rather less acute angle near its distal end, are formed by a completely different set of waves. These waves approach from the northeast along the Solent. They build ridges of material drifted round the sharp corner of the spit by waves with a southerly and southeasterly origin. These waves are likely to be very short and, therefore, to approach very obliquely. Hence they cause marked beach drift of material to the north and northwest. The sharp angle of the spit is due to the proximity of the Isle of Wight to the southeast, preventing the approach of powerful ridge-building waves from this direction.

All these variables and controls are built into the simulation model. It includes four different types of waves, the westerlies, the storm waves, the southeastern waves, causing beach drifting along the recurves, and the northeasterly waves, building the recurves. The depth factor and the refraction factor are also very important controls in the model. All these variables have been adjusted in turn to different values in the series of runs made with the model by varying the input data which control the different variables in the program.

The results show that the standard model produces the optimum combination of the different variables. The spit simulated with these values of the variables is remarkably similar to the form of the real spit. This close similarity illustrates the value of simulation models in testing the effect of different processes in the building of spits of this type.

One of the great advantages of this type of model is the speed with which different values for the process variables can be tested, as each run of the program, once compiled, only takes about 20 sec of c.p.u. time. When the spit has been correctly reproduced, its future development can be predicted. The standard model suggests that more recurves will be built onto the spit and that these will become longer and closer as the distal end of the main spit slows down its eastward movement as the depth becomes greater. Thus one of the major controls of the spit form is the increasing depth offshore. The length of the recurve on figure 4 does not take into account growth of salt marsh in the lee of the spit. Models of this type offer much scope for more sophisticated development. It would be possible, for example, to include the effect of tidal processes. These have been shown to be of negligible importance in the present instance by Lewis. It would also be possible to provide a map of a considerable area, including relief, by means of trend surface equations, which could provide a best-fit surface at any required level of complexity up to the octal surface. For regularly recurring features, such as a series of offshore banks, a Fourier-type analysis might supply a more suitable form of surface. The number of subroutines could also be increased and the range of random numbers enlarged. This relatively simple model, however, illustrates some of the possibilities of this method of approach to model building. It has the advantages of speed and flexibility, as well as those of simplification and control of the many variables that operate in reality to form geomorphological features, such as complex recurved spits.

REFERENCES CITED

Lewis, W. V., 1931, The effect of wave incidence on the configuration of a shingle beach: Geog. Jour. v. 78, p. 129–148.

——— 1938, The evolution of shoreline curves: Geol. Assoc. London Proc., v. 49, p. 107–127.

McCullagh, M. J., and King, C. A. M., 1970, Spitsym: a FORTRAN IV computer program for spit simulation: Computer Contribution 50. D. F. Merriam, ed., Univ. Kansas, Lawrence. 20 p.

Editor's Comments on Paper 17

While all of the preceding papers have been essentially scientific in nature, this final work is one of application in dealing with changing spit morphology. It is felt that a practical, field-oriented problem such as this is a fitting conclusion to the collected works presented here.

For years Ediz Hook has jutted out into the Strait of Juan de Fuca, undergoing the usual spit evolution described by our other authors. Trouble began when military and industrial development virtually covered the area and nourishment of the spit was cut off. With erosion and migration continuing and replenishment absent, the occupants faced the prospect of having the spit literally disappear from under their feet. The background to these events is outlined on pages 4 and 5 of the paper.

The whole approach at which this brochure is aimed is a relatively new one for the Army Corps of Engineers. Charged with the maintenance of public waterways, the Corps has in the past carried out its duties as it perceived them. With the current world-wide trend toward environmental and public concerns, however, the Corps has undergone a commendable change, and now solicits the opinions of the public and other governmental agencies. In this new approach all possible alternatives are considered and debated prior to selection of the most appropriate course of action to be taken. At present the Corps of Engineers is in the final discussion and planning stage of the Ediz Hook project.

PUBLIC BROCHURE

17

ALTERNATIVES AND THEIR PROS AND CONS

EROSION CONTROL, EDIZ HOOK PORT ANGELES, WASHINGTON

Authority for study: The Congress of the United States has directed the Corps of Engineers to make a study of the shores of Ediz Hook, Port Angeles, Washington, and such adjacent shores as may be necessary, in the interest of beach erosion control and related purposes.

*Purpose of this brochure: The purpose of this brochure is to concisely portray to the public and to agency officials the full range of alternatives available for reducing erosion control at Ediz Hook as well as the alternative of "no action."

*Methodology: This brochure is developed in an atmosphere of unrestrained or "fish bowl" planning. The alternatives depicted were suggested by local people and the Corps of Engineers through public meetings, correspondence, and personal contacts. Interested parties participate repetitively by making proposals (alternatives), defending them with factual advantages (PROS), and describing disadvantages (CONS) of other competing alternatives. The brochure is not a static document, but is continually under revision until all meetings and studies are completed and the final report is prepared. Changes are made as new comments are received from people, agencies, and associations, and as studies reveal new facts about alternatives. Individuals, as well as local, State, and Federal agencies, are urged to participate now, instead of waiting until after completion of the study to comment. The brochure is not intended as a device for obtaining votes favoring or rejecting alternatives. Selection of alternatives for final study must be based on consideration of all economic and environmental advantages and disadvantages for each alternative. We would appreciate your examination of alternatives set forth and invite your comments (PROS or CONS), suggested additions, or modifications of the alternatives.

*Changed, this draft.

SEATTLE DISTRICT, U. S. ARMY CORPS OF ENGINEERS
1519 ALASKAN WAY SOUTH, SEATTLE, WASHINGTON 98134

Fourth Edition - 13 April 1971

Early Congressional action: The existing Federal project is identified on the map on the opposite page. Congressional acts which ultimately authorized the existing Federal project are listed below:

Acts	Recommendations	Documents
2 Mar 1945	Report stated that measures for protection of Ediz Hook from the effects of storm waves could not be justified but recommended shoal removal to 30 feet in the vicinity of the ITT Rayonier, Inc., pier. The project has been classified "inactive."	H. Doc. No. 331, 77th Congress, 1st Session
3 Sep 1954	Report stated that protection of Ediz Hook from erosion action of storm was not justified as annual benefits were insufficient to justify the annual cost of constructing protective works, but recommended constructing a prorected mooring basin 15 feet deep adjacent to the existing basin. The project was completed in August 1959.	H. Doc. No. 155, 82d Congress, 1st Session

Previous study action: At the request of the city of Port Angeles, the Washington State Legislative Delegation petitioned the Public Works Committees of both the Senate and House of Representatives to authorize a survey of the shores of Ediz Hook, Washington. Resolutions authorizing such a study were adopted on 13 September and 8 October 1968 by the Senate and House Public Works Committes, respectively. In 1969 funds in the amount of $13,000 were authorized by the Congress to initiate the study. However, these funds were placed in budgetary reserve for use in 1970. In 1970 studies were initiated and additional funds allotted to complete the study by 30 June 1971, the end of the current fiscal year.

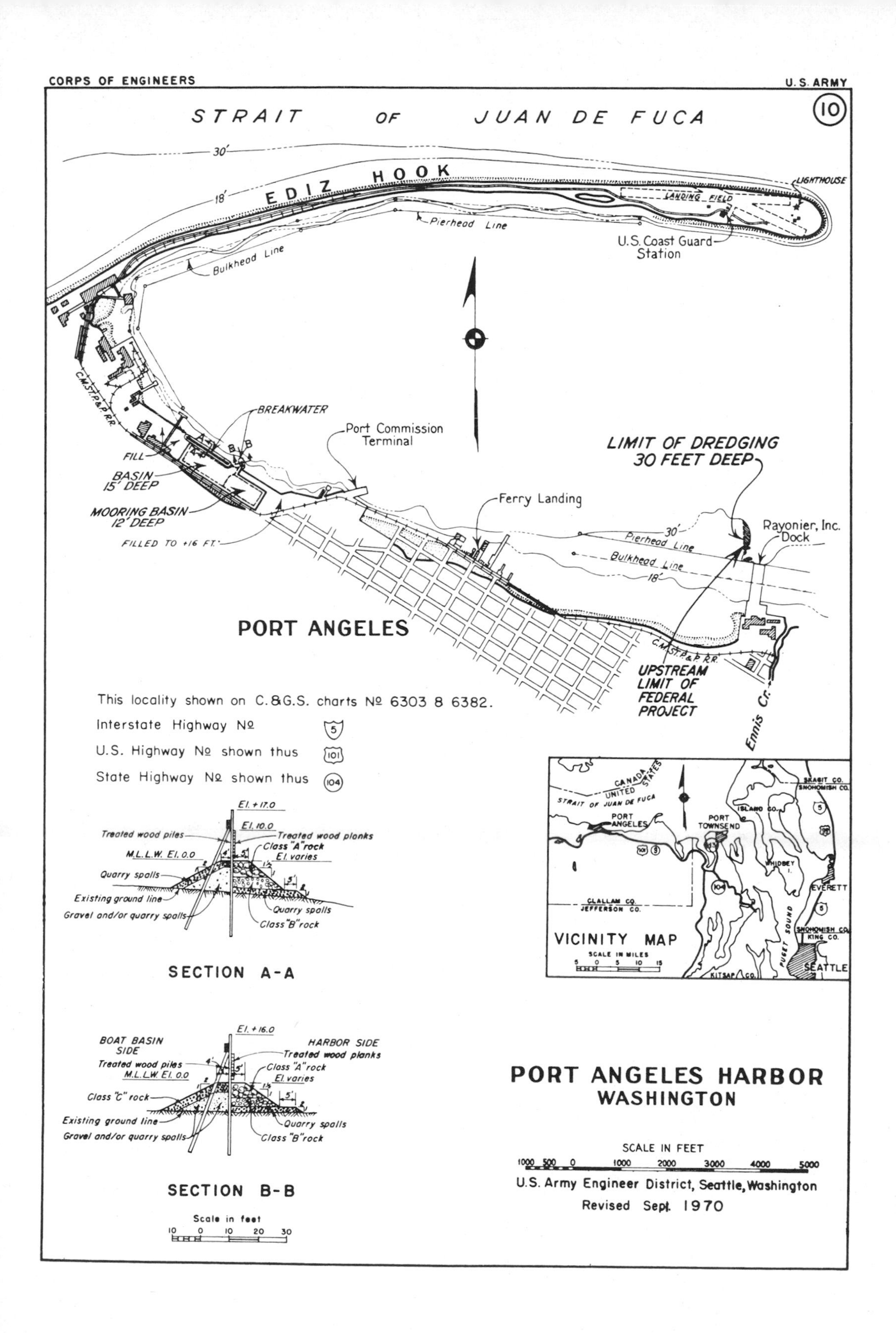
10
STRAIT OF JUAN DE FUCA
EDIZ HOOK
LIGHTHOUSE
LANDING FIELD
Pierhead Line
U.S. Coast Guard Station
Bulkhead Line
C.M.St.P.&P. R.R.
BREAKWATER
Port Commission Terminal
FILL
BASIN 15' DEEP
MOORING BASIN 12' DEEP
FILLED TO +16 FT.
LIMIT OF DREDGING 30 FEET DEEP
Ferry Landing
Rayonier, Inc. Dock
Pierhead Line
Bulkhead Line
PORT ANGELES
UPSTREAM LIMIT OF FEDERAL PROJECT
Ennis Cr.
This locality shown on C.&G.S. charts No 6303 & 6382.
Interstate Highway No
U.S. Highway No shown thus
State Highway No shown thus
El. +17.0
El. 10.0
Treated wood piles
Treated wood planks
M.L.L.W. El. 0.0
Class "A" rock
El. varies
Quarry spalls
Existing ground line
Gravel and/or quarry spalls
Class "B" rock
SECTION A-A
CANADA
UNITED STATES
STRAIT OF JUAN DE FUCA
PORT ANGELES
PORT TOWNSEND
ISLAND CO.
SKAGIT CO.
SNOHOMISH CO.
WHIDBEY I.
EVERETT
CLALLAM CO.
JEFFERSON CO.
PUGET SOUND
KING CO.
KITSAP CO.
SEATTLE
VICINITY MAP
SCALE IN MILES
El. +16.0
BOAT BASIN SIDE
HARBOR SIDE
Class "C" rock
SECTION B-B
Scale in feet
PORT ANGELES HARBOR
WASHINGTON
SCALE IN FEET
U.S. Army Engineer District, Seattle, Washington
Revised Sept. 1970

Preliminary investigations: Preliminary investigations indicated that the Hook was formed during very recent geologic times and is the product of gravel, sand, and cobbles being moved eastward, partly by currents, but mostly by waves and swells. The main source of these materials was erosion of the sea cliffs to the west, as well as material brought to the shoreline by the flooding Elwha River. As the eroding sea cliffs retreated and the spit built progressively eastward, the western and older section of the spit began to migrate southward by a process of breaching and overtopping during storms. The result is the surface landform we have today, consisting of a fat, easterly migrating distal or "bulb" end and a thin southerly-migrating central arm. At one time, the system was probably in equilibrium, with the majority of material being moved toward the easterly end of the spit. However, the apparent reduction in the source of material brought on by construction of the dams on the Elwha River and the pipeline protective works along the base of the sea cliffs has upset this balance. The graph on the next page, entitled "Ediz Hook Historical Changes," graphically shows the change. In addition, the concentration of wave energy along the neck (thin portion) of the hook by Crown Zellerbach mill is contributing to an offshore movement of material, with a corresponding reduction in the alongshore movement of materials. Thus, the seaward beach adjacent to the thin arm has progressively eroded to furnish the material for the easterly migration of the hook. The results of this erosion are the general lowering of the beach adjacent to the arm and base of the hook and the coincident concentration of wave energy directly against the thin arm of the hook. If this process is permitted to continue, it could result in ultimate destruction of the hook as we know it, and the formation of a submerged shoal in its place.

Subsequent to funding and study initiation, a public meeting, held on 13 October 1970, discussed alternative solutions contained in a public brochure mailed to interested parties on 5 October 1970. A second brochure, incorporating comments and suggested additional alternatives, was made available to all interested persons on 7 December 1970. A revised brochure summarizing pros and cons received from agencies and individuals commenting on earlier editions was distributed on 17 February 1971, prior to a second public meeting.

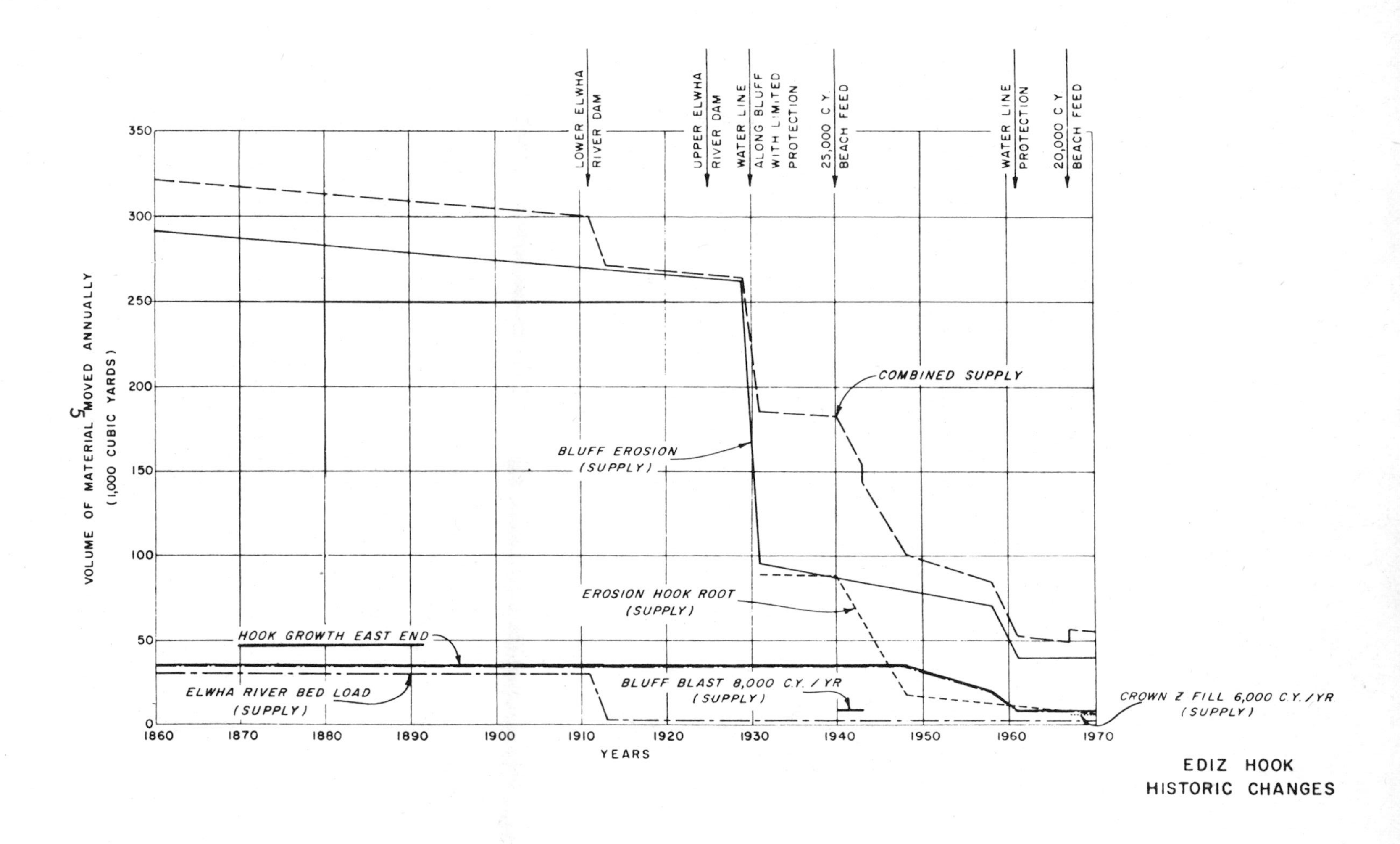
LOWER ELWHA RIVER DAM
UPPER ELWHA RIVER DAM
WATER LINE ALONG BLUFF WITH LIMITED PROTECTION
25,000 C.Y. BEACH FEED
WATER LINE PROTECTION
20,000 C.Y. BEACH FEED
VOLUME OF MATERIAL MOVED ANNUALLY
(1,000 CUBIC YARDS)
5
COMBINED SUPPLY
BLUFF EROSION (SUPPLY)
EROSION HOOK ROOT (SUPPLY)
HOOK GROWTH EAST END
ELWHA RIVER BED LOAD (SUPPLY)
BLUFF BLAST 8,000 C.Y./YR (SUPPLY)
CROWN Z FILL 6,000 C.Y./YR (SUPPLY)
YEARS
EDIZ HOOK
HISTORIC CHANGES

At a second public meeting (23 February 1971) the results of Corps of Engineers reconnaissance studies were presented, and a decision was reached on the alternative to be studied in detail. The proposal selected for detailed study is alternative 5, Rock Revetment with Beach Nourishment. The rationale for selection of this alternative, as well as the reasons for rejecting other alternatives, is presented on pages 37 and 38 of this brochure.

*Use to be made of this public brochure: This brochure summarizes the results of our studies and the pros and cons received to date on all alternatives presented in previous editions of this brochure, and includes the rationale for the elimination of all alternatives, except number 5, from further consideration. At the final public meeting, tentatively scheduled for May 1971, the results of detailed studies of alternative 5 will be presented prior to finalizing our report. The District Engineer's tentative recommendation on Federal financial participation will also be presented. A revised edition of this brochure, plus the study documents, will comprise the Seattle District Engineer's report. The recommendations will be reviewed by the Corps of Engineers Division Engineer in Portland, Oregon; Board of Engineers for Rivers and Harbors; and the Chief of Engineers in Washington, D. C. The Coastal Engineering Research Center will act as a consultant during the review. Comments will also be requested from various Federal agencies and the Governor of the State of Washington. The report will then be submitted to Congress. Comments on this brochure may be made at the forthcoming meeting, by letter to the District Engineer (address on the front page), or by informal comments to Seattle District personnel.

Federal participation: Federal financial participation in any alternative finally recommended after detailed study (except alternative 1, no action) would be contingent upon the following criteria:

a. The alternative must satisfy provisions of Federal participation by providing sufficient benefits, including economic, environmental, and social considerations, to offset the costs.

b. The local sponsoring agency, when required, must agree to fulfill certain requirements, which include, but are not limited to:

(1) Contribute in cash the local share of project construction cost.

(2) Provide without cost to the United States all necessary lands, easements, and rights-of-way.

(3) Hold and save the United States free from claims for damages which may result from construction and subsequent maintenance of the project.

(4) Assure continued public ownership or continued public use of the shore upon which the amount of Federal participation is based, and its administration for public use during the economic life of the project.

(5) Assure maintenance and repair, and local share of periodic beach nourishment where applicable, during the useful life of the works, as required to serve the project's intended purpose.

(6) Provide suitable borrow areas for beach replenishment and periodic nourishment where applicable, during the useful life of the project.

BASIC REFERENCES

a. Letter dated 6 January 1971 from United States Department of the Interior, Fish and Wildlife Service, Bureau of Sport Fisheries and Wildlife.

b. Letter dated 12 January 1971 from Environmental Protection Agency, Federal Water Quality Administration, Northwest Region.

c. Comment dated 9 December 1970 on Brochure mailed out 7 December 1970.

d. Letter dated 22 October 1969 from Manager, Port of Port Angeles.

e. Letter dated 15 December 1970 from Mr. Harvey Manning, Issaquah, Washington.

f. Letter dated 22 January 1971 from Washington State Department of Fisheries.

g. Letter dated 24 December 1970 from Interagency Committee on Outdoor Recreation.

h. Note dated 9 December 1970 from Mr. H. Engstrom, Port Angeles, Washington.

i. Letter dated 20 October 1970 from U. S. Department of the Interior, Bureau of Outdoor Recreation.

k. Testimony contained in the Record of the 13 October 1970 Public Meeting in Port Angeles, Washington.

l. Letter dated 13 October 1970 from Manager, Port of Port Angeles.

m. Testimony contained in record of public meeting held on 23 February 1971.

Note: Pros and cons without references are presented by the Corps of Engineers based on our reconnaissance studies.

INDEX

BASIC REFERENCE AND LIST OF INCLOSURES

ALTERNATIVE 1

No Action

Physical Structures. This alternative contemplates no additional erosion control structures along the seaward side of Ediz Hook. The existing revetment works would be allowed to deteriorate and would not be maintained.

Effects. The beach fronting the hook would continue to erode along the narrow portion of the hook. The hook would eventually breach permanently and become a shoal area. Protection of Port Angeles Harbor would be reduced.

Funding Costs. None if social costs are excluded.

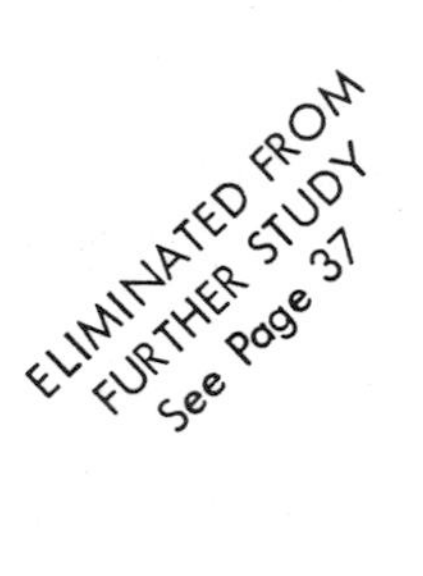

ALTERNATIVE 1

No Action

No.	PROS		Comments on Pros (CONS)
1.	Spit may heal itself if breached as has been observed in the past.	→	
2.	Will not disturb ecological systems.	→	
3.	No further investment required for protective measures.	→	
4.	Fish and wildlife resources would remain undisturbed.	→	
5.	Natural beauty of spit would remain unmarred by man-made structures.	→	
6.	Water quality would be improved inside Port Angeles Harbor (U.S. Bureau of Sport Fisheries, Ref. a).	→	Wave damage to inner harbor would occur if spit is permanently breached.
7.	Would have least impact on aquatic environment (FWQA, Ref. b).	→	

No.	Comments on Additional Cons (PROS)		Additional CONS
8.		←	Access road and utilities would eventually be destroyed.
9.		←	Lack of feed material would prevent beach from maintaining present slope.
10.		←	Industries located on spit would be abandoned.
11.		←	Access to beach and outer portion of hook would be impaired.
12.	Any change occurring naturally cannot destroy natural setting only alter it. Recreation loss is questionable (U.S. Bureau of Sport Fisheries, Ref. a).	←	Natural beach setting would be destroyed with subsequent loss of recreation opportunities.
13.		←	Cost to U.S. Coast Guard to relocate is estimated @ $2,850,000 exclusive of runway construction and moorage facilities for Coast Guard cutters (Cmdr. 13th Coast Guard District, Ref. c).
14.		←	Reduction in Port revenue is estimated at $450,000 annually if spit is breached permanently (Port of Port Angeles, Ref. d).
15.		←	Would be the end of Port Angeles as shipping and industry would be greatly affected. (C. D. Curran & H. Engstrom, Ref. K.)

ALTERNATIVE 2

Concrete Seawall

Physical Structures. Reinforced concrete wall in area most susceptible to erosion. The design concept provides for a reentrant lip to throw back the wave. Foundation would be buried with a weighted toe. Stairways to the beach would be spaced along the entire length of the wall.

Effects. This alternative would protect the roadway and utilities and maintain the integrity of the spit. However, the costs are higher than a rock revetment (Alternative 5) which would serve the same purpose.

Funding Costs	Original Investment	Annual Costs
1. Seawall Construction		
Minimum length (3,400')	\$3 - 4 million	\$171,000 - 228,000
Maximum length (10,000')	\$8 - 12 million	\$492,000 - 680,000

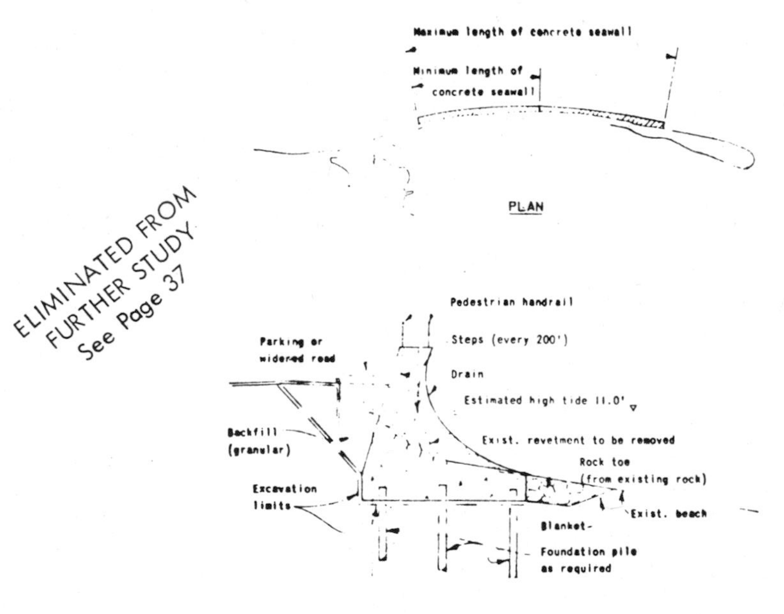

ALTERNATIVE 2

Concrete Seawall

PROS		Comments on Pros (CONS)
1. Protects road, utilities and maintains integrity of spit. (H. Engstrom, Ref. h.)	→	1. Would amount more or less to a complate loss of the natural integrity of the hook and its ecosystems (Harvey Manning, Ref. e).
2. Would not interfere with natural littoral drift movement.	→	2. Would alter gravel transport to the east (U.S. Bureau of Sport Fisheries, Ref. a).
3. Would dissipate wave energy and reduce freauency of overtopping.	→	3. Would alter wave attack patterns to beaches both east and west of seawall (U.S. Bureau of Sport Fisheries, Ref. a).
4. Provides safer access to the beach than under present conditions and would be architecturally pleasing.	→	4. ________________
5. Disturbance to ecological systems limited to construction site.	→	5. Effects on ecology would extend much farther than construction site (U.S. Bureau of Sport Fisheries, Ref. a).
6. Would have little impact on aquatic environment (FWQA, Ref. b).	→	6. ________________

Comments on Additional Cons (PROS)		Additional CONS
7. ________________	←	7. Would detract from natural beach setting.
8. ________________	←	8. Top of wall would be higher than road blocking view.
9. ________________	←	9. Existing protective works could not be utilized effectively.
10. ________________	←	10. Eliminates further optional changes.
11. ________________	←	11. Continued lowering of beach profile could undermine structure.

ALTERNATIVE 3

Offshore Breakwater

Physical Structures. Intermittent rock breakwater constructed approximately 400 feet offshore. Top layer of armor rock would be cemented with asphalt to reduce porosity and prevent excessive movement of rock by severe wave action. This alternative is designed to dissipate wave energy before it reaches the shoreline of the spit; however, it would require extensive model testing prior to design and construction.

Effects. Would act as a wave break to dissipate some of the wave energy before it reaches the upper shoreline of the hook. However, it would provide only limited protection to the roadway; would interfere with littoral transport from west and has a high cost when compared with other methods of erosion control without providing any special social or environmental advantages.

Funding Costs	Original Investment	Annual Costs
1. Breakwater Construction		
Minimum length (3,400')	$3 - 5 million	$171,000 - 284,000
Maximum length (15,000')	$10 - 14 million	$568,000 - 802,000

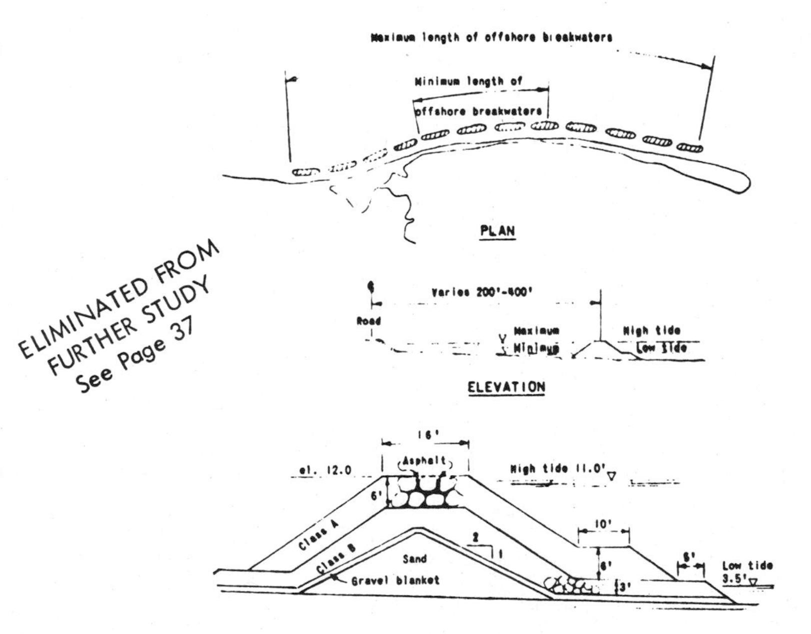

ALTERNATIVE 3

Offshore Breakwater

PROS		Comments on Pros (CONS)
1. Would provide only limited protection to road and utilities.	→	1. Would amount more or less to a complete loss of the natural integrity of the hook and its ecosystems. (Harvey Manning, Ref. e.)
2. Will provide habitat for sea life if modified to be more in the form of a reef (Wash. State Dept. of Fisheries, Ref. f).	→	2. ______________
3. Can be used in conjunction with existing protective works.	→	3. ______________

Comments on Additional Cons (PROS)		Additional CONS
4. ______________	←	4. Structure will detract from natural beach setting at low tides.
5. ______________	←	5. Could cause formation of irregular beach due to spacing between structures.
6. ______________	←	6. Could cause localized tide rips dangerous to swimmers. Would be a hazard to navigation.
7. ______________	←	7. Would interfere with littoral drift movement and increase hook erosion.
8. ______________	←	8. Would have some impact on aquatic environment (FWQA, Ref. b).

ALTERNATIVE 3A

Liberty Ships in Lieu of
Rubblemound Offshore
Breakwater

Physical Structure. This alternative would utilize Liberty Ships currently in moth balls, securely anchored in place, to serve as a breakwater. The ships could be used for storage, manufacturing or garbage disposal by constructing suitable roads on and off the ships.

Effects. The anchored ships would dissipate wave energy to a certain extent. However, to anchor the ships securely against all storm and tidal conditions would be costly. Also, the ability to maintain roads out to the ships is questionable. This method, used elsewhere, has not been too successful.

Funding Costs	Original Investment	Annual Costs
Not estimated		

ELIMINATED FROM
FURTHER STUDY
See Page 37

ALTERNATIVE 4

Groin Field with Periodic Beach Nourishment

Physical Structure. The groin field alternative consists of a series of treated-pile walls with horizontal planking, constructed at right angles to the beach. These walls would be spaced about 400 feet apart and would be about 500 feet long. The space between the walls would be filled with sand and, at the same time, the elevation of the road would be raised about four feet to prevent excessive overtopping. Once the spaces between the walls are filled, they would be nourished as needed by depositing material on the beach west of the spit and allowing the natural easterly currents to carry the material through the groin field. Modification of this concept, using a terminal groin is also a usable alternative.

Effects. A groin field along the entire length of the hook would provide only limited protection and would be ineffective in the narrow portion of the hook because of the offshore movement of material in this area. In addition, groins in this area would be detrimental to the beach to the east as they would intercept a greater percent of material and direct it offshore. However, a series of groins to the east of this area on the distal or bulb end would be acceptable as alongshore movement of material is greater in this area, but some type of protective work, such as a rock revetment, would be needed in the narrow portion of the hook.

Funding Costs	Original Investment	Annual Costs
1. Groin Field Construction		
Minimum project (8,000')	$2.0 - 2.5 million	$137,000 - 165,000
Maximum project (18,000')	$3.0 - 4.0 million	$207,000 - 273,000

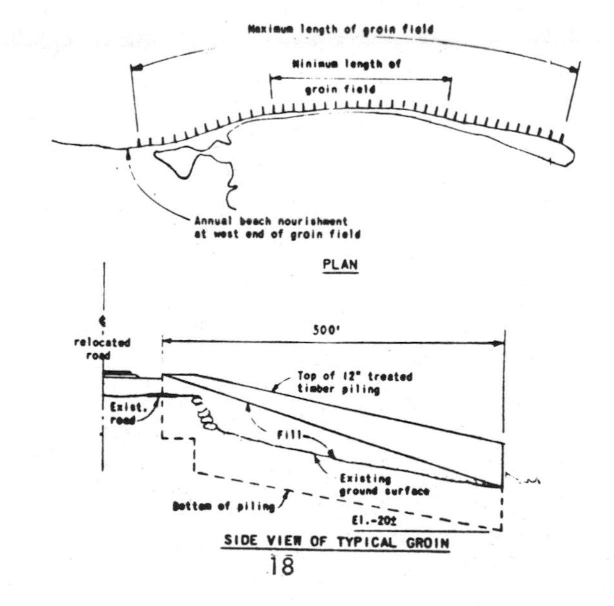

ELIMINATED FROM FURTHER STUDY See Page 37

ALTERNATIVE 4

Groin Field with Periodic Beach Nourishment

PROS		Comments on Pros (CONS)
1. Would provide only limited protection to the hook.	→	1. Would amount more or less to a complete loss of the natural integrity of the hook and its ecosystems (Harvey Manning, Ref. e).
2. Would be near natural form of protection (Wash. State Dept. of Fisheries, Ref. f).	→	2. ______
3. Would obtain necessary erosion control and maximize outdoor recreation potential. (Interagency Committee on Outdoor Recreation, Ref. g and Bureau of Outdoor Recreation, Ref. j.)	→	3. ______

Comments on Additional Cons (PROS)		Additional CONS
4. ______	←	4. Would tend to starve downdrift areas by forcing material into deeper water.
5. ______	←	5. Material in vicinity of Crown Zellerbach plant would be lost to offshore moving currents.
6. ______	←	6. Would have considerable impact on aquatic environment (FWQA, Ref. b).

ALTERNATIVE 5

Rock Revetment with Periodic Beach Nourishment

Physical Structure. The rock revetment alternative is a refinement of what is being done now by the city of Port Angeles. This concept consists of carefully placing various grades of granular material ranging from rock fragments or quarry spalls to heavy armor rock weighing several thousand pounds. An initial and periodic beach nourishment program would be instituted to maintain existing beach profiles and prevent undermining of structure. The top of the revetment would be about four feet above the roadway.

Effects. This alternative would utilize much of the work that has been accomplished and would provide the needed protection for the roadway and utilities as well as maintain the integrity of the spit.

Funding Costs	Original Investment	Annual Costs
1. Rock Revetment		
Minimum project (3,400')	$1.0 - 1.5 million	$58,000 - 88,000
Maximum project (9,000')	$3.0 - 4.0 million	$172,000 - 231,000

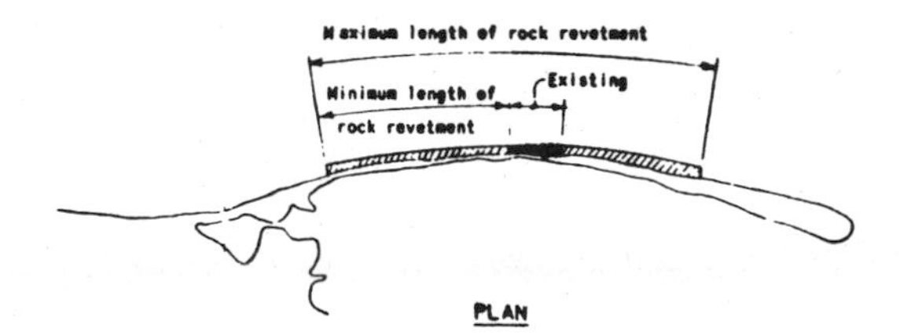

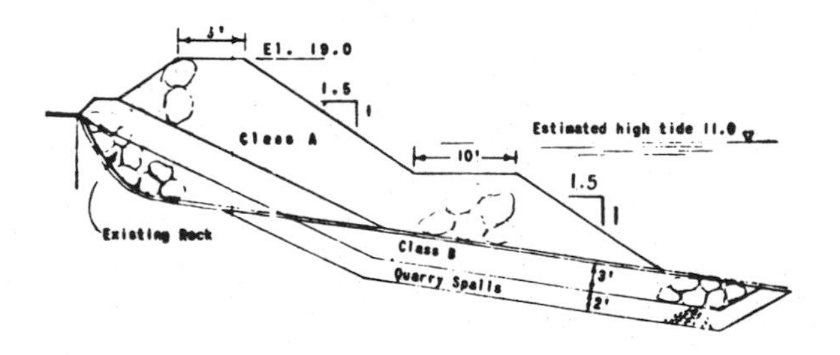

ALTERNATIVE 5

Rock Revetment

PROS		Comments on Pros (CONS)
1. Would protect road, utilities and maintain integrity of spit.	→	1. Would amount more or less to a complete loss of the natural integrity of the hook and its ecosystems (Harvey Manning, Ref. e).
2. Would not interfere with natural littoral movement.	→	2. Could alter gravel transport to the east (U.S. Bureau of Sport Fisheries, Ref. a).
3. Would dissipate wave energy and reduce frequency of overtopping.	→	3. Could alter wave attack pattern to beaches both east and west of revetment. (U.S. Bureau of Sport Fisheries, Ref. a).
4. Would readily adjust to any lowering of beach profile.	→	4. Continued lowering of beach profile could undermine and destroy structure in neck area.
5. Would not have serious effect on aquatic environment (FWQA, Ref. b).	→	5. ______________
6. Disturbance to ecology only in construction area.	→	6. Effects on ecology would extend much farther than construction site. (U. S. Bureau of Sport Fisheries, Ref. a.)

Comments on Additional Cons (PROS)		Additional CONS
7. ______________	←	7. Access to beach would be difficult over large rocks.
8. ______________	←	8. View would be impaired from roadway.
9. ______________		*9. Wave action would pulverize any small rock placed on the beach and turn it into sand. (Mr. Curran, Ref. m.)

ALTERNATIVE 6

Elevated Roadway

Physical Structure. This alternative consists of constructing on the harbor side of the spit a reinforced concrete roadway supported by concrete piers. The elevation of the roadway would be high enough to prevent overtopping. Utilities would be suspended beneath the roadway. This would eliminate the need for holding the spit in its present location and would allow natural forces to reshape the shoreward end of the spit. The easterly end would remain at approximately its present location for some time.

Effects. Would provide access to the Coast Guard station but would allow the spit to erode. The Coast Guard would be forced to provide protective works at the installation. Additional protective works would be required at the base of the hook to prevent erosion and outflanking of the elevated roadway. Also, Port Angeles Harbor would be exposed to storm wave action.

Funding Costs	Original Investment	Annual Costs
1. Construct Elevated Roadway	$8 - 10 million	$448,000 - 563,000

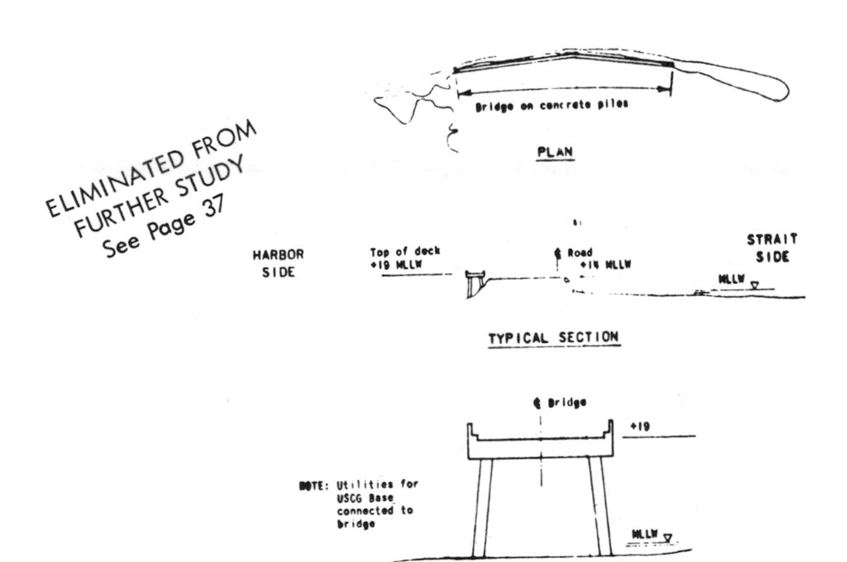

22

ALTERNATIVE 6

Elevated Roadway

PROS		Comments on Pros (CONS)
1. Would provide dependable access to Coast Guard Base.	→	1. Would amount more or less to a complete loss of the natural integrity of the beach and its ecosystems (Harvey Manning, Ref. e).
2. Would not appreciably affect existing ecology.	→	2. ______
3. Would have little effect on aquatic environment (FWQA, Ref. b).	→	3. ______

Comments on Additional Cons (PROS)		Additional CONS
4. ______	←	4. Hook would eventually breach causing possible wave damage in harbor:
5. ______	←	5. Industries on hook would probably be abandoned (F. M. Thompson, Ref. k).
6. ______	←	6. Would require additional protective works at Coast Guard Base.

ALTERNATIVE 7A

Beach Nourishment

Physical Structure. The beach nourishment alternative is similar to Alternative 4 - the groin field. This concept eliminates the use of the timber walls and envisions raising the beach profile to some height, as shown on the sketch below. An annual supply of feed material would be placed on the beach near the west end of the built-up area, to maintain the new beach elevation.

Effects. Beach nourishment would maintain the existing beach profile but would provide only limited protection for the roadway and utilities.

Funding Costs	Original Investment	Annual Costs*
1. Initial Beach Buildup		
Minimum project (3,400')	$1.5 - 2.0 million	$109,000 - 147,000
Maximum project (9,200')	$ 4 - 5 million	$248,000 - 314,000

*Includes annual nourishment costs of $25,000 - 35,000.

ELIMINATED FROM FURTHER STUDY See Page 37

ALTERNATIVE 7A

Beach Nourishment

PROS		Comments on Pros (CONS)
1. Would provide limited protection for road and utilities.	→	1. ________
2. Would maintain near normal integrity of spit. (Wash. Dept. of Fisheries, Ref. f.)	→	2. ________
3. Would increase available beach recreation area.	→	3. ________
4. Short-term effect on ecology.	→	4. Dumping of unbeachlike materials would detract from natural quality (Harvey Manning, Ref. e).

Comments on Cons (PROS)		Additional CONS
5. ________	←	5. Damage to road and utilities would continue to occur during storm activity.
6. ________	←	6. Would require a large amount of borrow material.
7. ________	←	7. Would have considerable impact on aquatic environment (FWQA, Ref. b).

ALTERNATIVE 7B

Surveillance Program with Option
for Structural Measures

Physical Structure. Alternative 7B is similar to 7A, but eliminates the initial beach buildup. This concept envisions periodically placing beach material along one mile of beach updrift from the erosion area. Existing protective works would be upgraded to protect roadway and utilities. Hydrographic and topographic surveys would be conducted quarterly for a period of one to two years to determine extent of deposition along the hook. Based on results of above surveillance program, one of the following courses of action would evolve:

a. Make a one-time fill with periodic nourishment, as considered in plan 7A.

b. Adopt one of the alternative structural measures, or a combination of these alternatives.

Effects. Would not provide protection to the hook during the surveillance period. Could result in serious damage to hook which could increase costs of taking corrective action at a later date.

Funding Costs	Original Investment	Annual Costs
Not available		

ELIMINATED FROM
FURTHER STUDY
See Page 37

ALTERNATIVE 7B

Surveillance Program with Option
for Structural Measures

PROS		Comments on Pros (CONS)
1. Would have least disruptive effect on natural setting (H. Manning, Ref. e).	→	1.
2. Would leave options open for other alternatives should they become necessary.	→	2.

Comments on Cons (PROS)		Additional CONS
3.	←	3. During interim period, deposited material could be lost to deep water and serious damage to spit would occur.
4.	←	4. Would not provide protection to the road and utilities.
5.	←	5. Would require continued maintenance of existing protective works until a final program is selected.
6.	←	6. Would have serious impact on aquatic environment (FWQA, Ref. b).

ALTERNATIVE 8

Relocate Industrial Water Supply Pipeline and Remove its Existing Protective Works

Physical Structure. This alternative suggests relocating the existing 10,000 feet of 57-inch concrete industrial water supply line from its present location along the base of the cliff into about 30 feet of water. The protective riprap and sheet piling would also be removed.

Effects. Removal of protective works would allow the sea cliffs to erode and supply material to the hook. However, continued erosion of the cliffs would eventually endanger homes located atop the bluff area as well as the landward end of the hook. Also, the eroded material would be insufficient to replace what has been lost and additional work would be required to protect the hook and roadway.

Funding Costs	Original Investment	Annual Costs
1. Relocate Waterline	$1.3 - 2.0 million	$80,000 - 120,000

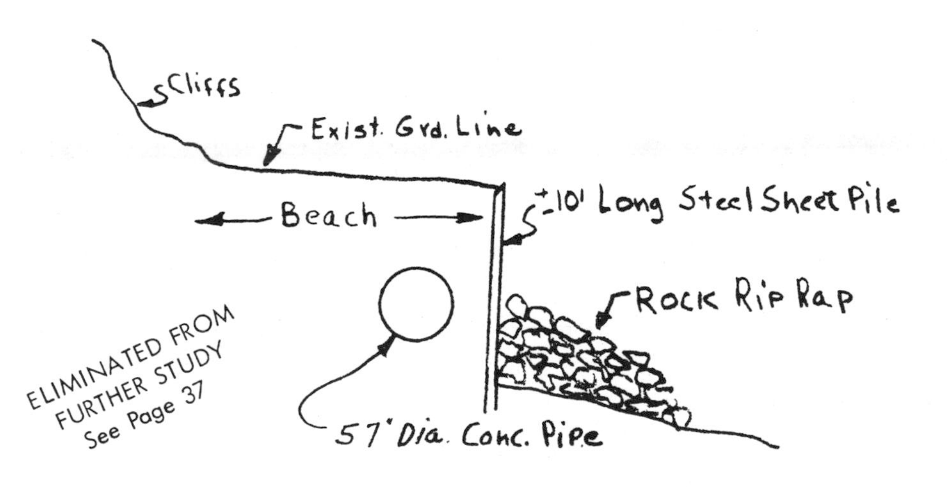

ALTERNATIVE 8

Relocate Industrial Water Supply Pipeline
and Remove its Existing Protective Works

PROS		Comments on Pros (CONS)
1. Restore area to natural condition.	→	1.
2. Increase natural supply of littoral material (A.M. Devine & A. W. Hare, Ref. k).	→	2.
3. Allow feed material from the west to move unrestricted.	→	3.

Comments on Cons (PROS)		Additional CONS
4.	←	4. Sea cliffs would erode and eventually threaten hook at its base and the homes situated atop the bluffs. *(Mr. Hassel, Ref. m.)
5.	←	5. Would not protect road and utilities without additional protective works along the hook.
6.	←	6. Would have considerable impact on aquatic environment (FWQA, Ref. b).
7.	←	7. Waterline is only 1 of several contributing factors of erosion along the hook.

ALTERNATIVE 8A

Removal of Pipeline and Dams

Physical Structures. This alternative suggests removal of all man-made structures including the two water supply dams on the Elwha River and the waterline along the base of the sea cliffs.

Effects. This would restore the natural feed material which has in the past kept the system in a relatively crude equilibrium. However, restoring these conditions would not save the roadway as the spit is attempting to migrate southward under natural conditions and would continue to do so if natural conditions were restored. Pros and cons of this alternative would be the same as of Alternative 8.

Funding Costs.

Cost of removal of dams not estimated.

ELIMINATED FROM
FURTHER STUDY
See Page 37

* ALTERNATIVE 8A

Removal of Pipeline and Dams

PROS

1. ______________________

2. ______________________

Comments on Pros (CONS)

1. ______________________

Comments on Additional Cons (PROS)

3. ______________________

Additional CONS

3. Dams are needed for industry (Mayor Wolff, Ref. m).

ALTERNATIVE 9

Porous-Walled Breakwater

Physical Structure. The porous-walled breakwater alternative, as shown below, would be similar to Alternative 3 except that the rubble-mound structure would be replaced by a reinforced concrete box structure with a porous wall facing seaward. This alternative is expected to dissipate wave energy before it reaches the shoreline of the spit and is a modification of the suggested alternative.

Effects. Would dissipate some of the wave energy before it reaches the upper shoreline of the hook. However, as in the case of Alternative 3, it would tend to block natural drift, would provide only limited protection for the roadway and has excessive costs when compared with other methods of erosion control without providing any special social or environmental advantages.

Funding Costs	Original Investment	Annual Costs
1. Construct Porous-Walled Breakwater	\$18 – 20 million	\$1.0 - 1.2 million

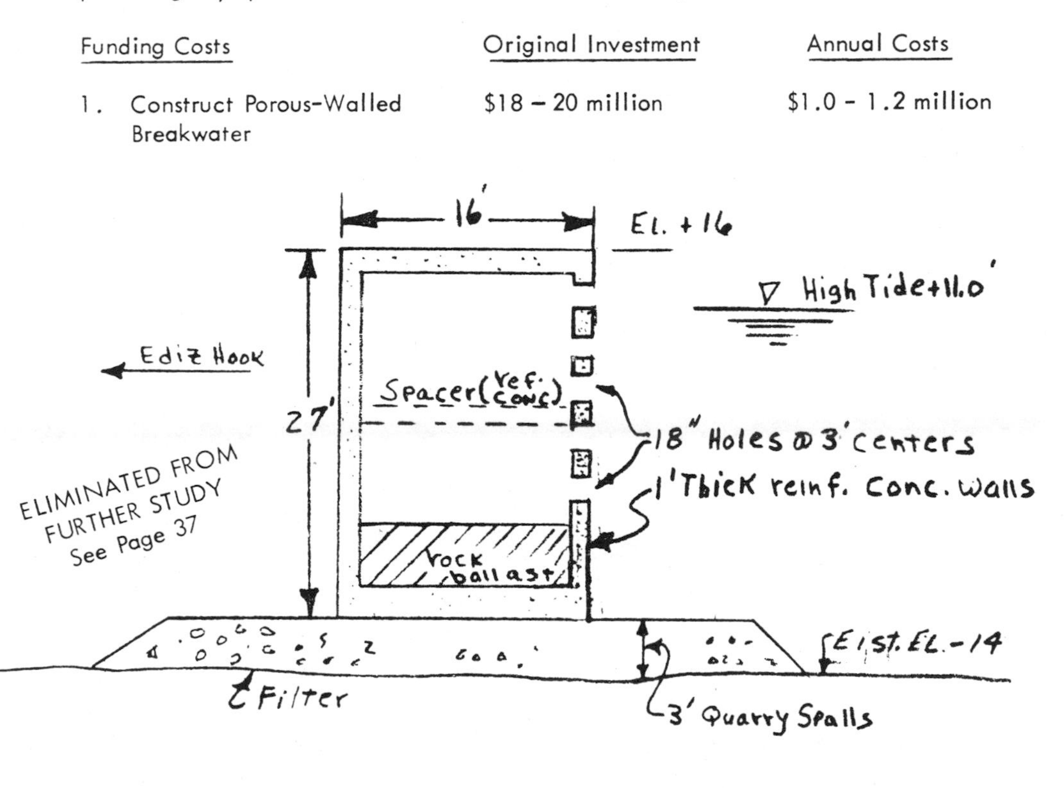

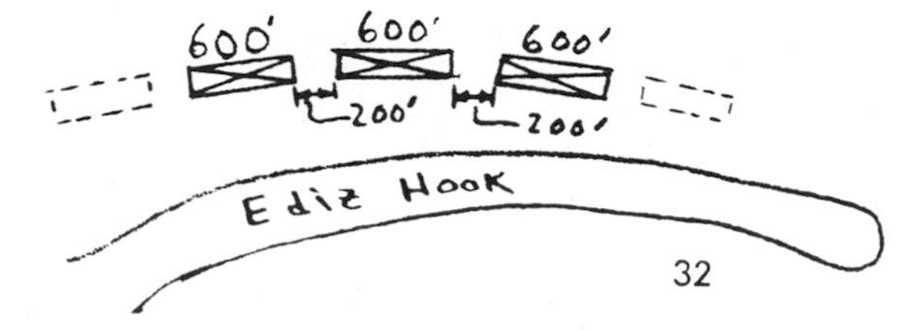

ALTERNATIVE 10

Industrial Fill - North Side of Ediz Hook

Physical Structure. This alternative, as shown in the sketch below, consists of a land fill out to the 30-foot contour. The land fill would comprise about 320 acres and require approximately 18.5 million cubic yards of fill material slope protection would be required along the north and west exposures.

Effects. Would provide protection for road, utilities, and would maintain integrity of the hook. However, protective works for the fill would be required as the shoreline would not be sufficiently reoriented to change the pattern of wave attack and the new shoreline would be subject to erosion. Also, costs would be high.

Funding Costs	Originial Investment	Annual Costs
1. Fill material only	\$18 - 20 million	\$1.0 - 1.2 million

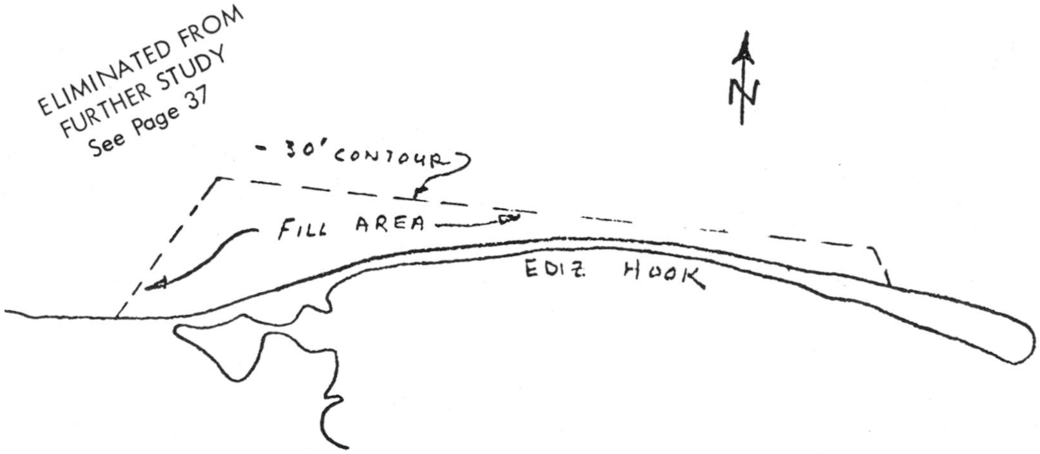

ELIMINATED FROM FURTHER STUDY See Page 37

30' CONTOUR

FILL AREA

EDIZ HOOK

N

ALTERNATIVE 10

Industrial Fill - North Side Ediz Hook

PROS		Comments on Pros (CONS)
1. Would provide additional waterfront lands for industiral and port development (Port of Port Angeles, Ref. k & l).	→	1. ______
2. Would maintain integrity of spit.	→	2. Would require additional protective works along Strait side of fill (U.S. Bureau of Sport Fisheries, Ref. a).
3. Additional recreation areas could be provided in conjunction with the industrial park.	→	3. Would alter quality and character of existing recreational usage (U.S. Bureau of Sport Fisheries, Ref. a).
4. Would enhance economic conditions and employment.	→	4. ______

Comments on Additional Cons (PROS)		Additional CONS
5. ______	←	5. Could cause changes in tidal and current patterns creating erosion or accretion problems else where (Interagency Committee for Outdoor Recreation, Ref. g).
6. ______	←	6. Would cover several acres of tidal and subtidal lands.
7. ______	←	7. Ecology would be drastically altered.
8. ______		8. Would amount more or less to a complete loss of the natural integrity of the beach and its ecosystems (H. Manning, Ref. e).

PRINCIPAL AGENCIES AND GROUPS URGED TO CONTRIBUTE FACTS TO THE STUDY

ORGANIZATION		Provided Additional Facts or Helped Edit Draft Brochure Mailed On Date Indicated			
LOCAL	POINT OF CONTACT	5 OCT. 70	7 DEC. 70	17 FEB 71	1 APR 71
Port of Port Angeles	Tom Neal, Port Manager	X	X		
City of Port Angeles	Don Herman, City Manager	X	X		X
Clallam County					
STATE - State of Washington					
Dept. of Ecology	John A. Biggs			X	
Dept. of Fisheries	Thor C. Tollefson		X		
Dept. of Game	Carl N. Crouse		X		
Dept. of Commerce & Economic Development					
Dept. of Natural Resources					
Planning & Community Affairs Agency					
Interagency Committee for Outdoor Recreation	Stanley E. Francis	X	X		
FEDERAL					
Dept. of Transportation					
U. S. Coast Guard	Cmdr. Clark	X	X		
Dept. of the Interior					
Bureau of Sport Fisheries & Wildlife	Edward S. Marvich		X		
Bureau of Outdoor Recreation	Fred J. Overly	X	X		
Environmental Protection Agency	Robert D. Zeller Ph.D.		X		
OTHER AGENCIES & ORGANIZATIONS					
Puget Sound Pilots					
INDUSTRIES					
Crown Zellerback Corp.	A. W. Gunlogsen	X	X		X
ITT - Rayonier, Inc.	John Gray				X
Penninsula Plywood					
INDIVIDUALS CONTRIBUTING					
Mr. Harvey Manning, Issaquah		X	X		
Mr. Everett Hassel, Port Angeles		X	X	X	X
Mr. Deven Curran, Port Angeles		X	X	X	X
Mr. A. M. Devine, Port Angeles		X	X		
Mr. H. Engstrom, Port Angeles		X	X		X
Mr. Gary M. Kass, Port Angeles		X	X		X
Mr. Larry Lack, Port Angeles					X

Alternative selected for detailed study: Based on information gained from reconnaissance studies and input from the public meetings and earlier editions of this brochure, alternative 5, Rock Revetment with Beach Nourishment, appears to be the most desirable plan for further study. The revetment works would provide protection to the roadway and utilities and establish a last line of defense against any severe winter storm. The nourishment material would maintain the beach profile, thereby minimizing scour of the revetment toe. In the winter, if accreted material from the nourishment program is lost, exposing the revetment toe, the danger of breaching is not imminent, as the weighted toe would adjust to the lowered beach profile and maintain the integrity of the spit. Also, radical disruption of the environment would be avoided.

Rationale for rejecting other alternatives: Each of the other alternatives, when compared with alternative 5, is less desirable for the following reasons:

a. Alternative 1 (no action). Current studies indicate that, without further protective works, the spit would be breached, resulting in a loss of the roadway atop the spit, land access to the Coast Guard Station, and the protection of Port Angeles harbor. There would also be a serious detrimental effect on industries which rely on the spit for their existence. The social and economic impact on the local community could be severe. While there is a possibility that the spit could heal itself without construction of protective works, the "no action" alternative was inconsistant with current and desired continued use of the spit.

b. The concrete seawall (alternative 2) would serve the same purpose as a rock revetment if beach nourishment were added. However, local and Federal costs would be much greater than for a rock revetment. No significant additional advantages were stated which were considered sufficient to justify the increased cost.

c. Offshore breakwaters or sunken ships (alternatives 3 and 3A) would provide some degree of protection to the hook and would not be overly disruptive to the environment. However, both alternatives would be more costly and would not insure against occasional breaching of the spit and roadway. Continued high maintenance costs of these facilities would be required.

d. The groin field with beach nourishment (alternative 4) is an acceptable alternative, providing adequate clean material for nourishment is available at all times. However, there are indications of a strong offshore littoral movement in the thin neck section of the spit, and groins would not assist appreciably in holding material at that location. The groins would also be detrimental to downdrift areas, by intercepting and directing offshore a greater portion of the littoral material than is now being lost. Use of groins farther east would be acceptable, as littoral movement conditions change toward the east end of the spit.

e. The elevated roadway (alternative 6) is costly ($8-10 million) and would have most of the social and economic disadvantages of alternative 1. This alternative was not selected for detailed study because it poses too great a risk to the integrity of the spit and harbor.

f. Beach nourishment programs (alternatives 7A and 7B) could stabilize the beach profile, but would allow storm waves to attack the upper beach, causing damage to roadway and utilities.

g. Removal of all manmade structures (alternatives 8 and 8A), such as the riprapped water line east of Ediz Hook and the dams on the Elwha River, is not a viable solution as the natural supply of littoral materials is insufficient to adequately restore the depleted beach areas. No Federal financial assistance would be available for this plan.

h. The porous-walled breakwater (alternative 9) has extremely high costs and would provide the same effects as alternatives 3 and 3A.

i. The industrial fill north of the hook (alternative 10) would be very beneficial to the economic climate by providing additional industrial areas. However, the proposal would be very detrimental to the environment, is costly, and would require protective works along the strait side of the fill.

Author Citation Index

Subject Index